AF580333

Introduction to Dynamic Systems Modeling for Design

David L. Smith

Prentice Hall
Englewood Cliffs, New Jersey 07632

Library of Congress Cataloging-in-Publication Data

Smith, David L. (David Lee)
Introduction to dynamic systems modeling for design / David L. Smith.
p. cm.
Includes bibliographical references and index.
ISBN 0–13–588344–X
1. Electric networks—Mathematical models. 2. System analysis.
I. Title.
TK454.2.S65 1994
620′.0042—dc20 93–3908
CIP

Acquisition Editor: Linda Ratts
Production Editors: Fred Dahl and Rose Kernan
Copy Editor: Rose Kernan
Designer: Fred Dahl
Cover Designer: Anne Ricigliano
Production Coordinator: Linda Behrens

© 1994 by Prentice-Hall, Inc.
A Paramount Communications Company
Englewood Cliffs, New Jersey 07632

All rights reserved. No part of this book may be reproduced, in any form or by any means, without the permission in writing from the publisher.

Printed in the United States of America
10 9 8 7 6 5 4 3 2 1

ISBN 0-13-588344-X

Prentice-Hall International (UK) Limited, *London*
Prentice-Hall of Australia Pty. Limited, *Sydney*
Prentice-Hall Canada Inc., *Toronto*
Prentice-Hall Hispanoamericana, S.A., *Mexico*
Prentice-Hall of India Private Limited, *New Delhi*
Prentice-Hall of Japan, Inc., *Tokyo*
Simon & Schuster Asia Pte. Ltd., *Singapore*
Editora Prentice-Hall do Brasil, Ltda., *Rio de Janeiro*

Contents

CHAPTER 2

Functional Analysis of a Multiport System, 26

CHAPTER 3

Physical Analysis of the System Inputs, 59

CHAPTER 4

Component Physical Analysis: The Fundamental Approach in the Time Domain, 90

Preface

This book is written as a textbook for junior-level engineering undergraduates. Its aim is to introduce the topic of dynamic systems modeling using ideas from two resources of critical interest to engineers: systems thinking and design orientation. Students in the mechanical, aeronautical, and electrical disciplines are the target audience, but the introductory systems training may also be useful to anyone who desires to numerically model the response of a dynamic physical system within the context of a changing, interacting environment.

Prerequisites for this study are first courses in linear algebra, physics, ordinary differential equations, computers, statics, and dynamics. Dynamics may be studied simultaneously, if desired.

Unfortunately, limited time and economy required concessions in the writing of this text. These appear mainly in three ways: first, in omitting extensive reviews of prerequisite material; second, in writing for scope over depth in the study; and third, in emphasizing the analogies in the system mathematical formulations.

The first concession is eased somewhat by writing the text so that less depth is required in mathematics and physics.

The second concession comes from a recognition that the student fresh from the above prerequisites is best served by studying the whole modeling process, even if the physical analyses are not very deep. In fact, the physical and mathematical analysis sections in this text are mostly a recasting of the prerequisites into systems terms, thus reinforcing both the previous studies as well as the present one.

The third concession rests on the idea that physical phenomena of different types are often analogous, both physically and mathematically, and these analogies can be effectively used to model the system dynamics. That is, for example, the way that a liquid flows in a pipe and the way that electrons flow in a wire are physically analogous and their model equations are very similar. Such modeling analogies are discussed in this text among the four basic engineering specialty areas: electronics, structural dynamics, thermodynamics, and fluid mechanics.

Further, such analogy modeling is *fundamental* to systems thinking. This is not to suggest that analogies are some "magic bullet" for all engineering modeling problems, but they do provide a consistently useful starting point for physical model building, they offer a powerful unifying and reinforcing theme to aid understanding of various physical phenomena, and emphasizing them helps to teach them.

However, the practicing system modeller soon finds that there are also times when a physical situation has no clear analogy. Such times should become apparent to the reader through further study in the areas of engineering specialization. In any case, the study of fundamental physical analogies in this text is designed to provide a systems framework for the integration of all such follow-on studies.

The text is organized into eight chapters, generally following the flow of the modeling process. Chapter 1 begins with a look at the engineering design process and how it gives context and focus for all modeling analysis efforts.

In Chapter 2, the idea of system definition and system functional analysis is presented. The goal there is to reduce a given complicated hardware system into a set of simpler, functionally connected components for physical and mathematical analysis. The system multiport analysis method in the text is chosen for study because of its conceptual simplicity and its power to organize what might otherwise be an overwhelming system analysis.

In Chapter 3, the idea of directing a component physical analysis according to the nature of its known or expected inputs is presented, as are methods for the analysis of the inputs.

Chapters 4 through 6 present component physical and mathematical studies to develop the fundamental modeling judgement of the reader.

Inevitably, components arise in systems which are too complicated to be analyzed from fundamentals. These components must be modelled with experimental studies, as discussed in Chapter 7.

Chapter 8 contains the methods used to assemble the component mathematical models into a system mathematical model, and those to predict the system performance for numerical design studies.

Each of the chapters contains homework problems which are designed to adequately cover the analysis material and bring out the design motivation where possible. The reader should do as many of the homeworks as time allows.

Each chapter in the text also has a bibliography. The references cited in the bibliographies are chosen for their readability, for their difficulty level in view of the reader's assumed background, and for their relevance for further in-depth study.

A complete instructor's guide for this text is available through the publisher. This guide includes recommended lecture schedules for teaching formats of one quarter, one semester, or two quarters. It also has coordinated lab suggestions for each teaching format, and expanded solutions to all the homework problems.

This text writing effort would not have been possible without the assistance of many people. At the top of the list must go my thesis and dissertation advisor, Dr. Karl Reid of the Oklahoma State University. He first taught me about systems thinking, and he's patiently encouraged me on as I've reached out to him through the years.

The entire editorial staff at Prentice Hall, Inc., have also been very helpful, but especially Doug Humprey. He shared the vision for this text right from the beginning and it's been a delight to work with him.

Many reviewers provided comments at various stages of the manuscript, each helping to fashion the text in their own way. This started with the students at the Naval Postgraduate School who provided many candid, insightful comments for the earliest draft material. Soon after the project was formally begun, Professor Robin Redfield of Texas A&M University went an extra mile in his constructive evaluations and I'm very grateful to him for his brotherly diligence. Professor Greg Starr of the University of New Mexico also provided many helpful comments at the early stages. Others offering helpful comments on various drafts of the manuscript were Alan Schneider of the University of California at San Diego, Stanley Johnson of Lehigh, Galip Ulsoy of the University of Michigan, and C. N. Bapat of the City College of New York.

This book would somehow be incomplete without some mention of the all-out encouragements of Dr. Bob Bose and Dr. Allan Kraus. I found them to be true design-oriented engineers, extremely effective teachers, and wonderful work-ethic role models.

The efforts of the research staffs at the Library of Congress, and the libraries of the University of Maryland and the George Washington University are all gratefully appreciated.

Last, but by no means least, I would like to thank my parents, Milo and Eunice, for creating a home for me, and my Father for generously providing for all my needs.

David L. Smith
Vienna, Virginia

Chapter 1

Overview: Top-Down Design, Top-Down Modeling

1-0 INTRODUCTION

Engineers create new or improved products (systems) using the complementary efforts of modeling and design. These efforts are "complementary" in the sense that models give insight to design decisions, and design objectives guide modeling decisions. Yet, the two efforts are distinct.

When designing, an engineer is *assembling* a hardware system to meet one or more design objectives. The tool that the engineer uses to measure success of the design is a mathematical model of the system performance. Thus, given a good system model, the engineer can make effective design decisions about a candidate product based upon accurate performance estimations. In this way, dynamic systems modeling is a tool which engineers use to improve their designs.

So, the modeling effort is aimed at finding an accurate *numerical* estimate of the system performance. The engineer does this by conceptually *disassembling* the complicated system hardware into easily analyzed parts, and then mathematically modeling those parts. Assembling the part models into a system model thus gives the numerical tool needed to evaluate the system performance.

Note that the level of detail in the design must guide the modeling effort, and the model must be suitable to answer design questions.

The goal of this text is to develop a useful, introductory-level understanding of those system modeling skills which, when properly used, help an engineer to accurately predict the dynamic performance of physical system designs. To better understand systems modeling, we must begin with a look at the process of systems design.

Systems Design. Systems design is based upon the functional hierarchy of systems organization. The levels of this hierarchy are as follows, from top to bottom:

- systems;
- subsystems;
- components; and
- elements.

All products (systems) can be subdivided using this hierarchy in much the same way as the United States can be subdivided into states, counties, and cities. (Notice that such hierarchies coexist functionally as well as physically. In the case of the United States, the governmental hierarchy is functionally organized along the same lines as the physical hierarchy.)

The top level of this hierarchy, the systems level, is where the system *design performance requirements* are specified by the customer. These requirements are numerical acceptability standards (speed, cost, availability dates, etc.) against which the design must be numerically measured before it is manufactured.

The system engineer's problem is to synthesize (design) a collection of hardware, software, and/or procedures (e.g., a system) that will meet or exceed the system-level requirements. While there are many ways to solve this problem, this text focuses on the *systems approach.* In this approach, the engineer conducts a functional design and a hardware design working down through the systems hierarchy, on a level-by-level basis. This procedure is called *top-down design.*

Three distinct stages of top-down engineering design can usually be identified. These are, in the order in which they are addressed:

1. Concept design (down through the sub-system level);
2. Preliminary design (through the components); and
3. Detail design (element level).

If the system (product) is very simple, then the engineering design stages tend to merge together and are often indistinguishable. For medium-complexity systems, the first two stages are likely to merge, and are sometimes referred to collectively as *advanced design.*

At each stage of design, the design must be numerically evaluated to test it against the performance requirements, to see if further design effort is merited. This is done by creating a numerical model which captures the essential physical nature of the design.

Systems Modeling. Three analytical steps are used in creating useful numerical system models:

1. Functional analysis;
2. Physical analysis; and
3. Mathematical analysis.

The *model* is the set of equations which emerge at the end of step two. Step three is necessary to check the proposed model against physical experience and intuition, and to compute the numerical performance. Collectively, these three steps are called *system modeling*.

Notice that each of the levels of system design has a corresponding level of modeling in terms of accuracy and analysis. For example, when new system concepts are first considered, only a very rough model is appropriate. At that time, few details are known well enough to permit an in-depth analysis. The system is cloaked in vague concepts of operation. Consequently, following the concept stage of design, such expressions as *back-of-the-envelope design*, or *rough-order-of-magnitude estimates*, or *single-digit accuracy* are descriptive phrases which are often heard during the discussion of modeling.

As the design evolves, more accuracy is possible and appropriate. However, do not be misled, zero error can never be achieved. The real world is hopelessly complex. The modeler is always forced to approximate. Thus, the accuracy of the model depends upon the stage of the engineering process.

These observations apply also to the analysis method. As more becomes known about the design, more rigorous analysis becomes possible.

This text presents analysis methods which are appropriate to preliminary design. The analyses are thus focused on the isolation and performance of the system *components*. At times, element behavior is analyzed, but not of such detail that the system becomes overmodeled. We are here interested in analyzing the essence of the operation of the system, at a preliminary level of analysis and accuracy.

Detail design modeling methods are discussed in other texts which are less concerned with *system* performance. Consequently, this text does not have methods for the detailed analysis of gears and bearings; or perhaps the analysis of heat exchanger fins; or even schemes for modeling rivet patterns and bolt dimensions. Such details are generally not necessary to understanding the dynamic performance of a preliminary system design.

In this text, it is assumed that the details of the design will be successfully addressed at a later stage of the design process so that advanced design engineers can focus on the overview problem of effective synthesis of the bigger parts of the product (e.g., which engine to put in the car or how many engines in the ship).

The remainder of this chapter takes a closer look at the design process, and overviews the modeling analyses which typically support the system preliminary design.

1-1 CREATING EFFECTIVE DESIGNS

In all but the simplest engineering design problems, there are choices to be made between competing options. For example, will an engine in a new car have four or six cylinders? Or will the body panels in a new airplane be made of aluminum or composite materials? Thus, design is called a process of *synthesis* since there are usually many such different options which must be effectively combined, or synthesized, to achieve the best overall system design. These options can only be sorted out through the use of a clear design objective, and an orderly design process.

The Design Objective. In starting the design process, an engineer must first establish a set of system-level design requirements, or *design criteria*, for evaluation of the candidate designs. The first list of these criteria usually comes from the customer, or from a marketing survey. The final list usually emerges after the concept modeling effort. In any case, the criteria may include, but are not limited to, the following:

1. Static performance thresholds.
2. Dynamic performance thresholds.
3. Cost maxima.
4. Producibility requirements.
5. Maintainability requirements.
6. Reliability requirements.
7. Survivability requirements.
8. Human operability standards.
9. Legal, political, and ethical considerations.
10. Marketability, public opinion, and aesthetics.

The company which employs the engineers could build all the candidate designs and test them before selecting the best design. In recent times, of course, this has become a cost-ineffective approach to engineering. Nowadays it is far more effective to conduct mathematical modeling to find the best of the candidate designs, then to build one or more hardware prototypes to prepare for production. Consequently, modern engineers use an iterative process which works back and forth between creative synthesizing of new parts into products (systems), and the modeling of those systems to measure their capabilities against the design criteria.

The engineer is finished with the development of the design when the company and customer is satisfied with the product. However, all products must be introduced into a changing marketplace which is characterized by technological breakthroughs and shifting customer needs. The profit motive of the company requires product adaptation. Thus, the engineer never really quits designing and modeling. In this way, product engineering is sometimes referred to as an *open-ended process*.

The Design Approach. New products move through *stages of engineering design* and are responsive to the design criteria. In order to demonstrate these stages of design, the evolution of a new engine system is discussed in the next few sections of this chapter. For this system, we will assume that the design is started under the motivation of a new carburetor and combustion technology breakthrough to achieve less pollution in the combustion process.

Based on this motivation, management may authorize the engineers to synthesize, say, eight engine system *concept designs* and a set of design criteria against which to measure the designs. These first design efforts must necessarily be very experience-intense in order to present a reasonable set of systems to choose from, and to avoid unnecessary redesign and analysis in later stages of engineering. Thus, engineering new-hires are seldom involved in product concept design.

The main products of concept design are an artist's rendering of the best design(s) and a very rough estimate of the expected performance of the new systems. The best concept designs are usually selected on the basis of a few main features from the criteria list.

At this early stage of design, little is known of the details of the new engine concepts, so the analyses must be very rough. One or two digits of analysis accuracy is often all that is reasonable. Extrapolated data from similar systems of the past are often used in order to make quick assessments of potential.

Figure 1-1 is representative of the historical and theoretical data which is often used to make the concept assessments. Such figures as this represent a great deal of previous systems engineering. They can thus be of great help to guide the engineer in deciding how much risk (and cost) may be associated with a new design concept by revealing a departure from previous successful systems. Notice that these initial comparisons are only on the basis of a few main features of the design (horsepower, number of cylinders, and cylinder radius in the example). Based on these comparative analyses, the engineers may be able to reduce their choices from eight to five viable concepts for the new engine.

The next stage of design is *preliminary design*. This stage is "preliminary" in the sense that the design emphasis is to complete the selection of the system components prior to the design of the details of the system. In the engine system example, the engineers would do this using a process of top-down design. That is, they first

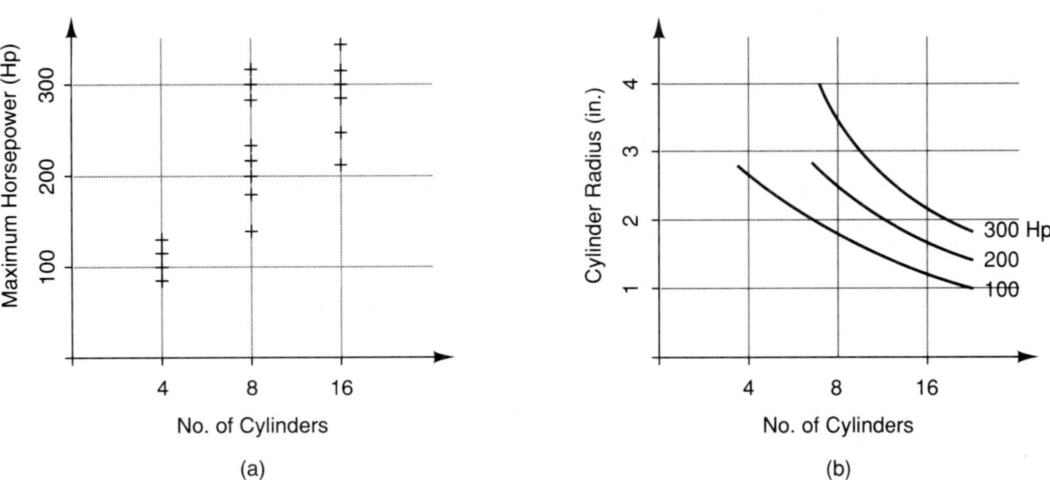

Figure 1-1. Engine Concept Design Data (hypothetical). (a) Past experience, (b) Theoretical maximum performance

express each engine system concept in terms of required supporting subsystems, they then express all the subsystems in terms of supporting components.

Top-Down Design. The system engineer begins preliminary design by applying a process of *functional requirements allocation* for each of the system concepts. This is done by first organizing each system concept into functional subsystems, and distributing all the system-level requirements among these subsystems.

By way of example, consider a computer design. If the computer is to cost $4,000 (a system requirement), then all of the part costs must not add up to more than $4,000 (the subsystem requirements). The designer can allocate the $4,000 in any fashion desired between the subsystems, but the total is fixed at the system level.

Similarly, if a car is to have a cruising speed of 200 MPH (a system functional requirement), this may require an engine with a torque of 1,000 ft-lb_f (the subsystem functional requirement). Note that some very crude analysis is usually required to accomplish these functional design allocations (torque-speed computations here and a survey of manufactured computer part costs above). Further, some preliminary assumptions must invariably be made (like the estimated mass of the car system and its projected friction losses here, and the best distribution of budgets above).

As implied in the examples above, there are many ways that the requirements could be grouped into the subsystems based on the assumptions for the allocations. Further, these functional groupings will ultimately be designed into hardware subsystems. Consequently, you can expect that there may be many different sets of hardware subsystems which could meet the system functional requirements. Thus, the main part of the hardware design problem is to sort through all the candidate subsystem collections to pick the best system.

The system engineers may continue through several layers of functional sub- and sub-sub-subsystems organization until coming to the smallest hardware pieces—the *elements*. The *components* are the functional groupings of the elements. (For example, capacitors and resistors are elements of tuner and transmitter components.)

The components are the lowest functional level for which the system engineers allocate design requirements. This is true since the elements are manufactured to standardized industry requirements and are not usually made to meet special requirements due to cost restrictions.

At the end of preliminary design, the engineers are able to show all the hardware components in the new product, and the design can be analyzed in enough detail for comparison against the criteria. In the previous engine example, this would mean that the engineers now have a good idea of the geometry and function of both the new carburetor and the new combustion chambers for the design.

However, it is not uncommon to find that there is no known analysis technique for the new components which are created in preliminary design. In some cases, engineers have been known to change their designs slightly in order to create a design which can be reliably evaluated by proven analysis methods. In other cases, designs are not built because of the extra risk associated with unreliable performance estimates. Laboratory research and testing are often used to help fill in the preliminary performance evaluations.

The major objective of this text is to present modeling methods which are appropriate to the evaluation of *dynamic physical performance* of electrical and mechanical systems in preliminary design.

In the last stage of design, *detail design*, the smallest details are designed and analyzed and a set of production drawings is created. Considerations during detail design seldom have any impact on the previous system configuration selections. The kinds of things which are configured at this time are:

- printed circuits;
- seals;
- pipes;
- cooling fins;
- fasteners;
- nuts and bolts;
- wiring.

Complementary Modeling and Design. So, the system engineer faces both a design problem and a modeling problem. The design problem is: given the performance requirements, what is the hardware system? The modeling problem is: given the hardware, what is the dynamic performance? These two problems are coupled together by the engineering question: does the proposed design meet the performance requirements?

Notice that, if the system already existed, it would likely be easier and cheaper to simply observe or measure the performance—there would be little need for modeling. A twist on this is that the system may exist, but an engineer is interested in how an improvement may affect performance. In any case, the problem considered in this text is to model the given hardware through a set of approximate equations which can acceptably predict the system response to input stimulation. The systems approach to this modeling problem is to begin with an analytical process called *functional reductionism*.

In reductionism, the given hardware system is first subdivided, or reduced, to its major functional subsystems. The subsystems are then, in turn, functionally subdivided. This process is repeated until the analyst has small enough pieces that physical modeling equations can be written. The technique of *multiport analysis* (Chap. 2) is an application of the process of functional reductionism to achieve a component-level functional systems model.

Note the similarity between functional reductionism in modeling and functional allocation in design synthesis. In applying reductionism, you attribute function *to* hardware. In applying synthesis, you create function *in* hardware. These efforts complement one another, and each lends insight to the other. Thus, design motivates modeling. But it is equally true that modeling motivates design since you can best know which designs to pursue through modeling. Consequently, the most effective engineers will be those who can achieve a balanced mastery of modeling and design.

HOMEWORK

1-1. An engineer wants to design a car which can accelerate uniformly from 0 to 60 ft/sec. in 3 sec. The mass of the car is to be 93 slugs. How much torque must the engine produce? Ignore the inertia of the drive train, the mass of the tires, and wind effects. Assume no slipping of the tires and a 1 ft. wheel radius. (hint, force $F = Ma$ and torque $Q = Fr$).

1-2. A house electrical system is in concept design. The system includes a power supply, a wiring network with outlets, and a light fixture attached to each outlet. It is estimated that the supply will yield 24 kilowatts. The supply is to provide current at a constant 120 v. (a) Estimate the current available at each outlet if there are to be 100 outlets in the house, all requiring the same amount of current from the supply, and all in operation at the same time? (hint, Power $= e\ i$) (b) What are the system and subsystems in this design?

1-3. As studied in physics, there are three types of models for solid materials; these are the point mass (*p*.), the rigid distributed mass (r. d.), and the flexible distributed mass (f. d.). Choose one of these three for the proper physical model for the following dynamic motions:

a) A bullet in flight;

b) A bullet at impact;

c) An airplane in steady, level flight through calm air;

d) An airplane in air combat maneuvering;

e) The planets orbiting around the sun;

f) The earth's rotation on its axis;

g) An automobile accelerating in a straight line;

h) An automobile cornering; and

i) An automobile crashing.

1-4. Using the two graphs in Fig. 1-1, (a) estimate the cylinder radius for a new internal combustion engine which will give a maximum power of 200 horsepower. (b) If a friend of yours wanted to design a four cylinder engine that produced 250 HP (he wants a simpler engine), what could you tell him about the risk in this based on the data of Fig. 1-1a?

1-5. An important part of concept design is understanding the functional basis of product competition. For a product of your choice, identify at least two primary functions which give that product an edge against its competitors. Also identify at least two secondary functions which are nice to have, but not critical.

1-2 SYSTEMS ENGINEERING IN INDUSTRY

In the discussion which follows, a medium- to large-company structure is used to demonstrate product engineering principles. Small engineering companies operate in much the same way, with fewer individuals performing more roles. In any case, all sizes of companies hire and train engineer specialists to develop their products. If more skill is required than the company staff has, then a highly skilled outside specialist is retained on a temporary basis to provide the needed analyses. Such consultants are usually college professors or recognized authorities from industry.

Many companies are organized with a simultaneous engineering hierarchy for *technology* management and a project hierarchy for *product* management. These two management structures work effectively together to create and support the company products.

Engineering Hierarchy. The description which follows assumes that you have a full-time job with an engine company. As an entry-level engineer, your first assignment might be a brief training assignment of six months to learn engineering skills for the company product line. You probably would start as an engineer at the bottom of the specialist's structure as shown at the left in Fig. 1-2. Your working group may be composed of you and four other engineers who all report to the same supervisor. As the junior person on the staff, you are engineer number five.

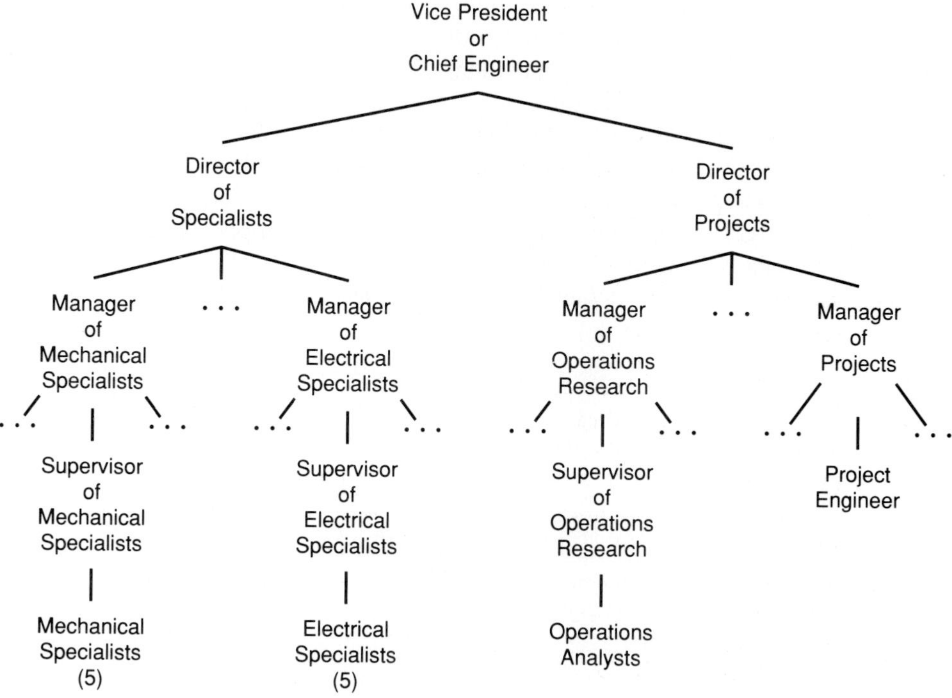

Figure 1-2. Company Engineering Management

Your supervisor is responsible for all hiring, firing, and promotions, and he ensures that the engineers in the group become proficient in the *technology* which supports the company product line ("technology" is the application of materials and methods, like aerodynamics and composites). The supervisor reports to a manager, who reports to a director, who reports to the chief engineer or vice president of engineering.

Project Hierarchy. Notice that Fig. 1-2 also has a branch to the right for project engineers. Project engineers report through a similar structure as the specialists, but they focus on products (systems) rather than technology. That is, they are generally the engineers who are in charge of product concepts, and they manage the system functional evolution which was discussed earlier in this chapter. The project engineers work closely with the customers, the top levels of the company, and the design engineers to make the value judgments necessary to find and develop worthwhile products.

Project Operation. Your first engineering assignment may be to a project under the control of a project engineer. A typical project structure is shown in Fig. 1-3. The specific members of the project are identified at the conclusion of concept design, and the project runs at least through preliminary design. If the concept survives preliminary design, then the project team is greatly expanded in order to complete the many detail design tasks which will be required for the creation of production drawings. Notice that the project team in this example includes an electrical engineer, a mechanical engineer, and an operations analyst, all of whom must work together to develop the preliminary design.

The *operations analyst* (OA) is a person who specializes in understanding the environment of the product during the future years in which it is to be manufactured and supported. In the case of a new engine concept, the OA might make estimates of the cost of fuel for the years that the engine is to be supported. The OA might also make estimates of the engine competition in those years. The engineers might make performance estimates of fuel consumption and mileage. Based on these estimates,

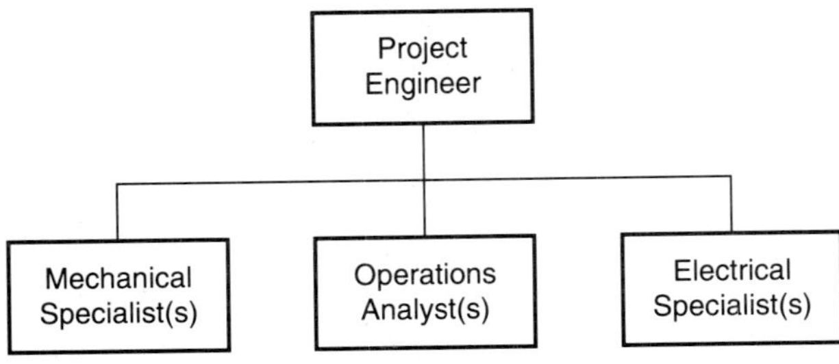

Figure 1-3. Typical Project Structure

the project engineer can begin to evaluate profitability and competitiveness of the engine concept.

The OA also evaluates how the product operates in its projected environment, except for performance. For example, returning to the design requirements list mentioned in the last section, the operations analyst may be responsible for evaluations of survivability and human operability. This does not mean that the engineers can ignore these areas in the creation of their designs. It does mean that the OA is the evaluation specialist in these areas.

The assignments of the five engineers in your work group may be as shown in Fig. 1-4. For example, suppose that you are assigned to work on engine concept five by your supervisor. The figure shows that you will be working under project engineer two for product development, who also has responsibility for concepts three and four. Your supervisor will monitor your progress and see that you get help if you need it. Meanwhile, project engineer one has responsibility for concepts one and two. Such a style of organization is called *matrix management*, and is very common in medium and large companies.

Your initial focus for work on engine concept five is to design and model an electrical or mechanical hardware subsystem to meet or exceed the static and dynamic performance thresholds established in concept design and allocated to your subsystem. You are told that some, but not all, of the engine components have been identified, and you may have to include one or more of these components in your subsystem design. Further, you are told what *interface requirements* you must meet with the other subsystems. These interface requirements ensure that the whole engine system remains viable, and they allow specialists on the project team to effectively work together.

In working on a very large project, many types of specialists are often present. Some common types of engineering specialization are shown in Table 1-1.

While the list of specializations shown in Table 1-1 is far from complete, it does show those which are used to design and analyze many different types of common products. This text presents introductory modeling methods in each of these specialization areas except for controller design, which is left as a follow-on study.

	Project Engineer 1		Project Engineer 2		
Specialist	Concept 1	Concept 2	Concept 3	Concept 4	Concept 5
Electrical	1	2	3	4	5
Mechanical	1	2	3	4	5
Operations Analyst	1	2	3	4	5

Figure 1-4. Typical Engineering Work Assignments, by Seniority

Table 1-1. Engineering Specializations

I.	Mechanical Engineering
	A. Structures
	B. Fluids
	C. Thermodynamics
II.	Electrical Engineering
	A. Electronic circuits
	B. Computers
III.	Controls
	A. Sensors
	B. Actuators
	C. Controller design

HOMEWORK

1-6. Write out a brief definition (at least two to three sentences) of the following terms:

a) Systems hierarchy;

b) Synthesis;

c) Top-down design;

d) Concept design;

e) Preliminary design;

f) Requirements allocation;

g) Detail design;

h) Project engineer;

i) Matrix management;

j) Operations analyst; and

k) Reductionism.

1-3 CREATING USEFUL MODELS

In light of the previous discussion, it is very important that engineers have modeling tools which they can use to accurately predict the dynamic performance of a given hardware design. To meet this need, engineers turn to the three-step systems approach to modeling:

1. Functional analysis of the design, followed by,
2. Physical analysis of the design to create an idealized model, and
3. Mathematical analysis of the model.

The goal of the preliminary-design functional analysis is to produce a diagram of the cause-and-effect relations existing between the hardware components. This diagram is very useful for both modeling and design, but perhaps its best feature is that it clearly shows *how* the components physically operate within the system.

The goal of the physical analysis is to find the equations of dynamics (e.g. the model) for the functional components. Such a numerical representation of the performance is important in assessing the merit of the design. For example, if an expensive part change is designed but a model shows only a negligible improvement in performance, then it is likely to be a highly questionable change. Such mathematical modeling analyses are aided by the systems approach to function and physical analysis.

Functional Analysis. The systems approach to modeling makes heavy use of cause-and-effect functional block diagramming, such as that shown in Fig. 1-5. This figure shows the subsystem-level *multiport diagram* of a general system (contained in dashed lines) as a power-source subsystem interacting with a load subsystem and influenced by one external input. Note that the arrows in the diagram indicate the cause-and-effect relationships. They show that the subsystems affect one another,

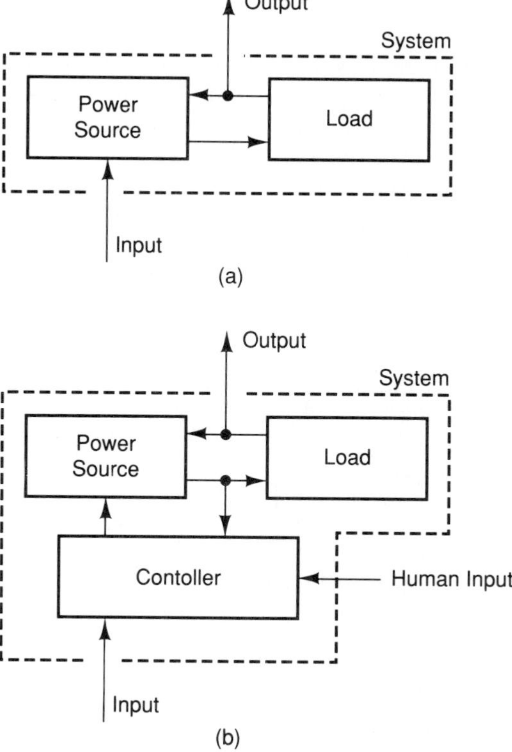

Figure 1-5. General Subsystem Multiport Diagrams. (a) The plant, (b) the controlled plant

and the system inputs and outputs. Here, the system output in the figure is a measurement of the load output, and the system input is the fuel. So, in a general sense, this plant may be thought of as a processor which changes fuel to an output.

A system may also have a *controller* by which it is made more useful to a human operator, with an arrangement such as that shown in Fig. 1-5b. Here, the controller (perhaps a computer) measures a source output and uses it, along with the human input, to control a source input to accomplish what the human desires. Thus, the controller operates to keep the system in a desirable operating condition. This arrangement is called a *controlled system* or a *controlled plant*.

In general, the *load* acts as an uncontrollable disturbance to the operation of the power source, and is often excluded from the system. This idea can be seen in the operation of an electric power generator as the electric load is varied (say, by switching on or off various connected appliances), or in the atmospheric disturbance of an airplane as it goes through various maneuvers. Sometimes, an electrical load is partially controlled during operation through various inputs. Similarly, a mechanical load may be partially controlled through transmission inputs.

In this text, the term *system* applies equally well to the plant, the controlled plant, or any part thereof. Control engineers build a model of the plant in order to design the controller. System design engineers build a model of the controlled system in order to better understand system design tradeoffs, say between competing component candidates.

So, the choosing of the system boundary is the first step of the functional analysis. If the system is chosen without a power source, then it would be a *passive system*. If the system contains a power source, then it is an *active system*. In all of these cases, the methods of this text will work quite well.

Figure 1-6 shows the first step in an internal combustion engine multiport analysis by showing the major *interacting external systems*. This is necessary to show the *environment* in which the system of interest (the engine) will perform. The

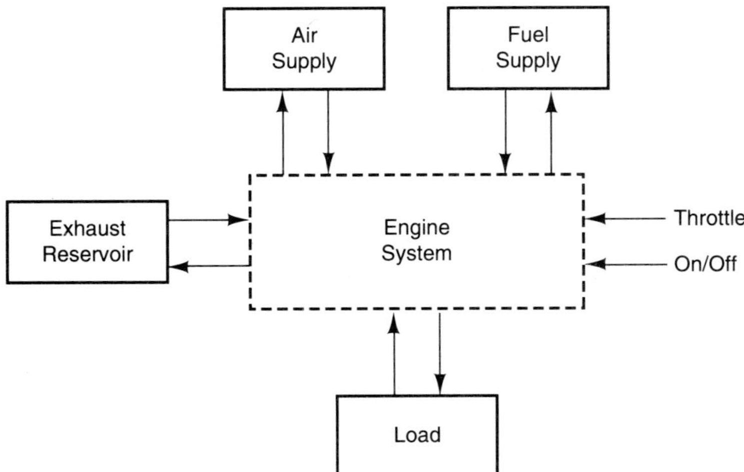

Figure 1-6. Engine System Multiport Diagram

interacting external systems are the air, fuel, exhaust, and load systems. The interaction of these systems with the engine is shown by the double arrow which connects them. There are certain conventions which guide in the labeling of the connecting arrows, these are discussed in Chap. 2 where the method of multiport functional analysis is presented in more detail. For the present, the arrows merely show the mutual interaction of the connected system parts through "ports" (arrow pairs) for power transfer. Since there are usually many ports in such a figure, the method is termed a "multiport" analysis.

Figure 1-7 shows the first level of engine system reduction. In this figure, the engine system is broken down into its major *subsystems*. Note that a "starter" subsystem is not modeled since starting dynamics are not of interest to the modeler. Again, in the figure, a double arrow is used to connect interacting subsystems.

Notice also the *idealized* engine interaction with exhaust, fuel, and air systems as shown through the one-way connecting arrows in Fig. 1-7. This is possible under the physical assumption of an infinite air source, an infinite exhaust reservoir, and an infinite fuel supply. Thus, the engine demands something from these supporting systems, but their capacity is assumed to be so great that this demand does not affect the supply to the engine. It is through such physical idealizations that the analysis is sim-

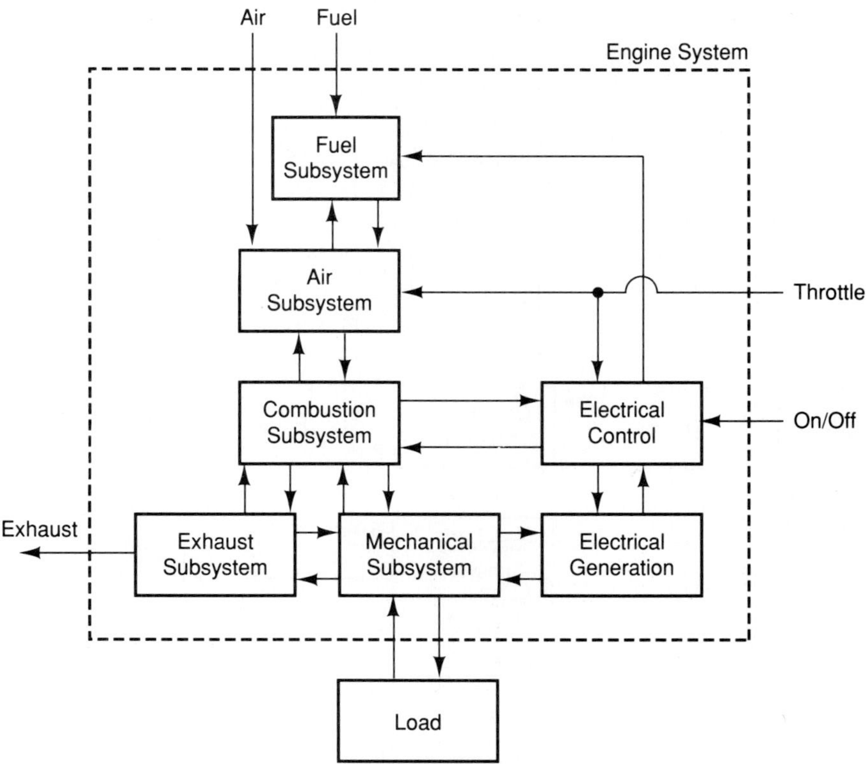

Figure 1-7. Engine Subsystem Multiport Diagram

plified and concentrated to focus only on that which is important, the dynamic response of the engine.

In Fig. 1-8, the air, combustion, fuel, exhaust, and electrical control subsystems have been reduced to the *component level*. Recall that it is at the component level that the physical parts become clear. That is, a subsystem figure shows functional relationships, while the component figure extends those relationships to the interaction of physical parts.

The multiport diagram created by the multiport analysis method (Fig. 1-8) thus offers several advantages:

1. It allows the modeler to express a complex system in terms simpler, more understandable physical components;
2. It shows cause and effect of the components;
3. It allows the modeler and the designer to focus their attention on the small pieces of the system during the physical modeling process; and

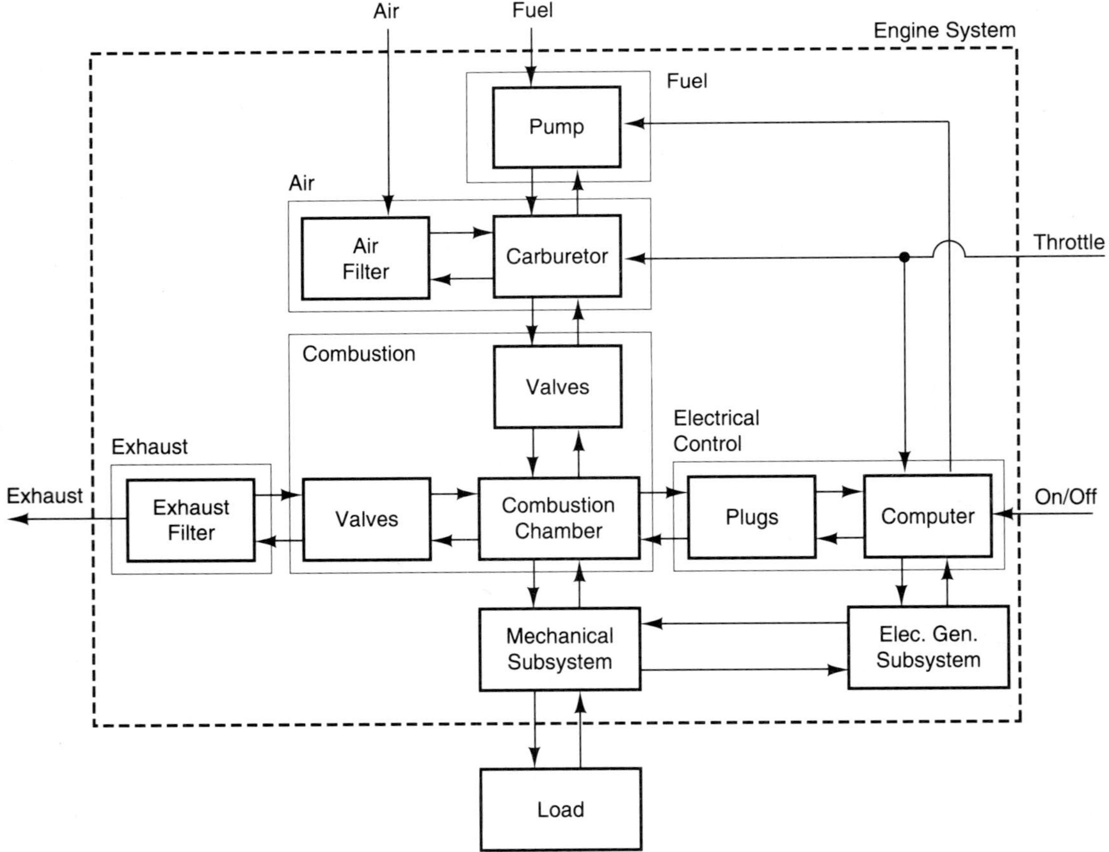

Figure 1-8. Engine Components and Subsytems Diagram

4. It provides a framework for the assembly of equations to form the overall mathematical system model.

The translation of the components in the functional diagram into equation-set models is accomplished through physical analysis.

Physical Analysis. Analysis of the system inputs offers the first guidance in the physical analysis. For example, designers often use *worst-case conditions* in terms of environment and inputs for the evaluation of their designs. This is done under the assumption that adequate performance under the worst case will yield a design that is adequate for any performance requirement. So, engineers evaluate electric generators while powering the maximum number of devices, they analyze automobiles over the most demanding terrain, and they model aircraft during the most severe maneuvers.

Further, designers know that the system inputs dictate what type of physical components are needed to create the desired system outputs. And modelers know that the inputs also thus dictate the approach to component modeling. (The physical inputs and their models are considered in detail in Chap. 3.)

Given a proper analysis of the system inputs, the component physical modeling approach then uses one of two methods: either analyzing behavior from physical principles (called the "fundamental" approach) or measuring the component's input/output empirical relationships. Newton's Second Law, $\Sigma F = \mathrm{M}\, dV/dt$, is a fundamental model of particle dynamics. The Ideal Gas Law, $P/\rho = \mathrm{R}\theta$, is an empirical model measured in laboratories. Often, during the physical analysis of a single complicated system, both modeling methods are needed to complete the analysis.

Sometimes, more detail than the component level is needed to perform the analysis. This is often the case when modeling from first physical principles. In such cases, the *element* level is the next step in reductionism. An example of this is the reduction of an engine fuel-pump component (Fig. 1-8) into its various elements for analysis (gears, pistons, electronic elements, motor elements, etc.).

Chapters 4 and 6 discuss physical modeling using a menu of fundamental building-block elements. This systems approach to fundamental physical models allows the modeler to readily choose a simple model to capture the physics of many types of common components.

Chapter 7 examines the experimental and theoretical methods for the input/output empirical modeling of components. The danger here is that the component may be modeled by too much or too little detail, or by wrongly assuming a physical form. However, the chapter also presents guidelines to sustain the modeling effort towards a simple-yet-accurate model.

HOMEWORK

1-7. Given the following hardware list, identify those parts that could be modeled as components and those which should be modeled as elements:

a) A resistor;

b) A simple radio;

c) An automobile alternator;

d) A capacitor;

e) An inductor;

f) A fuel pump; and

g) A spring.

1-8. Given that a system is composed of three subsystems (A, B, and C) which all physically interact with each other. Show this interaction through a subsystem-level multiport diagram.

1-9. What are the parts of a plant in a subsystem-level multiport diagram?

1-10. An electronics engineer has estimated that two damaging voltage spikes may occur during switching of the load lines for his new electronic system. One spike has a peak of 200 volts, the other has a peak of 310 volts. The engineer decides to analyze his system hardware only for the worst spike of 310 volts. What principle has been used in making this decision?

1-4 SYSTEMS MATHEMATICAL ANALYSIS

The purpose of the system mathematical analysis is to accomplish three tasks:

1. To combine all the component equation sets into one set of equations to model the system;
2. To evaluate the suitability of the system model; and
3. To use the system model to find the numerical predictions of system performance for the given system inputs.

The mathematical analysis begins with the component models found from physical analysis.

Component Models. A component model appears as a set of equations, one for each of the component's outputs. For example, a component on the edge of a system may appear in a functional diagram as shown in Fig. 1-9. Here, the component is stimulated by two system-*internal* inputs (U_1 and U_2), and two system-*external* inputs (U_1^* and U_2^*). The component responds with two system-*internal* outputs (Y_1 and Y_2), and one system-*external* output, Y_1^*. Notice that the system-external output Y_2^* is assumed to be negligible, as indicated by the dashed arrow shown in the figure.

In modeling a component, you create steady and dynamic equations. A steady-model equation has no time derivatives. A dynamic-model equation may or may not have time derivatives.

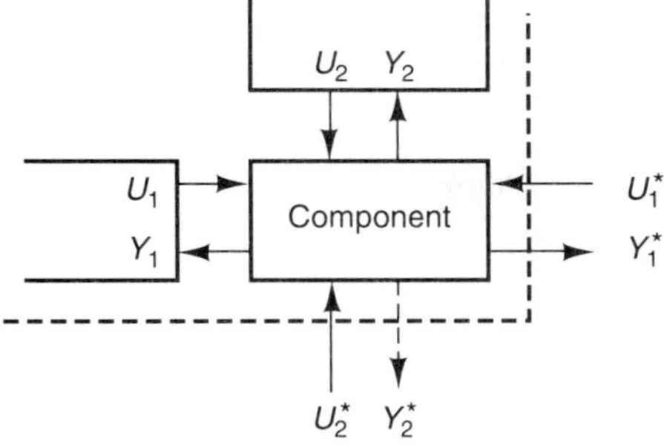

Figure 1-9. General Component Diagram

Thus, a steady model for the component of Fig. 1-9 is

$$Y_1 = f_1\,(U_1, U_2, U_1^*, U_2^*), \tag{1-1}$$
$$Y_2 = f_2\,(U_1, U_2, U_1^*, U_2^*), \tag{1-2}$$
$$Y_1^* = f_3\,(U_1, U_2, U_1^*, U_2^*). \tag{1-3}$$

These algebraic equations can be expressed in the more general vector form

$$\boldsymbol{Y} = \boldsymbol{f}\,(\boldsymbol{U}, \boldsymbol{U}^*), \tag{1-4}$$

where $\boldsymbol{Y} = (Y_1, Y_2, \ldots, Y_N, Y_1^*, Y_2^*, \ldots, Y_L^*)^T_,$ (1-5)

$\boldsymbol{f} = (f_1, f_2, \ldots, f_M)^T,$ (1-6)

$\boldsymbol{U} = (U_1, U_2, \ldots, U_N)^T,$ (1-7)

and $\boldsymbol{U}^* = (U_1^*, U_2^*, \ldots, U_L^*)^T.$ (1-8)

Notice that each component's outputs are its neighbor's inputs. While this may seem confusing, it is important to realize that the multiport diagram expresses these interconnections to ensure that there are enough equations, with enough unknowns, to give a sufficient equation set at the system level. This means that you need only focus on each component as you model. All component interactions are taken into account through the multiport diagram.

Equation 1-4 is the general form of an *algebraic* component model. This form may correspond to either a steady or a dynamic model. However, before dynamic models can be discussed further, some systems terminology must first be presented.

Terminology. If a component has a single input with a single output, then it is called a *SISO* component. Otherwise, it is grouped with the multiple input, multiple output *MIMO* components.

Simple SISO components were extensively analyzed during the formative years of systems analysis. Consequently, there is a large body of unified reference material for SISO analyses. In contrast, MIMO components are much more difficult to model in general, so the modeling theory is more fragmented. The most common analyses for MIMO components are in the areas of computer analysis of input/output data histories, and modeling from first physical principles, both of which are discussed in this text.

All components are either analog or quantizing. *Analog* components produce output histories which can take on any value in the design range. *Quantizing* components have discrete levels to their outputs, and are also called time-sampled components. They are usually associated with computers or computer parts in the component. Notice that it is very likely that there will be both quantizing and analog parts in modern systems. Modeling methods for both these types of components are discussed in this text.

Given models for these different types of components, it becomes necessary to combine them into a system model. This combination is very easily done using the multiport diagram, as discussed in Chap. 8.

After the component equations are combined, they are transformed through a change of variables into a more suitable form for numerical processing. The variables used in this transformation are very useful in avoiding the problem of overmodeling the system, so modelers using the system approach seek them out in every modeling situation. They are called *state variables*.

System State Models. The *state* of the system is synonymous with the fundamental condition of the system. This fundamental condition is expressed through a set of characteristic variables called the *state variables,* or simply *states*, indicated by the vector symbol $\boldsymbol{X}$. These variables are "characteristic" in the sense that they come from the characterizing differential equations in the model of the system response, as will be shown in Chap. 5.

Given the state variables, all other variables of interest can be readily computed using state transformation equations. Thus, the modeler needs only to predict the state variables in order to express the condition of the system at any time desired.

The system state model is said to be *predictable* if all variables in the system can be calculated for all $t \geq 0$. This requires three things:

1. A mathematically sufficient equation set;
2. The initial conditions for all differential equations; and
3. Knowledge of the expected time histories of all external inputs ($\boldsymbol{U}^*$).

The system equation set is sufficient if all the component equation sets are sufficient. This, in turn, is true if all the component outputs are properly modeled by equations, and if the multiport analysis method is used (it ensures correct mathematical coupling of the components).

The other two requirements for predictability are related to solving the system differential equations, and are necessary to project the initial conditions forward through time in response to the external input stimulation.

At the end of modeling, the system model appears as a set of *state equations* to predict the states, $\boldsymbol{X}$. This set is composed of first-order, linear or nonlinear, differential equations in terms of the states and the external inputs. These equations are often expressed in the general vector form

$$dX/dt = \boldsymbol{G}(\boldsymbol{X}, \boldsymbol{U}^*), \tag{1-9}$$

where

$$dX/dt = (dX_1/dt, dX_2/dt, \ldots, dX_k/dt)^T. \tag{1-10}$$

This is very convenient since equations in this form are ideally suited to processing on a digital computer. Consequently, the state equations are extensively used in numerical solution methods.

The integration of the system state equations may be possible using the analytical theory of differential equations. However, in all but the simplest linear equations, adequate mathematics simply do not exist to do this. Instead, modelers routinely rely on computer-based methods to integrate the equations and predict the response of the system to the inputs of interest. This computer-based prediction is called *simulation*. It is a powerful tool, based on the science of numerical analysis, for analyzing the dynamic performance of new or existing systems. This text discusses the solving of differential equation models using both analysis and simulation.

Thus, the engineer can predict a response time history for all variables of interest in a given system. Armed with such a predicted system response, the engineer is in a position to evaluate the system preliminary design with respect to the criteria of dynamic performance.

HOMEWORK

1-11. Briefly define the following component descriptive terms in one or two sentences:

a) SISO

b) MIMO

c) Analog

d) Discrete

1-12. Define the following modeling terms:

a) State

b) Simulation

c) Predictable system model

1-13. An engineer has designed a SISO component which has the following differential equation model,

$$C\, dY/dt + 3Y = U^*,$$

with $U^* = 0$ for $t \leq 0$, and $U^* = 6$ for $t > 0$, and $Y(0) = 0$.

The engineer has two elements to choose from for the component, as represented by the constant C in the above equation. These elements can be represented by either of two values as follows,

$$\text{for element 1: } C = 3,$$
$$\text{and for element 2: } C = 6.$$

a) Which element will give a higher value of Y at $t = 1$ sec.? (Hint: solve the equation with element 1's value and compare it to the solution with element 2's value).

b) Which element gives a higher value of Y at $t = \infty$ sec.?

1-14. A SISO component has the input/output equation,

$$Y = (U^*)^2/18.$$

a) Plot the output of the above component on the same Y-time axes as the output for the "element 1" component ($C = 3$) from the previous problem. Subject the components to the same input and initial condition.

b) Which component reacted quicker to the input according to your plot?

1-5 SUMMARY, BIBLIOGRAPHY, AND REVIEW PROBLEMS

Chapter 1 was an overview of system design and modeling using the systems approach. In this chapter, the motivation and guidance for modeling was presented as coming from the system design process, where engineers move ideas from concept design, through preliminary design, and into detail design. And each of these stages requires a suitable model. However, the preliminary-design model was selected for study in this text since management uses it to make a decision on whether to continue into detail design or to drop the concept based upon performance estimations.

The many performance requirements typical of a preliminary design were briefly discussed, and the dynamic performance was selected for modeling from the requirements list as the focus for this text.

So, the preliminary-design modeling procedure to be followed in the text is composed of three steps:

1. Functional analysis of the system. This is done by building a multiport diagram down to the component level. This translates the given system hardware into a correctly connected set of functional components. Methods for this analysis are discussed in Chap. 2;
2. Physical analysis of the components. This begins with an analysis of the system inputs, as discussed in Chap. 3. It continues with a study of fundamental physical component modeling techniques in Chaps. 4 and 6. It is completed in Chap. 7, where the methods for input/output modeling of components from experimental

data are presented. The result of these analyses is the mathematical equations which model the component responses;

3. Mathematical analysis of the system. The first component mathematical analyses in the text appear in Chap. 5 where the properties of differential equations are studied from a systems modeling point-of-view to gain insight for physical model building. In Chap. 8, the multiport diagram is used to mathematically combine the component equations to find the system model, which is then used to predict the dynamic system response.

BIBLIOGRAPHY

Beam, W.R., *Systems Engineering*, New York, N.Y.: McGraw-Hill Book Co., 1990.

Blanchard, B.S. and W.J. Fabrycky, *Systems Engineering and Analysis*, Englewood Cliffs, N.J.: Prentice-Hall, Inc., 1990. This is a very readable text which covers the process of system engineering and the many support disciplines required for its practice.

Bronikowski, R.J., *Handbook of Product Design for Manufacturing*, New York, N.Y.: Van Nostrand Reinhold Co., 1986.

Christiansen, D. (ed.), *Engineering Excellence*, New York, N.Y.: IEEE Press, 1987.

Dasgupta, S., *Design Theory and Computer Science*, New York, N.Y.: Cambridge University Press, 1991.

Dieter, G.E., *Engineering Design*, New York, N.Y.: McGraw-Hill Book Co., 1991.

Eisner, H., *Computer-Aided Systems Engineering*, Englewood Cliffs, N.J.: Prentice-Hall, Inc., 1988. This is one of the few design-oriented texts which discusses the national standards of design that are usually imposed by government authorities.

Jacoby, S.L.S. and J.S. Kowalik, *Mathematical Modeling with Computers*, Englewood Cliffs, N.J.: Prentice-Hall, Inc., 1980.

Kamm, L.J., *Successful Engineering*, New York, N.Y.: McGraw-Hill Book Co., 1989.

Kantowitz, B.H. and R.D. Sorkin, *Human Factors: Understanding People-Systems Relationships*, New York, N.Y.: J. Wiley & Sons, 1983.

Middendorf, W.H., *Design of Devices and Systems*, New York, N.Y.: Marcel Dekker, Inc., 1990.

Mittra, S.S., *Structured Techniques of System Analysis, Design, and Implementation*, New York, N.Y.: J. Wiley & Sons, 1988.

Pahl, G. and W. Beitz, *Engineering Design*, New York, N.Y.: Springer-Verlag, 1984.

Parsons, H.M., *Man-Machine System Experiments*, Baltimore, Maryland: The Johns Hopkins Press, 1972. This book reports on a series of human factors experiments to explore the man-machine interface.

Sanders, M.S. and E.J. McCormick, *Human Factors in Engineering and Design*, New York, N.Y.: McGraw-Hill Book Co., 1987.

Smith, D.B. and G. Rowland, *Systems Engineering and Management*, Reading, Mass.: Addison-Wesley Publ. Co., 1974.

Turner, W.C., Mize, J.H., and K.E. Case, *Introduction to Industrial and Systems Engineering*, Englewood Cliffs, N.J.: Prentice-Hall, Inc., 1987. This is an excellent management-oriented text.

REVIEW HOMEWORK

1-15. An engineer wants to design a lighting system for a remote location as shown in Fig. 1-10. The system is to be composed of a 100 watt power source, $\mathcal{P}$, which supplies a light bulb of resistance R. The engineer must choose from three off-the-shelf bulbs made by three manufacturers:

R_1 = 100 ohms, standard size, cost = \$0.80,
R_2 = 120 ohms, non-standard size, cost = \$1.20,
R_3 = 150 ohms, standard size, cost = \$1.00.

The system is to be designed for lowest cost, such that the current in the wires, i, is 0.85 amps ± 10%. Which bulb should the engineer choose? (Hint: $e = i\text{R}$, $\mathcal{P} = ei$).

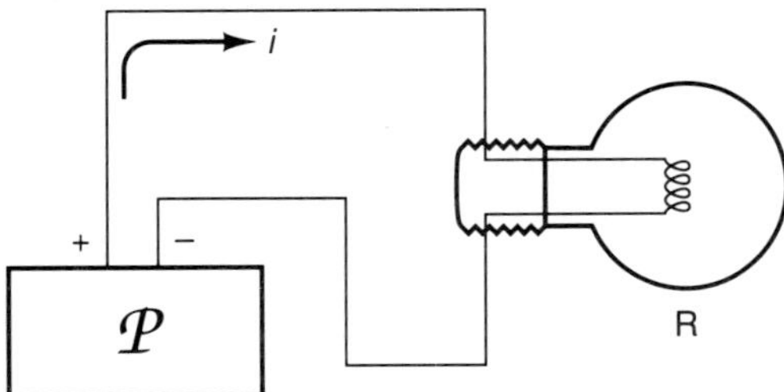

Figure 1-10. Schema for Prob. 1-15

1-16. An engineer wants to design a machine (a system) for removing the square sealing nut from the top of unpressurized cylinders (shown in Fig. 1-11a). To do this, a set of measurements of the loosening force using a set of wrenches is first made and plotted, as shown in Fig. 1-11b.

The engineer's company wants the loosening torque motor to supply a torque at least 10% greater than the largest measured torque, at the least cost. The engineer has found four motors as shown in the list below. Which motor should the engineer choose?

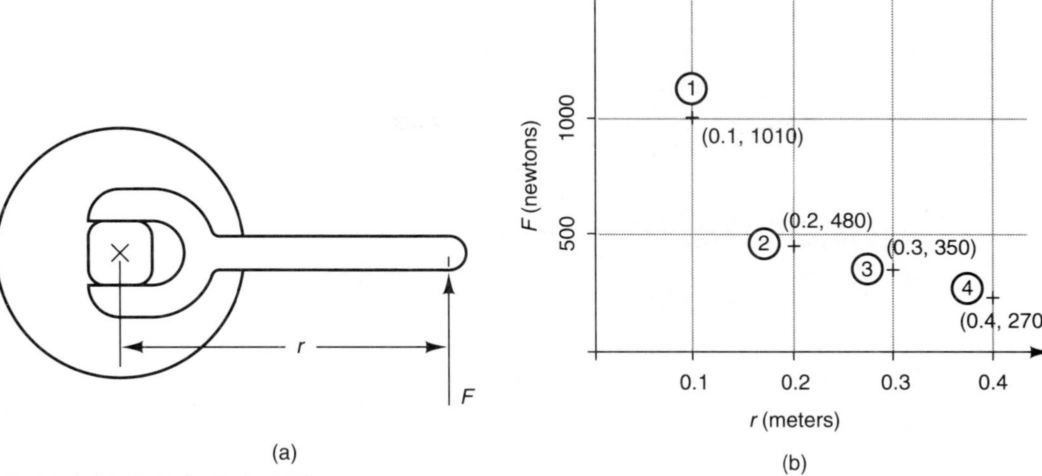

Figure 1-11. Data for Prob. 1-16

	Loosening Torque	*Cost*
Motor 1	110 (*n-m*)	$100
Motor 2	115	$200
Motor 3	120	$300
Motor 4	125	$400

1-17. Indicate whether each of the following is used in either modeling or design:

a) Multiport diagram

b) Reductionism

c) Synthesis

d) Simulation

e) Requirements allocation

f) Performance requirements

g) System states

1-18. Name and briefly describe the three stages of design for complex systems.

1-19. A new computer concept, based on a new microchip, is being considered. Briefly describe why it is *not* appropriate to try to analyze how the printed circuit boards will be mounted onto the computer frame before the first computer system concept is evaluated for performance?

1-20. Briefly describe why design is called a process of system synthesis.

1-21. Name and briefly describe the three steps of engineering systems modeling.

1-22. Briefly describe why engineering systems modeling is called a process of reductionism.

Chapter 2

Functional Analysis of a Multiport System

2-0 INTRODUCTION

Multiport analysis derives its name from its origin in the electronics industry where a pair of terminals was first called a *port*. That is, electrical engineers thought of a pair of terminals as a port through which electrical energy or power could enter and leave an electrical device. Many devices had more than one port, hence they were called *multiport* devices. However, systems engineers have since generalized these concepts to study many types of systems, not just electrical ones.

As shown in the previous chapter, the modeler uses multiport functional diagramming to organize and segment a system dynamic analysis based on component function. This method allows the modeler to later focus on the physical modeling of the small parts of the system (the components and elements) with the assurance that the connections of these parts are properly modeled. The modeler achieves this assurance by being very careful to consider the functional *connectivity* and *causality* of the parts as the system is reduced down to the component level. Here, the connectivity of a system has to do with *what* parts are connected together, the causality has to do with *how* they are connected.

The objective of this chapter is to present the multiport analysis method for system functional block diagramming, and especially to develop an understanding of the principles of system connectivity and causality.

The functional modeling begins with the *isolation of the system* to be studied. Thus, a system boundary is drawn to limit the analysis to the system of interest in preliminary design. The free body diagrams of statics and dynamics are applications of this system-isolating concept.

When the system boundary is drawn, connections are cut between the system (whatever is inside the boundary) and its environment (whatever is outside the boundary—conceptually shown in Fig. 2-1).

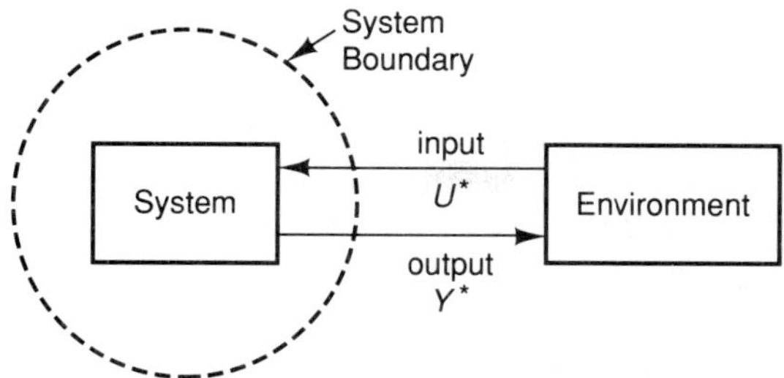

Figure 2-1. A System with One Port to the Environment

Each system-environment connection which is cut must then be represented by an external port to show this interaction. Each external port has one or two system inputs (inward-directed arrows labeled U^*) and one or two outputs (outward-directed arrows labeled Y^*). Thus, each external port graphically represents the essential interactions at that location. Further, since the "system" may be chosen arbitrarily (as far as systems theory is concerned), you can correctly expect that the internal component-connecting ports are similar to the external ports.

In this text, four types of ports are discussed:

1. Structural;
2. Electrical;
3. Fluid; and
4. Thermal.

These four port types are adequate to analyze a wide variety of engineering systems, they will serve you well in the study of external and internal connectivity, and in system modeling in general.

In introducing this topic of system functional analysis, this chapter is organized into four sections: the first two deal with defining the external-connective ports, the third presents a discussion of reductionism and internal connectivity, and the fourth presents the topic of causality and the selection of the specific port variables.

2-1 EXTERNAL CONNECTIVITY—STRUCTURAL AND ELECTRICAL PORTS

The study of external connectivity begins by reviewing the concept of *structural isolation* which is necessary for the analysis of structural systems. This concept is then generalized for use in identifying the other types of physical system-environment connections.

The *free body diagrams* of statics and dynamics are usually the first rigorous examples of system isolation to which the engineering student is exposed. In those diagrams, all significant external interactions with the body are represented by forces, moments, and kinematic constraints acting on a sketch of the body. In this way, the real physical situation is abstracted so that the modeler can focus on the body/system of interest. In statics, the known forces are used to determine the unknown forces; in dynamics, the forces are used to determine the motion.

In multiport functional analysis, the system boundary also identifies the free body for structural analysis. And it identifies all ports for power ($\mathcal{P}$) into and out of the system of interest, both structural ports and otherwise.

The three types of *structural ports* for structural power flow are shown in Fig. 2-2 (note the dashed lines which show the system boundary). Note that the structural ports are defined in such a way as to account for translational power separately from rotational power, if possible. This is done since translational motion is functionally different from rotational motion. They can exist separately or at the same time. Therefore, two ports are provided for the two different *functions* of translation (ST) and rotation (SR).

The ST port is used to account for the structural translational power. Translational power ($\mathcal{P}_T$; units: ft-lb$_f$/s, n-m/s) occurs wherever the structure has both net force ($\boldsymbol{F}$; lb$_f$, n) and velocity (V; ft/s, m/s) along the same direction. This power may be computed using the inner product of the force and the velocity,

$$\mathcal{P}_T = \boldsymbol{F} \cdot \boldsymbol{V}. \tag{2-1}$$

The SR port is used to account for the structural rotational power. Rotational power ($\mathcal{P}_R$) occurs wherever the structure boundary has both moment ($\boldsymbol{Q}$; ft-lb$_f$, n-m) and angular velocity ($\boldsymbol{N}$; rad/s, rad/s) about the same axis. Thus,

$$\mathcal{P}_R = \boldsymbol{Q} \cdot \boldsymbol{N}. \tag{2-2}$$

Note in Fig. 2-2 (1a) that the boundary cuts through a pin joint without friction. No moment is possible at such a joint, therefore, the rotational power is always zero there. The function of the joint is thus to restrain translation without applying moment, and the horizontal motion gives an ST port. In Fig. 2-2 (1b) the boundary cuts through an axis of rotation where a pure moment (the double-headed arrow) is applied—this is an SR port.

The SC port is provided for those times when the two kinds of function cannot be separated.

Notice that the examples shown in Fig. 2-2, and the problems in this text, are chosen to demonstrate the *concept* of the ports, rather than exhaustive applications. So, it is not uncommon to find structural ports in three-dimensional motion problems. And the use of structural ports in such problems rests on the expansion of the inner products of Eqs. 1-1 and 1-2. That is, the ports are applied on a dimension-by-dimension basis, by assigning a port for each dimension (an example of this is presented later in the text). For now, it is only important to realize that the structural ports can accommodate any dimension of motion.

Also, the ports of Fig. 2-2 are used to model the *essential* nature of the structure/environment interaction. That is, a two-dimensional pin joint similar to Fig. 2-2 (1a) may in reality have some rust on it, thus providing some negligible moment about the pin. So, some physical modeling may be used at the start to simplify the analysis. Thus, it is assumed that the pin's *primary* function is to restrain translational motion, so the ST model is used rather than the SC model for the port.

Further, the ST port can be used to show the effect of forces on the structure without contact. The ST port thus provides a way for the structure to translate in response to gravity. Similarly, the SR port can be used as a way to account for rotational power entering the structure due to applied moments without contact. In

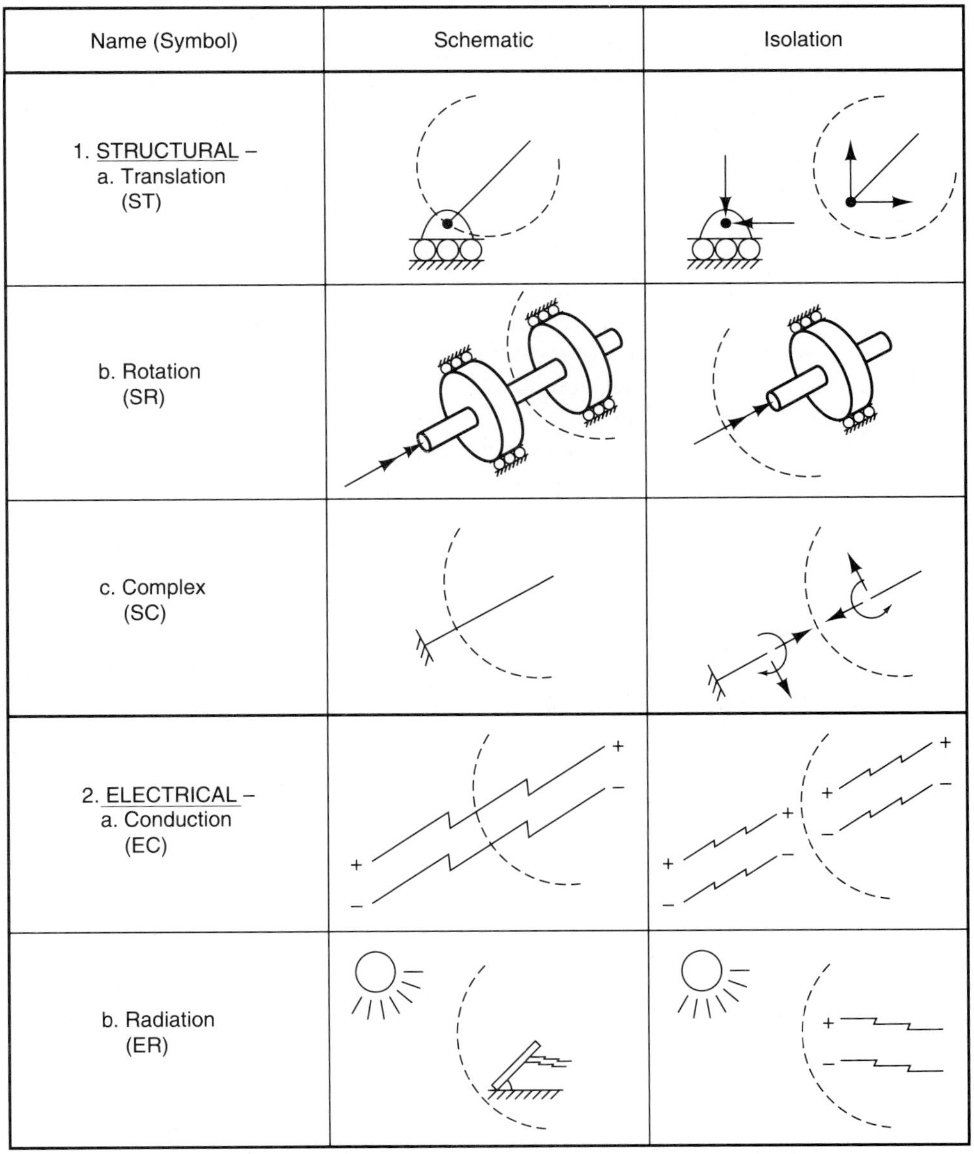

Name (Symbol)	Schematic	Isolation
1. STRUCTURAL – a. Translation (ST)		
b. Rotation (SR)		
c. Complex (SC)		
2. ELECTRICAL – a. Conduction (EC)		
b. Radiation (ER)		

Figure 2-2. Structural and Electrical Ports

fact, many different types of field force effects can be modeled through the structural ports.

The *electrical ports* operate to provide access for electrical power to enter the system. The electrical power ($\mathcal{P}_E$; watts) is a product of the voltage difference (E; volts) and current (I; amps) in the wires which are cut at the system boundary,

$$\mathcal{P}_E = EI. \tag{2-3}$$

The two electrical ports which are considered in this text are shown in Fig. 2-2 (2a and b).

The conduction port (EC) is for those times when the system boundary cuts through connecting leads of any type. The most common case is two conducting leads which supply either DC power or low-power, single-phase AC circuits, but other conductors are possible. Again, the port merely represents the *concept* of conduction.

The electrical-radiation port (ER) is provided for those cases when the sun's photoelectric energy is directly converted into electrical power.

A Robot Example. Consider, for example, the simple robot arm of Fig. 2-3a. (This example will be followed throughout this chapter to illustrate the principles of multiport analysis.) This robot manipulator has an electric motor and a position sensor at each joint. It has been designed to manipulate the deadweight load to any location within the planar working area shown in the figure, and is designed for gravity-free space environments. A modeler wants to find the dynamic response of the arm to a given load and a given motor input history.

The first thing the modeler does is to select the boundary for the system of interest. In this, the motors are included, but not whatever supplies their inputs; the sensors are included, but not their output attachments; and the arm's foundation/base support is excluded. In other words, the system boundary cuts through the motor lead wires, through the sensor wires, and through the base pin joint, as shown in Fig. 2-3a. The modeler decides not to include the load in the system, so the boundary cuts through the load/arm connection.

The modeler begins by constructing a diagram which shows the structural connections/ports between the system and the environment, as in Fig. 2-3b. Notice that the figure shows two environmental systems which are structurally interacting with the system of interest; namely, the base and load systems. All ports are generally indicated by two arrows, facing in opposite directions, and labeled with the port type.

The port to the base is an SC port, showing the combined action of the pin joint at the base and the motor torque applied about that joint. The port to the load is also an SC port, showing the moment-carrying action of the stiff wrist. Notice that since the system is designed to operate in space, no ports are needed to show the effects of gravity. If gravity is operating, then ST ports must be shown to account for this effect as well.

In the final version of the preliminary multiport diagram, Fig. 2-3c, the modeler sketches in the electrical connections/ports, one port for each set of wires that the boundary has cut.

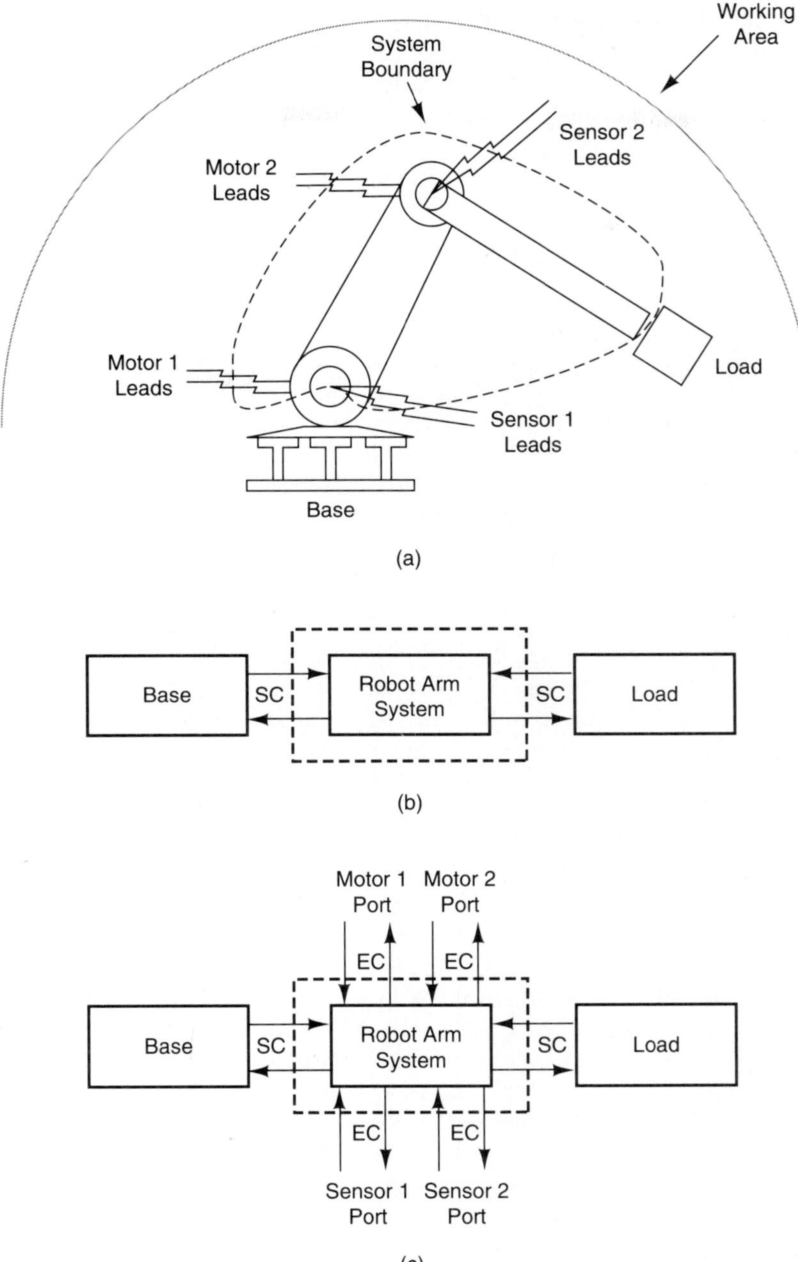

Figure 2-3. Robot Manipulator. (a) Schematic diagram, (b) mechanical parts, (c) completed system-level diagram

From this point onward, the modeler reduces the multiport diagram to ever-finer functional resolution until the components of the system are defined. At that time, physical modeling is begun in earnest. But, for the moment, the system of interest has been successfully isolated.

HOMEWORK

2-1. An engineer wants to analyze a vehicle suspension system design as shown in Fig. 2-4. Components that are not shown restrain the motion of the wheel axle to the plane of the page. The single rigid link b-d-a is pinned to rotate at b, and is rigidly attached to the axle a-a' at the right. The link c-d contains a spring to support the car body.

a) Given the suspension system boundary in the figure, show the connectivity between body, suspension, and roadway in a multiport diagram. Ignore the effect of gravity on the suspension system parts.

b) Redraw the system boundary to include the entire car (four wheels), and use a multiport diagram to show the connectivity between the car and the roadway. Don't forget the effects of gravity.

2-2. A communications satellite engineer wants to analyze a satellite solar array design as shown in Fig. 2-5. The sole external input to the satellite is $R(t)$, the incoming radiation from the sun. The solar array is physically connected to its positioning arm and to the storage batteries. Using the system boundary shown in the figure, sketch a multiport diagram which shows the connectivity between the array and the sun, the batteries, and the arm.

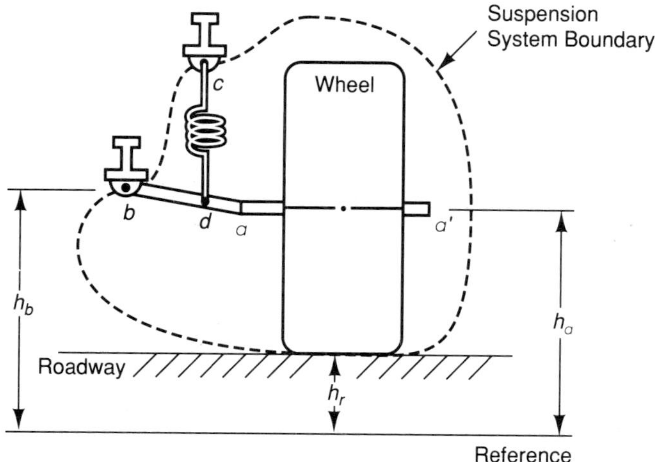

Figure 2-4. System for Prob. 2-1

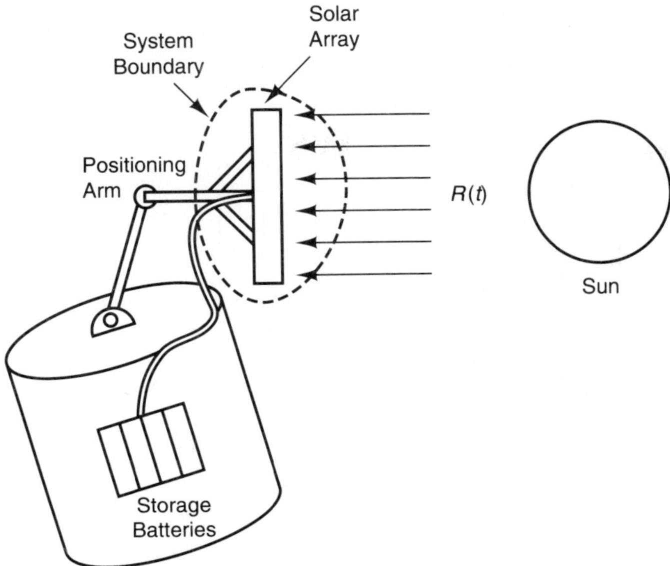

Figure 2-5. System for Prob. 2-2

2-3. An engineer has found a model of a suspension system that has the following equation for $h_a(t)$, the height of the axle:

$$h_a = 0, \qquad t < 0 \text{ sec.},$$
$$5\, dh_a/dt + 10h_a = 20 \qquad t \geq 0 \text{ sec.}$$

Plot h_a versus t over the time interval $(-5, 3.0)$ sec.

2-4. Draw a multiport diagram showing the structural connectivity of an auto engine to the motor mounts and the transmission.

2-2 EXTERNAL CONNECTIVITY—THERMAL AND FLUID PORTS

Thermal and fluid ports are very similar to structural ports in that they model distinct physical phenomena, but they can be more confusing since several types of these ports often occur simultaneously. Again, you must focus on the primary dominant functions occurring at the external connections during port modeling. The various types of thermal and fluid ports are shown in Fig. 2-6.

The power through a thermal port ($\mathcal{P}_{TH}$; Btu/s, kg-cal/s) is the same as the heat flowrate (H),

$$\mathcal{P}_{TH} = H. \tag{2-4}$$

There are three types of thermal ports for this heat flow; these represent conduction (TC), convection (TV), and radiation (TR). Each of these types arises in those cases where there is a substantial difference in the temperature (Θ; °F, °C) between system

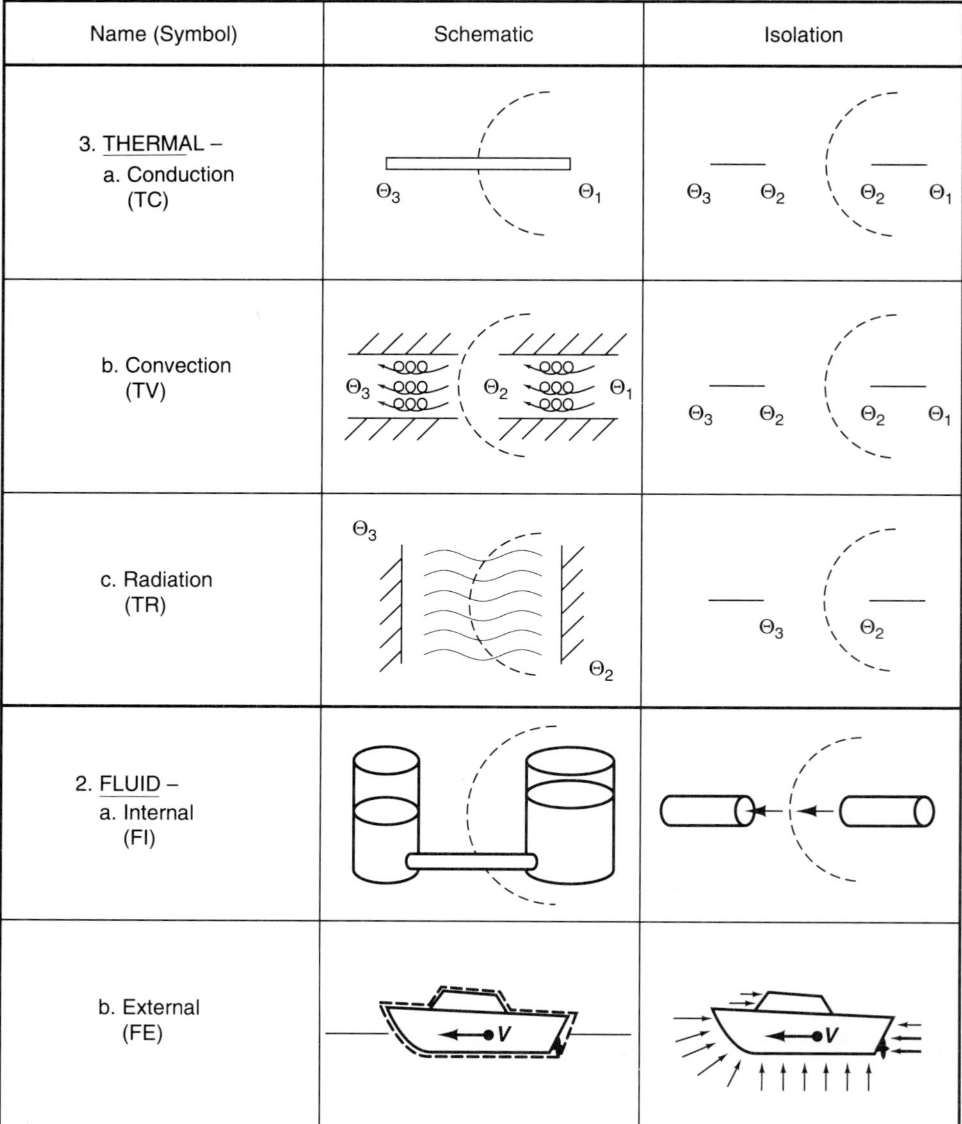

Figure 2-6. The Thermal and Fluid Ports

parts. This temperature difference works to cause heat flow in much the same way as voltage difference works to cause current flow. This heat flow must be accounted for at all system boundaries. (The similarity of heat flow to electrical current flow is one which will be exploited in detail later. It is mentioned here to help you recognize heat flow ports.)

In *thermal conduction*, the heat is transmitted through a solid material, or a *nonmoving* liquid. Thermal conduction through gases is generally negligible. Therefore, at those places where the system boundary cuts through a path of significant heat conduction, you must supply a thermal conductive port to represent that path/connection to the environment. Conduction occurs in an automobile engine through the engine block between the hot cylinders and the cooler outside surfaces. Heat conduction also occurs through a metal spoon with one end immersed into boiling water and the other held in a cooler stirring hand.

Sometimes, heat is carried along by a moving fluid. This is known as *thermal convection*. Convection can occur in both liquids and gases, but never in solids. Convection *always* occurs when gases and liquids are heated, and it results in both heat transfer and fluid motion. This effect is what gives rise to the activity in boiling water, and the turbulent updrafts and downdrafts in the atmosphere, where uneven heating of the Earth's surface creates "hot spots" in the atmosphere which are more buoyant than cooler surrounding areas. This is called *free convection*. *Forced convection* appears in home central heating-ventilating-air-conditioning systems (HVACs) where the conditioned air is forced along by fans to the living spaces through ducts to provide a comfortable atmosphere within the home.

The third type of thermal energy transfer is *thermal radiation*. The main driving force behind thermal radiation is a substantial temperature difference between two surfaces (say, about 1,000 °F). This effect can be felt when standing close to a hot fire, such as a furnace or other hot, open flame. Solar radiation may also give rise to significant radiant heating.

The two types of fluid ports are those for internal-fluid flows (FI) and those for external-fluid flows (FE). Figure 2-6 (4a) shows the system boundary cutting through a pipe connecting two reservoirs. The function of all such pipes and ducts is to provide an *internal flow* passage for the fluids in them. The fluid power through an internal-fluid port ($\mathcal{P}_F$; ft-lb$_f$/s, n–m/s) is a product of the fluid pressure (P; lb$_f$/ft^2, n/m^2) and the fluid-volume flowrate (Z; ft^3/s, m^3/s),

$$\mathcal{P}_F = PZ. \tag{2-5}$$

Figure 2-6 (4b) shows the *external fluid flows* which occur around a small ship which is underway at sea. In the system isolation, the fluids have been removed and replaced by their equivalent effects on the ship. This approach is used in the study of airplane flight, ship motions, and in those cases where a fluid is flowing around the system of interest. However, generally speaking, the external flows are too complex for this introductory text. They are presented here only for the sake of completeness.

A major point of confusion often arises in distinguishing between an internal fluid-flow port (FI) and a convection heat-transfer port (TV). This occurs since both involve the flow of a fluid in a duct as the transport medium. This confusion can be resolved by considering the *primary function* of the port. That is, in a fluid-mechanical analysis, it becomes an FI port; while in a heat-transfer analysis, it becomes a TV port. Convection ports (TV) are designed to account for the power from *both* the fluid flow and the heat transfer, but internal flow ports (FI) neglect all temperature effects as secondary.

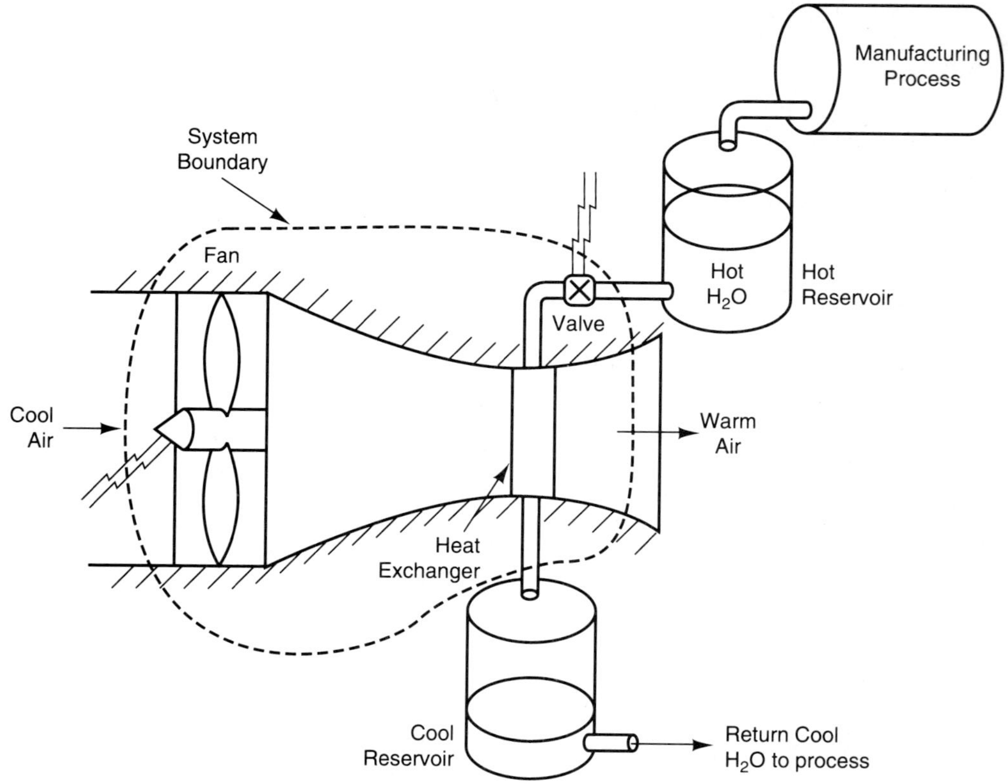

Manufacturing Process
System Boundary
Fan
Hot H_2O
Hot Reservoir
Valve
Cool Air
Warm Air
Heat Exchanger
Cool Reservoir
Return Cool H_2O to process

(a)

Hot Water
Warm Air
Cool Water
Cool Air

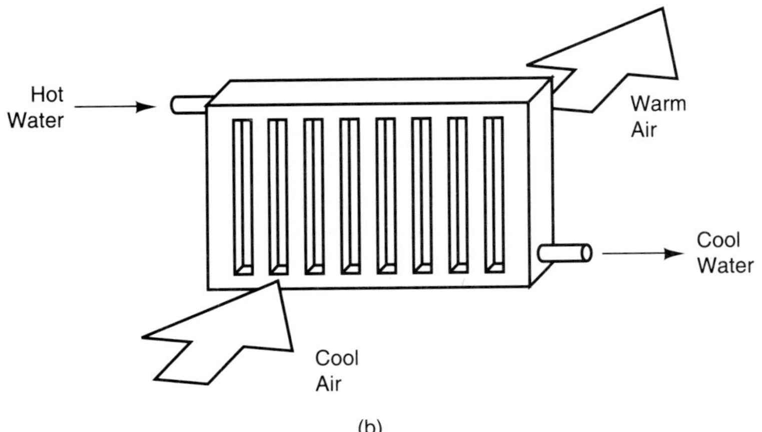

(b)

Figure 2-7. A Water Cooling System. (a) System schematic, (b) heat exchanger schematic

A Cooling System Example. Consider the cooling system shown in Fig. 2-7. This figure shows a design for the production of cooling water to be used in a certain manufacturing process. The manufacturing process produces hot water which is dumped to a reservoir outside the system. A valve is used to control the flow of the gravity-fed water through a heat exchanger and out to a cool reservoir, also outside of the system. The manufacturing process draws from the cool water reservoir to meet its process needs.

The cooling of the water is accomplished through the use of a liquid-to-air heat exchanger as illustrated in Fig. 2-7b. Here, hot water enters the manifold at the top of the heat exchanger where it is distributed to the vertical cooling channels of the device. As the water flows down through tubular, finned cooling channels, it gives up some of its heat through a combination of conduction through the channel walls, and convection to the cool air. Exiting the exchanger then is cooler water and warmer air.

The air flow of the cooling system is shown in Fig. 2-7a. The geometry of the duct for the air is precisely designed to make the fan as effective as possible. The fan draws in the cool air, blows it across the heat exchanger surfaces, and exhausts the resulting warm air to the atmosphere.

The system boundary of Fig 2-7a is chosen to isolate the water cooling system. In the first analysis, the modeler decides to focus on the heat transfer within the system since that is the primary function of the system. The modeler's first problem is thus to translate the system isolation into a proper multiport diagram, which is done in Fig. 2-8a.

The multiport diagram of Fig. 2-8a shows the system with six ports to the environment: two for electrical energy (EC), and four for thermal convection (TV). But why aren't there any ports for internal fluid flow since several pipes were cut by the system boundary? The answer to this question lies in the modeler's analysis of the *function* of the system, that the system is modeled to function as a heat-transfer machine. Therefore, the modeler chooses the convective TV ports rather than the fluid-flow FI ports for the analysis. Of course, an electrical conduction port must be supplied for the valve since its leads were cut by the system boundary. Similarly, an electrical conduction port is needed to represent the severed leads of the fan.

In Fig. 2-8b, a multiport diagram is shown of the heat exchanger alone. The assumption behind this figure is that heat flow to other places than the fluid volumes (i.e., the support structure) is negligible compared to the primary function of the heat exchanger: to transfer heat from the liquid to the air. The air and water volumes are shown to function as collection sites for the heat flow (TV) to/from other parts of the system.

In any case, during the preliminary design of the water conduction system, the fluid mechanics could be considered independently of the heat transfer. This is possible since, during the rough sizing of the pipes and in the selection of the valve, the temperature of the water makes little difference. So, the modeler could well have constructed the multiport diagram of Fig. 2-9 for the preliminary analysis of the fluid-flow design. Note that this figure makes use of the fluid-flow FI ports for

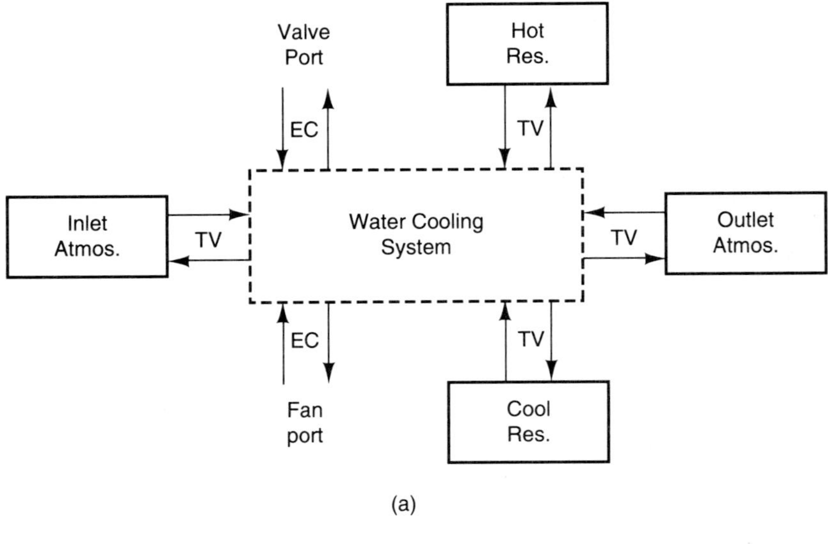

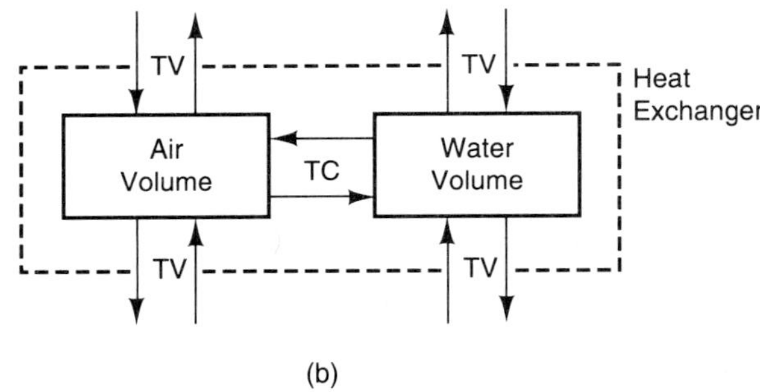

Figure 2-8. The Water Cooling System. (a) System multiport diagram, (b) heat exchanger multiport diagram

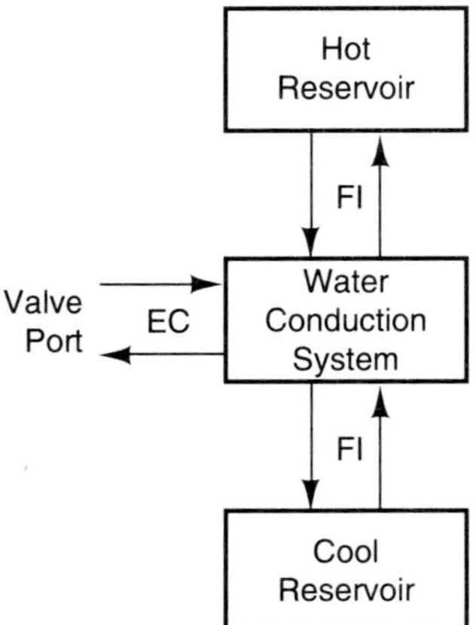

Figure 2-9. The Water-Flow Multiport Diagram

the evaluation of the fluid problem apart from the thermal problem. It also highlights the notion that the modeler must be careful to understand, and focus on, the *function* of the system under analysis.

HOMEWORK

2-5. A designer's scheme for an improved hot water storage design is shown in Fig. 2-10. In this system, the designer wants to enclose a glass container for the hot water within two layers of new insulating materials, Z-1 and Z-10, as shown. The modeler has cut out a 1 ft^2 section of the storage wall for analysis. The inside liquid is assumed to be unmoving. The outside air is heated by conduction. Draw a multiport diagram showing the isolation of the 1 ft^2 wall system.

2-6. A mechanical engineer has designed the hydraulic-assisted pump for the supply of electrical power to a remote village as shown Fig. 2-11. Draw a multiport diagram which shows the isolation of the pump system.

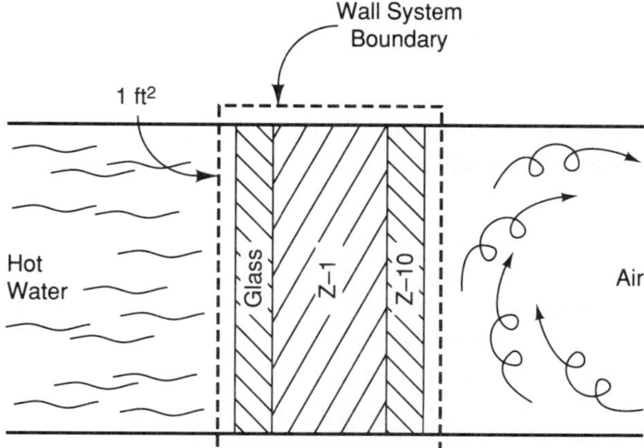

Figure 2-10. Schematic for Prob. 2-5

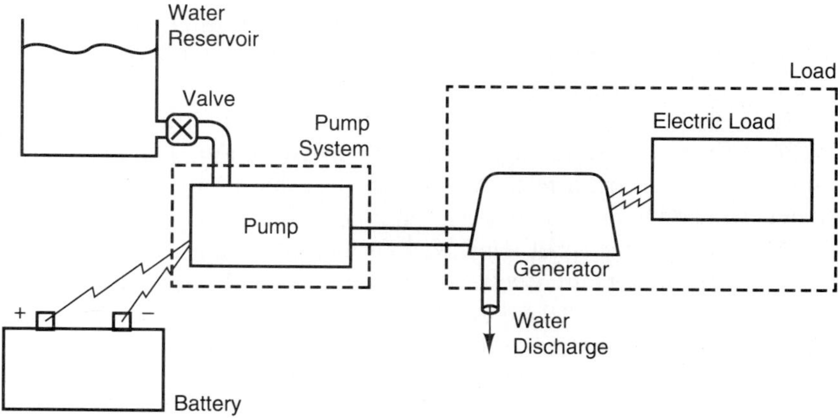

Figure 2-11. Schematic for Prob. 2-6

2-7. Draw a multiport diagram which isolates the radiator on a water-cooled automobile engine. When you look at the radiator, you will see two large rubber hoses connected to it, these are the hoses which carry coolant to and from the engine. You will also see one small rubber hose which is connected to an overflow reservoir. In addition, you may also see two small metal tubes connected—these are the tubes which carry the coolant to and from the automatic transmission. Be sure you account for all the connections when you draw your diagram.

2-8. In the physical modeling of a certain water reservoir design, the modeler has found the following equation for the pressure at the bottom of the reservoir tank:

$$d^2P/dt^2 + 4\ dP/dt + 13P = 13.$$

Find the mathematical solution of the above equation for the pressure, P, if the following initial conditions are applied at time $t = 0$ sec.:

$$P = 0.0, \text{ and}$$
$$dP/dt = 0.0.$$

2-3 APPLIED REDUCTIONISM AND INTERNAL CONNECTIVITY

Recall from Chap. 1 that the modeler's objective in reductionism is to refine the multiport diagram to the point where physical components are shown for further modeling. The guiding idea that the modeler uses in this effort is to *segregate function* in ever more detailed levels of the hardware. Also, recall from Chap. 1 that this idea was used to reduce an example engine system to its functional subsystems (ignition, fuel, mechanical, exhaust, air, and electrical). While these subsystems are not specific physical parts, they do have specific tasks. This is generally true since the subsystems are collections of components which operate as a group to perform the given subsystem tasks. Consequently, it is often very hard to build a physical model of each subsystem directly. Such an observation indicates that the system is not yet diagrammed at the component level.

The Robot Example. Returning to the robot system of Fig. 2-3, the interplay of the subsystems for the robot arm can be identified as shown in Fig. 2-12. Note that this figure shows the motors, the linkage, and the sensors grouped together to form the three subsystems of the robot. However, further discussion with the designer would indicate that the motors are electrical torque motors with the shaft connected to one link, and the housing connected to a neighboring link. The motors thus perform two functions: they supply torque to the links, and they add mass to the links. The sensors are attached to the links in much the same way as the motors.

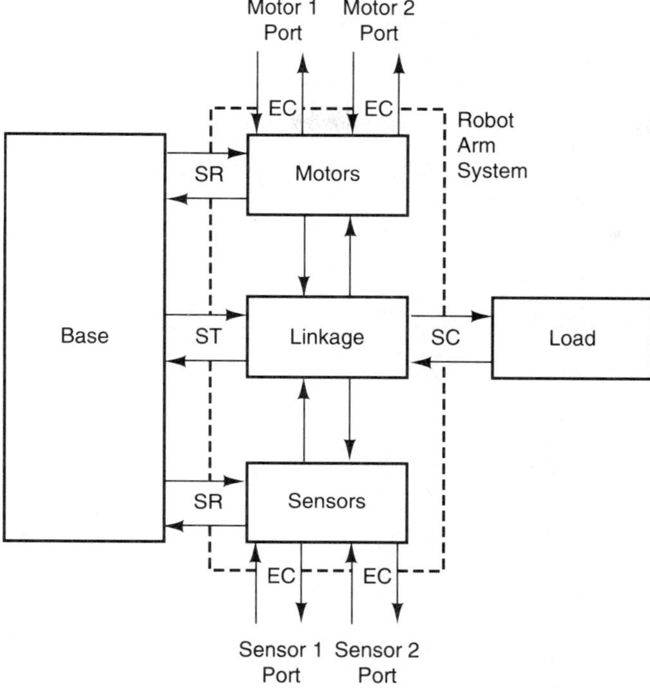

Figure 2-12 Robot Subsystem-level Diagram

Consequently, these subsystems are not yet at a component level since most of the blocks in the diagram are not yet specific enough for physical modeling. Here, the "sensors" must have more than the one port shown in order to interact with the two physical links. Also, the "motors" produce two torques to the two links, they thus require two ports. Another step in reduction is indicated.

In Fig. 2-13, the physical action of the individual motors has become apparent; the same is true for the links. Now, the analysis methods of dynamics could be used to build physical models of the links and the load. Later in the text, methods will be introduced to model these, and other physical parts. Figure 2-13 is thus a component-level multiport diagram, but it is not yet complete.

The Completed Multiport Diagram. Four things must be done to complete the component-level multiport diagram:

1. The physical components of the system must be shown and labeled as interacting boxes;
2. All ports for power transfer must be shown;
3. The modeling assumptions must be listed; and,
4. All input/output arrows (port variables) must be shown and labeled for the coming physical analysis.

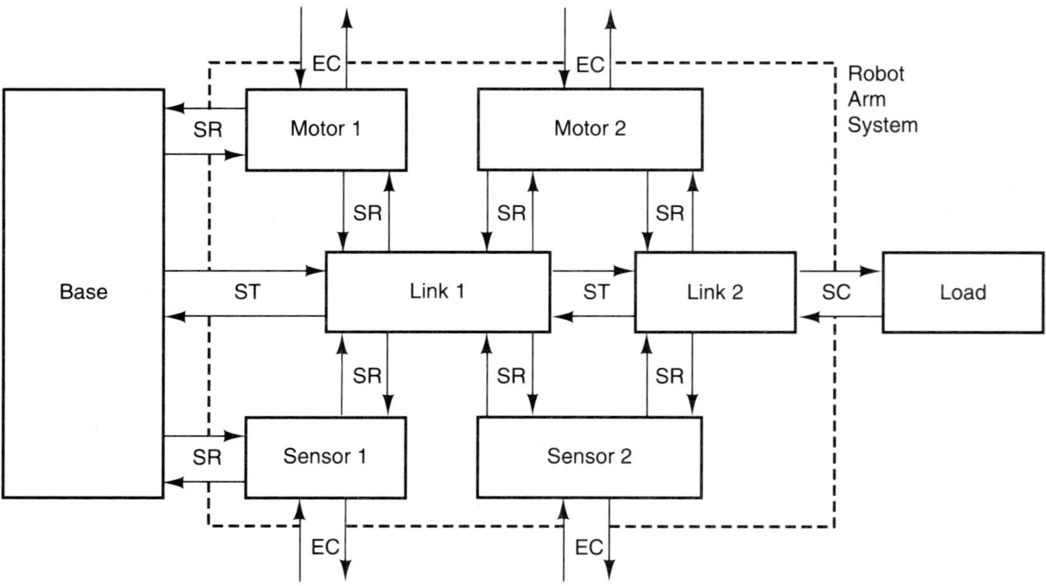

Figure 2-13. Robot Component-level Diagram

This section of the text (Sec. 2-3) deals with the first three steps in the component-level diagram (connectivity), the next section (Sec. 2-4) deals with the last step (causality).

Since the components in the robot example have been identified, the next step is to list the modeling assumptions; these are shown in Table 2-1. The assumptions shown are those which would normally be made in a first dynamic analysis. Friction is usually ignored unless it is known to have a substantial effect on the system (about 10% of the forces or more). The assumption of rigid links is often a good approximation, even for motions of flexible links. Combining the load with link 2 further allows some simplification in not having to deal with a separate mass for the load. Or, from another point of view, link 2 and the load *function* as a combined mass, therefore should be combined in the figure and in the analysis. The last two listed assumptions are drawn from discussions with the robot designer in order to simplify, yet analyze fully enough, for preliminary design evaluations.

Table 2-1. Robot System Assumptions

1. Gravity does not act on the components.
2. Ignore friction.
3. All links are rigid.
4. The load may be combined with link 2. These two parts function as one rigid body.
5. The motors supply pure torque; their inertia may be lumped with the link inertias.
6. The sensor inertias may also be lumped with the links.

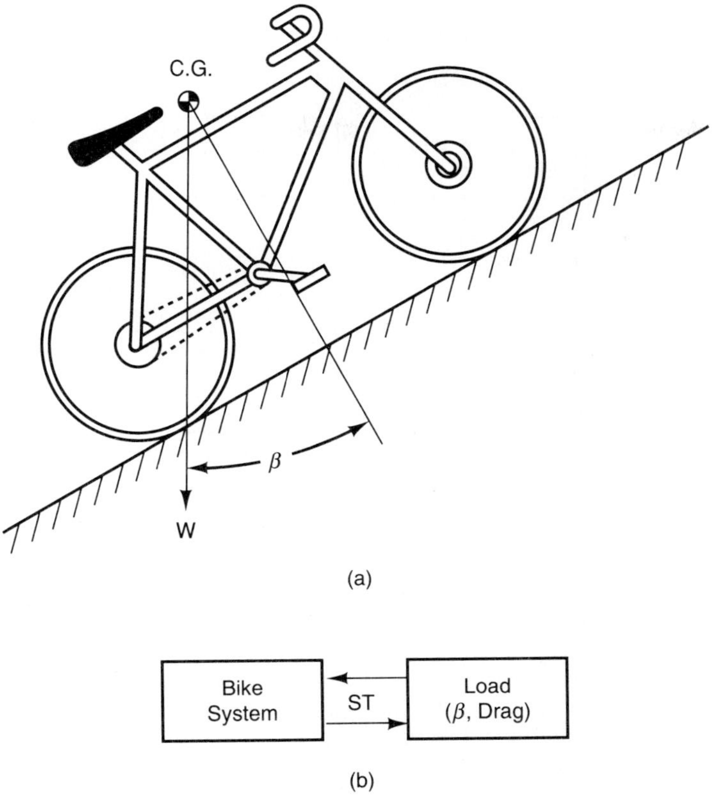

Figure 2-14. Bike System-level Diagram (rider not shown. (a) Schematic, (b) multiport Diagram

A Bike Example. By way of further example, consider the bike/rider system shown in Fig. 2-14. A modeler wishes to evaluate the speed potential (V) of the new bike design over a certain terrain, represented by the angle β, as shown in Fig. 2-14a.

The modeler begins with the very simple diagram of Fig. 2-14b. Here, the "load" in the figure is assumed to be made up primarily of terrain, weight, and aerodynamic drag considerations. The modeler knows that the weight and aerodynamic drag operate directly on the rider; but, functionally, the rider's power works against the load forces only through the structure of the bike. Consequently, all loads are grouped together as shown. The ST port for structural power flow is provided to account for the power needed to translate the bike system.

The modeler's next level of reduction is shown in Fig 2-15, where a set of functional subsystems is shown. This subsystem-level bike diagram shows that the rider is seen to function as a power source which outputs vertical translational power to the pedals. The structure subsystem transforms this power into the forward translation of the bike system.

The modeler now asks if this is a component-level diagram, from which a physical analysis may be started. But, on further inspection of Fig. 2-15, the modeler is not content with the detail in the structure subsystem.

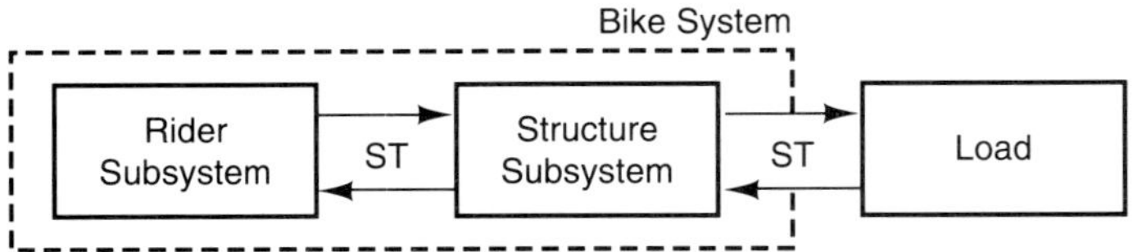

Figure 2-15. Bike Subsystem-level Diagram

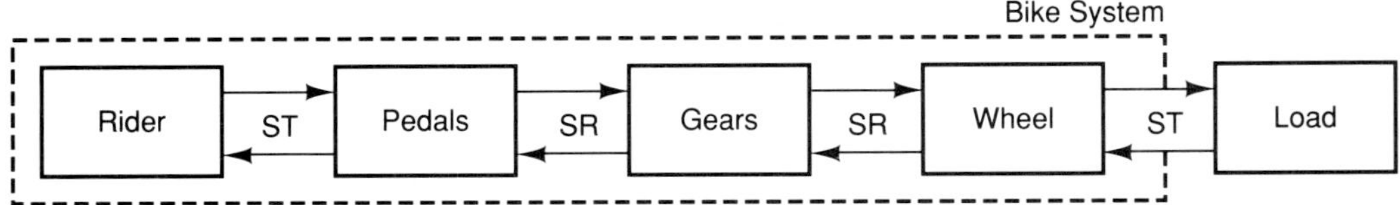

Figure 2-16. Bike Component-level Diagram

Also, the rider must exert power not only on the pedals, but also on the steering, and control through the gears. While these are all important functional components, the modeler decides to neglect steering and control power for the first analysis in order to focus on the propulsion mechanisms.

The next diagram in reduction is shown in Fig 2-16. In this diagram, the pedals are modeled as devices to transform the rider's translational foot power into rotational power for the gears. This is indicated by the typical pin joint which mounts the foot piece to the pedal lever. Gears are always used to modify rotational power (they are discussed more fully in Chap. 4), hence they are always shown between two SR ports. The wheel acts to convert the rotational power from the gears into the translational power needed to move the bike system.

Now the modeler can recognize distinct hardware parts in each box of the component-level diagram and is ready to begin physical modeling.

The assumptions that go with Fig. 2-16 are shown in Table 2-2. Again, the same assumptions on friction, rigidity, and C.G. (Center of Gravity) are used as in the robot example, in order to simplify the first analysis. The last three assumptions in the table came out of the development of the diagram.

Table 2-2. Bike System Assumptions

1. Neglect friction.
2. All rigid parts.
3. Ignore motion of Center of Gravity (C.G.) with respect to the structure.
4. Neglect steering power.
5. Neglect shifting power.
6. A rider model is available in this input/output form.

HOMEWORK

2-9. A rocket engineer has sketched the component-level hardware assembly as shown in Fig. 2-17 for a high-altitude rocket. The engineer is now interested

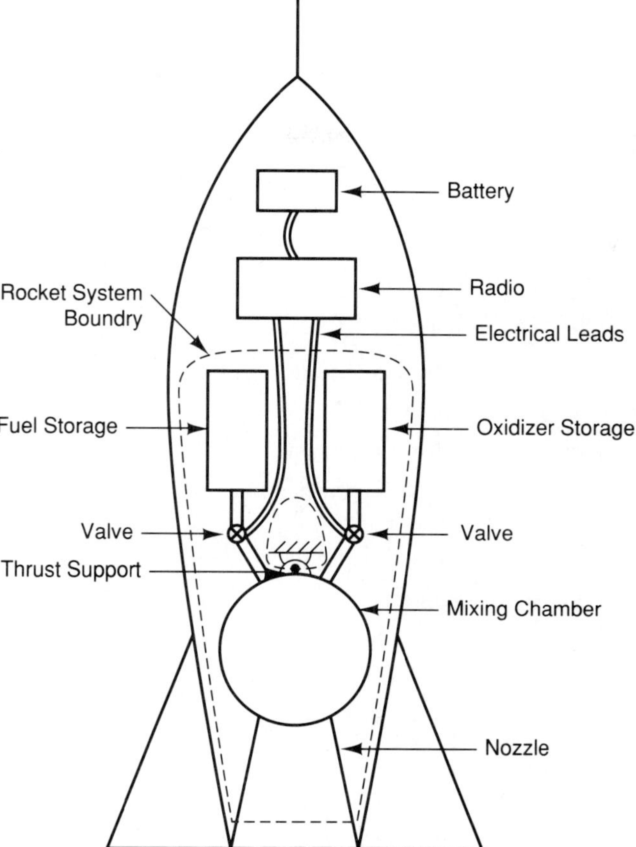

Figure 2-17. Rocket System for Prob. 2-9

in a dynamic performance analysis of the thrust production for the rocket system, and wishes to begin this with a multiport diagram of the rocket function.

The components in the rocket function in the following ways. The radio receives engine firing commands from the ground station. The ground commands are relayed from the radio to open the valves and mix the pressurized fuel and oxidizer in the mixing chamber. The hypergolic (self-igniting) fuel ignites when it is exposed to the oxidizer, they then produce thrust when they expand through the nozzle. Heat convection occurs throughout the fluid flow.

The engineer has arranged the components into these subsystems:

Propulsion subsystem
- Mixing chamber
- Nozzle

Structural subsystem
- The rocket frame and all inertias
- The thrust support pin joint

Fuel subsystem
- Oxidizer storage
- Fuel storage
- Fuel valves

Electrical subsystem
- Battery
- Radio
- All electrical leads

The system boundary shown in the figure includes the propulsion and fuel subsystems only. Complete the following:

1. Sketch the system isolation multiport diagram.
2. Sketch the subsystem multiport diagram.
3. Sketch the component-level multiport diagram.
4. State the modeling assumptions which you'd make in a first analysis regarding:
   a) The rigidity of the rocket structure.
   b) The atmospheric exhaust pressure.
   c) The action of gravity.

2-10. Select a system of your own choosing and complete the first two steps in a multiport diagram of that system. Some suggested systems are:

An automobile drive train;

An AM radio;

A PC computer printer;

A city power plant;

A small ship; or

A small airplane.

This problem may take some research on your part to find the functioning subsystems and components of your system. Also, it will be easier to identify the system levels in a simple system, but choose a system which has at least three levels in hierarchy (system, subsystems, and components).

2-4 CAUSALITY AND THE SYSTEM VARIABLES

The final step in multiport diagramming requires that all input/output arrows (port variables) are labeled for the coming physical analysis. The complete and correct set of port variables is a natural product of the component port selections discussed previously and it has to do with specifying *how* the components affect one another. In assigning this *causality*, the modeler chooses and labels the inputs and outputs at every port, guided by a small menu of possible port variables.

Once completed, the set of all the input and output variables is collectively called the *system variables*. The physical analysis which follows the diagramming effort is thus expressed in terms of these system variables.

The Menu of Port Variables. The list of possible port variables comes from considerations of port causality for each type of port. We begin with the electrical ports and follow with the other port types.

Consider first the simple electrical system shown in Fig. 2-18a. This figure shows an electrical system composed of a single resistor and two connecting leads. The system isolation has cut the leads, so the modeler knows that an electrical-conductive port (EC) must be supplied. The modeler must now determine the causality of the port.

Since a port is chosen, in general, to model the power flow at a boundary point, the *causality* of the port is determined by the variables comprising the power at that port, and by their fundamental relationship to each other.

Thus, the variables for the electrical port must be the voltage, E, and the current, I, since these are the two variables which make up the electrical power. The assignment of the variables as input or output depends on their fundamental relationship.

Clearly, the resistor must obey Ohm's Law,

$$E = \mathrm{R}I. \tag{2-6}$$

Here, the value of the resistance, R (ohms), is a known constant which relates the current, I, to the voltage drop across the resistor, E. Thus, the system can be modeled to accept either E or I as an input, but not both as inputs. Both cannot be inputs since the system must be allowed to respond in some way. So, if the modeler calls I the input, then E is seen as the system response, or output, and vice versa.

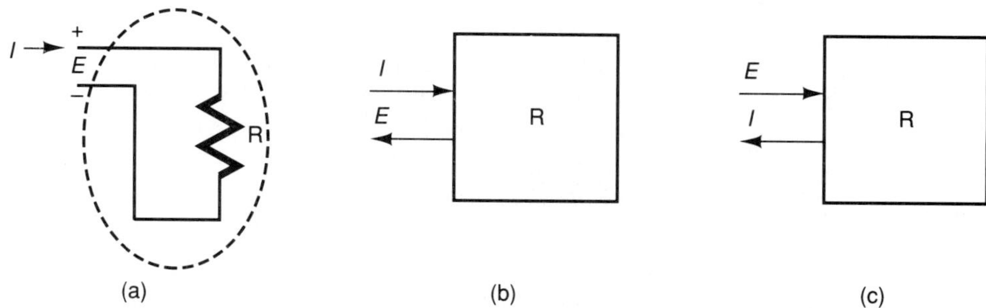

Figure 2-18. A Simple Electrical System. (a) Schemetic, (b) I as input, (c) E as input.

Similarly, the electrical system cannot have both E and I as outputs since some input is necessary to provoke a response from the system.

Multiport diagramming thus *forces* the modeler to choose an assignment of input and output. Later modeling may well prove that a wrong choice is made, but many times it simply makes no difference how the input/output assignment is made, *as long as the correct variables are used.*

When in doubt, the modeler should first talk to the designer to identify the causality of a port connection. If that is not possible or not productive, then the modeler must make some assignment of input/output, with the understanding that it may need to be corrected later.

As far as the resistor system is concerned, the modeler may choose either forcing shown in Fig. 2-18b or 2-18c. The correct choice may well have to do with the nature of the system which is separated to the left Fig. 2-18a. For example, if a current source were "cut off" on the left from the resistor, then the input/output of Fig. 2-18b would be appropriate. If a voltage source such as a battery were cut off, then the forcing of Fig. 2-18c would be appropriate.

In any case, E and I are the electrical input/output port variables regardless of the nature of the severed electrical system parts. This is true since these two variables give the power in electrical ports. One has to be the input, the other has to be the output.

These electrical-port causality ideas may be extended into the realm of structural ports, starting with the power at the ST port,

$$\mathcal{P}_T = \boldsymbol{F} \cdot \boldsymbol{V}. \qquad \text{(2-1 repeated)}$$

Thus, the general structural-translating port variables are $\boldsymbol{F}$ and $\boldsymbol{V}$. Also, since these general port variables are vectors, each vector ST port must be redefined by assigning one port for each axis of translation in the problem of interest. In the final diagram, each ST port thus accommodates one force component and a similar velocity component. One of these vector components must be an input, and one must be an output. (Several examples of this vector-port expansion are presented following the discussion of causality.) The results so far are summarized in Fig. 2-19, along with the remaining port variables.

The *structural-rotating port* (SR) is very similar to the structural-translating port, and is composed of the general vectorial quantities of angular velocity, $\boldsymbol{N}$, and torque, $\boldsymbol{Q}$. Again, these general ports must be expanded by assigning one SR port for each axis of rotation. The input and output for the expanded ports are the torque and velocity components about the axis of interest.

The *structural-complex port* (SC) is simply a combination of the ST and SR port variable assignments. It, too, must be expanded to account for the power using the vector components.

The *thermal-conduction port* (TC) is very similar to the electrical-conduction port. That is, when the system boundary cuts perpendicular to the axis of a thermal conductor (a metal bar, for example), a surface is exposed which has an average tem-

PORT TYPE Sym.	Name	PORT VARIABLES	
EC ER	Electrical – Conduction Electrical – Radiation }	Voltage, E	Current, I
ST	Structural – Translating	Linear Velocity, V	Force, F
SR	Structural – Rotating	Angular Velocity, N	Torque, Q
SC	ST + SR	–	–
TC TR	Thermal – Conduction Thermal – Radiation }	Temperature, Θ	Heat Flowrate, H
FI	Fluid – a. Incompressible	Pressure, P	Volume Flowrate, Z
TV	b. Compressible, or Thermal – Convective }	Pressure, P	{ Mass Flowrate, W Temperature, Θ

Figure 2-19. The Port Variable Menu

perature, Θ, and a net rate of heat flow through it, H. Thermodynamic studies show that temperature difference causes heat flow. Thus, even though the heat flowrate is the same as the power, the "systems" causality convention is to assign one of these variables as the input, and the other as the output at every thermal port.

The *internal-fluid port* (FI) occurs when the system boundary cuts through a pipe or duct which has a fluid flowing in it, and where heat transfer is not of interest. The fluid may be either a liquid or a gas. The difficulty here is in recognizing those gas flows which may be successfully modeled as incompressible. This difficulty arises in accounting for the latent heat changes, and corresponding temperature changes, which are naturally associated with gas compression/expansion. All liquid flows may be safely regarded as incompressible for the preliminary analyses in this text.

Gas flows may be regarded as incompressible as long as they are not subjected to too large a pressure change across any component element. For the introductory purposes of this text, *compressible* gas flows are defined as those which have,

$$\bar{P}_U/\bar{P}_D \geq 1.7, \tag{2-7}$$

in the ratio of sustained upstream-absolute pressure ($\bar{P}_U$) to sustained downstream-absolute pressure ($\bar{P}_D$) across any fluid-flow restriction in the component.

When in doubt about the nature of the gas flow, an incompressible flow model is usually assumed. After a preliminary simulation of the system response, the pressure ratios are then used to check this flow assumption. Note that gas-flow components are usually designed so that the critical equality in the pressure ratio (Eq. 2-7) is not closely approached or exceeded, thus avoiding incompressible behavior.

For the FI port with an *incompressible* flow, the port variables are instantaneous absolute pressure, P, and instantaneous volume flowrate, Z. These are the power variables which arise when the system boundary plane cuts through a pipe with this type of flow in it. So, across the plane of intersection, the flow has some instantaneous average pressure and some instantaneous net flowrate; again, one of these is the input, and one is the output.

For the FI port with a *compressible* flow, the flowrate must be supplemented with the temperature. Thus, the mass flowrate (W; slugs/s, kg/s) and the temperature are used instead of the volume flowrate so that the modeler can account for the heat changes due to compressibility effects. Consequently, the FI ports for compressible flows have three arrows, one for each of pressure, P, mass flowrate, W, and temperature, Θ.

Further, the three compressible-flow port arrows must be labeled according to the *thermal-fluid convention*: the mass flowrate arrow and the temperature arrow are *always* shown in the same direction since the flow convects the heat (indicated by the temperature). Thus, if the mass flowrate is an input, so is the temperature, and the pressure must be the output. The converse of this is true as well.

The final port type is the *thermal-convective port* (TV). This port is used to model a fluid which has external heat interactions, as in a home HVAC system. Such a port is modeled in exactly the same way as a compressible-flow FI port, using the variable set P, Θ, and W. Again, the thermal-fluid convention must be followed: the two arrows for temperature and flowrate are assigned in the same direction, the pressure is directed in an opposing manner.

The Bike Example Completed. By way of example, consider the bike component-level diagram previously shown in Fig. 2-16, which now can be completed as shown in Fig. 2-20.

The figure shows that the rider is thought to be a force source (force output) as a function of foot/pedal speed (speed input). Note that the forces and torques in the figure all proceed from left to right, while the velocities proceed from right to left. This sort of causality is common in a system with many machinery structural ports. In fact, this "flow" of inputs/outputs is typical of all passive systems (no sources in the system) with many ports of the same type, whether the system is structural, electrical, fluid, or thermal.

In the bike example, all the velocities and forces are one-dimensional, so the vectorial aspects of the ports do not come out. In the next examples, the structural ports are multi-dimensional to demonstrate the treatment of vectorial ports.

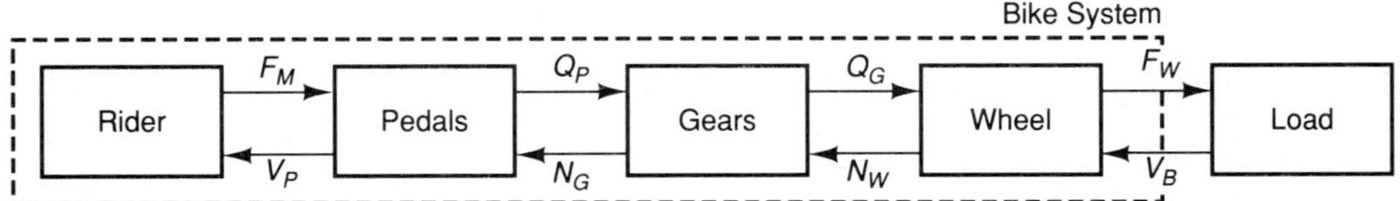

Figure 2-20. The Completed Bike Multiport Diagram

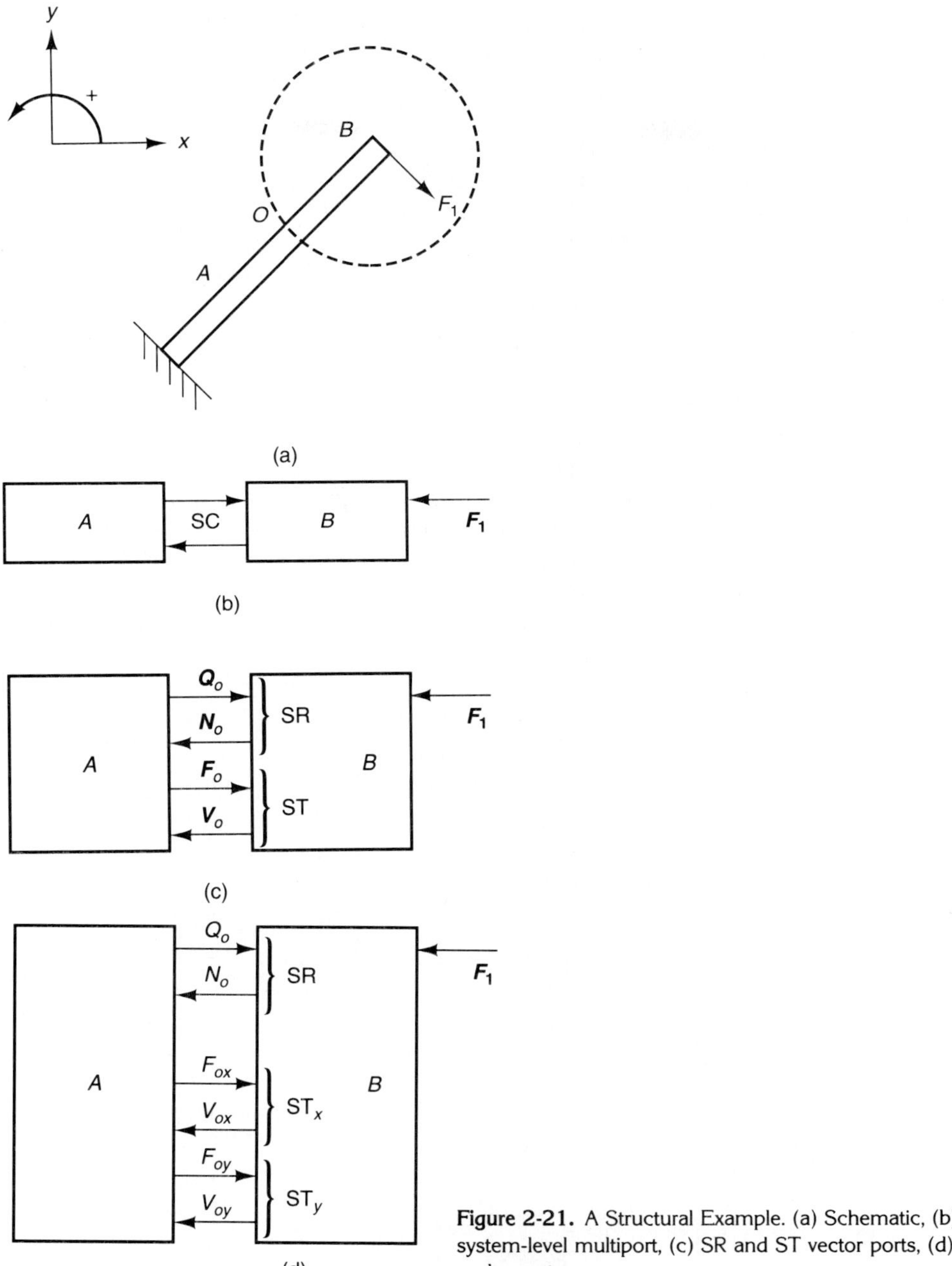

Figure 2-21. A Structural Example. (a) Schematic, (b) system-level multiport, (c) SR and ST vector ports, (d) scalar ports

A Structural Example. Consider the structural system of Fig. 2-21. The system comes from a flexible beam which is rigidly implanted into a wall, as shown in Fig. 2-21a. The modeler draws the system boundary as in the figure, such that it cuts through the beam at point O and separates the beam into two parts, A and B. The system of interest is beam part B.

Figure 2-21b is a system-level multiport diagram of the system under consideration. Here, the modeler correctly shows the interacting system parts, and the structural-complex port (SC) which connects them. The modeler is next concerned with interpreting the SC port.

Figure 2-21c shows the expansion of the general SC port into two ports: a rotational SR port and a translational ST port, both of which are general vectorial ports. The modeler also assigns the causality of the vectors as shown in the figure.

In Figure 2-21d, the modeler expands the vectorial ports into their scalar equivalents. The modeler is now ready to start the physical analysis, once the assumptions are listed which govern the modeling.

The Robot Example Completed. By way of further example, the robot analysis begun earlier (Fig. 2-13) is completed by making the port causality assignments. The result is shown in Fig. 2-22. While this figure is quite complicated, it is also very instructive, if studied carefully.

For example, the role of the Base as a boundary/support structure is functionally shown. It receives the forces from Motor 1, Sensor 1, and Link 1. If it is properly designed, it will produce a stable, non-moving support platform for the robot arm with all output velocities, $\boldsymbol{N}$ and $\boldsymbol{V}$, equal to zero. Such a platform is critical to the precise, repeatable performance of the arm.

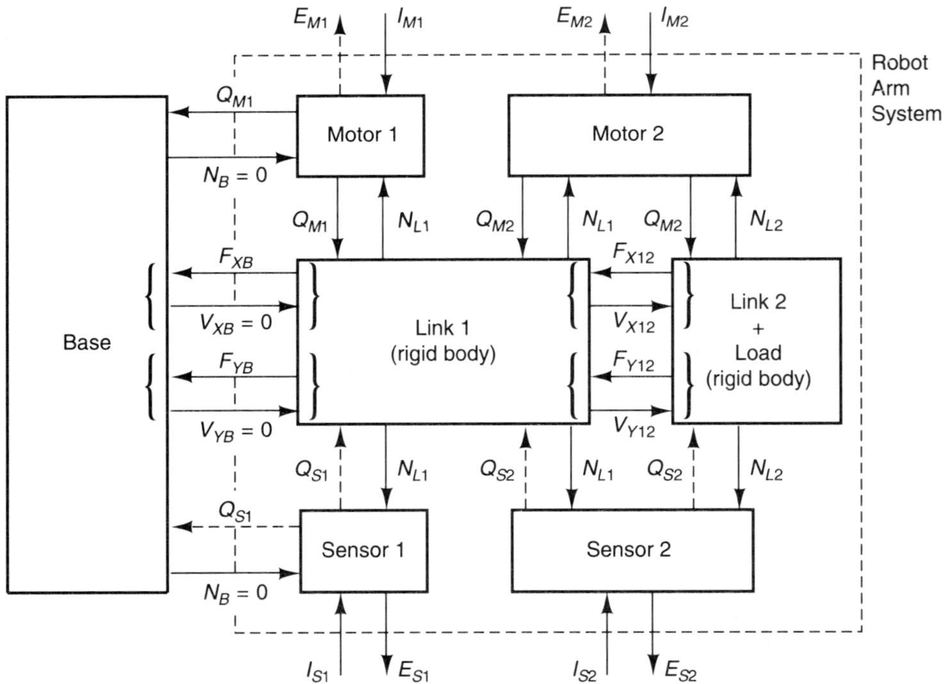

Figure 2-22. The Completed Manipulator Multiport Diagram

The function of the motors is also clearly displayed. Such motor-component diagrams are developed from conversations between the designer and the modeler. Here, the designer describes how the motors are carefully designed to isolate the motor electrical input, I_{M1}, from the motor electrical output, E_{M1}, for control effectiveness reasons. Thus, the modeler shows each motor electrical port with inputs only (the outputs at the motor ports are shown as dashed lines in Fig. 2-22 merely to show a complete port for each motor).

The important idea here is that for each motor, its *input is shown inward regardless of its sign* as a variable. This is true because the input *causes* the motor to turn. The direction of the motor turning is irrelevant to this discussion, as is the sign of the input. The motors are thus seen as devices which have current and rotational velocities as inputs, and torques to the links and base as outputs.

Notice that the structural-translational ports to the links have been expanded to account for the two-dimensional nature of the motion of the arm. Also notice that the Load has been combined with Link 2 in keeping with the assumption made earlier in the analysis that they function as a single rigid body and they should be combined.

The *kinematic chain* of the robot is also clearly shown in the "flow" of the velocities from left to right in the figure, from the non-moving Base out to the tip of the arm. This chain is seen regardless of how many links there are in a mechanism.

The *load path* of the robot follows the forces as they move from the Load to the supporting Base. This concept is a little confusing in this case because it also includes the torques through the motors. That is, the load path includes not only the translational forces but the torques necessary to maintain the position of the mechanism.

The manipulator sensors are designed to produce a position-measuring output without placing significant torque on the mechanism. This is true of all well-designed sensors: they measure without interfering with the process being measured. Thus, the sensors are shown with dashed lines for their torques. The modeler plans to ignore these sensor torques in the coming physical analysis.

HOMEWORK

2-11. Complete the rocket system multiport diagram that was started in homework Prob. 2-9. Treat the fluid storage units as flow sources, and remember the "flow" of system variables as in Fig. 2-20. Treat the "structure" as a translating mass which moves only in the vertical direction (Hint: remember Newton's Second Law.)

2-12. Complete the multiport diagram on the system of your own choosing that was begun in homework Prob. 2-10.

2-5 SUMMARY, BIBLIOGRAPHY, AND REVIEW PROBLEMS

Successful dynamic analysis of complex, modern engineering systems requires a proper functional analysis before a physical analysis is attempted. Such an evaluation of function is often supported with discussions between modeler and designer. These discussions are important but difficult when attempted during preliminary design since some of the system parts may not be well understood at that time. In fact, such discussions can lead to improvements in preliminary designs once the system function is clearly diagrammed.

In this chapter, the method of multiport diagramming was introduced to provide a graphical representation of the function of a system. The objective was to present a method for finding an accurate component-level multiport diagram for the system of interest. The two issues of particular interest in this method were those of connectivity and causality. These issues were addressed in the method through an orderly progression from system isolation, through reductionism and finding the components, to assigning input/output variables at the component ports.

Four types of ports were examined in the chapter, these being the electrical, structural, thermal, and fluid port types. These same ports were shown to occur at every level of the multiport diagram, from the system isolation down to the component connections. They are, in fact, the only ports that need to be considered in the vast majority of engineering systems.

Given a complete and accurate multiport diagram, the physical analysis of the system is properly segmented to allow you to concentrate on modeling only one component at a time. The multiport analysis method is thus shown to both simplify and organize the analysis. It simplifies the analysis because it allows you to model physics at a low level of complication, and it organizes the analysis because it shows you how to assemble the simple physical models once they are completed.

BIBLIOGRAPHY

Cassell, W. L., *Linear Electric Circuits*, New York, N.Y.: J. Wiley & Sons, 1964. This book has a very clear discussion of two-port electrical analysis. This type of analysis led to the multiport analysis method.

Critchlow, A.J., *Introduction to Robotics*, New York, N.Y.: Macmillan Publ. Co., 1985.

Karnopp, D. and R.C. Rosenberg, *Analysis and Simulation of Multiport Systems*, Cambridge, Mass.: M.I.T. Press, 1968. There are two excellent bibliographies and discussions in the front of this book describing the onset of system graphing methods for engineering systems. Although multiport diagrams are not specifically mentioned, they were used at this time as well.

Newcomb, R.W., *Linear Multiport Synthesis*, New York, N.Y.: McGraw-Hill Book Co., 1966.

REVIEW HOMEWORK

2-13. A heating and ventilation engineer has designed a passive heating system for a home as shown in Fig. 2-23. In this system, the sunlight passes through the glass covering and the air in the heater without noticeable effect on them. However, when the sun hits the black surface, the surface is warmed substantially by radiation from the sun. The heat from the black surface goes to the air in the heater by means of conduction. The hot air at the surface then becomes more buoyant than the surrounding air and is convected out the top of the heater while cool air is convected in the bottom. A significant amount of heat is lost by conduction out through the glass. Draw a system-level multiport diagram for the heater showing all system-environment interactions, the modeling assumptions, and the port variables.

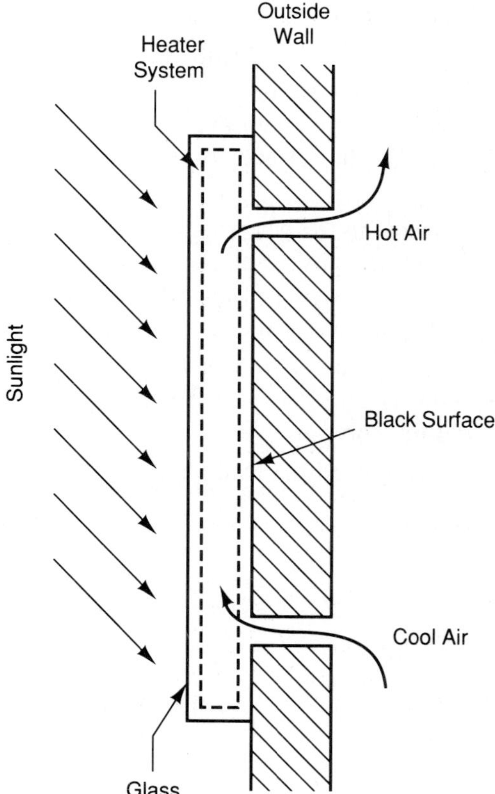

Figure 2-23. The Heating System for Prob. 2-13

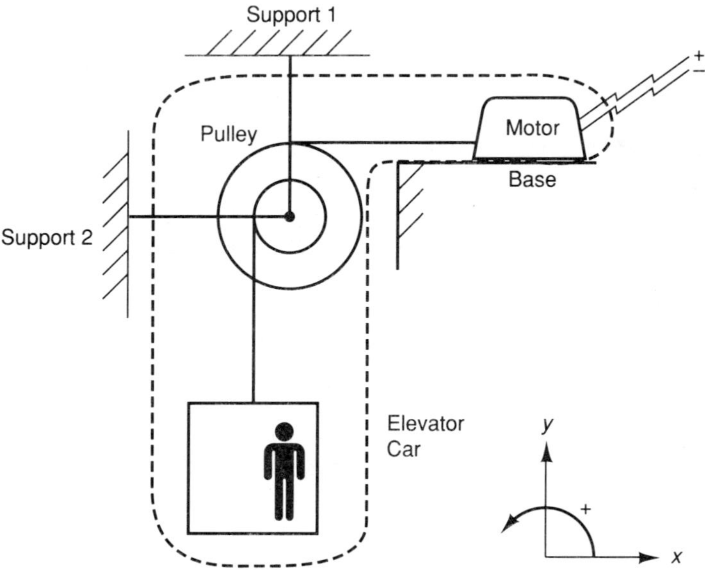

Figure 2-24. The Elevator System for Prob. 2-14

2-14. An elevator engineer has designed the system in Fig. 2-24, and is now interested in a functional analysis. Draw the completed multiport diagram for this system. In this you should assume that the motor inputs are voltage and speed of the pull cable. Also, assume that the car acts as a rigid, deadweight mass and that all cables are inextensible.

2-15. Automotive engineers are interested in the performance of their new connecting rod design, which fits between the piston and the crank, as shown in Fig. 2-25. In order to start the analysis, you have sketched the system boundary such that it cuts the pin joints at positions 1 and 2.

a) Draw the system-level multiport diagram which goes with the system boundary in the figure showing all system-environment interactions, the modeling assumptions, and the port variables.

b) Comment on the kinematic chain and the load path.

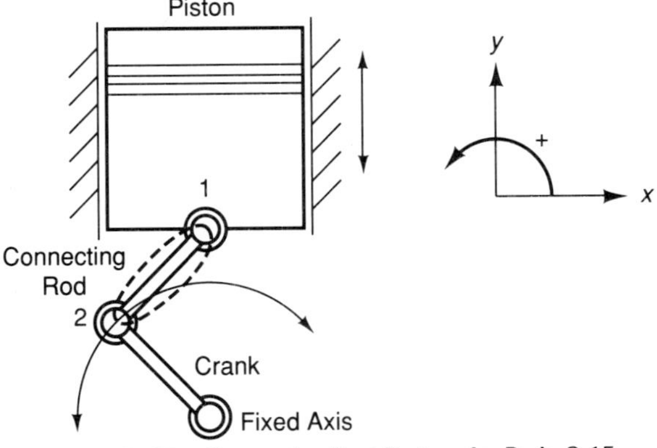

Figure 2-25. The Connecting Rod System for Prob. 2-15

2-16. A heating and ventilation engineer and an electrical engineer have co-designed the heating system shown in Fig. 2-26. The system has a new fin design for the heat exchanger, combined with a new electrical pump design to generate the optimal fluid flow for heating. The electrical engineer has designed the heater and pump to operate on an input of current, I. Draw the completed multiport diagram for this system.

2-17. Often in the design of structural systems (i.e., vehicles and buildings), a need to dampen motions arises. The device which is used to accomplish this damping is called a *damper*, as shown Fig. 2-27. The damper works by trapping a liquid between two sides of a ram, and forcing it to flow through a precise orifice. The pressure difference thus created gives opposing forces on either side of the ram to resist the applied force and its corresponding motion. Draw the completed multiport diagram for this system. In your functional analysis of the damper system, assume that the force, F, is the input to the system, and that the ram and the driving rod comprise a single mass. Ignore friction around the ram, and combine the orifice and chamber 1 components.

2-18. An electrical engineer is beginning a new power station design. The engineer has identified the three components of the load as shown in Fig. 2-28. The output of the station is to be its voltage.

a) Sketch the completed multiport diagram for the load.

b) Comment on the flow of variables which your diagram displays.

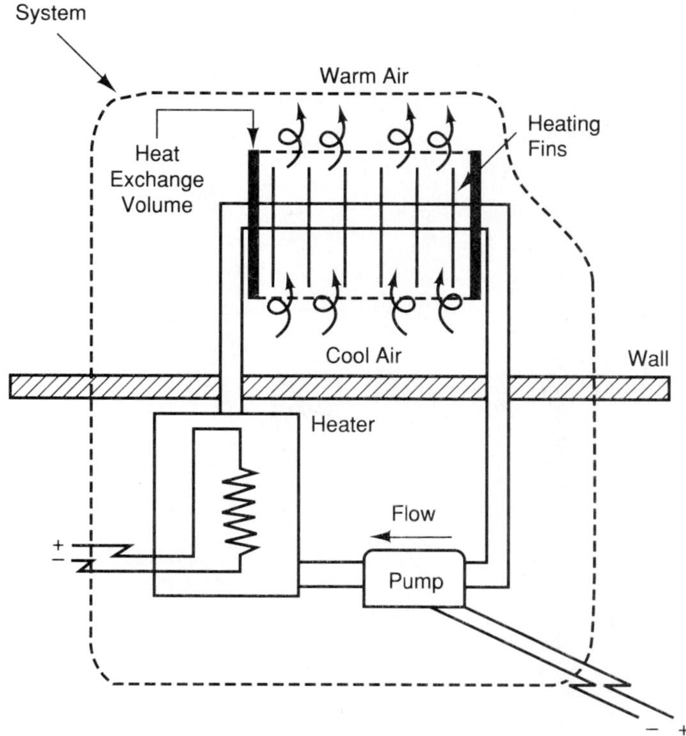

Figure 2-26. The Heating System for Prob. 2-26

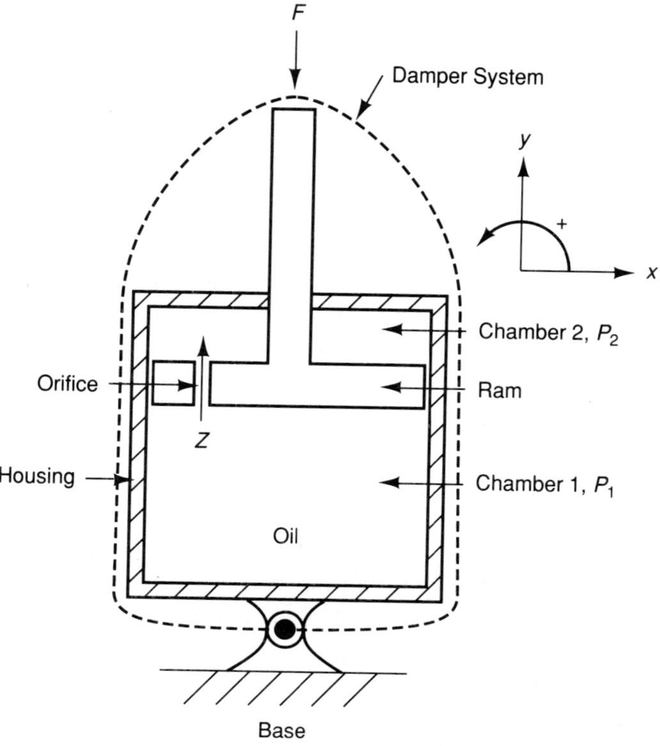

Figure 2-27. The Damper System for Prob. 2-17

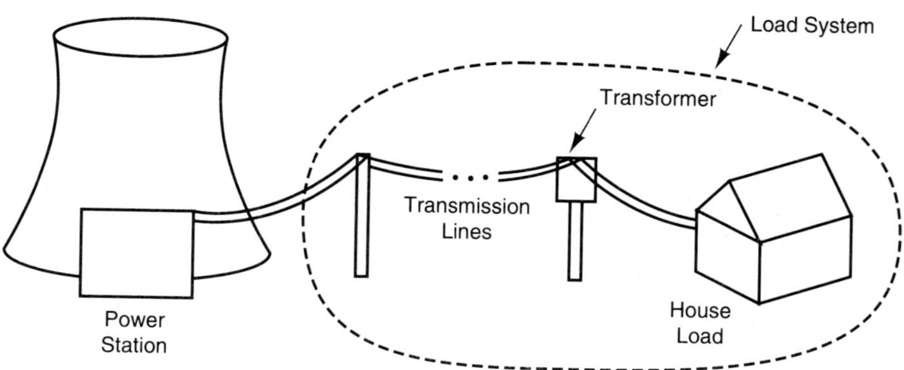

Figure 2-28. The Load System for Prob. 2-18

Chapter 3

Physical Analysis of the System Inputs

3-0 INTRODUCTION

Recall from Chap. 1 that the three sequential steps of systems modeling are: functional analysis, physical analysis, and mathematical analysis. Having developed methods for the first modeling step in the previous chapter, discussion of the physical analysis of the preliminary design is begun in this chapter.

So, given a preliminary system design, the time histories of the expected system inputs, and the multiport functional diagram, engineers accomplish the following sequential modeling and design tasks:

1. Create a set of *idealized inputs* from the given inputs. This is the first task of physical analysis,
2. Create an *idealized system model* guided by the system inputs and the multiport diagram. This is the second task of physical analysis,
3. Compute the idealized model response to the idealized inputs—this response is called the *characteristic response* of the system. This is a task of mathematical analysis,
4. Create *design improvements* by comparing the characteristic response of the given system to the design requirements, and drawing on knowledge of other characteristic responses to guide the design-change process. This is a task of design.

The idealization of the given system inputs begins with the observation that the inputs must be either periodic or aperiodic. If the system inputs are periodic, then the modeler expects to eventually express the model in the *frequency domain*, with frequency as the independent variable. This is true because a design which processes periodic inputs is best analyzed using frequency-dependent models.

If the system inputs are aperiodic, then the modeler expects to express the model in the *time domain*, with time as the independent variable.

Choosing a modeling domain based upon system inputs is called an *input-directed* approach to modeling. Often, however, the system designer requires a model to be built in a certain domain to assist further design work, regardless of the inputs. This is called the *designer-directed* approach to modeling.

Only the input-directed approach is presented in this text, because a thorough understanding of this approach develops the necessary background to meet the designer-directed need when it arises.

Notice that all physical inputs must really occur in the time domain, as do all physical outputs. An example of an aperiodic input and output is shown in Fig. 3-1. (Note that the time, t, is on the horizontal axis.) Thus, your first component models are built in the time domain, if at all possible. This approach allows you to draw on your fundamental physical intuition early in the model-building process. In addition, you have powerful model analysis tools in the time domain which can deal with all sorts of complications during system modeling and system response predictions.

However, many systems are designed and built to process frequency information from frequency-rich inputs. In these cases, you can rightly expect that a frequency-domain analysis is more insightful than a time-domain analysis for design improvement. The approach to this is to recognize from the inputs when such frequency-domain problems are occurring, then to build the frequency-domain model. Often, you simply build a time-domain model and then transform it into the frequency domain.

In this chapter, some guidelines for recognizing time-domain inputs and frequency-domain inputs are first developed. Then methods for idealizing the inputs in each of the two modeling domains are presented.

3-1 CHOOSING THE ANALYSIS DOMAIN

System inputs may be either periodic or aperiodic variables. A *periodic* input is one which is characterized by a repeating cycle of uniform period, T (sec.), and uniform frequency, ω (rad/sec.). An *aperiodic* input is one which has no uniformly repeating cycle.

A building, for example, might be designed to withstand an input wind gust of 300 ft/s (an aperiodic input since it only occurs once), or an input ground-vibration spectrum due to an earthquake (a spectrum is a frequency range of periodic inputs).

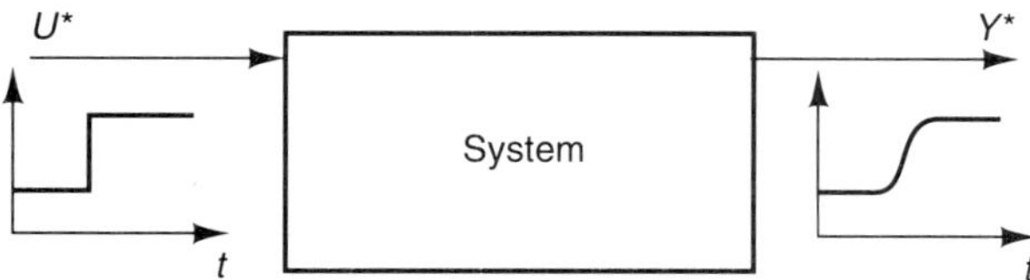

Figure 3-1. System with a Time Input and a Time Response

So, hardware is specifically designed to do engineering tasks in either of the two input domains, and a classification of the hardware can give some help for initial model building.

Classification of Hardware. Table 3-1 lists some of the common hardware/analysis areas in the two input domains—any given system model may involve some or all of these areas, and all are discussed in the chapters which follow.

Table 3-1. Engineering Hardware and the Input Domain

Time Domain	*Frequency Domain*
DC electronics	AC electronics
1D structural dynamics	1D structural vibrations
1D fluid mechanics	1D acoustics
1D heat transfer	

Table 3-1 shows that there are many one-dimensional analyses (1D) in this introductory study. This approach deals more clearly with the interconnection issues of different types of components within a system without overcomplicating the hardware dynamics—2D and 3D dynamic analyses are taught in other specialization texts. However, the multiport analysis method of the previous chapter is fully capable of providing for the higher dimensionality, as demonstrated by the vectorial-structural ports discussed earlier.

Note that each row of Table 3-1 is for hardware of a similar type. The first row is for electronics, the second for mechanics, the third for fluids, and the fourth for heat transfer. Notice also that the heat transfer row has no frequency-domain hardware. This is because heat-transfer systems are rarely designed to dynamically respond in the frequency domain.

Direct Current (DC) electronic systems are designed to operate subject to aperiodic inputs. Alternating Current (AC) electronic systems use AC currents because they are useful for processing frequency-domain inputs. In this, the modeler must be careful to distinguish between *supply* and *input*. In personal computers, the supply is an AC wall connection, but the input is an aperiodic keyboard signal. In portable radios, the supply is a DC battery, but the inputs come, in part, from the periodic radio-wave spectrum, and, in part, from the aperiodic tuning buttons. Consequently, you would rightly expect to do both a time-domain and a frequency-domain analysis for the radio, and for the shaking building mentioned earlier, since both have the two types of inputs. This, in fact, is the usual pattern of modeling and analysis. Modelers begin in the time domain since there is almost always one or more time domain input(s).

The other types of engineering hardware follow the same pattern that is established in the analysis of electronics. In structures, a time-domain approach is used to model the usual 1D Newtonian dynamics, while the frequency domain is used to study the motion of waves in structures (the study of vibrations). In fluids, the usual

1D pipe and duct flows are analyzed in the time domain, while the frequency domain is used to study the motion of pressure waves through the fluid (the study of acoustics).

Consider for example the system with one external input and one external output shown in Fig. 3-2a. The modeling approach for this system depends upon the form of the external input, U^*.

In Fig. 3-2b, an expected *aperiodic* input is shown at the far left of the figure. By virtue of this input, the modeler knows that the modeling is conducted in the time domain, and so are the response predictions of the output, y^*. Thus, the modeler builds the *idealized time-domain input*, u^*, for the design/model evaluations.

(Note that *all* idealizations are indicated by lower-case variables. This convention is adopted throughout this text to readily indicate all idealized models.)

Often, the idealized time-domain inputs come directly from the designer who is interested in evaluating the design response to idealized, *worst-case* system inputs. These standardized, idealized inputs are used by modelers and designers to reveal

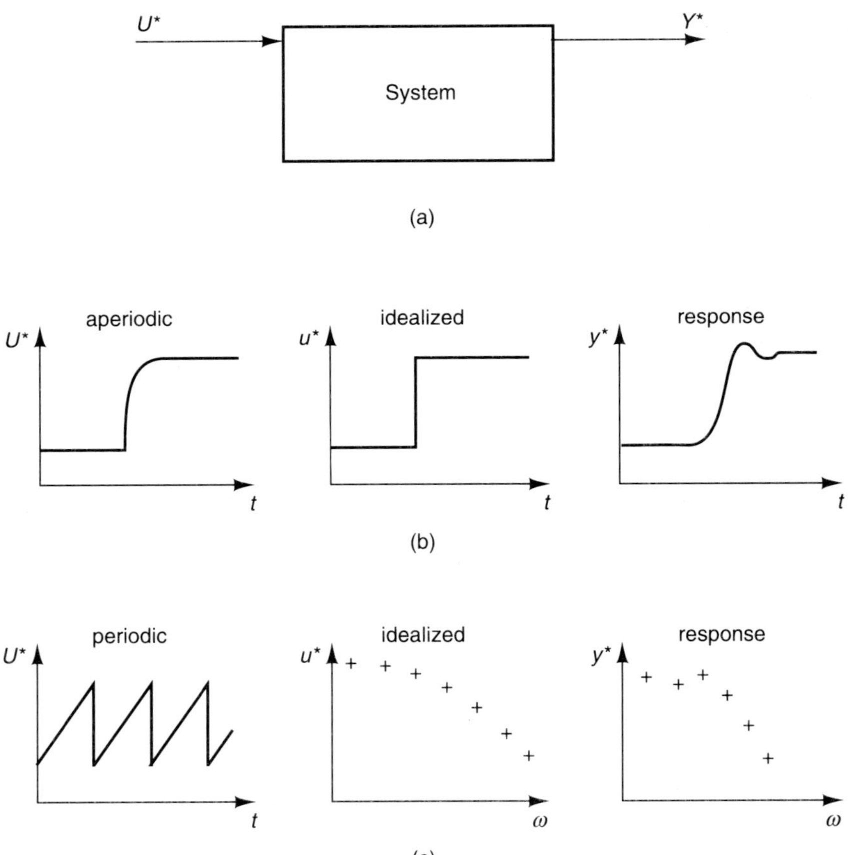

Figure 3-2. System Response Modeling. (a) System diagram, (b) time-domain input, (c) frequency-domain input

the model's *characteristic response* so that the system designer can achieve some comparative insight for design improvements. The idealized inputs do this by testing the response of the math model to input singularities in slope and value, hence they are called *singularity inputs*.

Figure 3-2c shows an expected *periodic* input to the system, and the modeler's treatment of the input. Here, the idealized input comes from a *transformation* of the expected input from the time domain into the frequency domain—note the change of independent variable from time, t, to frequency, ω. The modeler does this to prepare for the output response computation, which is also in the frequency domain, as shown at the far right of the figure. This is very reasonable since the primary information content of the input/output is in the frequency domain. Therefore, the model/design should be analyzed in the frequency domain.

The frequency-domain system model which is used to compute the output is usually obtained from a transformation of the time-domain model. Models are built directly in the frequency domain only if nothing is known about the physics of the component, and only then if there is frequency domain input/output data. Otherwise, the modeler always models in the time domain and *transforms* the model into the frequency domain for output computations. If the modeler wishes to do so, the frequency-domain output response may be transformed back into the more intuitive time domain for further evaluation.

Methods for the transformation of the *inputs* and *outputs* between the two domains are presented later in this chapter. Methods for the transformation of *models* between the two domains are studied in Chap. 6.

Complications. Complications in the analysis-domain selection can arise when a component is MIMO and/or nonlinear.

MIMO components have more than one input, hence they may have more than one input domain. The radio component shown in Fig. 3-3 is an example of this sort of complication.

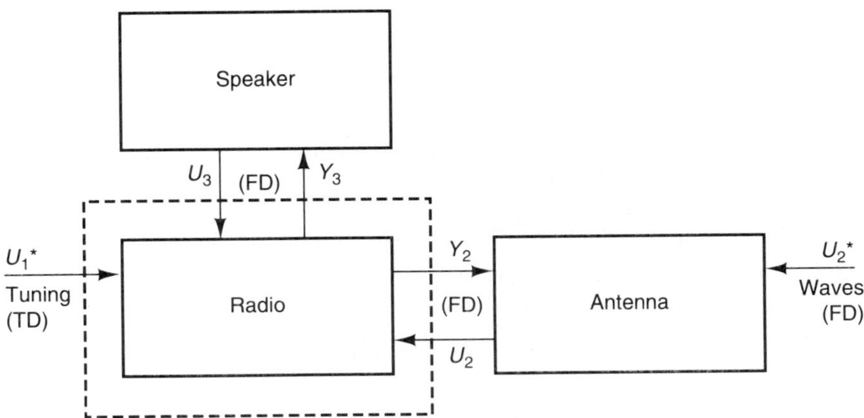

Figure 3-3. A MIMO Radio with Mixed Inputs

The radio waves which enter the system at the far right of Fig. 3-3 are clearly in the frequency domain (FD). Conversely, the tuning which is shown at the far left of the figure is done in the time domain (TD). The modeler is thus faced with the difficulty of choosing the model domain.

The general MIMO modeling approach to simultaneous inputs is to model *either* the time response using a time-domain model *or* to model the frequency response while all time inputs are held steady.

A further difficulty arises in that a MIMO frequency-domain model may only be created if the component input/output physical relationships are *linear*, or very nearly so. The component must not have any undifferentiable nonlinearities (this makes them unlinearizable) since the frequency-domain model relies so heavily upon the attributes of linearity. But, linearization usually gives an acceptable model for most electronic components due to careful element design and manufacture. In mechanical components (structural or fluid) this may be a realistic approximation only if small amplitude input oscillations are allowed, which is usually the case in vibrations and acoustics.

Modelers often work hard to idealize and avoid nonlinearities in their models, but sometimes this is not possible. Occasionally, the modeler may have to build a preliminary-design model for the express purpose of studying a new nonlinear component or element(s), especially if the main features of the system response are dictated by the nonlinearity. In any case, linearization and idealization are discussed more fully in later chapters on component modeling, as are proper methods to deal with the inescapable nonlinearities when they arise.

HOMEWORK

3-1. Classify each of the following inputs as to whether it is in the time domain (TD), or in the frequency domain (FD):

a) The force exerted on a computer frame when the computer is dropped onto concrete.

b) The shaking exerted on DC electronic structural frames in aircraft and space vehicles.

c) Electromagnetic television signals.

d) Manual television tuning button changes.

e) The sound that shatters an opera glass.

f) The sudden drop in blood pressure due to a wound.

3-2. What are the three conditions under which a frequency-domain model of a MIMO component may be built?

3-3. Given the dynamic model of a ball falling vertically in oil,

$$\mathrm{M}\, dV/dt + \mathrm{B}V = \mathrm{W},$$

where M is the mass of the ball, B represents the drag constant of the oil on the ball, and W is the weight of the ball. Is this model in the time domain or the frequency domain? Why?

3-4. If a system has an external input of 3sin5t, in what domain would you expect to build a model? Why?

3-2 SINGULARITY INPUTS: IDEALIZATIONS IN THE TIME DOMAIN

The system modeler's approach to modeling time-domain inputs is to turn to a set, or family, of *idealized* time functions which are used singly or in combination to model the actual inputs. This set of idealized inputs is called the set of *singularity functions* since they all involve discontinuities (singularities) in slope or value. The most common singularity functions are the impulse, step, and ramp as shown in Fig. 3-4a, b, and c, respectively.

The magnitude of the impulse is measured by its area, the magnitude of the step is measured by its height, and the magnitude of the ramp is indicated by its slope. Note the indication of the magnitudes in Fig. 3-4. Hence, the singularity functions are also called "unit" functions since their magnitudes are all equal to one, or unity.

The Unit Impulse. The essential idea behind the impulse is that of a very short duration pulse to the system; as a hammer blow in a structural system or a voltage spike in an electrical system. The duration is usually so short, in fact, that its width cannot be well represented on a graph. Instead, the idealized impulse arrow is used as in Fig. 3-4a, with the magnitude indicated at the arrow head.

The unit impulse is expressed mathematically as

$$U(t) = u_i(t - t_s). \tag{3-1}$$

The units of the unit impulse are input-time; for example, if the unit impulse is in force, then the units are lb_f-sec. The time shift, t_s, is used to set the start time of the unit impulse. For example, if $t_s = 0.0$ sec., then the impulse would occur at $t = 0.0$. This input would be written in the following form:

$$u(t) = u_i(t), \tag{3-2}$$

to represent a unit impulse occurring at $t = 0.0$ sec., and having a magnitude of *zero everywhere else*.

If the unit impulse were to occur at $t = 3$ sec., then the proper expression would be:

$$u(t) = u_i(t - 3). \tag{3-3}$$

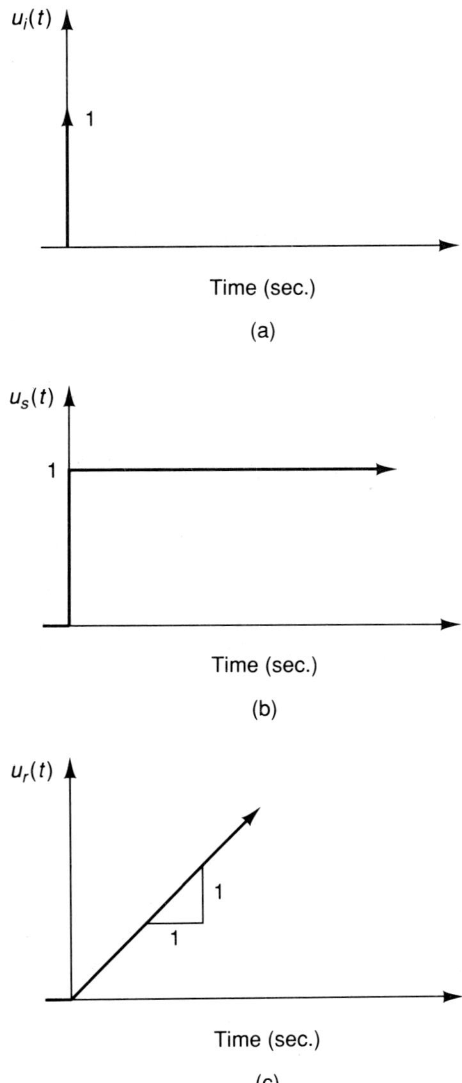

Figure 3-4. Common Singularity Inputs. (a) The unit impulse, (b) the unit step, (c) the unit ramp

To obtain an impulse of any other magnitude than unity, you would simply multiply the unit impulse by the desired magnitude. For example, the impulse function

$$u(t) = 6.0\ u_i(t - 5.0) \text{ amp-sec} \tag{3-4}$$

is an impulse which occurs at $t = 5.0$ sec. and has a magnitude (area) of 6 amp-sec.

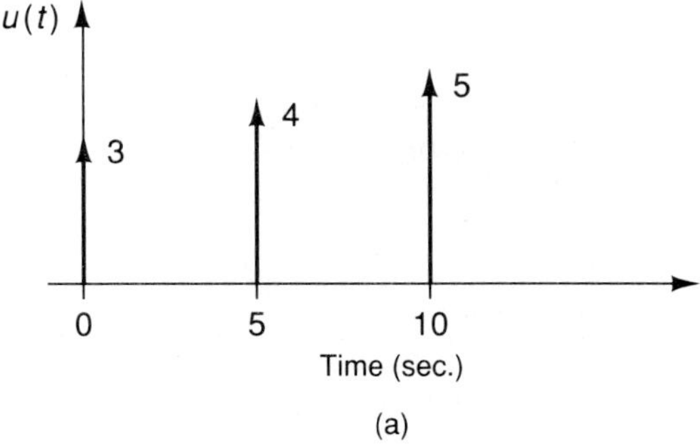

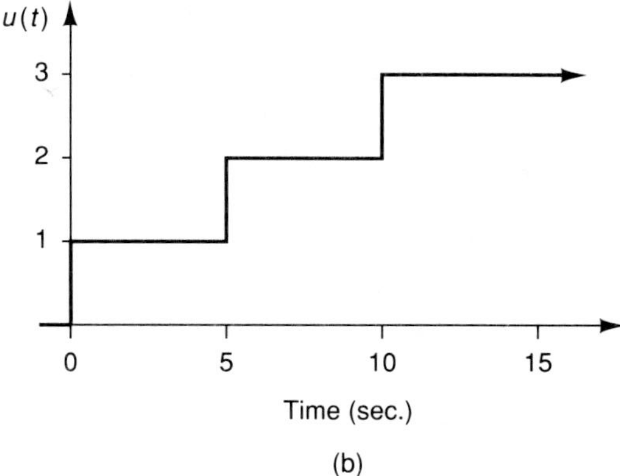

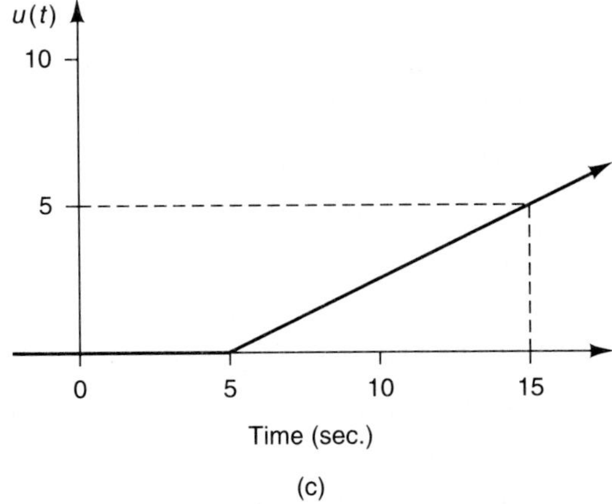

Figure 3-5. Built-up Idealized Inputs. (a) An impulse series, (b) a step series, (c) a delayed ramp

To represent a series of idealized functions, you add the scaled unit functions together. Consider, for example, the series shown in Fig. 3-5. For the impulse series shown in this figure, the correct representation would be:

$$u(t) = 3.0\ u_i(t) + 4.0\ u_i(t - 5.0) + 6.0\ u_i(t - 10.0) \text{ amp-sec.} \tag{3-5}$$

Notice that each of the impulses has zero value except at its start time, thus making built-up idealizations easy to assemble. This type of impulse representation is often used to study the time response of computer systems to changes in sampled data which the computer is measuring.

The Unit Step. The unit step is used to represent the sudden switching on, or switching off of inputs. Such inputs may be seen in the form of firing a rocket engine, or turning on an electrical switch. These actions may be graphically idealized in the form of Fig. 3-4b. Mathematically, the unit step is expressed in the following form:

$$U(t) = u_s(t - t_s), \tag{3-6}$$

which has the following values:

$$\text{if } (t - t_s) < 0, \text{ then } u_s = 0.0,$$

$$\text{if } (t - t_s) \geq 0, \text{ then } u_s = 1.0.$$

Consequently, the expression for the unit step function shown in Fig 3-4b is:

$$u(t) = u_s(t). \tag{3-7}$$

In other words, the start time, t_s, is used in precisely the same manner as the unit impulse to turn on the step. For a step down from 0 to –5 lb_f, occurring at 3 seconds, you would use:

$$u(t) = -5.0\ u_s(t - 3.0) \text{ lb}_f. \tag{3-8}$$

For another example, consider the series of steps shown in Fig. 3-5b. This function can be expressed mathematically by adding steps together in the following way:

$$u(t) = u_s(t) + u_s(t - 5.0) + u_s(t - 10.0) \text{ lb}_f. \tag{3-9}$$

The unit step is a very useful and common input for the design and modeling of engineering systems. The sudden change from an old steady input to a new steady input is often encountered in physical situations, and it is often an easily created experimental stimulation for design and model validation purposes.

The main focus of this text has to do with using step input functions in the modeling process.

The Unit Ramp. In the modeling of some physical systems, the unit step is too severe an input to be realistic or desirable. This occurs, for example, in the case of engine input fuel flow changes, or in aircraft pressure input changes due to altitude. In such cases, the ramp input can be useful.

A unit ramp is shown in Fig. 3-4c. It has the mathematical form:

$$U(t) = u_r(t - t_s), \tag{3-10}$$

where the start time, t_s, may again be set as desired by the modeler. The unit ramp represents the following:

$$\text{for } t < t_s,\ u_r = 0.0,$$

$$\text{for } t \geq t_s,\ du_r/dt = 1.0.$$

Consequently, if the modeler desires to start a ramp input at 5.0 sec., which is to have a slope of 0.5 volts/sec. (as in Fig. 3-5c), then the following would be used:

$$u(t) = 0.5\ u_r(t - 5.0) \text{ volts.} \tag{3-11}$$

The unit ramp is seldom used in model development work, but is often seen in the evaluation of completed models, for comparison to experimental data.

The Singularity Family. The unit impulse, unit step, and unit ramp are sometimes referred to as a *family* due to their relationships with one another. For example, the unit step is *defined* as the derivative of the unit ramp, and the unit impulse is *defined* as the derivative of the unit step. Similarly, the unit step is the integral of the unit impulse, and the ramp is the integral of the step.

These functional relationships may be similarly extended to higher- and lower-order singularity functions (e.g., second order, third order, etc.), in order to define an infinite family (or set) of singularity functions.

These definitions will be very helpful later in the text when components which act to differentiate or integrate their inputs are studied. Thus, a unit step may be physically input to a differentiating system, but the system response to a unit impulse must be analysed.

Complications. MIMO components require the modeler to express the idealized input in a vectorial form,

$$\boldsymbol{u}(t) = \begin{bmatrix} u_1(t) \\ \vdots \\ u_k(t) \end{bmatrix} \tag{3-12}$$

The entries in this vector, $(u_1, \ldots, u_k)^T$, are computed by idealizing the inputs one at a time, and then formulating the vector.

For example, if all the inputs from Fig. 3-4 were to be input to a system in the following fashion:

u_1 = unit impulse at t = 0.0 sec.,
u_2 = unit step at t = 0.0 sec., and
u_3 = unit ramp at t = 0.0 sec.

The idealized input vector for this example is thus,

$$\boldsymbol{u}(t) = \begin{bmatrix} u_i(t) \\ u_s(t) \\ u_r(t) \end{bmatrix}$$

HOMEWORK

3-5. Two sets of laboratory input measurements are shown in Fig. 3-6 which need to be idealized. Use the singularity functions to idealize the two data sets, and write out your idealized time-domain models for F and I. (Note that an impulse is a poor model for the start of the I steps since the small wiggle is negligible compared to the primary activity in the step.)

3-6. Use the integral definition of the singularity family to find the second-order singularity function (the unit ramp is the first-order function). Write out the definition of the second-order unit function and do not forget the time shift constant.

3-7. A certain MIMO component has $\boldsymbol{U} = (E, Q)^T$, with the two expected input histories shown in Fig. 3-7. Note that the area of the voltage impulse history has been approximated by the dashed lines shown in the voltage figure. Write out a MIMO idealization of $\boldsymbol{U}$.

3-8. Explain why it is better to use an idealization of E, as in the last problem, rather than the actual history of $E(t)$ for model and design improvement studies during preliminary design.

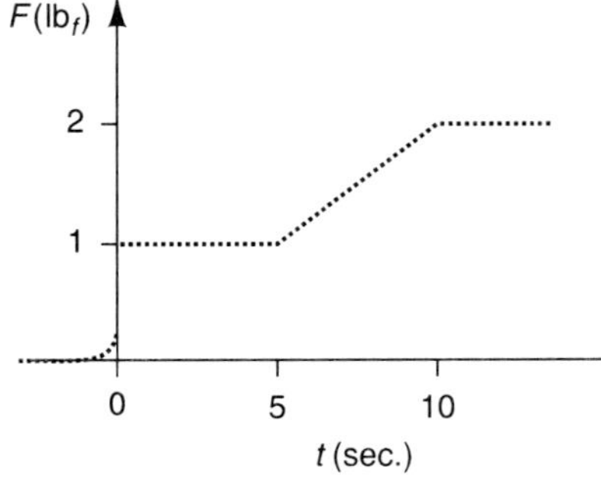

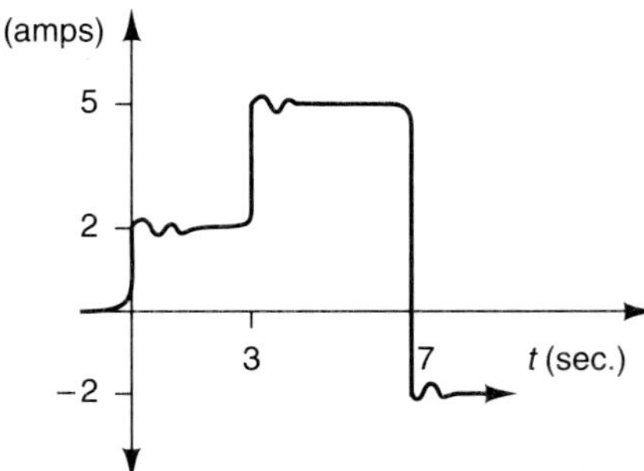

Figure 3-6. Time Histories for Prob. 3-5

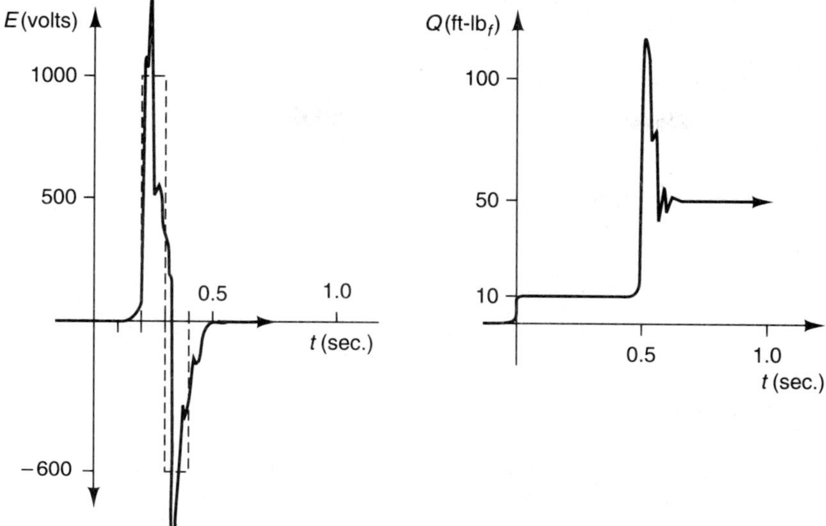

Figure 3-7. Time Histories for Prob. 3-7

3-3 PERIODIC INPUTS: IDEALIZATIONS IN THE FREQUENCY DOMAIN

Sometimes a component or system is subjected to inputs which occur in uniform periodic cycles. Examples of this are occasionally seen in propeller torque oscillations due to a ship's passage through the waves in a seaway, or sometimes in the vibrations felt in a car due to the tires passing over a wavy roadway. As mentioned earlier, these inputs are really occurring in the time domain, yet the modeler needs to focus on their frequency-rich content. This is done by *transforming* the periodic time-domain inputs into the frequency domain.

In this section, three commonly idealized periodic inputs are considered: sinusoids, step waves, and ramp waves (sawtooth waves). The tools to transform these time-domain waves into the frequency domain are introduced.

Sinusoidal Inputs. Occasionally, a system input is a simple sine (or cosine) function such as that shown in Fig. 3-8. This basic wave has the following mathematical representation:

$$U_1(t) = C_1 + C_2 \sin(\omega t + \phi), \tag{3-13}$$

where the *frequency*, ω(rad/sec.), is related to the *period*, T(sec.), through the following expression:

$$\omega = 2\pi/T. \tag{3-14}$$

The *phase angle*, ϕ(rad), is used to start the sine function at the proper point in its cycle when $t = 0.0$ sec. (Notice that a positive phase angle shifts the sine curve to

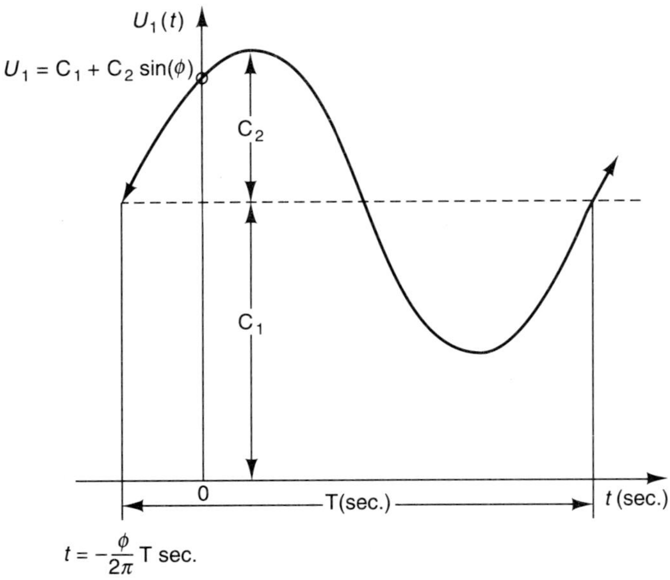

Figure 3-8. A Pure Sine Input

the *left* in Fig. 3-8) The constant C_2 indicates the *amplitude* of the input wave, and the constant C_1 indicates the *offset* of the wave from $U_1 = 0.0$. C_1 and C_2 have the same units, according to the nature of the input wave (i.e., volts, or torque, or whatever).

Thus, while the mathematical expression for the sine input is in terms of the independent variable time, t, the expression also has in it the three quantities that are used to characterize the frequency domain—these being the frequency, the amplitude, and the phase.

Broadly stated, the modeler's plan is to transform *any* periodic time-domain input into a combination of pure sinusoidal waves. Thus, the modeler can extract a wealth of frequency, amplitude, and phase information for frequency-domain modeling and design studies.

The solution to the transformation problem which follows is first developed theoretically for step waves and ramp waves. Often, however, the frequency domain inputs are more complicated then these simple waveforms. So, in the next section of the text, numerical methods are developed to deal with more complicated waveforms.

Square-Wave Inputs. Consider the example square wave shown in Fig. 3-9. This example square wave is composed of an infinite series of periodic steps between 4 volts and 6 volts, occuring at one second intervals. The modeler could choose to mathematically represent this input with an infinite series of steps after the time domain manner of the previous section:

$$u(t) = 6.0u_s(t) - 2.0u_s(t - 1.0) + 2.0u_s(t - 2.0) - \cdots \tag{3-15}$$

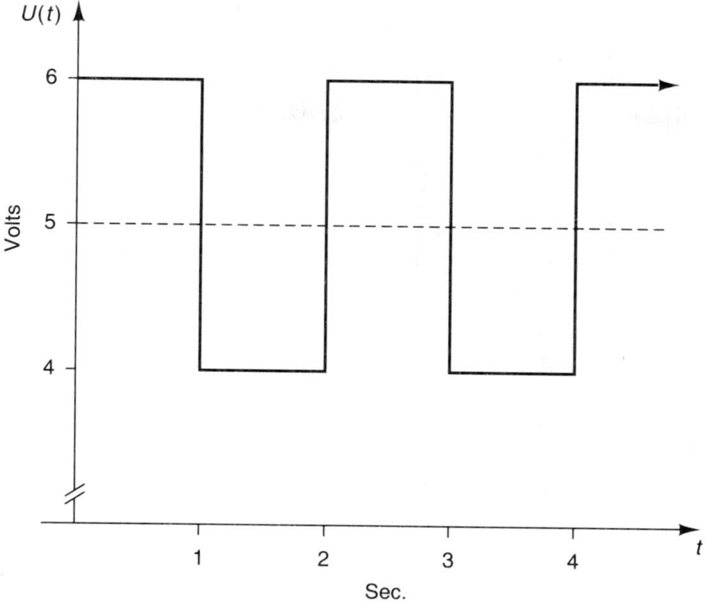

Figure 3-9. A Square-wave Input

$$= 6.0u_s(t) + \sum_{n=1}^{\infty} (-1)^n \, 2.0u_s(t - n). \qquad (3\text{-}16)$$

On the other hand, the modeler may have an insight that the periodic step is similar to a sine wave, as shown in Fig. 3-10.

So, the period of the example step wave is 2 seconds, its amplitude is 1 volt, the magnitude oscillates between 4 and 6 volts, it has an offset of 5 volts, and zero phase angle. Thus, the modeler may achieve a first (but very poor!) idealization of the step wave using the following form:

$$u_1(t) = [5.0 + \sin(\pi t)] \; u_s(t) \text{ volts.} \qquad (3\text{-}17)$$

Note that the frequency of both waves in Fig. 3-10 is π rad/sec, this is known as the *fundamental frequency* of the waves.

The modeler has thus achieved the reduction of an infinite series of inputs (Eq. 3-16) to a two-term model (Eq. 3-17). But, being a precise professional, the modeler desires a more accurate model of the stepped wave. To achieve this, the modeler has a second insight that it may be possible to add in higher frequency sine and cosine waves to better approximate the sharp corners of the steps.

At this point, the modeler has begun to evolve a *Fourier Analysis* of the step wave, which was first developed by J.B.J. Fourier, a French mathematician. The exact representation of the series as developed by Fourier is:

$$U(t) = \mathrm{U}_o + \sum_{n=1}^{\infty} \mathrm{a}_n \cos(n\omega_o t) + \sum_{m=1}^{\infty} \mathrm{b}_m \sin(m\omega_o t) \qquad (3\text{-}18)$$

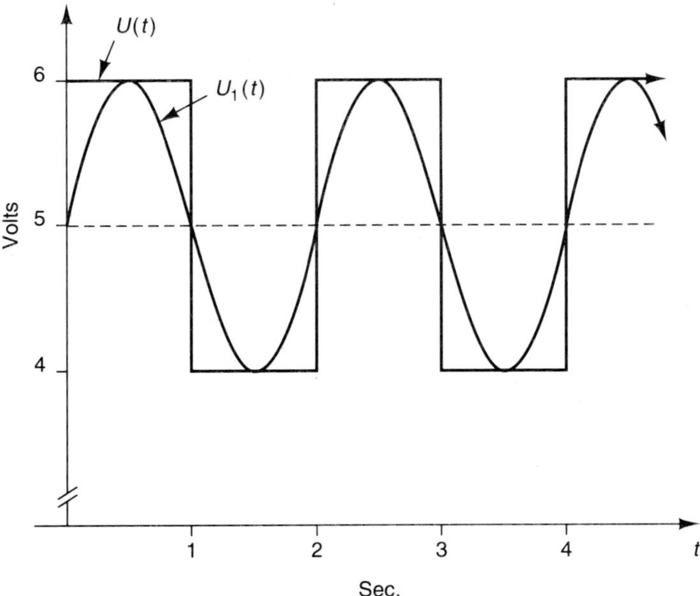

Figure 3-10. First Approximation of a Square Wave

This series representation can be applied to any *periodic* waveform.

The Fourier coefficients are evaluated from the following formulas:

$$U_o = \text{waveform offset},$$

$$a_n = \omega_o/\pi \int_{t_0}^{t_0+T} U(t) \cos(n\omega_o t)\, dt \tag{3-19}$$

$$b_m = \omega_o/\pi \int_{t_0}^{t_0+T} U(t) \sin(m\omega_o t)\, dt. \tag{3-20}$$

Notice that the above formulas begin with $n = m = 1$, which gives terms in sine and cosine at the fundamental frequency, ω_o. The proper higher frequencies to fit the desired waveform corners are called *harmonics*, and appear in the formulas as integer multiples of the fundamental ($n\omega_o$ or $m\omega_o$). The start time of the integrations, t_0, corresponds to the start of any single period, and the integration extends over one period. The start time may be selected for the convenience of the analyst.

Now, for exact accuracy, the modeler must choose between Eq. 3-16 and Eq. 3-18, both of which are infinite series. However, the modeler also realizes that there will necessarily be approximations in the remainder of the model which will set the accuracy level of the input modeling effort.

With this in mind, the modeler decides on an idealization which is a five-term Fourier series approximation of the example square wave. The following constants are used for this idealization:

$$U_o = 5 \text{ volts},$$

$$T = 2.0 \text{ sec.}, \ \omega_o = 2\pi/T = \pi \text{ rad/sec.},$$

$$t_0 = 0.0 \text{ sec.}$$

The integrals in the general expressions for the constants are next evaluated,

$$a_n = \omega_o/\pi \int_0^2 U(t) \cos(n\omega_o t)\, dt, \tag{3-21}$$

$$= \int_0^1 6 \cos(n\pi t)\, dt + \int_1^2 4 \cos(n\pi t)\, dt, \tag{3-22}$$

$$= (6/n\pi) \sin(n\pi t) \Big|_0^1 + (4/n\pi) \sin(n\pi t) \Big|_1^2, \tag{3-23}$$

and

$$a_n = 0, \text{ for all } n \geq 1. \tag{3-24}$$

Since the a_n's are all zero, Eq. 3-18 indicates that there are no cosine terms in the approximation for $U(t)$.

The sine coefficients are evaluated next,

$$b_m = \omega_o/\pi \int_0^2 U(t) \sin(m\omega_o t)\, dt, \tag{3-25}$$

$$= \int_0^1 6 \sin(m\pi t)\, dt + \int_1^2 4 \sin(m\pi t)\, dt, \tag{3-26}$$

$$= -(6/m\pi) \cos(m\pi t) \Big|_0^1 - (4/m\pi) \cos(m\pi t) \Big|_1^2. \tag{3-27}$$

So, when m is an even number, $b_m = 0$ from the above integral. However, when m is odd,

$$b_m = -(6/m\pi)\,(-1 - 1) - (4/m\pi)\,[1 - (-1)], \tag{3-28}$$

$$= (12 - 8)/m\pi, \tag{3-29}$$

and

$$b_m = 4/m\pi, \text{ for } m \geq 1, m \text{ odd only}. \tag{3-30}$$

The desired five-term model is found by using Eqs. 3-18, 3-24, and 3-30, and letting m equal one through seven in the Fourier series. This gives

$$u_5(t) = 5.0 + (4/\pi)\sin\pi t + (4/3\pi)\sin 3\pi t + (4/5\pi)\sin 5\pi t + (4/7\pi)\sin 7\pi t, \quad \text{for } t \geq 0.0. \tag{3-31}$$

Comparison between the earlier two-term model of the square wave (Eq. 3-17) and the above five-term expression (Eq. 3-31) leads to several conclusions. First, the similarity of the first two terms in each series is evident (offset and fundamental frequency terms). The only difference is in the amplitude of the fundamental frequency term, it being slightly larger in the Fourier expression. This is due to the fact that u_1 was not found from a rigorous analysis, it was simply proposed as a model. In contrast, all the terms in the Fourier model are those which have the least overall error when compared to $U(t)$.

The second conclusion to be reached is that a step wave can now be thought of as being built up from an infinite set of sine waves. In fact, any component upon which these waves act cannot tell at first that it is "seeing" the beginning of a periodic input, it simply sees many frequencies at once. And, if the fundamental frequency of the step wave is reduced to zero in some limiting sense, then the result is an approximate simple step.

In the limit then, the simple step can be thought of as containing all the frequencies. This notion is very important in understanding the relationship that exists between the time domain (step inputs) and the frequency domain (Fourier-equivalent step inputs). This idea is also very important to the study of engineering system behavior, and it will be seen again in several places later in this text.

Sawtooth-Wave Inputs. The Fourier series for a sawtooth wave may also be computed using the integral definitions of the Fourier method. However, these difficult integrations are more easily computed by taking advantage of the relationship of the sawtooth wave to the step wave. Consider, for example, the sawtooth and square waves of Fig. 3-11.

As Fig. 3-11 shows, the derivative of the sawtooth wave is a square wave, and the integral of the square wave is a sawtooth wave. Note, however, that the two waves have different amplitudes and offsets—these must be recognized in the analysis. The sawtooth wave can be expressed as

$$U_{SAW} = \mathrm{U}_{OSAW} + 2\mathrm{C}_2\left[\int U_{SQ}\,dt - 1/2\right], \tag{3-32}$$

where U_{OSAW} is the offset of the sawtooth wave, and U_{SQ} is a *zero-offset square wave* of unit amplitude. C_2 is the desired sawtooth amplitude, and the integral is an indefinite integral beginning on an up-pulse of the square wave. The 1/2 must be subtracted to correct the offset due to the integration (see Fig. 3-11).

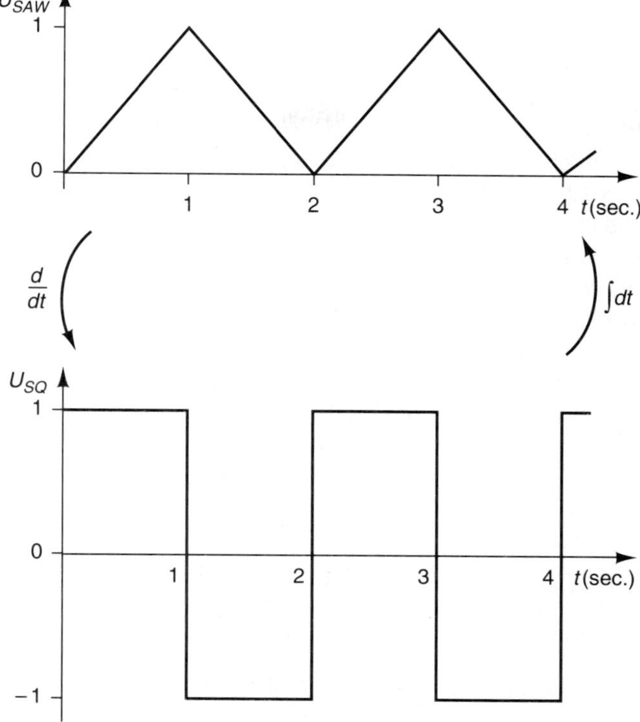

Figure 3-11. Sawtooth and Square Wave Relationships

The sawtooth offset is computed using

$$\mathrm{U}_{OSAW} = (1/\mathrm{T}) \int_{t}^{t+\mathrm{T}} U_{SAW}\, dt, \tag{3-33}$$

which is the formula for the time-averaged value of U over the period, T. (The offset and the time-averaged value are always the same.)

In the case of Fig. 3-11, $\mathrm{U}_{OSAW} = 1/2$, and the integral term for Fig. 3-11 is computed using the time-varying part of the square wave analyzed earlier (Fig. 3-9 and Eq. 3-31). So, for the sawtooth of Fig. 3-11, $\mathrm{C}_2 = 1/2$,

$$U_{SAW} = 1/2 + 2(1/2)\left[\left[\int U_{SQ}\, dt - 1/2\right]\right],$$

$$= \int_0^t [(4/\pi) \sin \pi t + (4/3\pi) \sin 3\pi t + \cdots]\, dt,$$

and $$U_{SAW} = 1/2 - (4/\pi)[(1/\pi) \cos \pi t + (1/9\pi) \cos 3\pi t + \cdots].$$

While this method is convenient, there are three points at which you must exercize great care. The first point is to be certain that a *zero-offset* square wave of the *same period* as the sawtooth is used in the integration.

The second point for care is that the *amplitude* of the triangle wave is used to scale the integration. Unit-magnitude square waves are often seen in tabulations of these functions for this reason. You then simply multiply the unit-amplitude result by the amplitude of interest, according to Eq. 3-32.

The final point for concern is that the *phase* of the two waves is properly selected for the relationship of interest. In the example of Fig. 3-11, the asymmetric square wave (about $t = 0$ sec.) is matched with the symmetric sawtooth. The integral determines this phase, and begins on the up-pulse of the square wave.

Given a computed Fourier model of a triangle wave, it can be scaled in frequency (or period), in amplitude, and/or in phase (e.g., time-shifted) to model any sawtooth wave. The same is true for the square wave.

HOMEWORK

3-9. The plot in Fig. 3-12 is for a step wave which is symmetrical about $t = 0.0$ sec. Use this data to:

a) Compute the Fourier series for this wave, through the second harmonic.

b) Compare your solution to the text solution for the asymmetrical wave regarding the frequency domain variables (amplitude, frequency, and phase).

c) Compute the Fourier series for the sawtooth wave which has corners at (−0.5, 0), (0.5, 1), and (1.5, 0). (Hint: compare to Fig. 3-12.)

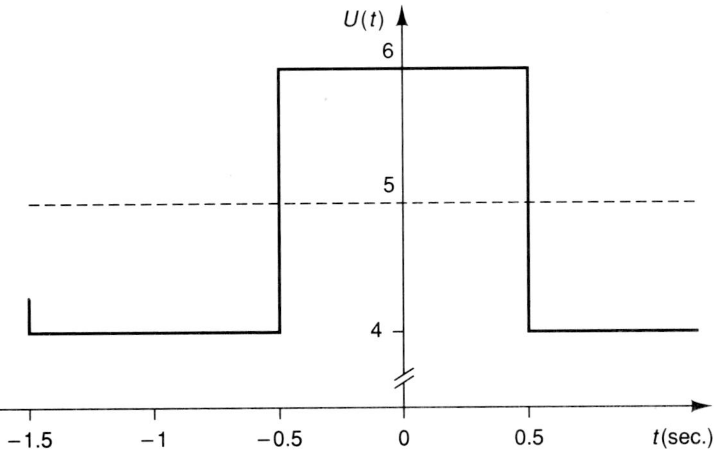

Figure 3-12. Symmetrical-Square Wave for Prob. 3-9

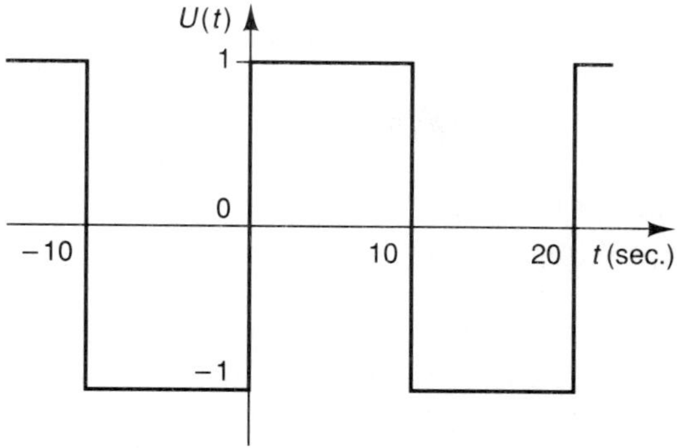

Figure 3-13. Asymmetrical-Square Wave for Prob. 3-11

3-10. Given the function U* = 3 cos(2t − 45°) amps, what are the values of the three frequency-domain variables which describe this input?

3-11. Use the asymmetrical step wave of Fig. 3-13 to find the following:

a) Compute the Fourier series for this wave out through the first two harmonics,

b) Compare your answer to the text answer for the asymmetrical step wave regarding amplitude, frequency, and phase.

3-12. An idealized input model for a preliminary design has only the 3 lowest harmonics present. Why has the modeler excluded the higher harmonics?

3-4 FREQUENCY DOMAIN COMPLICATIONS AND DATA PLOTS

For many periodic inputs, the theoretical integrations for the Fourier coefficients are simply too difficult. Thus, in order to deal with more complicated periodic functions than simple sinusoids, or stepped and sawtoothed waves, you must find additional tools to compute the required time-to-frequency transformations. In these cases, system modelers resort to the digital computer and use the methods of *numerical analysis* to perform the Fourier transformations.

Numerical Wave Analysis. Numerical analysis is a branch of mathematics which offers the modeler a means to approximate the continuum of mathematics in which the difficult Fourier integration formulas are expressed. This has become an increasingly useful branch of mathematics in the era of the digital computer since it translates infinitesimal, continuous formulae into finite-interval, discrete equations which are ideal for computer processing.

The basic idea of numerical analysis is the notion of *finite approximation*. In this, the infinitesimal differential variables are approximated in the following manner,

$$dt \sim \Delta t = (t_2 - t_1), \tag{3-34}$$

and

$$d\omega \sim \Delta\omega = (\omega_2 - \omega_1). \tag{3-35}$$

This is the reverse of the original differential-limiting process, where the infinitesimal variables are obtained by letting the delta quantities go, in the limit, to zero. In fact, it is upon this notion that the definition of differentiation is based,

$$\frac{dU}{dt} = \lim_{\Delta t \to 0}\left[\frac{\Delta U}{\Delta t}\right] \tag{3-36}$$

In numerical analysis, the modeler simply takes Δt small enough to allow for an accurate approximation. That is,

$$dU/dt \approx \Delta U/\Delta t. \tag{3-37}$$

An integration may also be approximated in a similar way. This applies directly to the study of the Fourier integrations since this idea returns the integral to its summation origins,

$$\int_{t_1}^{t_2} U(t)\,dt \approx \sum_{k=0}^{N-1} U(t_1 + k\Delta t)\,\Delta t. \tag{3-38}$$

Recall that the fundamental operation of integration embodied in Eq. 3-38 is to compute the area under U on the time plot. One way to numerically approximate this area is to use an addition of rectangular areas of uniform width, Δt, and uniform height, U_k. Such an approach is called *rectangular integration*. An example of rectangular integration is shown in Fig. 3-14 for a specific history of U, and an eight-interval integration ($N = 8$). Notice that some of the interval areas are too tall, and some are too short. On the whole, however, the net approximation is acceptable if the interval, Δt, is properly chosen.

There are, of course, many other ways to approximate the area underneath $U(t)$, and rectangular integration is only one possible approach. However, these other methods are presented in advanced studies in numerical analysis, they are not needed in this introductory study.

In any case, all numerical methods are strongly dependent on the *numerical error bounds*. That is, you might be tempted to believe that if Δt were to go very close to zero, then the theoretical integration could be exactly reproduced. However, this means that the computer must compute a nearly infinite number of additions and multiplications. Such a large number of arithmetic operations magnifies the tiny errors in each computation due to the inexact representation of the numbers being pro-

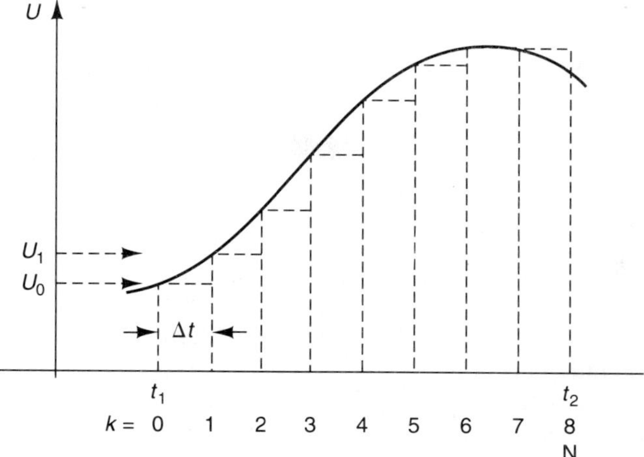

Figure 3-14. Rectangular Integration of *U*

cessed. In fact, this magnification can destroy the accuracy of the approximation. So, too many computations is not acceptable, Δt must not be chosen too small.

Further, if too few intervals are chosen (Δt too large), then you fail to meet the spirit of the limiting process that is necessary for the derivative and the integral. Again, erroneous results are achieved and too few computations is not acceptable either.

The best choice of Δt lies between these extremes. The interval must be chosen suitable to the integration problem under consideration to stay within the error bounds.

For the problem of the Fourier coefficients, the fundamental period, T, must be divided into N uniform intervals, each Δt wide, so that $t = t_1 + k\Delta t$, $k = 0, 1, \ldots, N$.

The Fourier terms thus become

$$U_o = \text{Offset in } U = \text{Time-averaged } U(t),$$

$$= \frac{1}{T}\int_{t_1}^{t_1+T} U(t)\, dt \tag{3-39}$$

$$\approx \frac{1}{T}\sum_{k=0}^{N-1} U(t_1 + k\Delta t)\, \Delta t. \tag{3-40}$$

Similarly, the other Fourier coefficients are transformed into the following forms:

$$a_n \approx A_n = \frac{\omega_o}{\pi}\sum_{k=0}^{N-1} U(t_1 + k\Delta t)\cos(n\omega_o k\Delta t)\, \Delta t, \tag{3-41}$$

$$b_n \approx B_m = \frac{\omega_o}{\pi}\sum_{k=0}^{N-1} U(t_1 + k\Delta t)\sin(m\omega_o k\Delta t)\, \Delta t. \tag{3-42}$$

Combining these results gives

$$u(t) = U_o + \sum_{n=1}^{\infty} A_n \cos(n\omega_o t) + \sum_{m=1}^{\infty} B_m \sin(m\omega_o t). \tag{3-43}$$

The general rule-of-thumb for error bounds in the rectangular integration of a Fourier series is

$$T/5000 \leq \Delta t \leq T/200, \tag{3-44}$$

where T is the period of the *fundamental* frequency.

The underlying rule which is used to develop the interval guideline is that at least 10 intervals are required to integrate the fastest significant frequency. This fastest frequency generally occurs before the twentieth harmonic. Thus, if 10 intervals are taken for the twentieth harmonic, the intervals will be 1/200th of the period of the fundamental.The lower limit is chosen to give a reasonable working range, without allowing too many intervals.

If the input history is a very "spikey" periodic function, an interval may also be estimated by taking one-tenth of the time width of the narrowest spikes (these will generally correspond to the highest frequencies).

Example 3-1

In order to demonstrate the numerical Fourier computations, consider the square-wave plot in Fig. 3-15. Note that the example wave has been divided into 10 equal intervals for numerical analysis. The problem is thus to estimate the a_4 coefficient using Eq. 3-41.

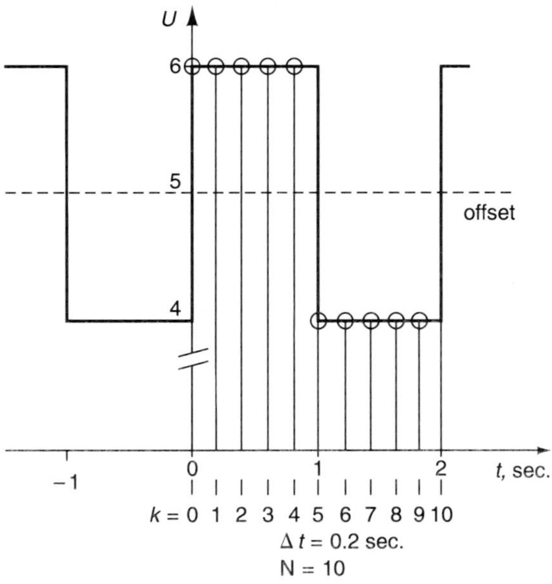

Figure 3-15. Square Wave Integration Example

Solution

The numerical parameters in the analysis are, T = 2 sec., $\omega_o = 2\pi$ rad/2sec. = π rad/sec.,

$$N = 10,\ \Delta t = N/T = 0.2 \text{ sec},\ n = 4.$$

You may then compute the entries of Table 3-2 according to Eq. 3-41.

Table 3-2. Computation of A_4 for Ex. 3-1

k	$k\Delta t$	$U(k\Delta t)$	$\cos(4\pi k\Delta t)$	$U(.)\cos(.)$
0	0.0	6.0	1.000	6.000
1	0.2	6.0	−0.809	−4.854
2	0.4	6.0	0.309	1.854
3	0.6	6.0	0.309	1.854
4	0.8	6.0	−0.809	−4.854
5	1.0	4.0	1.000	4.000
6	1.2	4.0	−0.809	−3.236
7	1.4	4.0	0.309	1.236
8	1.6	4.0	0.309	1.236
9	1.8	4.0	−0.809	−3.236
				Σ = 0.0000

Comments

The earlier theoretical analysis that was done on the offset-square wave of Fig. 3-15 (e.g., Fig. 3-9 and Eq. 3-31) showed that all of the cosine terms are equal to zero, which is confirmed by this analysis. Still, it is interesting that the *exact* result has been computed in only ten intervals in light of the interval guidelines given above.

The reason that rectangular integration worked so well in this case is that the product of the square wave and the cosine wave which is computed in the A_4 coefficient is also a well-behaved periodic wave. The selection of 10 intervals was exactly T/10 for the product wave, thus giving exact cancellation of the numerically added terms.

Bode-Diagram Plots. Frequency data are often presented on a frequency-domain figure called a *Bode diagram*. Such a diagram is shown in Fig. 3-16.

In the Bode diagram, the horizontal axis is a log-scale plot of either the frequency ratio, ω/ω_o, or the frequency, ω. The vertical scale is a uniform scale of the decibel value (db) of the function of interest. For an arbitrary periodic function, f, the decibel value is defined as,

$$db = 20 \log_{10} |f|. \tag{3-45}$$

That is, the db value of the function, f, is twenty times the base-10 logarithm of the absolute magnitude of the sinusoidal amplitude of the function. The offset value of f does not enter into the computation. If f is not a periodic function, then it has no decibel value.

The decibel value originated in the study of sound propagation (acoustics) but was later generalized for the study of AC electronics and other types of frequency-based systems. The diagram is named after H.W. Bode, who invented it as a tool to study the design and response of electronic amplifiers while he worked at Bell Laboratories.

Figure 3-16 is a Bode diagram of the transformed step wave, u_5, which was analyzed in the previous section (e.g., Eq. 3-31). Notice that the fundamental frequency (π rad/sec.) and the lowest harmonics all appear simultaneously as points on the diagram. Thus, each pure sinusoidal wave in the Fourier expansion of a periodic function has its own data point on the diagram, according to the frequency and amplitude of that part of the function. In this way you can graphically represent the many distinct frequencies in a periodic wave as a set of points on a single Bode diagram.

Conversely, the implication of the points on a Bode diagram is that they represent sinusoidal contributions which must be added together in order to reconstruct the function plotted.

Notice also that the general trend of the data in Fig. 3-16 goes from the upper left of the figure to the lower right. This is a trend which is encountered over and over again in physical data. In fact, this trend serves as a good general check on both instruments and data.

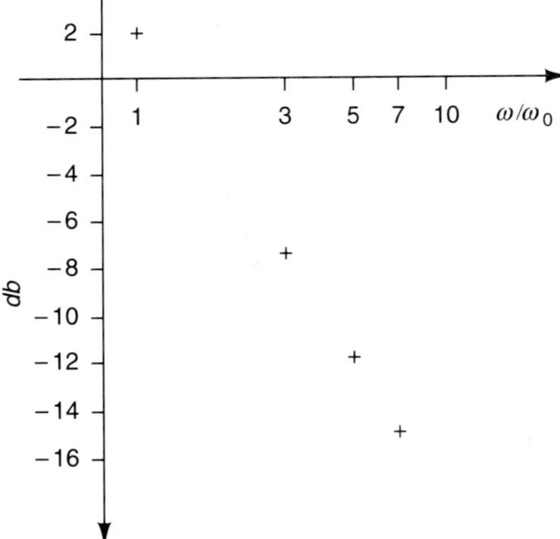

Figure 3-16. Bode Diagram for a Periodic Input

HOMEWORK

3-13. Write a computer program to compute the Fourier coefficients out through the tenth harmonic. Write your program to compute the value of U based on the assumption that you know the analytic time function for U (rather than read ing U from a file). Validate your program with the square wave coefficients that were computed in the text.

3-14. Use the computer program that you validated for the previous homework problem to do the following:

a) Compute the Fourier transformation for the triangle waveform given in Fig. 3-17.

b) Plot your result on a Bode diagram.

3-15. Given $U^* = 3\sin(2t + 15°)$. What is the frequency of a third harmonic of this wave?

3-16. Given $U = 10\cos(10t) + 5\cos(20t) + 2\cos(30t)$. What is the *maximum* time interval for the rectangular integration of this wave that you would recommend?

3-17. Write a computer program to integrate $U = \sin\pi t$ over the range $t = [0, 1]$ sec. using rectangular integration. Compute the error in percent [(numerical-exact)/(exact) × 100] for the cases, N = 2, 3, 5, 10, 100, 1000, 10^4.

a) Plot the error as a function of N on a log scale.

b) Comment on the error curve shape with regard to the minimum error and the lower error bound (N small).

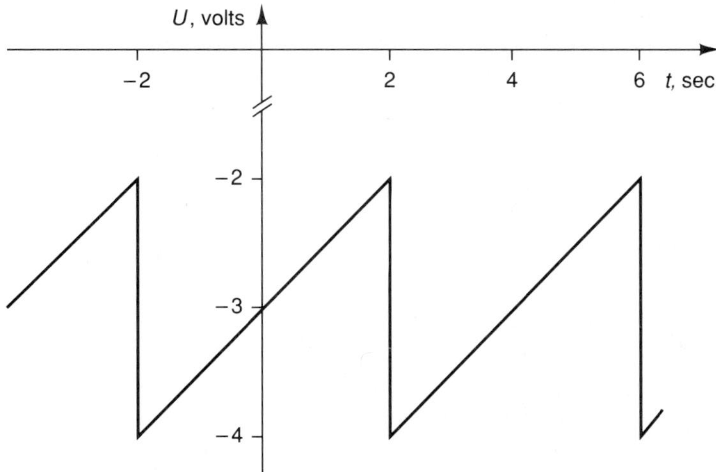

Figure 3-17. Periodic Input for Prob. 3-14.

3-5 SUMMARY, BIBLIOGRAPHY, AND REVIEW PROBLEMS

All physical inputs behave as if they are either aperiodic or periodic functions. Aperiodic inputs are modeled in the time domain, with time as the independent variable. Periodic inputs are modeled in the frequency domain, with frequency as the independent variable. So, this chapter presented physical analysis of system inputs with a view towards system-model domain selection and input idealization for preliminary-design studies.

Input idealization in the time domain was accomplished using a family of singularity inputs. Three members of this family are used much more commonly than the rest, these being the unit impulse, the unit step, and the unit ramp. Several examples of idealizations using these three common singularities were given.

Input idealization of periodic functions was accomplished using waveform analysis. This was first demonstrated for the simplest frequency-domain input, the pure-sine wave, which was shown to present all three of the frequency-domain variables: amplitude, frequency, and phase. For more complex periodic waveforms, the analytic Fourier method was used to transform a periodic time-domain input into a combination of pure sines. This transformation was demonstrated for the square wave and for the sawtooth wave.

For the time-to-frequency transformation of very complex periodic inputs, a numerical version of the Fourier integrals was developed and demonstrated.

The modeling of inputs discussed in this chapter is important for the coming component physical analyses because, generally speaking, the system processes the inputs and creates like outputs. Thus, if the inputs are in the time domain, then the outputs tend to be there as well. If the inputs are in the frequency domain, then the outputs are likely to be there as well. So, the domain of the inputs helps you to decide on the final form for the system model since the domain of the model usually matches the domain of the inputs.

BIBLIOGRAPHY

Bracewell, R.N., *The Fourier Transform and its Applications*, New York, N.Y.: McGraw-Hill Book Co., 1978.

Cartwright, M., *Fourier Methods for Mathematicians, Scientists and Engineers*, New York, N.Y.: Ellis Horwood, 1990.

Goldstein, H.H., *A History of Numerical Analysis from the 16th Through the 19th Century*, New York, N.Y.: Springer-Verlag, 1977. When did Newton live and what numerical methods did he invent? This is interesting reading.

Lee, J.A.N., *Numerical Analysis for Computers*, New York, N.Y.: Reinhold Publ. Co., 1966.

Maron, M.J. and R.J. Lopez, *Numerical Analysis*, Belmont, Calif.: Wadsworth Publ. Co., 1991. This is a large, thorough text, but advanced reading in some places.

Shearer, J.L., Murphy, A.T., and H.H. Richardson, *Introduction to System Dynamics*, Reading, Mass.: Addison-Wesley Publ. Co., 1967. This is a classic text on system dynamics. It has a good discussion of singularity functions and analytical Fourier analysis.

Thompson, W.J., *Computing in Applied Science*, New York, N.Y.: John Wiley & Sons, 1984. This text covers numerical methods for Fourier computations.

REVIEW HOMEWORK

3-18. Given the system and data shown in Fig. 3-18, answer the following:

a) Write out the preliminary-design idealization of U^*.

b) In what domain should this system be modeled to study its response to U^*.

c) In what domain is Y^*?

3-19. Use the input histories of the four inputs in Fig. 3-19 to answer the following:

a) Which of the inputs is periodic?

b) What is the idealization of U_a?

c) What is the idealization of U_c between 0 and 2.0 sec.?

d) If you were willing to ignore the high frequency, small amplitude waves in U_b, what model would you use to approximate the remaining waveform?

e) Idealize U_d over $(0 \leq t \leq 3.0)$ using sines and steps.

3-20. Use the information in Fig. 3-20 to answer the following:

a) What domain(s) of analysis would you expect to use for this component? Why?

b) What is the Fourier model of U_2?

c) What is the time-domain model for U_3, for $t \geq 0$?

d) Compute the frequency-domain model for U_3 using the analytical integral form of the Fourier equations (as opposed to the numerical form)?

e) Plot your results for (d) above on a Bode diagram, and compare them to the homework result that you got for Prob. 3-14.

f) Assuming that you wish to conduct a frequency analysis of this component, what should you do about U_1?

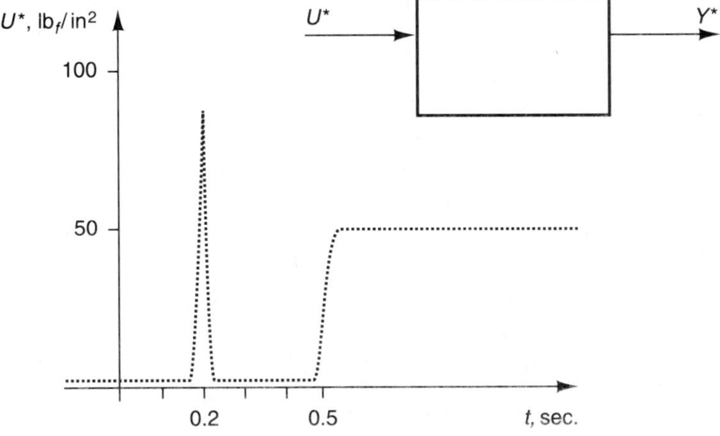

Figure 3-18. Aperiodic Input for Prob. 3-18

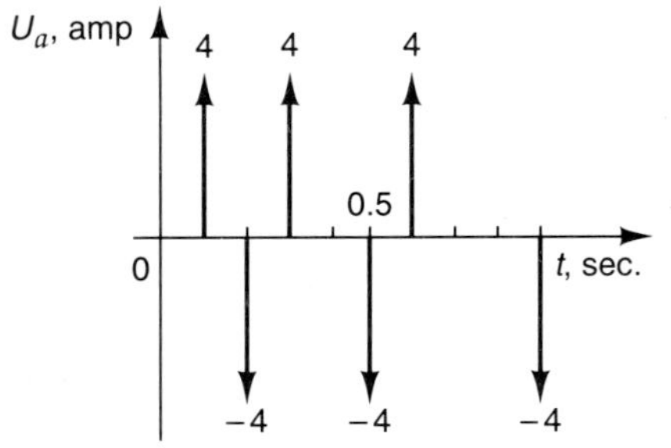

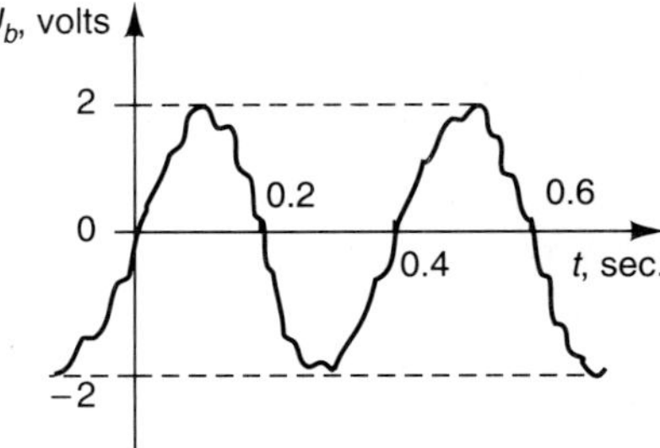

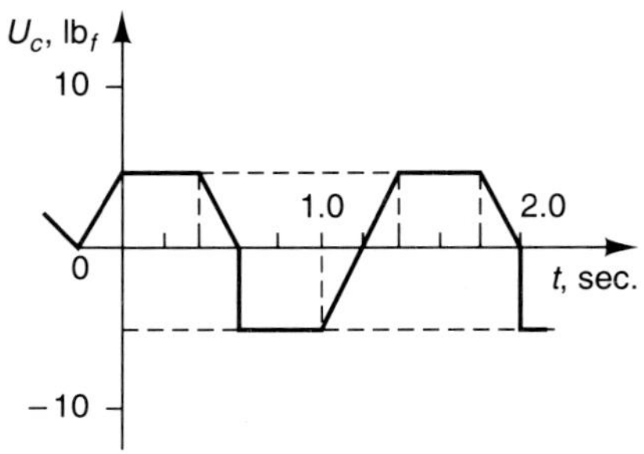

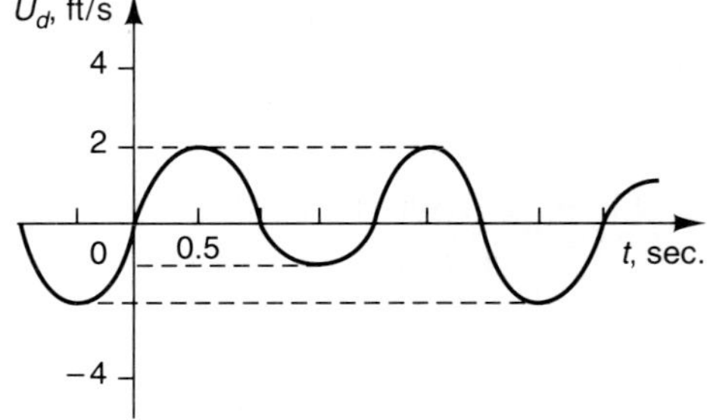

Figure 3-19. Inputs for Prob. 3-19

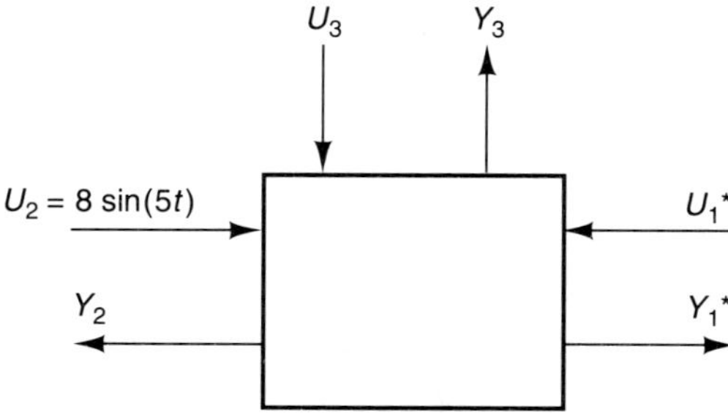

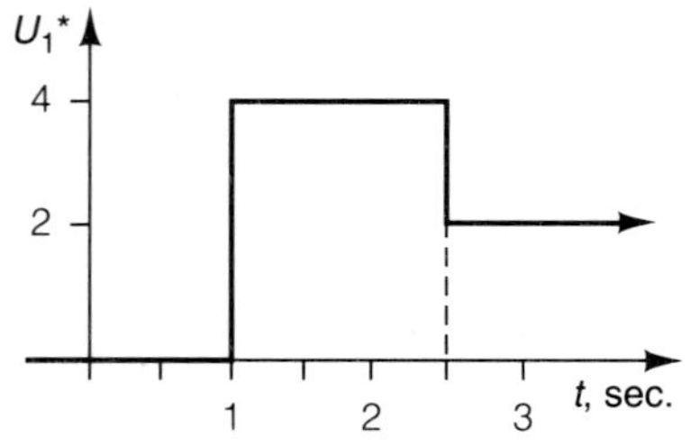

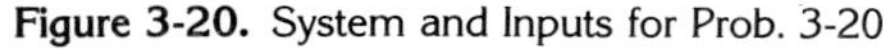

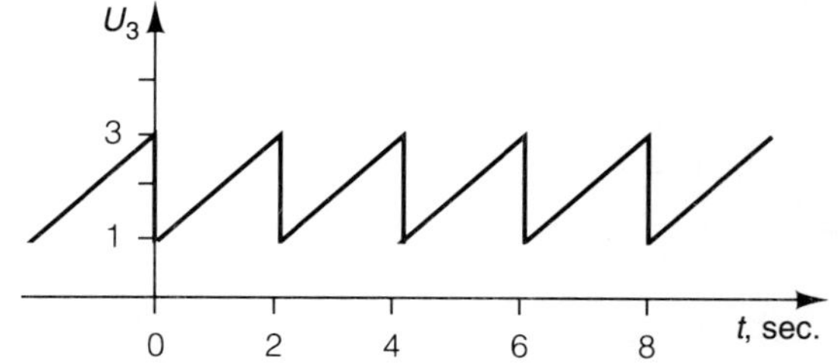

Figure 3-20. System and Inputs for Prob. 3-20

3-21. Given that $U(t) = 2\sin 3\pi t \cos \pi t$,

a) What is the fundamental frequency of $U(t)$?

b) How many intervals would you choose for numerical integration of $U(t)$? Why?

c) Write out the formulas for a Fourier series approximation of $U(t)$ using numerical integration of the first ten harmonics.

d) Compute the analytical Fourier series for $U(t)$ using the integral formulas developed by Fourier.

Chapter 4

Component Physical Analysis: The Fundamental Approach in the Time Domain

4-0 INTRODUCTION

In this chapter, it is assumed that time domain system inputs exist and have been idealized, and the multiport diagram for the system is completed. The modeler now selects a component from the system diagram for time domain modeling which has a known arrangement of physical elements. This knowledge lets the modeler conduct a fundamental analysis of the physical behavior of the elements to account for the output response of the component. Hence, this is called the *fundamental approach* to component modeling.

When the elements within the component are *unknown,* the modeler is forced to approach the component physics from the viewpoint of overall input/output relationships. That is, the component is then treated as a type of "black box," which performs some process to transform the inputs into outputs. Consequently, this alternative approach is called the *input/output approach* to component modeling; it is discussed in Chap. 7.

When digital components occur as a part of the system under study, they are usually modeled with input/output models. This approach is taken since digital components are very complicated in their design, yet their performance is very simple to analyze. In fact, this is the precise aim of their element constructions, to make their component performance simple and reliable. Thus, input/output models are highly effective for digital components.

So, the analysis of electrical physics is divided into two parts: the study of digital components and the study of analog-electronic components (all non-digital electrical hardware). Digital analysis is discussed in Chap. 7; DC-electronic analysis is discussed in this chapter; AC-electronic analysis is in Chap. 6.

For all non-digital system components, you should always attempt a preliminary fundamental analysis in preference to an input/output analysis since a great deal of component performance insight is gained through the attempt, even if unsuccessful. As stated in Chap. 1, the primary motivation for analysis is design insight.

However, given knowledge of the elements, fundamental preliminary systems modeling rests on the concept of *idealized elements* and their use to develop first models of preliminary designs. These ideal elements have their attributes lumped into *pure* types, and they have *linear* input/output mathematics. These modeling idealizations help you to understand the overall cause-and-effect of the component's physics, at a detail level appropriate to preliminary design.

Consequently, as a fundamental modeler, you are always dealing with two versions of the component: the actual component and its idealized model. You are trying to capture the *essence* of the actual response through an assembly of idealized elements, each of which captures the fundamental physics of some part of that response.

This chapter begins the study of fundamental modeling by looking at the idealized electronic elements which are used to model real electronic components. Such a component model is shown in Fig. 4-1, which is a simple electronic filter composed of one resistor, R, and one capacitor, C. The physical model is shown inside the component box on the multiport diagram. Note that the multiport diagram shows a causality which is not apparent in the physical model alone. Also note that the model is an idealization of the physics; this is indicated through lower-case i and e variables. The actual, non-idealized port variables are shown in upper case on the multiport diagram.

After discussing the creation of the physical model, this chapter goes on to show how to find a mathematical model in the *time domain*. Such a model for this particular component happens to be

$$\mathrm{RC}\, di_L/dt + i_L = \mathrm{C}\, de_L/dt - i_R, \tag{4-1}$$

and

$$e_R = e_L - i_L \mathrm{R}. \tag{4-2}$$

This model is recognized in the time domain because all of the component equations are in terms of the independent variable, t, the time. This type of simple, linear equation always arises when idealized elements are used in the modeling. This is well in keeping with the primary modeling objective of keeping the preliminary-de-

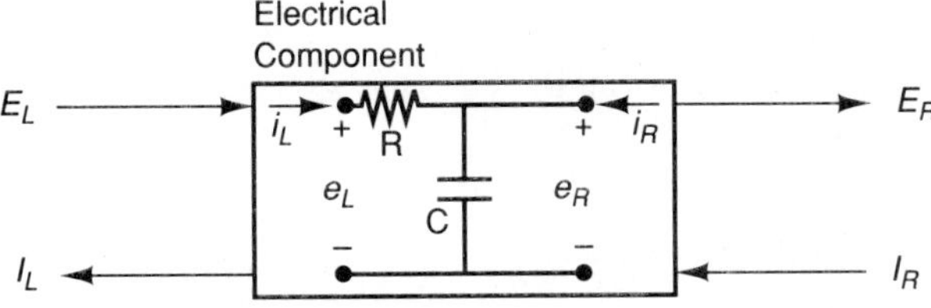

Figure 4-1. An Ideal Electronic Component

sign model as simple as possible. The models that are developed in this chapter are such as those shown in Eqs. 4-1 and 4-2.

In the first of the two model equations (Eq. 4-1), an idealized output, i_L, is expressed in terms of the two idealized inputs. In any physical analysis, the inputs are always assumed to be known or specified. Thus, once the input idealizations are obtained, the equation for the idealized output, may be solved. Consequently, the study of inputs in the previous chapter serves you well when solving such equations as these in later chapters.

The fundamental, physical analyses in this chapter are generalized among the component types through the powerful tool of *analogy*. That is, there is some basic similarity (or analogy) between the physical phenomena of electrical-current flow, structural-load path, fluid-flow path, and heat-flow path. An analogous "something" follows a "path" in a predictable way in each of these cases. As this predictability is studied, the physical parallels are seen in similar equations, and these similarities are used to find insights into the dynamics of various problems. Consequently, this chapter presents the fundamental, physical relationships of the four basic types of elements (electronic, structural, fluid, and thermal) using the powerful system-analysis tools of generalization and analogy.

In all cases, the strength of the analysis rests in the accuracy of the idealizations. Thus, the study of fundamental, preliminary modeling is begun with a study of the concept of idealization of physical elements.

4-1 FUNDAMENTAL ELEMENT IDEALIZATIONS

The fundamental relationships of physics may be modeled into system-element form using the idealization principles of *purity*, *lumping*, and *linearization*. In this way, the system modelers create a menu of idealized elements from which they may assemble an idealized-component model. In this section, the three principles of idealization are studied to see how they are used to capture the fundamental physics for the idealized elements.

Pure Elements. Consider the real, physical, electronic capacitor shown in Fig. 4-2a. The aptly named flat-plate capacitor in this figure consists of two flat-metal plates separated by a dielectric material, and equipped with lead wires attached to the plates. A voltage difference, E, is input across the leads of the capacitor and results in a current flow, I, through the device, as shown in Fig. 4-2a.

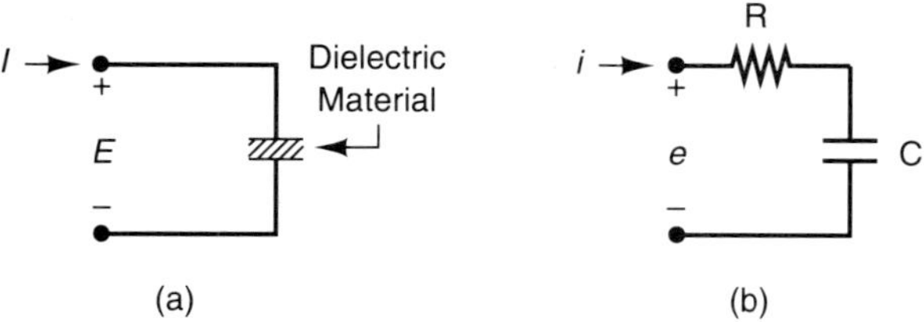

Figure 4-2. Electronic Capacitor Model. (a) Actual construction, (b) pure-element model

Experimental data shows that this device exhibits two *pure,* fundamental, input/output characteristics: *resistance* to current flow and *capacitance* of electrical charge in the plates.

A modeler thus separates the two pure physical effects and creates the idealized model in Fig. 4-2b. (Note the *lower* case variables to indicate the idealized model.) Here, the resistance effect is idealized into the pure element, R, and the capacitive effect is represented by the pure element, C. (These elements are "pure" in the sense that the ideal capacitor has no resistive response, it is pure capacitance. Much more will be said about the symbols and mathematics of these two elements in the next section. They are used here only to present the idea of using pure elements in idealized modeling.)

The idealized modeling problem thus centers first on recognition of significant, pure, fundamental physics, and using the pure elements to represent those effects.

Lumped Elements. Real physical effects are always distributed over some area or volume, however small. Yet such distributed effects are effectively idealized by using their lumped, net effect in the modeling effort. As an example of this, consider the diving board of Fig. 4-3. A modeler wishes to build a model to study the dynamic response of the right end of the board, V_R, to an input force, F, as shown. Here, the

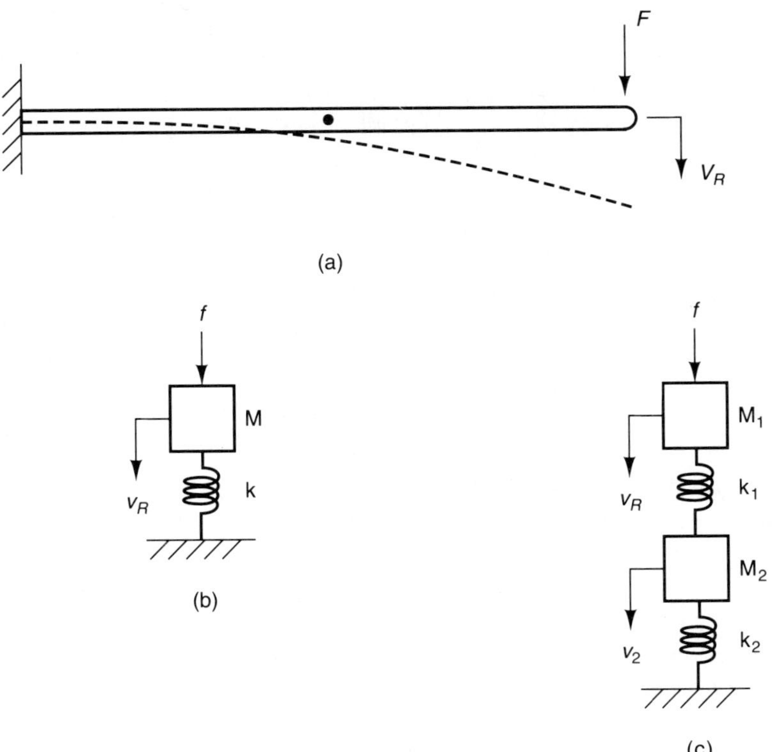

Figure 4-3. Lumped Diving-board Models. (a) One-lump model, (b) two-lump model

mass and spring-rate of the board are distributed between its two ends. When this sort of *distributed physics* arises, the fundamental continuum physics gives rise to *partial differential equation* models. This type of model is beyond the scope of this text. Fortunately, this difficulty is avoided by using *lumped models* to idealize the distributed physics. Such lumped models generate *ordinary differential equations* and thus simplify the analysis.

So, the modeler idealizes by lumping the entire mass of the board into a single pure-mass element, M, while the spring effect is lumped into the single pure-spring element, k, as in Fig. 4-3b. Note the change to lower-case variables in the figure indicating an idealized model. For design studies, the M and k values depend mostly on the dimensions of the board and the material from which it is made. The designer can thus "rough in" the board dimensions and material through the use of the one-lump model using M and k elements after considering their effects on the predictions of the dynamic response of the board.

If the designer needs predictions from a more distributed idealized model, then the board's physics may be lumped into two interacting pieces, as in Fig. 4-3c. This model may be more insightful than the one-lump model, but its solution takes more work since there are more model constants which need to be evaluated, and there are twice the number of output equations.

In the limit, using an infinite number of lumps, the modeler approaches the result of a distributed-physics model. However, a distributed-physics model is rarely, if ever, used for preliminary design since its detail is simply not appropriate at that stage of design. The idealized modeling problem thus centers on determining how many pure, lumped elements give acceptable accuracy for the modeling problem under consideration.

Linear Elements and The Principle of Superposition. The third, and final, characteristic of idealized elements is linear input-to-output relationships. The consequence of this linearity is that added (superposed) inputs create added (superposed) outputs. Superposed inputs were studied in the last chapter, superposed outputs are studied in the next chapter. In this chapter, the creation of a linear model which offers the advantage of superposition of outputs is studied.

We begin with a statement of the *Principle of Superposition*.

> If an input to a linear element is composed of several functions added together, then the output from that element is the same as if the functions were input separately, and their corresponding outputs were mathematically added together.

In order to demonstrate the idea superposition, consider the input/output plot of steady data shown in Fig. 4-4. This plot of f represents the data taken from an electrical resistor such that

$$E = f(I). \tag{4-3}$$

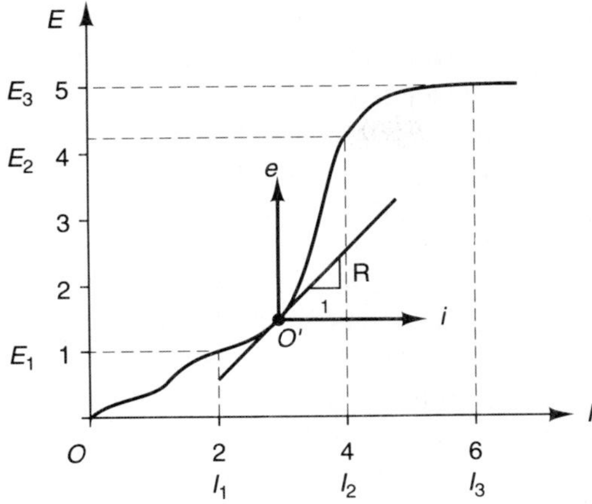

Figure 4-4. An Electronic Resistor Characteristic.

Superposition requires that the following be true for the resistor:

$$\text{if} \quad E_1 = f(I_1), \tag{4-4}$$
$$\text{and} \quad E_2 = f(I_2), \tag{4-5}$$
$$\text{then } E_1 + E_2 = f(I_1 + I_2). \tag{4-6}$$

However, the data yields:

$I_1 = 2, E_1 = 1,$
and $I_2 = 4, E_2 = 4.3,$
and $I_3 = 6, E_3 = 5.0.$

Thus, superposition has failed because $E_3 \neq E_1 + E_2$ for $I_3 = I_1 + I_2$. That is, the outputs did not superpose for the superposed inputs.

Still, the failure was not awful, and a linear model may be acceptable, especially in certain ranges of I (e.g., for the e-i axes located at O'). This experimental method is a common manner of checking the linearity of a system, component, or element prior to modeling.

Note that the origin of the axes (I, E in this case) must be set at $(I, E) = (0, 0)$ to check the overall *global linearity*, of the device, and all transient changes must be allowed to pass before any data is taken. With the axes chosen in this way, doubling any steady input doubles the steady output in a globally linear device.

Another way to check the linearity using steady data is to visually look for a straight-line (linear) relationship between the input and output over some *region* of the steady response. This checks for *local linearity*, such as that near O' in Fig. 4-4.

The two types of superposition which are checked above are *algebraic superposition;* they correspond to *algebraic linearity*. This type of linearity is demonstrated by the linearized e-i plot in Fig. 4-4,

$$e = i\text{R}. \tag{4-7}$$

Notice that this type of linearity always presupposes a point about which the linearization is computed and enforced (point O' in Fig. 4-4). The tangent to the curve at O' has a numerical slope to give value to the constant R. Thus, different linearization points give different values for R, but the *functional* form of the linear relationship remains the same near any linearization point.

Unless specifically mentioned, the assumed linearization point is *always* at (0, 0).

The other type of linearity with which the modeler must be concerned has to do with *rate linearity;* it comes out in the model *differential equations*. This type of linearity is much harder to detect in experimental data since it is only seen in the transient part of the data. However, nonlinearities in rate terms are usually ignored in preliminary-design system models.

In this text, *linear ordinary differential equations* are defined as those which have the general form:

$$\sum_{j=1}^{m} C_j \frac{d^j y}{dt^j} + C_0 y = \sum_{k=1}^{n} B_k \frac{d^k u}{dt^k} + B_0 u + g(t) \tag{4-8}$$

where,

1. The dependent variables (u and y) are functions of the independent variable, t, only;
2. The coefficients (B's, C's, etc.) can vary with time, but not with u or y;
3. The right-hand-side function, g, may vary in any manner with time, but it may not involve nonlinear expressions of u *or* y; and
4. No products of the dependent variables or their derivatives are allowed.

Ordinary differential equations which follow these rules obey the Principle of Superposition. And, if a model is created using only idealized elements, then any model differential equations which arise are linear ordinary differential equations.

The importance of idealization by linearization thus rests in the notion that linear elements obey the superposition of outputs, while nonlinear elements do not. Furthermore, as the elements go according to superposition, so goes the system. If there are all linear elements in a component, then the component is called *linear*. If all the components are linear, then the system is linear. Conversely, if there is even one nonlinear element in a component, then that component is nonlinear, and superposition fails. If there is even one nonlinear component, then that system is nonlinear.

One advantage to linear modeling is that linear systems are easy to analyze since there is a large body of powerful linear-analysis theory. Furthermore, the response of linear systems is easy to generalize for design purposes. The experienced linear modeler can thus readily anticipate a change in response due to a change in hardware. This is not the case for a nonlinear model.

However, it is also very important to realize at the outset that *all real, physical elements are nonlinear*. This ***Fundamental Truth of Modeling*** is often forgotten by designers and modelers alike, yet it is the most consistent source of modeling error. The modeler's idealization problem is partly to recognize when linearization yields an adequate representation of the real nonlinear system.

Idealization is thus shown to be a powerful ally in the analysis, modeling, and design of systems. The ideal system model consists entirely of elements which are pure, lumped, and linear. Such elements create linear system models for analysis. From time-to-time, however, the modeler finds that there is no choice but to include an element or two which are pure and lumped, but which are not linear, thus giving rise to a nonlinear, nonideal system model. In this text, such models are dealt with as a complication to the basic problem of system idealization for preliminary design.

As you begin to quantitatively model fundamental physical responses, you soon find that your first concern needs to be in the area of consistent units for your analyses. You also find that the time you spend preparing in this way is very helpful when you start evaluating your output prediction results.

Units. Two unit sets are used in this text, as shown in Table 4-1. This table is given only for an overall contrast of the unit sets; a much more complete units table is given in the Appendix, which also includes the conversion factors between the two sets.

In this text, the Standard International (SI) unit set has no Kilogram-force (kg_f), nor does the English set have a Pound-mass (lb_m). These are simply confusing units. The units shown in Table 4-1 require no proportionality constants for their use.

Table 4-1. Systems of Units

Variable	*Standard International, (SI) Unit*	*English Unit*
Mass	Kilogram, kg	Slug, sl
Force	Newtons, n	Pound force, lb_f
Distance	Meters, m	Foot, ft
Time	Seconds, s	Seconds, s
Translational speed	Meters/second, m/s	Feet/second, ft/s
Torque	Newton-meters, n-m	Foot pounds, ft-lb_f
Rotational speed	Radians/second, r/s	Radians/second, r/s
Pressure	Newtons/meter^2, n/m^2	Pounds/foot^2, lb_f/ft^2
Mechanical power	Newton-meter/second, n-m/s	Foot-pounds/second, ft-lb_f/s
Mechanical energy	Newton-meters, n-m	Foot pounds, ft-lb_f
Current	Amperes, amp	Amperes, amp
Voltage	Volts, v	Volts, v
Electrical power	Watts, w	Watts, w
Electrical energy	Joules, j	Joules, j
Heat	British thermal unit, Btu	British thermal unit, Btu
Temperature	Centigrade, °C	Farenheit, °F

For example, the weight, W, of one slug of mass can be directly calculated as:

$$\begin{aligned} W &= \mathrm{Mg} \\ &= (1\ \mathrm{sl})\ (32.2\ \mathrm{ft/s^2}) \\ &= 32.2\ \text{pound-force}\ (\mathrm{lb}_f) \end{aligned}$$

If you come across pound-mass units, your first job should be to convert those units to slugs by dividing by 32.2 (pound-mass/slug). Many good analyses have been confounded by a failure to apply this conversion properly and quickly.

HOMEWORK

4-1. What is the Fundamental Truth of Modeling and how is that truth reflected in the element model for a resistor, $e = i\mathrm{R}$?

4-2. Convert the following to SI units:

a) 210 lb_m

b) 26 lb_f

c) 15 ft

d) 35 MPH

e) 100 slugs

f) 50 kg

4-3. Classify each of the following equations as to whether they are linear or non-linear,

a) $y = \sin u$

b) $dy/dt = u \sin t$

c) $(dy/dt)^2 = 10t$

d) $y\ du/dt = \sin t$

4-4. A heating, ventilation, and air conditioning (HVAC) engineer is analyzing the distributed heat that is spreading into a room. The engineer's first model of the room is as a single heat capacitor, without heat-flow resistance. What principle(s) of idealization is the engineer using?

4-2 THE IDEAL ELECTRONIC ELEMENTS

There are three types of fundamental, electronic phenomena; the first is pure resistance, the second is pure inductance, and the third is pure capacitance. In this section, each of these is modeled using a distinct, ideal electronic element. These elements are called *passive* since they represent an electronic phenomenon which does not increase the net electrical energy of the component. In other words, these ideal elements are used to model the physics of storage or dissipation of electrical energy. Elements which originate electrical energy (such as batteries) are called *active* elements or *power sources,* or simply, *sources*; they are studied in the next section.

The Electrical Variables. The variables of interest in all electrical elements, whether digital or electronic, are the current, I, and the voltage, E.

The *current*, I, is a measure of the amount of charge (coulombs) which flows past a given point in an electrical *circuit* per unit time. The units of current are thus coulombs/sec, or *amperes*. Note that current (or charge) will not flow unless there is at least one complete circuit for it to follow. An electrical circuit, a combination of circuits, or a combination of non-circuited electronic elements is called an *electrical network*.

The *voltage*, E, is the electrical potential which may exist between any two points in an electrical component. It measures the amount of work that could be done to move a point-charge around a circuit, if the circuit were closed. Voltage is measured in *volts*, and is always a relative quantity. That is, it is always the voltage *difference* across an element which forces the current through the element. In this text, the notation for a voltage difference follows the convention, $E_{12} = E_1 - E_2$.

The *constitutive relationship* for each of the three fundamental, electronic phenomena is shown in Fig. 4-5. Notice that each of the linearizations for the constitutive, electronic relationships is about the $E_{12} = 0$ operating point. Also notice that the figure shows various current functions along the horizontal (input) axes, with the corresponding voltage drop along the vertical (output) axes. These inputs/outputs, correspond to various experiments which are used to characterize the elements. For the resistor, the current is the input. For the inductor, the time-rate-of-change of the current is the input. For the capacitor, the integral of the current (the charge) is the element input.

Notice that each of the element curves demonstrates a *saturation nonlinearity* if the input is allowed to get too large. At the far right of the figure the idealized element representations are shown.

Note, again, that the fundamental elements are *passive*; they do not produce energy. Inductors and capacitors are used to store electrical energy, and resistors are used to dissipate energy.

The real elements which correspond to these idealizations are carefully designed to exhibit the ideal-constitutive relationship over a wide operating range. Thus, through proper combinations of real forms of these ideal elements, electronic networks are very effectively designed to perform many useful engineering tasks.

The Ideal Resistance. The electronic phenomenon of resistance occurs whenever current flows (Fig. 4-5a). Still, in your preliminary-design models, you are not usually concerned with parasitic resistance, such as that in the connecting wires. Instead, you focus on those elements which are designed to demonstrate some form of *primary* resistance.

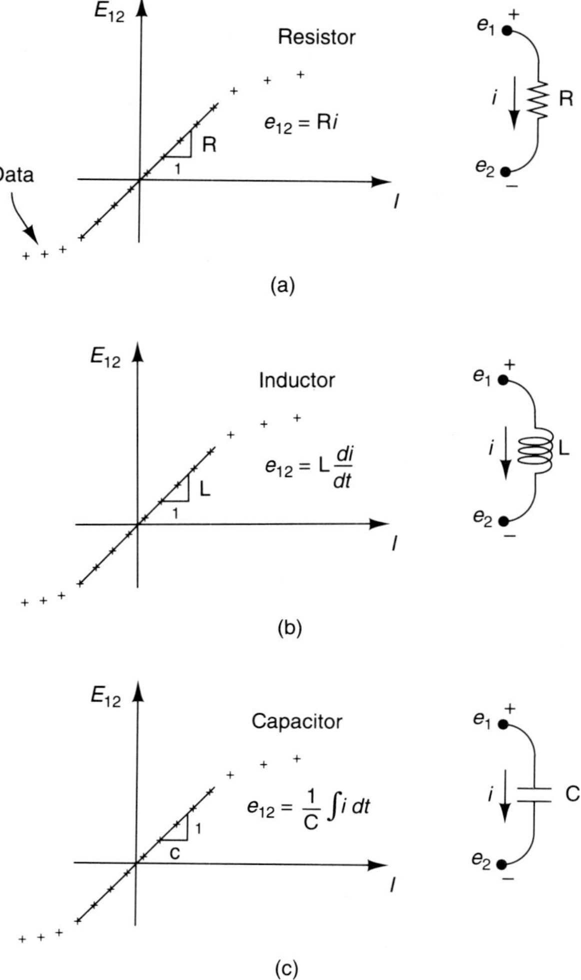

Figure 4-5. The Electronic Elements. (a) The resistor, (b) the inductor, (c) the capacitor

Resistance is demonstrated in two main classes of designs: those giving high temperatures using high resistance (ovens, heaters, some incandescent lights, and electrical appliances); and those giving low temperatures using precision resistance (electronic instruments and measuring apparatus). Thus, you see resistance in the first class as an element in a circuit which is used specifically to give off light and heat. You see resistance in the second class as a commercially available precision product.

All resistances work by resisting the current flow through a circuit. That is, when the electric potential (volts) is applied to a resistance, the free electrons in the material are accelerated. These electrons travel through the free space within the material until they strike more stationary atomic matter. These collisions result in heat and, perhaps, light. The collisions also act to slow down the electron flow and establish a rate of charge flow or current. Thus, the amount of resistance is determined by the type of material, as well as its physical dimensions. For example, more area in a conductor gives more electrons and more current flow; a longer conductor means a longer path to offer collisions and less current flow. Consequently, the resistance obeys the empirical relationship

$$i = e\left(\frac{a}{\ell\rho}\right), \tag{4-8}$$

where a is the crossectional area of the conductor, ℓ is the conductor length, and ρ is the resistivity of the material. Some common values for the resistivity are shown in Table 4-2.

Table 4-2. Electrical Resistivity

Material	*Composition*	*ρ, $10^{-6}\Omega$-cm at 20°C*	*Melt, °C*
Copper	Cu	1.724	1,082
Aluminum	Al	2.655	660
Bronze	Cu, Zn	4.2	1,015
Tungsten	W	5.523	3,370
Magno	Ni 0.955, Mn 0.045	20.0	1,425
Advance	Cu 0.55, Ni 0.45	40.0	1,210
Nichrome V	Ni 0.80, Cr 0.20	108	1,400

Notice that tungsten is often used in resistive-heating devices due to its very high melting point.

Thus, the modeler uses Eqs. 4-7 and 4-8 to compute the resistance for an ideal-resistor element model. Since

$$e = i\text{R}, \tag{4-7 repeated}$$

the consistent units are

$$\text{volts} = (\text{amperes})\ (\text{ohms}).$$

And, combining the previous equations, the ideal resistance is

$$\text{R} = \left(\frac{\ell\rho}{a}\right). \tag{4-9}$$

Resistance is often given the symbol Ω to represent its units, ohms.

Example 4-1

A common DC lighting appliance is the 25 watt incandescent light bulb. When this bulb is screwed into a DC light fixture and turned on, it is subjected to 12 volts DC from the house power. Compute the area of a 1 inch-long Magno filament for this bulb (Magno is a common filament material).

Solution

The bulb rating was given in watts, which is a unit of electrical power, $\mathcal{P}$. The power rating of the bulb gives the amount of current flowing

$$\mathcal{P} = ie, \tag{4-10}$$

or

$$\begin{aligned} i &= \mathcal{P}/e \\ &= 25 \text{ watts}/12 \text{ volts} \\ &= 2.08 \text{ amps}. \end{aligned}$$

The resistance is thus

$$\begin{aligned} \mathrm{R} &= e/i \\ &= 12 \text{ volts}/2.08 \text{ amps} \\ &= 5.77\ \Omega(\text{ohms}). \end{aligned}$$

This value of resistance can now be used to compute the geometry of the 1-inch Magno lighting element. From Eq. 4-9

$$\begin{aligned} a &= (\ell\rho)/\mathrm{R} \\ &= (1 \text{ in.})(2.54 \text{ cm/in})(20 \times 10^{-6}\ \Omega\text{cm})/5.77\ \Omega \\ &= 8.8 \times 10^{-6} \text{ cm}^2. \end{aligned}$$

This area corresponds to a wire diameter of

$$\mathrm{d} = 0.001674 \text{ cm}.$$

Comments

This wire size is, obviously, very small. Much of the cost of the light bulb, therefore, can be expected to occur in the manufacture of consistent filament material which will give the required 25 watt rated-power consumption in this application. Fortunately however, this resistance does not have to be very precise, which helps to keep the cost down.

In those applications where high-precision resistors appear (e.g., instruments and measuring apparatus), either the resistive length, ℓ, is shortened to increase the resistive area, or the resistivity is reduced so that the length may be increased. The first approach gives resistors appearing as small cylindrical shapes, usually about 1/2 in. long, which are manufactured in standardized values, and labeled with color-coded stripes to indicate their resistance. The second approach gives rise to the common form of the wire-wound resistor, such as shown in Fig. 4-6.

The wound resistor of Fig. 4-6a is created by winding an insulated wire N times around a cylinder of radius r. An insulated wire is used to minimize or eliminate the magnetic influence of adjacent windings. The area of the wire conductor (not including insulation area), is a. The length of the conductor for the resistance formula is thus approximately

$$\ell = 2\pi \mathrm{r} \mathrm{N}, \tag{4-11a}$$

which may be substituted into Eq. 4-9 to compute the resistance of the wire-wound resistor

$$\mathrm{R} = 2\pi \mathrm{r} \rho \mathrm{N}/a. \tag{4-11b}$$

The idealized network symbol for this resistance is shown at the right of Fig. 4-6a.

In Fig. 4-6b, the idea of the wire-wound resistor is extended to include a sweeper mechanism which is used for changing the resistance, and thus tuning a sys-

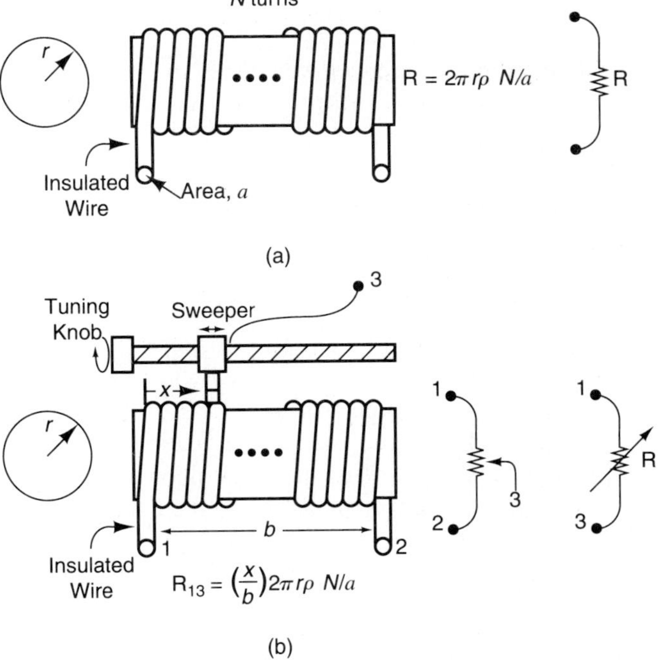

Figure 4-6. The Precision Wound Resistor. (a) Fixed resistor, (b) variable resistor

tem. Along the area where the sweeper comes into contact with the windings, the insulation is removed. This allows the designer to divide the current flow through the resistor according to the position of the sweeper. The resistance value thus becomes

$$R_{13} = (x/b)\ 2\pi\, r\rho\, N/a. \tag{4-11c}$$

Where b is the length of the *coil*, as shown in the figure. The two common ideal network symbols for this variable resistance are shown at the right of the figure.

Real resistors are put into electrical networks to tailor the component response in a predictable manner. Ideal resistors are used to model the power losses of real resistors and/or other significant resistance effects when they occur.

The Ideal Inductor. The electronic phenomenon of inductance occurs when an uninsulated conductor is coiled or looped in such a way that the *magnetic field* surrounding the conductor reinforces itself. When this happens, the current flowing through the coils exhibits a reluctance to change, such that a voltage must be generated in proportion to the desired rate-of-change of the current. The experimentally observed inductor (e.g., Fig. 4-5b) thus demonstrates the input/output relationship

$$E_{12} \sim dI/dt. \tag{4-12}$$

And the constitutive relationship for the *ideal inductor* is

$$e_{12} = L\ di/dt. \tag{4-13}$$

The consistent units of the inductor relationship are

$$\text{volts} = (\text{henrys})\ (\text{current})/(\text{sec}).$$

Recall that resistors are made by coiling *insulated* wire, but inductance arises by coiling *uninsulated* wire. In fact, a precision inductor is often made by winding lacquer-coated copper wire around a core. The lacquer coating prevents direct-contact conduction between the coils without significantly inhibiting the magnetic field effects. So, real inductors exhibit resistive effects due to their wire content, and this effect must be considered in selecting an idealized model.

The construction of a precision inductor is shown in Fig. 4-7. When a current flows through such an uninsulated conductor, a magnetic field is established in the magnetic-permeable core material. This magnetic field is often used to operate solenoid switches or to trip circuit breakers. Or, the inductor may be used as a precise electronic element to store electrical energy in its magnetic field.

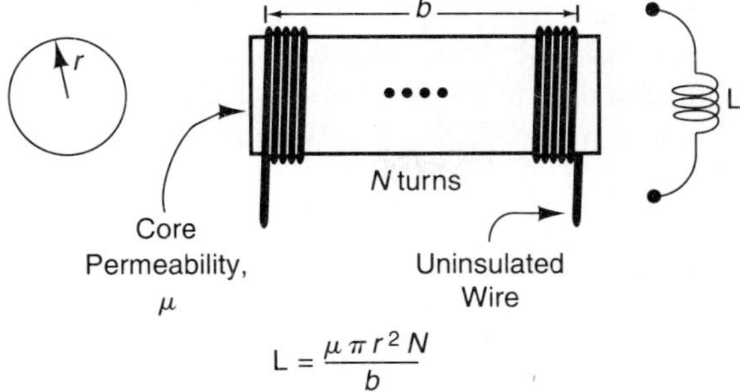

Figure 4-7. The Wound Inductor

The inductor strength, L, is observed to be proportional to the area of the core, the number of turns about the core, and the permeability of the core. It is inversely proportional to the length of the core. Thus, for a closely wound coil

$$L = \pi r^2 N^2 \mu / b, \tag{4-14}$$

which has the units of henrys. Again, r and b are the radius and the length of the *coil*, μ is the magnetic permeability of the core material, and N is the number of wire windings.

For an inductor core of air (or a vacuum), the permeability is approximately

$$\mu_{AIR} = 4\pi \times 10^{-7} \text{ henrys/meter},$$

and the linear model of Eq. 4-14 is quite good over a very large operating range.

If a ferromagnetic material is used for the inductor core, then near $E_{12} = 0$,

$$\mu_F = 40\pi \text{ henrys/meter}.$$

The non-air type of inductor typically begins to saturate at a fairly low value of voltage and is often modeled using a nonideal element.

The most important things to remember about the inductor are: first, when current flows through it, a magnetic field exists in its core; and second, if the current is unchanging, then there is no voltage across the device. The inductor is thus seen as a device which functions to *store energy* by virtue of the *current* through it. This energy is stored in the magnetic field. It takes an input potential (a voltage) to increase this energy storage by increasing the current flow.

In this text, the ideal inductor is used only to model real inductors. General inductance, when it occurs, is a very complex phenomenon and must be modeled with advanced methods.

The Ideal Capacitor. Electrical capacitance occurs whenever electrical charge accumulates. Some everyday examples of this are seen in overhead power-transmission lines, in static electricity in carpets and clouds, and in the charge residue which sometimes is felt on the screen of a television. However, in this text, electrical capacitance is analyzed and modeled only in the form of precisely manufactured *capacitors* (also called condensers) which are used to supply capacitance in electronic designs.

A precision capacitor is constructed by placing a *dielectric material* between two conducting surfaces. A dielectric material is one which permits the passage of electrostatic lines of force without transmitting the charge. That is, electromotive force (voltage) is passed, but current is blocked through the dielectric. Once the voltage is reestablished in the output lead, however, the current flows again if a circuit exists. This situation gives rise to a net-positive charge on the current-input side of the capacitor, and a net-negative charge on the current-output side.

The experimental relationship for the capacitor (e.g., Fig. 4-5c) is of the form

$$E_{12} \sim \int I\,dt. \tag{4-15}$$

So that the *idealized capacitor* model becomes

$$e_{12} = (1/C) \int i\,dt. \tag{4-16a}$$

This is more often expressed in the equivalent form

$$i = C\, de_{12}/dt. \tag{4-16b}$$

The consistent units of the capacitor are

$$\text{amperes} = (\text{farads})\,(\text{volts})/(\text{sec}).$$

The computation of the capacitance, C, comes from the experimental observation that the capacitance is proportional to the area of the dielectric, the kind of dielectric material, and is inversely proportional to the separation of the plates. Thus, the capacitances are computed as shown in the three cases of Fig. 4-8.

In the case of the *flat-plate capacitor*

$$C = \frac{\varepsilon A}{d}, \tag{4-17}$$

with the symbol for a fixed capacitor as shown at the right margin of Fig. 4-8a. The *dielectric constant*, ε, of various materials is often measured relative to the dielectric value of free space, ε_o. This measured value is

$$\varepsilon_o = 8.854 \times 10^{-12} \approx 10^{-9}/36\pi \text{ farads/meter}.$$

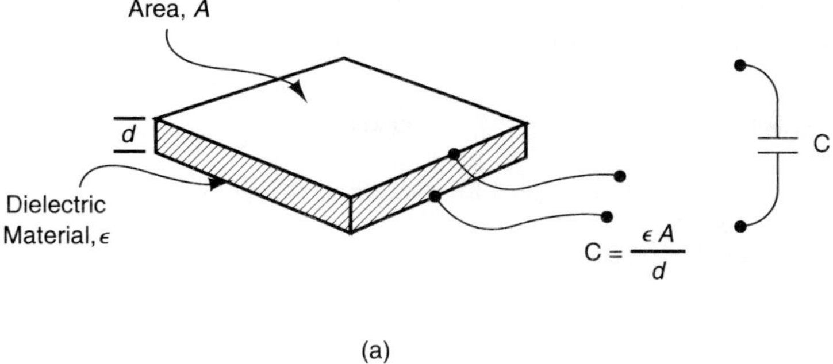

(a)

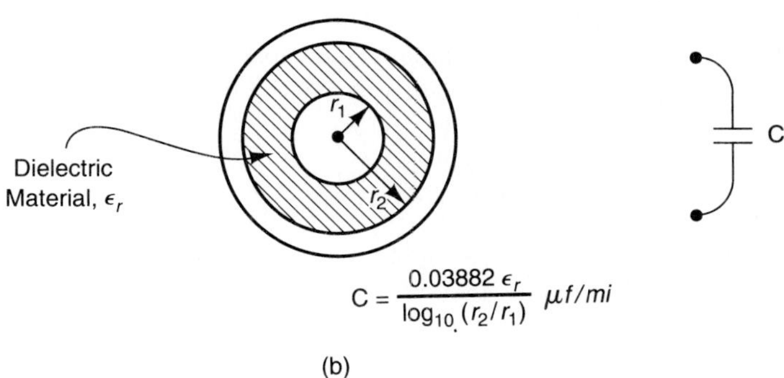

(b)

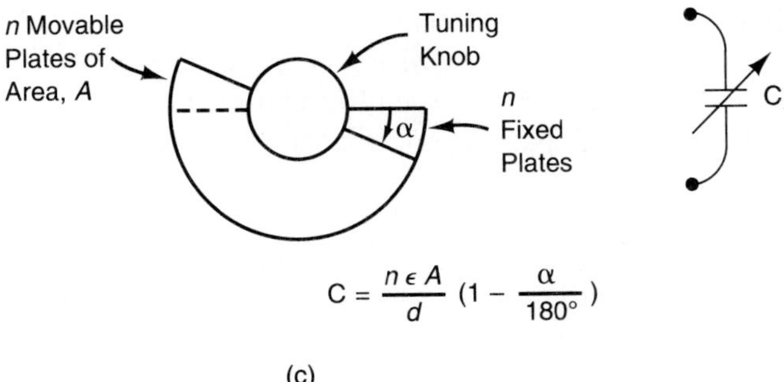

(c)

Figure 4-8. The Capacitor. (a) Flat-plate, (b) coaxial, (c) variable

In the tabulation of various dielectric values, the *relative dielectric constant*, ε_r, is sometimes given. This constant is computed as follows

$$\varepsilon_r = \varepsilon/\varepsilon_o. \tag{4-18}$$

Values for the relative dielectric constant are given in Table 4-3.

Table 4-3. Relative Dielectric Constants (dimensionless)

Material	ε_r
Paper	1.7-2.6
Ebonite	1.9-3.5
Bakelite	4.5-5.5
Polyvinyl Chloride	6.5-12.0
Indian Mica	7.07-7.90

In the case of the *coaxial capacitor*, Fig. 4-8b, a slightly more complicated expression for the capacitance must be used. This case is often seen in the problem of power-transmission lines, hence the units of the capacitance are in *micro*-farad/mile (10^{-6} farads/mi.). The expression which is used to compute the capacitance of an overhead-transmission line is

$$C = \frac{0.03882\ \varepsilon_r}{\log_{10}\left[\dfrac{r_2}{r_1}\right]}\ \text{micro-f/mi.} \tag{4-19}$$

The expression which is used to compute the capacitance of a shorter coaxial cable (such as an oscilloscope lead) of length, ℓ, cm, is

$$C = \frac{0.2171\ \varepsilon_r\ \ell}{9 \times 10^5 \log_{10}\left[\dfrac{r_2}{r_1}\right]}\ \text{micro-f.} \tag{4-20}$$

A *tunable capacitor* is easily made using the configuration of Fig. 4-8c. Here, a set of *movable*, evenly spaced, semicircular plates is attached to an axle. Evenly spaced between these is a set of *fixed* semicircular plates. Thus, as the axle is rotated using a tuning knob, the capacitance is changed according to the following expression

$$C = \frac{n\varepsilon A}{d}\left(1 - \frac{\alpha}{180°}\right). \tag{4-21}$$

This *variable plate capacitor* has the symbol shown at the right-hand margin of Fig. 4-8c.

All capacitors *store energy* by virtue of the *voltage* across their dielectric. This is true because the primary function of the capacitor is to accumulate charge or voltage.

The Network Model. The three ideal electronic elements may now be used to model a real electronic network. In doing this, you may see a variety of resistors, capacitors, and inductors which are easy to idealize, but you also have to look carefully for more general resistive effects.

Consider, for example, the flashing-light network shown in Fig. 4-9. When the switch between 1 and 2 is closed, the current flows around the circuit, lighting the bulb between 2 and 3, charging the inductor between 3 and 4, and charging the flat-plate capacitor between 4 and 5. The light flashes because the designs of the coil and

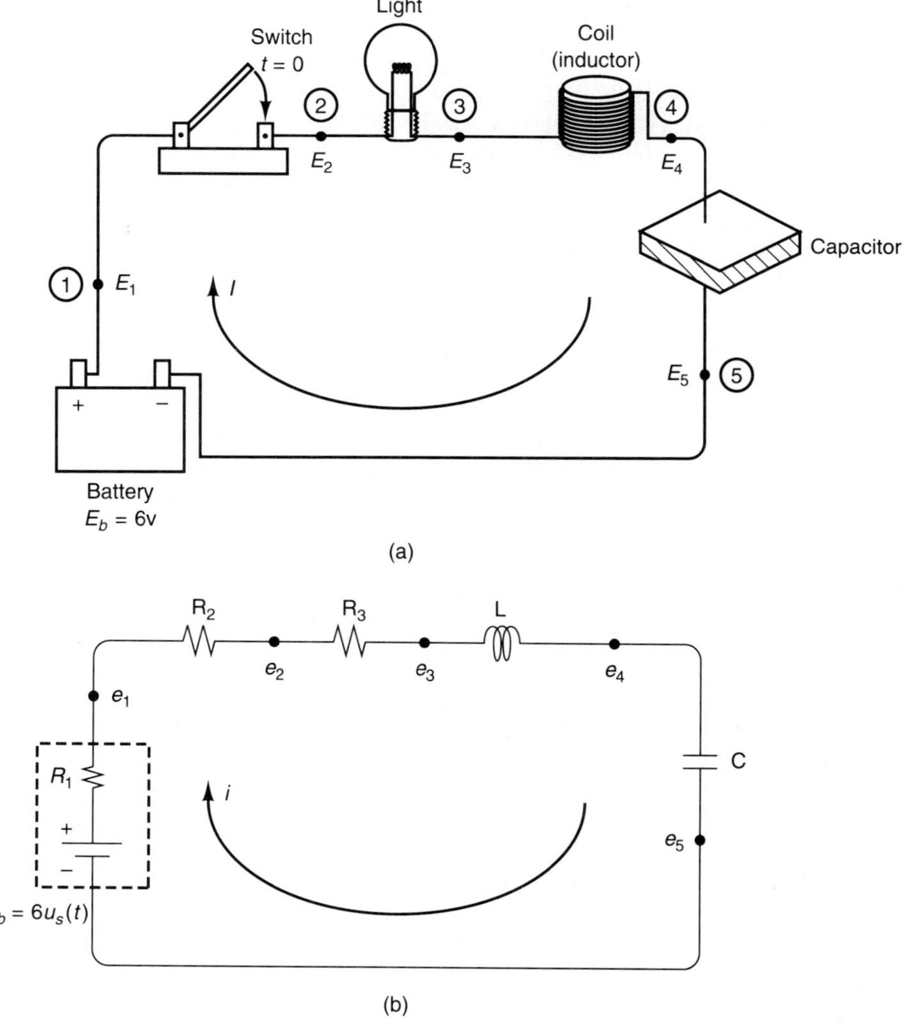

Figure 4-9. A Simple Electronic Circuit. (a) Physical Elements, (b) idealized elements

the capacitor are such that their stored energy goes back and forth between them, creating an oscillating current. The voltage is symbolically indicated at various points in the circuit by the subscripted E variables.

Consider now the idealized model of the circuit as shown in Fig. 4-9b. Here the various physical effects are lumped into their ideal R, L, and C elements. Notice first that all the variables (the current and voltages) are shown in lower-case symbols to identify this as an idealization. Notice also that the switch and the light are modeled as resistances, using ideal resistors. The battery is modeled with a resistance placed inside dashed lines with an ideal source (more about this in the next section). The input, on-off action of the switch is modeled using a step function on the battery strength.

The connecting wires and the secondary resistances of the inductor and capacitor are assumed to be negligible resistances when compared to the primary resistances in the battery, switch, and light. If you desire to put these explicitly in your model, then you must assign them an ideal resistor.

The idealized variables of current, i, and voltage, e, are shown on the idealized network for the coming mathematical analysis. The observed direction of current flow is shown as an arrow on the figure, labeled i. The voltages, e, are shown at points in the figure which separate all the elements.

Given such an idealized physical model, it is possible to find corresponding equations, which is addressed later in the text.

HOMEWORK

4-5. An inductor is made of coated Bronze wire after the manner of Fig. 4-7. It is closely wound on a cylinder of reinforced paper, such that it has an air core. The following parameters are used in the construction

Core	*Wire*
d = 2 in	d = 0.01 in
b = 1 in	$\rho = 4.2 \times 10^{-6}$ ohm-cm
N = 100 turns	

Answer the following for this inductor

a) What is the value of inductance, L?

b) If this inductor is used in a circuit with a 100 ohm resistor, does it contribute a significant (± 10%) resistance?

4-6. Compute the capacitance of an oscilloscope lead if the following construction is used

$$\text{dielectric material} = \text{Polyvinyl Chloride } (\varepsilon_r = 10.0)$$

$$\text{length of lead} = 91.44 \text{ cm (3 ft)}$$

$$\text{outer radius} = 1 \text{ cm}$$

$$\text{inner radius} = 0.5 \text{ cm}$$

4-7. A resistor is made of Advance alloy. It has the following dimensions

$$\ell = 3 \text{ in}, \; a = 0.01 \text{ cm}^2.$$

If there is a current of 3 amps in the resistor,

a) How much power is the resistor consuming?

b) How much power is the resistor storing?

4-8. A physical inductor is modeled using an ideal resistor element and an ideal inductor element. The modeler has used what idealization principle(s) in doing this?

4-3 COMMON DC-ELECTRONIC COMPONENTS

Electrical subsystems often need components to supply power, to raise the power level, or to change power into electrical form from another type (structural, fluid, or thermal). In this section, some common forms of these components are discussed. Note that these are components, as opposed to elements. These are common devices which usually contain many elements to produce a more-or-less ideal type of operation. Thus, their models are easily created directly at the component level.

The components which supply electrical power are called *sources*, those which amplify power are called *amplifiers*, and those which change power into electrical form are called *transducers*.

The DC-Power Sources. The two common DC-power sources are the voltage source and the current source.

DC-voltage sources are found in the form of *batteries* and *DC power converters*. (The latter converts AC voltage into DC voltage for equipment and instrument needs. Most personal computers have an internal power converter to change the AC wall-socket power into the DC power needed for the digital elements inside the computer.)

The circuit symbol for a typical battery is shown in Fig. 4-10; note the upper-case variables indicating a real battery, and the lower-case variables indicating the idealizations. The real DC battery is represented using a symbol with parallel lines of alternating long and short lengths, as shown in Fig. 4-10a. The positive direction of current flow is assumed to be *out* of the positive terminal of the battery, as indicated by the small " + " and " − " symbols on the diagram and the arrow for current.

The input/output data for the battery is shown in Fig. 4-10b. Note that the battery shows some droop in its output voltage as more current passes through the device. This is assumed to be a linear droop of constant slope, $-R$.

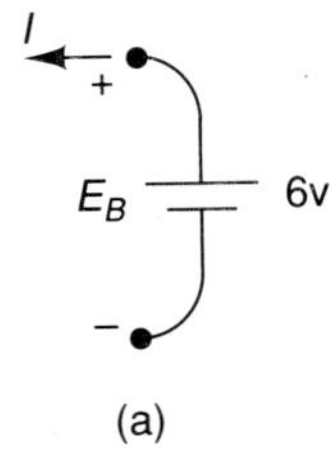

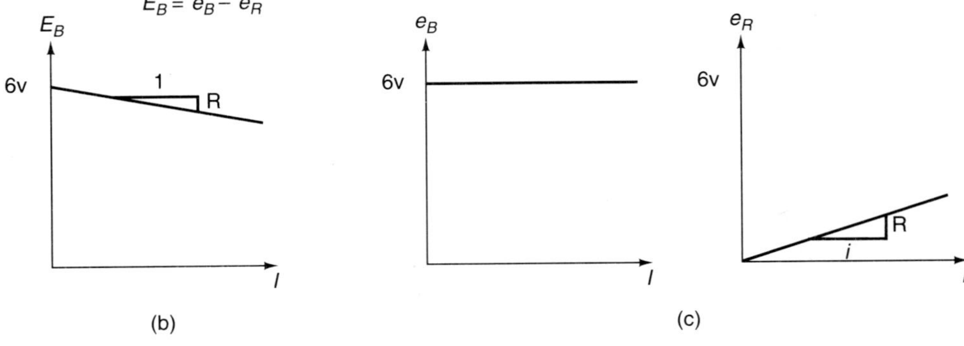

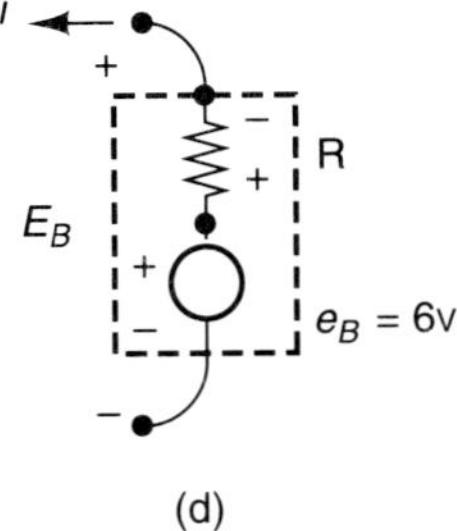

Figure 4-10. The Battery. (a) Schematic symbol, (b) output data, (c) idealized data, (d) ideal model

In Fig. 4-10c, the battery voltage is modeled using two idealized voltages according to

$$E_B = e_B - e_R. \tag{4-22}$$

Where e_B is a constant 6 volts coming from an *ideal* source (one that is unaffected by the current through it). And e_R is a resistor-like element which accounts for the droop in E_B.

The idealized battery performance is thus shown by the electrical network in the dashed box of Fig. 4-10d. Here, the idealized battery source, e_B, has the network symbol of a circle placed in the network, with its positive and negative terminals properly labeled in the same sense as E_B, and with the constant strength of the source indicated nearby.

The ideal resistor which accounts for the real battery droop is taken from the data of Fig. 4-10c. Its positive and negative ends show a *voltage drop* in the direction of positive-current flow since it models the losses in the battery; that is, it opposes the battery.

Example 4-2

Consider the voltage-output characteristic of the common 48-watt photovoltaic module shown in Fig. 4-11. This module is made up of many small photovoltaic cells wired together in such a way that they produce the net input/output curves shown in the figure when they are exposed to full sunlight normal to their surfaces. Notice the loss in voltage output as the temperature increases. This is due, in part, to the rise in internal resistance which occurs as the temperature goes up; and it is due, in part, to the loss of sunlight-conversion efficiency of the cells at higher temperatures.

Suppose that a modeler wanted to create an idealized model of this voltage source for the study of a new remote powering system, to be run at 25°C.

Solution

The modeler needs to find the amount of droop in the curve as current is increased. To do this, the modeler creates a linear model of the curve near $I = 0$, using an intercept (e_o) and slope (m) formula for the module voltage

$$E_m = \text{m}i + e_o. \tag{4-23}$$

The slope, m, is found from a two-point expression, using the two points *a* and *b* shown in the figure,

$$\text{m} = \Delta E/\Delta I \tag{4-24}$$

Or,

$$\text{m} = (E_b - E_a)/(I_b - I_a). \tag{4-25}$$

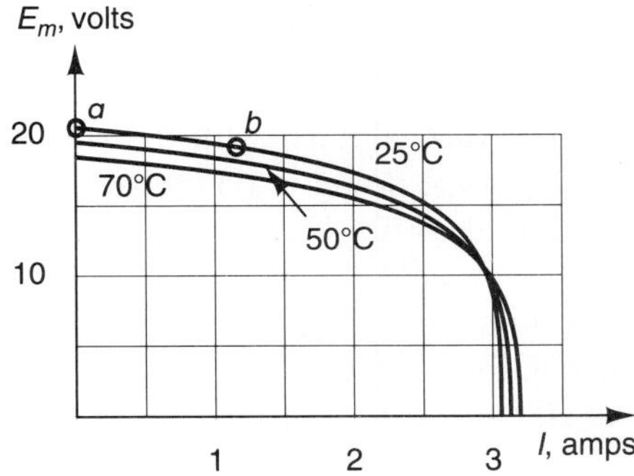

Figure 4-11. Output of a Photovoltaic Module

In terms of the data given,

$$\begin{aligned} m &= (19.0 - 21.0)/(1.1 - 0.0) \\ &= -1.82 \text{ v/a.} \end{aligned}$$

By inspection, the intercept is

$$e_o = 21.0 \text{ volts.}$$

Thus, the modeler plugs into Eq. 4-23 to find the idealized module relationship

$$E_m = -1.82i + 21.0 \text{ volts.}$$

The right-hand side of this last expression is composed of two terms. The first term is the resistor-like loss term, the second term is the ideal-source term.

Thus, the idealized source model is

$$e_m = 21.0 \text{ volts,}$$

with an internal resistance of

$$R = 1.82 \text{ ohms.}$$

The network diagram for this source model is exactly as shown in Fig. 4-10d.

Comments

Both the modeler and the designer need to be very careful regarding the true nonlinearity of this device when using the predictions from this linear model. Returning to Fig. 4-11, notice the catastrophic, nonlinear loss in output as the load current approaches 3 amps. Fortunately, this nonlinearity can be avoided through proper design, so that it never occurs in application. So, though the linear model is easy to find and use in a system model, it always implies a limit of application.

DC-current sources are treated in a manner very similar to that used for voltage sources. These devices also appear as *DC-power converters*, supplying DC current in this case.

The modeling of a common DC-current source is shown in Fig. 4-12. Note that the current output is now plotted on the vertical axis, with the voltage load on the horizontal. Again, the actual device has some droop in its output characteristic, and the modeler is faced with the problem of idealizing the source. The linear droop is assumed to have the slope -1/R.

The approach to idealizing the current source is very similar to the approach taken to the voltage source, and is shown in Fig. 4-12c. Here, the *current* is used to create an ideal source and a resistor-like element to account for the droop, according to the equation

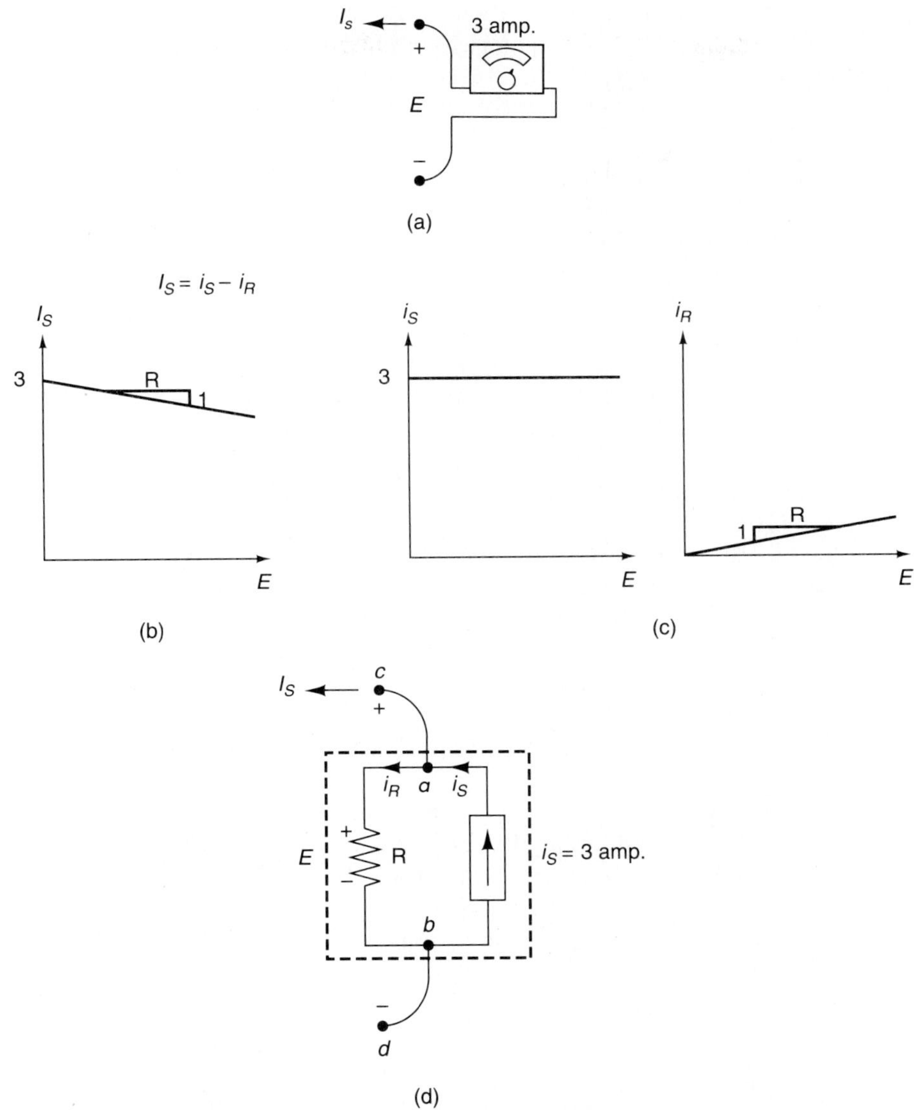

Figure 4-12. The Current Source. (a) Schematic symbol, (b) output data, (c) idealized data, (d) ideal model

$$I_S = i_S - i_R. \tag{4-26}$$

The idealized source is again shown as two ideal elements in a dashed box on the network diagram, as in Fig. 4-12d. The ideal source is represented by a box with an arrow in it to show the forced current in the same direction as the real source, and with the source strength indicated nearby. The confusion here usually has to do with placing the droop-modeling resistor in a network sketch.

The correct resistor location comes from physical interpretation of the current equation (Eq. 4-26). For example, consider point *a* in the idealized-source model of Fig. 4-12d. The sum of all currents flowing into this point must total zero, as will be shown in the next section of the text. Thus, the idealized current, i_S, must equal the sum of the output current, I_S, plus the resistor current, i_R. This relationship is rearranged to give Eq. 4-26.

Further, the resistor must be located as shown in Fig. 4-12d in order to give the same voltage sense as in the real component of Fig. 4-12a. If the resistor were in-line with the ideal source, then the voltage drop would have the opposite sense as the real case. So, the resistor must be located as shown in Fig. 4-12d.

The DC Transducers. Transducers are used by engineers to change power from one type or form (structural, electrical, fluid, or thermal) into another type. For example, a DC generator is a structural-to-electrical transducer which changes structural-rotational power into DC-electrical power. Similarly, a DC motor is a transducer which changes DC-electrical power into structural-rotational power.

High-power transducers (like motors and generators) are quite different from low-power transducers (also called *sensors*), and they result in very different models. This is true for several reasons. First, it is very common to encounter sensors with outputs to electrical ports since electricity offers such an easy way to transmit and manipulate data (e.g., as a basis for computer-based controllers). Sensors are thus manufactured to performance standards to ease the design problem. Second, the best sensors are designed to not interfere with the sensed environment, nor to introduce significant dynamics into the system. For these reasons, sensors are very simple to model.

This point may be amplified by returning to the robot example of the second chapter, and reconsidering the sensor multiport relationship, as in Fig. 4-13. Here, two complete ports are drawn, illustrating the fact that the sensor transduces between a rotational port and an electrical port. The dashed arrows in the figure represent the two variables which need to be present to complete the ports, but which are less

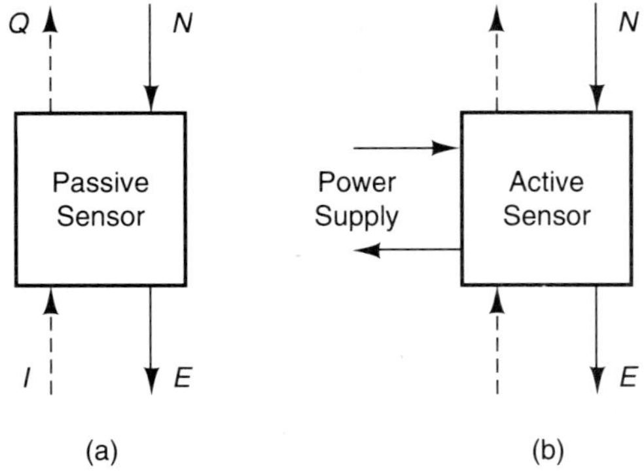

Figure 4-13. Sensor Multiport Diagrams.
(a) Passive, (b) active

important to the physics. Q, the sensor torque, is less important since the sensor is designed to offer negligible friction to the joint motions. The current input, I, is less important because it is only a constant value supplied by the robot computer to activate the sensor. However, the computer does need the sensor voltage output, E, to measure the arm position.

Sometimes, the passive sensor of Fig. 4-13a becomes the active sensor of Fig. 4-13b. This change is accomplished by giving the sensor its own power supply. In fact, this changeover to an active sensor is usually accompanied by the incorporation of intelligent signal processing at the sensor level. This is due to the advent of low-cost, micro-miniaturized transistors which have led to a whole generation of *smart sensors*. Such sensors incorporate local-computer processing of the sensed variables, including such factors as noise filtering, sensor fault detection, and preprocessing of data.

The detailed modeling of active or passive sensor dynamics is beyond the scope of preliminary-design modeling. They are modeled here with simple algebraic models which are expressions of their idealized functions. For the robot sensor of Fig. 4-13a, such an ideal model is, $e = \mathrm{k}n$, where k is a constant of proportionality.

Much different are the high-power transducers (motors and generators), whose dynamics are very interactive with their loads. However, the discussion of the performance of these components must be postponed until after the ideal structural elements are presented in a later section of this chapter.

The DC Amplifiers. Many electrical component designs require DC-electrical power to be boosted, or amplified, to a higher level so that useful work can be accomplished. Such amplifiers come in many different types, but two types are of special interest: the voltage-isolation amplifier and the current amplifier.

These amplifiers are designed to sense a voltage in a low-power network. They then faithfully reproduce it in an amplified form as either voltage or current in a high-power network, while isolating the two networks from each other. This is very important since low-power electronics are used to process sensor signals, which must be converted to high-power levels to perform useful work, such as the powering of a motor to rotate the shaft of a piece of heavy machinery. Furthermore, high-power networks are often very noisy, and can cause spikes and pulses which are very damaging to low-power networks.

The modeling of the *voltage-isolation amplifier* is shown in Fig. 4-14a, note the isolating action of the amplifier as indicated by the dashed current arrows. The figure is drawn this way since the amplifier creates no load on the low-power side by drawing negligible current, I_{in}. And it creates no demand on the high-power side by drawing negligible current, I_{out}. Furthermore, it is designed to produce a proportional voltage regardless of the elements which are attached to its terminals, so the figure is always drawn this way. The voltage-isolation amplifier is often found within communication systems and those computer applications which require a high, variable-voltage output.

The idealized voltage-isolation amplifier symbol is shown in Fig. 4-14b. Note

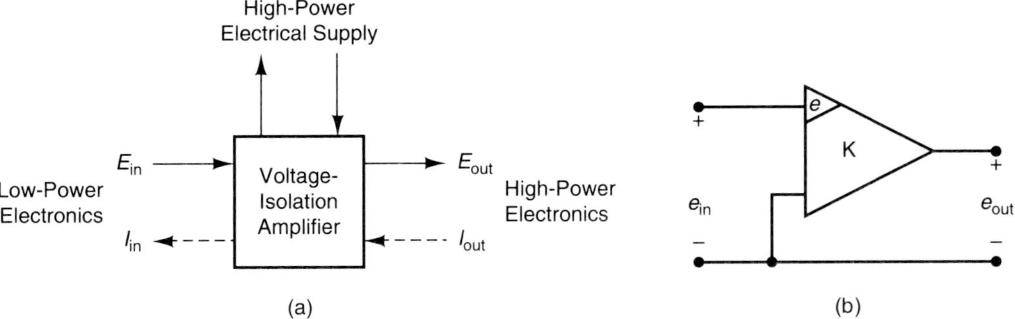

Figure 4-14. The Voltage-isolation Amplifier. (a) Multiport, (b) idealization

the "*e*" which is inset in the upper corner of the amplifier triangle, the multiplication constant "K" in the center of the triangle, and the absence of a connection to a power supply.

An example of modeling using a voltage-isolation amplifier is shown in Fig. 4-15, where an amplifier is being used to sense and amplify the voltage across an inductor. This voltage is varying as the resistor, R_2, varies—perhaps due to a person tuning the component.

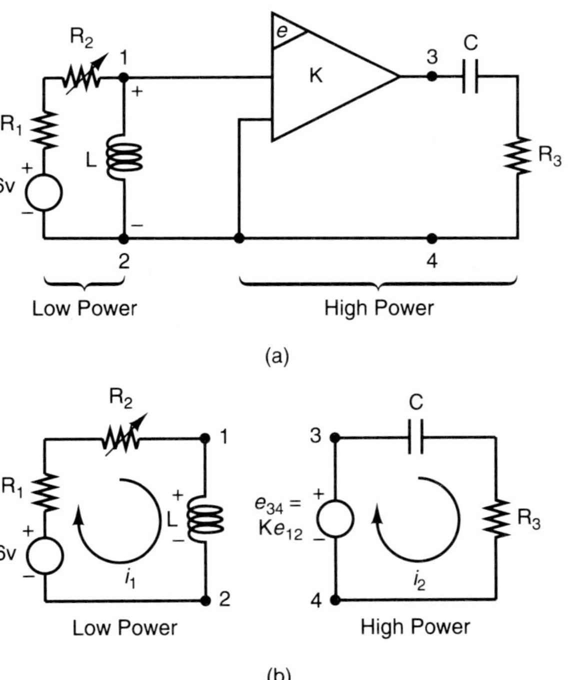

Figure 4-15. Voltage-isolation Network. (a) Idealized newtork, (b) equivalent networks

Since the voltage-isolation amplifier draws negligible current from the low-power circuit, it does not effect the operation of that circuit. Thus, you can draw the two equivalent idealized networks of Fig. 4-15b to model the single network of Fig. 4-15a. Note that the source that models the amplifier in the high-power network is of the same voltage sense as that assumed for the inductor.

The operation of the *current amplifier* is very similar to that of the voltage-isolation amplifier, and is shown in Fig. 4-16. Here, the multiport shows that the sensed variable is still voltage in a low-power network, but the amplifier output is now a *current* in a high-power network. This kind of operation is very useful in taking the low-power voltage output of a computer and amplifying it into a high-power current to drive an electric motor, or a variety of electro-hydraulic components.

The idealized symbol for the current amplifier is shown in Fig. 4-16b. Note that this amplifier has an "i" in the top of the triangle, it has a "K" in the triangle, and it has a current as the output.

An example of the current amplifier in a network is shown in Fig. 4-17, where the amplifier appears between a low-power circuit and a high-power circuit. The amplifier senses the voltage across the resistor at 1-2, which varies due to changes in the tuning capacitor shown at the top left of the figure.

The two equivalent idealized circuits for mathematical modeling are shown in Fig. 4-17b. Note the ideal current source in the high-power circuit. Here, the ideal source-current direction corresponds with the assumed positive sense of the voltage across the resistor at 1-2, which is carried over to 3-4.

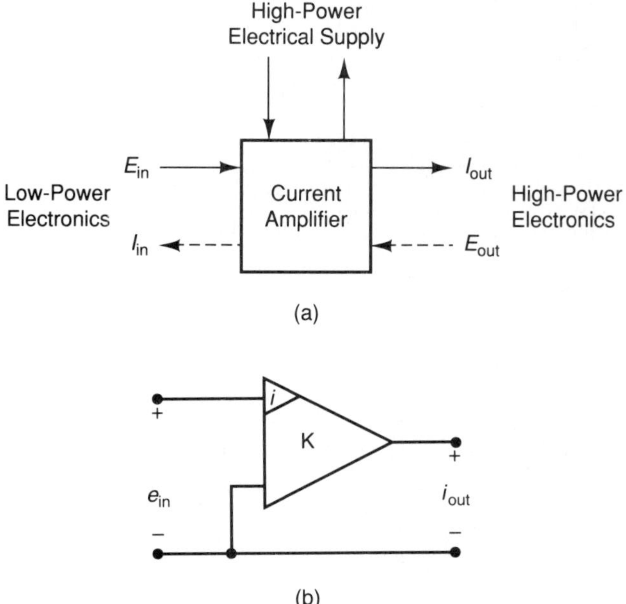

Figure 4-16. The Current Amplifier. (a) Multiport, (b) idealization

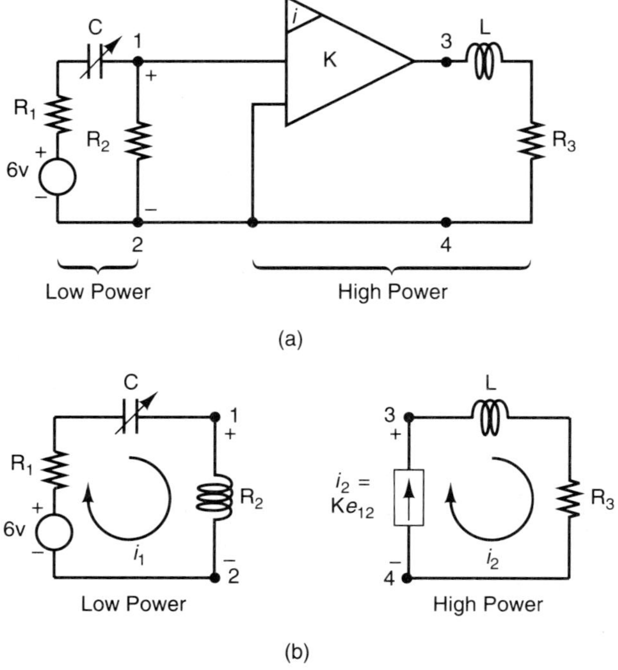

Figure 4-17. Current-isolation Network. (a) Idealized network, (b) equivalent networks

HOMEWORK

4-9. Using the data for the source given in Fig. 4-18, compute the following:

a) Approximately, what value of internal resistance does the data indicate?

b) Sketch the idealized circuit model for this source.

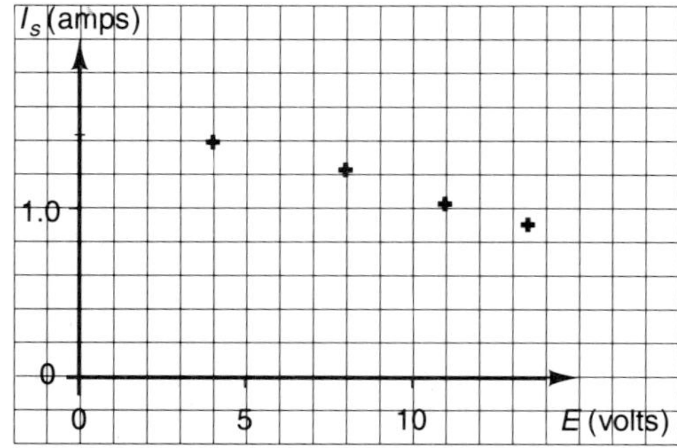

Figure 4-18. Current Data for Prob. 4-9

4-10. Given the following model for a real battery:

$$E = 12 - 100I, \quad (E \text{ in volts, } I \text{ in amps})$$

a) What strength of ideal battery source is indicated by this model?

b) What size of ideal resistance should be used to account for the real battery droop?

4-11. Use the multiport in Fig. 4-19 to complete the following,

a) Label the input/output arrows in the figure, except for the "amp. power" port,

b) Draw the idealized circuits for the figure, assuming that there is no significant droop in the battery.

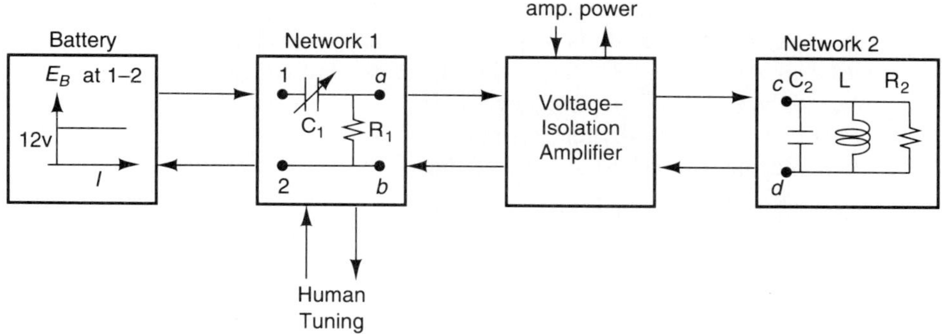

Figure 4-19. Multiport for Prob. 4-11

4-12. A real current source has the following input/output model:

$$I = 0.3 - 50E, \quad (I \text{ in amps, } E \text{ in volts})$$

a) What ideal current source strength is indicated by this model?

b) What ideal resistance may be used to model the real source droop?

4-13. **a)** What are the two types of amplifiers discussed in this section?

b) How are they alike?

c) How are they different?

4-4 ELECTRONIC NETWORK ANALYSIS: THE MATH MODEL

The goal of this section is to present a five-step analysis method which is used to translate a given network diagram and a given multiport diagram into a set of mathematical equations. This method is based on Kirchhoff's Laws of electronic networks.

The resulting model equations are linear if only idealized elements are in the network, they are nonlinear if there are one or more nonlinear elements. In either case, the method works quite well to give the network's mathematical model.

The Five-Step Method. *Step 1* is to set the ground rules and the analytical notation. This is where you recognize and assign the independent voltages and currents in the network. *Step 2* is the application of Kirchhoff's Voltage Law. This law shows you how to write a set of independent, voltage-based equations for the network under analysis. *Step 3* is the application of Kirchhoff's Current Law. This law shows you how to relate the independent currents in the network. In *Step 4* the general element and component models from previous sections of the text are expressed using the notation of the network under analysis. In *Step 5* you use equation manipulations to find the proper component output equation(s).

As the five analysis steps are discussed below, the network of Fig. 4-20a is used as a first example. Such SISO networks are called *planar networks* since they can all be flattened out and drawn on the plane of the page. The analytical method of this section also works for non-planar networks, but this text restricts your attention to planar networks for the sake of clarity.

Step 1: Notation and Definitions. The first thing to do is to assign an ideal input source to represent each input at every port. The polarity of these inputs must be chosen the same for both components connected at a port. If a polarity is not known, then one must be assumed at this point.

Notice that there is no "right" and "wrong" choice for the polarity assumption. Whatever positive variable assignments you make here or later in this step, the sign of the solution must be interpreted in light of the polarities assumed. Thus, if you assumed that e_{ab} is positive as shown in Fig. 4-20b, but it should be the reverse, then your solution will show e_{ab} is negative in value. The same is true for the assumed current directions.

Further, notice that there is no necessary relationship between the multiport input and the $+/-$ orientation. "Inputs" are inputs because they cause a response, no sign is mandatory from the multiport diagram. However, the same input/output polarity assignments for the separated component at the port must always be chosen. This is necessary for proper system assembly which comes after the separated component modeling is complete.

An idealized voltage-source input, e_{ab}, is thus assigned in the example of Fig. 4-20b due to the indicated causality from the multiport diagram. However, the $+/-$ orientation of the input is arbitrarily chosen as shown since no indication is given of the correct orientation, say by some solution from the port-matched component.

Once chosen, the polarity of the port-matched component variables are fixed as well. So, if the orientation for the output, i_{ab}, is known from the port-matched component, it would also be assigned at this point. In Fig. 4-20b, it is assumed that the positive current enters at a and leaves at b. Thus, the current in the port-matched component (not shown in Fig. 4-20a) must leave at a and enter at b, while the voltage across the port is "+" at a and "−" at b.

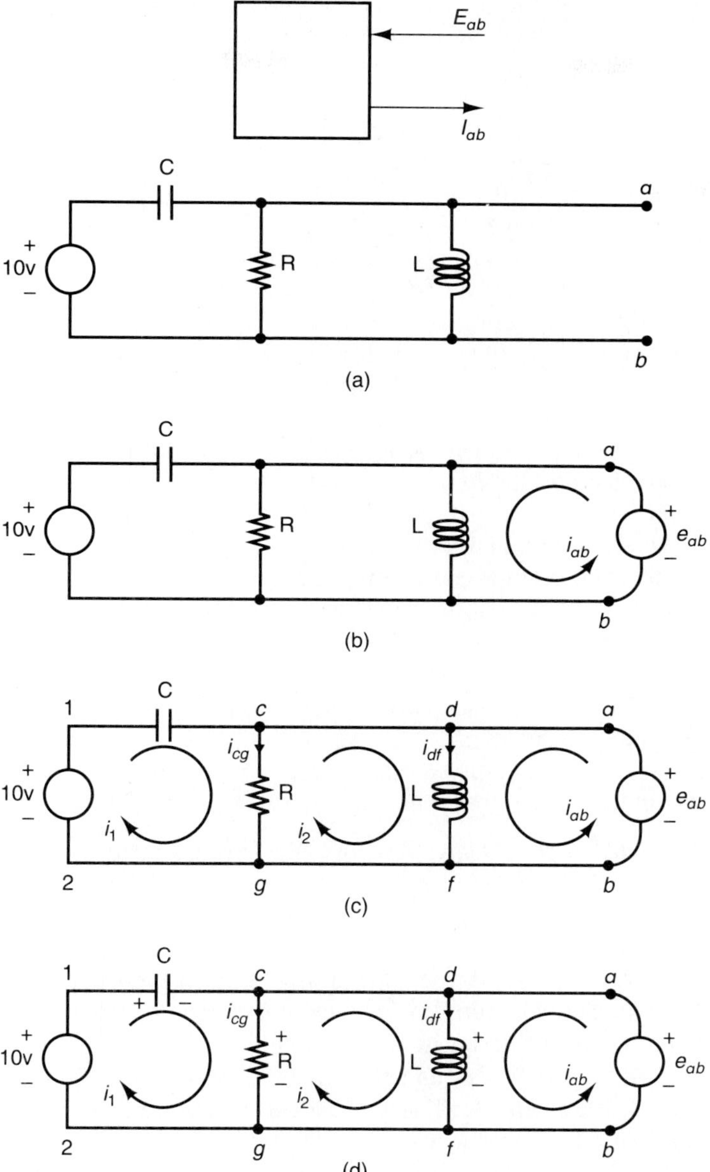

Figure 4-20. SISO Component Variable Assignments. (a) Given diagrams, (b) input/output assignments, (c) current conventions, (d) voltage conventions

The next thing to do is to recognize the *independent circuits* in the given network. These are recognized by the enclosed areas in the network diagram. Each enclosed area is assigned an *assumed* positive-current direction, which is indicated by a curved arrow drawn in the area, and labeled with a distinct current name (i_1, i_2, etc.). An example of this is shown in Fig. 4-20c.

The enclosed areas in the networks are also called *meshes* by systems engineers. Each independent circuit (or mesh) has one independent equation which may be used to solve for the network output(s). Note that there are three meshes in the example of Fig. 4-20.

Also at this time, all points at the ends of elements should be distinctly labeled (c, d, f, etc.), as in Fig. 4-20c. Note, however, that a-c-d are really all the same point since there is no intervening element between these points (the resistance of the idealized connecting wires is usually ignored). Consequently, you could recognize them all as point a. Similarly, b-f-g-2 is really one point b. In this example, however, these points are kept distinct for clarity.

Such points of circuit intersection, and all similar points separating elements are called *nodes* by systems engineers. Like the meshes, they are used to give independent relationships to solve for the network output(s). Note that there are three nodes in the example of Fig. 4-20; these are at 1, a, and b.

The final notation to be made on the network is to show the *assumed voltage drops* on all the elements excluding the sources. These are indicated by the "+" and "−" sign pairs on all the elements, as shown in Fig. 4-20d. Wherever obvious, these voltage assignments should be consistent with the assigned mesh currents. For example, the capacitor voltage drop should be as shown in the figure. But the assigned voltage drops over R and L are arbitrary, and may be chosen as shown or opposite to these choices.

The important things to remember about sign conventions is to *make them* at the beginning, *show them* on your network diagram, and *use them* in your analysis.

Step 2: Kirchhoff's Voltage Law (KVL). This law states that,

> *The total change of voltage around any closed circuit is zero.*

In other words, if you start at any point in the electrical network, and move around any closed circuit, you eventually come back to the point where you started.

For example, designate a start point as zero voltage and take a trip around a circuit. As you move around the circuit, you add or subtract voltage as you pass through each element. KVL states that you should have the same voltage value at the end of your circuit that you started out with, namely zero.

This is very much like making an excursion in hilly country. The place that you start has some altitude; designate this as the reference altitude. No matter what route you take through the hills and valleys, you always arrive back at the same reference altitude when you get back to the start point.

One way to ensure that you achieve independent equations in your analysis is to apply KVL to the *meshes* of your network, including the ports. It is not theoretically necessary to write KVL relations for this exact set of circuits, but it is procedurally desirable since it creates a consistent method resulting in equations which are easy to check.

For the example of Fig. 4-20d, you get three equations,

For mesh 1, starting at 2, $10\text{v} - e_{1c} - e_{cg} = 0,$

For mesh 2, starting at g, $e_{cg} - e_{df} = 0,$

For the i/o port, starting at f, $e_{df} - e_{ab} = 0.$

Note that these last two equations can be combined to give

$$e_{cg} = e_{df} = e_{ab}.$$

Such an arrangement where the voltage is the same across a number of elements is one which is called a *parallel arrangement* by systems engineers. These elements are thus said to be "in parallel."

Step 3: Kirchhoff's Current Law (KCL). This law states that,

> *In any network, the algebraic sum of the currents in all the wires that meet at a point is zero.*

This law has to do with the conservation of charge that is always taking place in an electronic network. It applies to the wires that connect elements. It says, basically, that the current into a point in the network must equal the current out of that point.

For example, consider point 1 in Fig. 4-20d. Clearly, the current into point 1 must equal the current out of point 1. There is no reason for the current (or charge) to accumulate at that point. While this looks like a trivial result, you will find that you get much use out of applying KCL to the nodes of the network.

The application of KCL to other nodes in the example allows you to see how these independent currents are related to each other through the common node points, as in Fig. 4-21. Note that the relationships of this figure show you what cur-

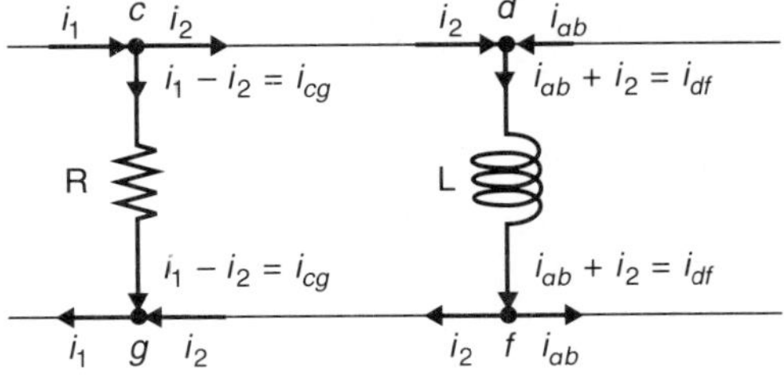

Figure 4-21. KCL at Nodes in Fig. 4-20

rent is flowing through any element in the network in terms of the independent mesh currents. Also note that the 10v source and the capacitor in mesh 1 both have the same current flowing through them. This is called a *series arrangement* of elements by systems engineers and they are said to be "in series" with each other.

Step 4: Element Relationships. In this step you take the element constitutive relationships and apply the specific variables of the problem under analysis. You must do this for every element in the network.

In the example problem there is a capacitor, a resistor, and an inductor. Thus, using the voltage and current definitions from the previous steps, you write

$$i_1 = C\, de_{1c}/dt,$$

$$e_{cg} = R\,(i_1 - i_2),$$

and

$$e_{df} = L\, d(i_{ab} + i_2)/dt.$$

Step 5: Component Output Equations. In this step, the element relationships of Step 4 are substituted back into the KVL or KCL expressions, which are then manipulated to find the mathematical equations for the component output(s) in terms of inputs only.

This step is made vastly easier by keying on the sources in the network. Thus, wherever there is a *voltage* source in a mesh, write the KVL expression for that mesh first, and your goal is to find the mesh *current*. Wherever there is a *current* source attached to a node, the KCL expression for that node is used to begin the procedure, and the goal is to find the node *voltage*.

For the example problem of Fig. 4-20 you start with the mesh 1 KVL expression because of the voltage source there

$$10 - e_{1c} - e_{cg} = 0.$$

Next, recognize that the resistor is "in parallel" with the input voltage. This allows you to write

$$10 - e_{1c} - e_{ab} = 0.$$

Now you can substitute for the capacitor element

$$10 - (1/C) \int i_1 dt - e_{ab} = 0.$$

In general, you will find it a good idea to eliminate any integrals as you go. This is done by taking the derivative of both sides of the equation

$$i_1 = C\, d(10 - e_{ab})/dt.$$

But, since 10 is a constant value

$$i_1 = -C\, de_{ab}/dt,$$

and you have found the first mesh current.

You may find the second mesh current directly from the resistor expression

$$e_{ab} = R\,(i_1 - i_2).$$

Solving this for the second mesh current gives

$$i_2 = i_1 - e_{ab}/R.$$

And substituting for the first mesh current gives

$$i_2 = -C\, de_{ab}/dt - e_{ab}/R.$$

Since you now know both mesh currents, you are in a position to solve for the network output, i_{ab}. This can be achieved directly from the inductor equation

$$e_{ab} = L\, d(i_{ab} + i_2)/dt.$$

Solving for the output gives

$$di_{ab}/dt = e_{ab}/L - di_2/dt.$$

The final result is found by substituting for i_2 from above

$$\boldsymbol{di_{ab}/dt = C\, d^2e_{ab}/dt^2 + (1/R)\, de_{ab}/dt + e_{ab}/L.}$$

This last equation only involves the single, idealized, SISO-network, output variable, i_{ab}, the idealized-network input variable, e_{ab}, and the constant parameters of the network (R, L, and C). Thus, this is the *output equation* of interest—you are done manipulating the equations.

Also note that the output in all its derivatives is shown on the *left side* of the equation, and the input in all its derivatives is shown on the *right side*. Further, the derivative terms are arranged in descending order, with the highest on the left. This attention to formatting helps you to quickly check your final result.

In general, there must be *one equation for each output* of the network, in terms of one or more of the network inputs and the network parameters. For this SISO example there is only one output and one input, but more of each is possible in a MIMO network.

So, given the inputs and the values for the parameters, you may solve the output equations. Given the outputs, you may then back-substitute for any variable in the network using the mesh current and node voltage equations from Step 5 of the analysis procedure. However, this is best done after the model is thoroughly checked.

Checking the Model. Three checks on the model are usually made. The first two are design checks which compare the output equation(s) to the network diagram, the third is a mathematical check of the model form.

The first design check has to do with the disappearance of any network parameter from the output equations. This may be telling the modeler that there are unnecessary elements present, or it may be a mistake in equation writing or manipulation.

For the example problem, the disappearance of the 10v supply from the output dynamic equation is of concern. In other words, is that supply really necessary to this design? Note that the supply may be used to establish and maintain the proper initial conditions for the physical transients, and would thus appear in the initial conditions which are used to solve the equation of output dynamics. In this case, you should first recheck your derivations, then talk with the designer.

The remaining two checks are related to a linear MIMO model which is expressed in terms of L outputs, y_i, *each* of which is of the form

$$C_n\, d^n y_i/dt^n + C_{n-1}\, d^{n-1} y_i/dt^{n-1} + \cdots + C_o y_i = \sum_{m=1}^{N} (D_{kmi}\, d^k u_{mi}/dt^k + \cdots + D_{omi} u_{mi}). \tag{4-27}$$

Thus, if there are L outputs of a linear network, then there must be L linear equations in the network model, each in the form of Eq. 4-27 ($i = 1, \ldots, L$), and each potentially in terms of all the inputs ($m = 1, \ldots, N$), as shown.

In the second design check, the highest order of output derivative for the set of output equations for the component should equal the number of energy storage elements in the network diagram. Again, if this is not true, you should first recheck your derivations, then talk with the designer.

In the second check of the example, the network diagram shows two energy-storage elements (one capacitor and one inductor). The second-order derivative in the output equation thus checks with the network diagram.

The third check of the network model is on a port-by-port basis. In real physical ports, the order of the highest derivative of the output is greater than or equal to the order of the highest derivative of the port-matching input. This is checked on an equation-by-equation basis, by checking the highest derivatives of input/output variables with matching subscripts. If the check fails, then the fix is quite simple, as described below.

Returning to the example SISO-problem model, the third check shows that the order of the highest input derivative is greater than that of the output derivative. The cure for this is recognizing that the *causality* was improperly assigned during the multiport analysis, and it merely needs to be reversed. This may entail a discussion with the designer to verify the proper causality of the port in question.

The reversal of the causality is easily accomplished graphically by reversing the port arrows on the multiport diagram. It is accomplished mathematically by merely reversing the port input/output in the equation,

$$\mathbf{C}\, d^2e_{ab}/dt^2 + (1/\mathbf{R})\, de_{ab}/dt + e_{ab}/\mathbf{L} = di_{ab}/dt.$$

Sometimes, in MIMO systems, you find that the output equations consist of an algebraic equation which is coupled to one or more differential equations. This case is examined in the next example, along with the difficulty of a MIMO component having a nonlinear element.

Example 4-3

The component shown in Fig. 4-22 has two ports for electrical power and a nonlinear resistor, R_2; note the special symbol given to the idealized form of this nonlinear resistor. The problem here is to find the output equation(s) for this component based on the causality shown in the figure.

Solution

As in the previous example, the five-step Kirchhoff approach is used to solve the problem.

Step 1: The notation is selected as in Fig. 4-22b.

Step 2: The KVL relationships are as follows:

Left port, $e_L - e_{fg} = 0$,

Mesh 1, $e_{fg} - e_{f1} - e_{12} = 0$,

Mesh 2 and the right port, the current source and the resistor are in parallel

$$e_{34} = e_R.$$

Step 3: The KCL relationships are as shown in Fig. 4-22c. Note that the mesh 2 current is equal to the current source strength.

Step 4: The element relationships are

$$e_{fg} = (i_L - i_1)R_1,$$

$$e_{f1} = L\, di_1/dt,$$

$$e_{12} = (i_1)^2R_2,$$

$$i_S = Ke_{12},$$

$$e_{34} = (i_R + i_2)R_2.$$

Step 5: Beginning with *mesh 1* since there is a known voltage input at the *f-g* location,

$$e_{fg} - e_{f1} - e_{12} = 0.$$

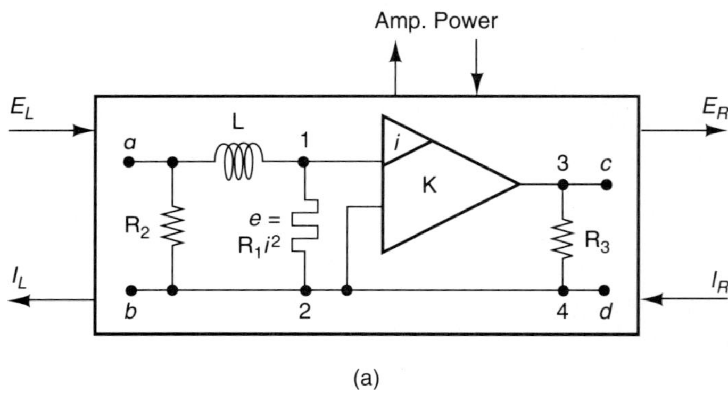

(a)

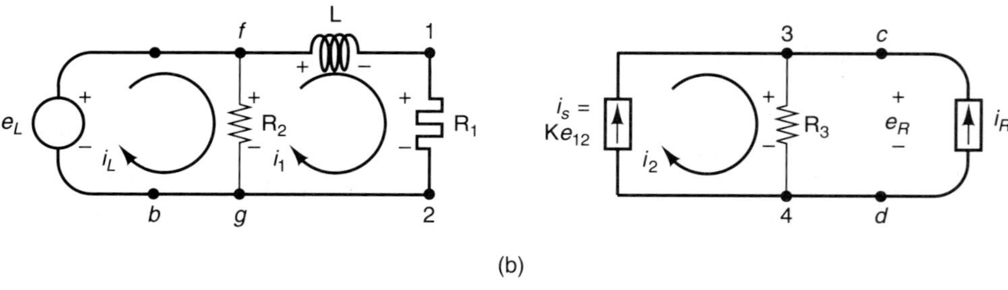

(b)

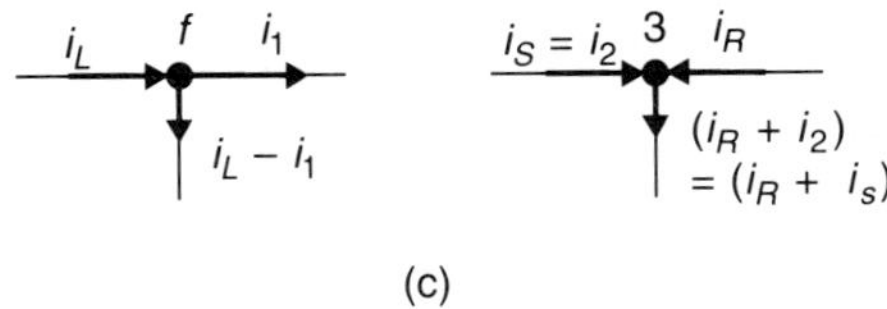

(c)

Figure 4-22. A Nonlinear MIMO Component. (a) Given diagrams, (b) mesh conventions, (c) node analyses

Substituting the input voltage gives

$$e_L - e_{f1} - e_{12} = 0.$$

Now, substituting the element relationships gives

$$e_L - \mathrm{L}\, di_1/dt - (i_1)^2\, \mathrm{R}_1 = 0.$$

This expression is solved for the mesh 1 current

$$\mathbf{L}\, \boldsymbol{di_1/dt} + \boldsymbol{(i_1)^2}\, \mathbf{R_1} = \boldsymbol{e_L}.$$

which is a nonlinear differential equation.

There is not much more you can do with mesh 1 at this point other than to realize that the solution to the mesh current is expressed in the above equation.

To find the output of mesh 1, you must turn to the unused resistor equation,

$$e_L = (i_L - i_1)\,\mathrm{R}_1,$$

which can be solved for the output current

$$\boldsymbol{i_L = i_1 + e_L/\mathrm{R}_1}.$$

In the analysis of mesh 2, you start by recognizing that there is both a current source and a current input to node 3. Thus, KCL yields the element current directly, as in Fig. 4-22c, which gives an easy solution for mesh 2. First for resistor 2,

$$e_R = \mathrm{R}_3\,(i_R + i_s).$$

Then substituting for the current source strength,

$$e_R = \mathrm{R}_3\,(i_R + \mathrm{K}e_{12}).$$

Finally, substituting for the amplifier voltage gives

$$\boldsymbol{e_R = \mathrm{R}_3\,[i_R + \mathrm{K}(i_1)^2\mathrm{R}_2]},$$

which is a nonlinear algebraic output equation.

The three checks of the output equations are next applied. First, all of the network elements appear in at least one of the model equations, and the first check passes. Second, the highest output derivative is one, which corresponds with the single energy storage element, the inductor in mesh 1—the second check passes as well. And third, there is only one equation with a derivative in it. In that equation the output is differentiated, but the input is not. So the third check passes.

Comments

Note first that there are now *three* equations in this nonlinear component model, even though there are only *two* outputs. This occurs since you are unable to solve explicitly for the mesh 1 current, i_1, due to the nonlinearity.

HOMEWORK

4-14. Find the equivalent total resistance, R_T, to the resistor networks of Fig. 4-23.

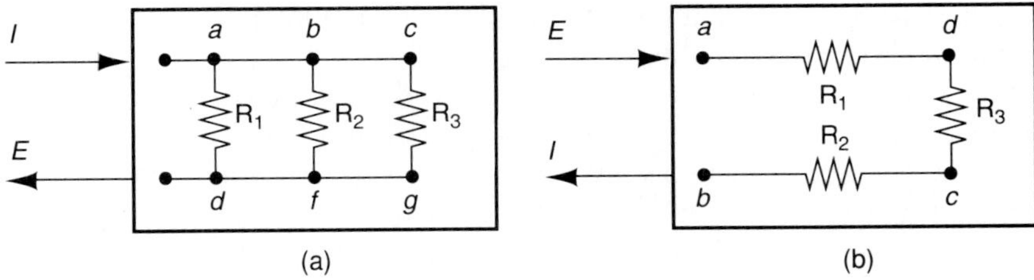

Figure 4-23. Resistor Networks for Prob. 4-14

4-15. Define the following system concepts in one or two sentences:

a) Mesh, and

b) Node.

c) Why are these two system concepts important?

4-16. Solve for the output equations of Fig. 4-1.

4-17. A network has four nodes; *A*, *B*, *C*, and *D*. This network has two meshes: *A-B-C* and *B-C-D*, such that side *B-C* is common to both. In side *A-B* there is a voltage source. Between nodes *C* and *D* is a current source.

When you begin writing equations, should you use KVL for

a) Circuit *A-B-C* (why?),

b) Circuit *B-C-D* (why?)

4-5 THE IDEAL STRUCTURAL ELEMENTS

In structural components, unlike electronic components, you often find that you need to sort out intermixed physical effects for ideal models. You also find that the structural physics often involve significant nonlinearities with which you must contend. Consequently, you need to be careful in your idealized-structure modeling to remember the Fundamental Truth of Modeling—that these effects are nonlinear. You are only approximating, and the range of accurate idealization may not be very large.

However, like electronic components, structural components are usefully modeled with idealized elements and components, some of which are passive and some of which are active. In this section, the three passive structural modeling elements are presented. These elements are designed to represent the fundamental physical properties of mass, spring-rate, and friction.

The Structural Variables. This text only discusses problems in *one-dimensional* structural dynamics in order to develop the principles of physical analysis without getting lost in the complexities of the elements. Thus, all masses are allowed to move in *either* translational *or* rotational motion, but *not both* at the same time. Further, the translational motion is allowed only in a straight line, otherwise it is multidimensional and beyond the scope of this text. Similarly, the rotational motion is only allowed about a fixed axis.

While these requirements may seem overly restrictive, you will soon find that a number of interesting and useful problems fall into this realm. You will even find that many structural components which exhibit more complicated motions can be well-approximated by these simple elements. So, these restrictions lead to a useful preliminary understanding of structural system dynamics, which is the objective.

The dynamic variables which are used to describe one-dimensional *translational* motions are the *force*, F (lb_f, n), and the *translational speed*, V (ft/s, m/s). Often, the *net* force is of interest—this quantity is called the *resultant force*, R_F, such that

$$R_F = \Sigma F, \tag{4-28}$$

and the directions of the forces and the resultant are understood to be along the axis of motion.

The translational motion of any point in the structure is described by a family of *kinematic variables*. This family is comprised of the position, X, (ft, m), the speed, V, and the acceleration, A (ft/s^2, m/s^2). These variables are related by derivative in the following way: $dX/dt = V$, and $dV/dt = A$. Thus, even though you focus on the speed in the discussion which follows, you may readily compute any other kinematic variable using the derivative relationships.

Note that the kinematic variables must be measured with respect to an inertial (non-accelerating) reference frame. Further, the speed of any point in the structure must not exceed 0.2 times the speed of light in order to avoid relativity effects in mass modeling.

The variables which are used to describe one-dimensional rotational motions are the *torque*, Q (ft-lb_f, n-m), and the *angular speed*, N (rad/s). The *resultant torque*, R_Q, is computed such that

$$R_Q = \Sigma Q, \tag{4-29}$$

where the directions of the torques and the resultant are understood to be about the axis of rotation.

The rotational motion about an axis in the structure is also described by a family of *kinematic variables*. This family is comprised of the angular position, β (rad), the angular speed, N, and the angular acceleration, α (rad/s^2). These variables are related by derivative in the following way: $d\beta/dt = N$, and $dN/dt = \alpha$.

It is the convention of this text to only show forces and torques which are *externally* applied, in the same manner as on a free-body diagram. Furthermore, only the positive sense of the speed is indicated on the diagrams, the positive senses of the other variables are understood to be the same.

The Pure Mass. Of course, one of the most common physical properties is that of mass, which is thoroughly discussed in the study of dynamics. The pure mass is one which demonstrates pure inertia effects (without spring or friction effects) according to *Newton's Second Law of Motion*. This law may be stated in two parts, first for *translational* motion,

> If an unbalanced resultant force acts on a particle, it will accelerate in proportion to the magnitude and in the direction of the resultant force.

Then for *rotational* motion,

> If an unbalanced resultant torque acts on a body with a fixed axis of rotation, it will experience an angular acceleration in proportion to the magnitude and in the direction of the resultant torque.

Note that according to Newton, the *input* to the mass is *always* the *resultant*. The mass responds with acceleration. It is *never* the other way around, no matter how deceptive a problem seems. The consequence of this is that the input to a mass on a multiport diagram must *always* be the force, never the speed.

The constant of proportionality in the translational law is the *mass*, M (sl, kg), of the particle. The constant of proportionality in the rotational law is the *inertia*, J (sl-ft^2, kg-m^2). Thus, the pure masses with one-dimensional kinematics appear as in Fig. 4-24.

The experimental data for the mass follows the Newtonian response, which is idealized in the linear model

$$r_F = \mathrm{M}\ dv/dt. \tag{4-30}$$

This equation has the units in the English system

$$\mathrm{lb}_f = (\mathrm{sl})(\mathrm{ft/s^2}).$$

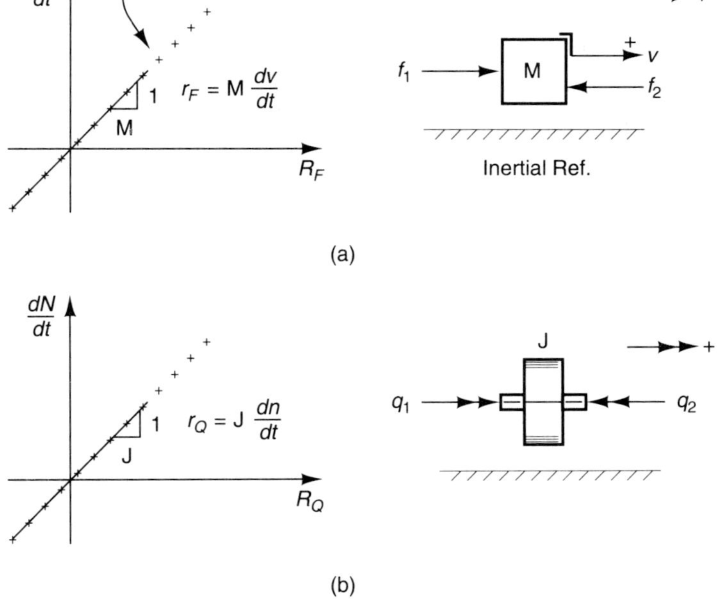

Figure 4-24. The Ideal Mass. (a) Translating mass, (b) rotating inertia

And in the SI system

$$n = (kg)\ (m/s^2).$$

The measurement of the mass often comes from weight measurements. That is,

$$M = W/g, \tag{4-31}$$

where W is the weight, and g is the acceleration of gravity (32.2 ft/s^2, or 9.81 meters/s^2, at the earth's surface).

The mass may also be computed from measurements of volume, $\mathcal{V}$ (ft^3), and density, ρ (sl/ft^3), according to

$$M = \mathcal{V}/\rho. \tag{4-32}$$

The symbol for the idealized mass is shown at the right of Fig. 4-24a. In this, the assumed positive directions of the applied forces and the motion are shown on the figure. Note that the positive-force direction and the positive-motion direction must be the same; thus $r_F = (f_1 - f_2)$. The inertial reference is shown at the bottom of the figure to indicate the need for such a restriction.

In Fig. 4-24b, the data for the *inertia* is presented. Again, the data conforms to the Newtonian model of a rotating mass. This time the idealized model is

$$r_Q = J\ dn/dt, \tag{4-33}$$

with the English units

$$\text{ft-lb}_f = (\text{sl-ft}^2)\ (\text{rad/s}^2).$$

The SI units for this equation are

$$\text{n-m} = (\text{kg-m}^2)\ (\text{r/s}^2).$$

The inertia, J, is computed from the moment of mass about the axis of interest,

$$J = \int d^2\ dM, \tag{4-34}$$

where d is the distance of dM from the axis of interest. J is thus sometimes called the *mass moment of inertia*. For very complicated mass shapes, this integral may be approximated with the numerical formula

$$J \approx \Sigma\ d_i^2\ \Delta M_i, \tag{4-35}$$

as done in the study of Fourier series approximations in the previous chapter. Here, d_i is the distance of mass element, ΔM_i, from the axis of interest.

Fortunately, J has been tabulated for many common shapes [see Baumeister and Marks, for example]. In this text, you will deal mostly with the disk as the inertia shape. The disk has

$$J_{DISK} = (1/2)M\, r_D{}^2, \tag{4-36}$$

where r_D is the disk radius.

The idealized inertia symbol is shown at the right of Fig. 4-24b. Note that the torque is shown with a double-headed arrow. This arrow shows the assumed direction of the positive net torque in a *right-handed* sense. That is, if you point your right thumb in the direction of the double arrow, then your curved right-hand fingers point in the direction of positive torque, and $r_Q = (q_1 - q_2)$. The idealized mass speed, n, is also an angular variable, and its positive direction is shown by a curved arrow, also interpretted in a right-handed sense.

Again, the rotational kinematic variables (angular position, velocity, and acceleration) must be measured with respect to an inertial reference, which is shown by the reference symbol at the bottom of the figure.

All mass elements are energy *storage* elements. They store energy by virtue of their *speed* (e.g., translational kinetic energy is $\frac{1}{2}MV^2$).

The Ideal Spring. Spring rate may be measured experimentally, or it may be computed theoretically. However, theoretical computations of spring-rate are very complicated, even for simple structures. In this text, it is assumed that the spring rate is known from experimental data or previous computations. Again, there are two types of structural spring effects: the translational and the rotational types. The *translational spring* is used to model two main effects: hardware manufactured to produce springiness, or that due to secondary flexibility effects. The former is seen in precisely manufactured devices which are used to achieve a designed spring rate, as in automobile suspension or diving boards. The latter appears in a structure designed to accomplish some other primary task than springiness, as in a bridge or building subjected to a lateral wind loading.

A typical model of the translational spring is,

$$F = k(X_1 - X_2), \tag{4-37}$$

where X_1 and X_2 are measured from the *unstretched* positions of the spring ends. This model may be easily idealized into the form,

$$f = k(x_1 - x_2), \tag{4-37a}$$

where x_1 and x_2 are now measured from the *initial* positions of the spring ends. The conversion between these two formulae is accomplished using constant δ's to equate the two positions,

$$x_1 = X_1 + \delta_1, \tag{4-37b}$$

$$x_2 = X_2 + \delta_2, \tag{4-37c}$$

and

$$f = F + \mathrm{k}(\delta_1 - \delta_2). \tag{4-37d}$$

You will usually find that the ideal spring model is much more convenient to use since the initial positions of the spring ends are easily located, but the unstretched ends are not. Further, springs are often put into machinery under compression or in tension in order to avoid difficulties when the spring changes between these two cases. However, be sure to account for the initial force in the spring following idealized modeling.

It is the convention of this text to idealize about the point $\delta_1 = \delta_2 = 0$. That is, the initial positions of spring ends are usually assumed to be their unstretched positions. The contrary case to this convention will be indicated.

The idealized spring model in terms of speed is thus

$$f = \mathrm{k} \int v_{12}\, dt. \tag{4-38}$$

The English units of the spring model are

$$\mathrm{lb}_f = (\mathrm{lb}_f/\mathrm{ft})\ (\mathrm{ft/s})\ (\mathrm{s}),$$

and the SI units are

$$n = (n/m)\ (m/\mathrm{s})\ (\mathrm{s}).$$

The modeling of pure spring effects is thus as shown in Fig. 4-25. The response demonstrated by this spring is very characteristic of translational spring rate. That is, the data shows the commonly observed nonlinear stiffening of the spring as deflection grows.

Notice that the force in an idealized spring must be equal and opposite at both ends since it is a pure, *massless* element. Any imbalance of the force cannot be allowed since this would create an infinite acceleration of the massless element. Thus, the force f is said to pass through the element instantaneously. The forces on an idealized spring thus always exist as oppositely directed, collinear forces of equal magnitude.

The symbol for the idealized translational spring is shown at the right of Fig. 4-25a.

By convention, a positive force for the translational spring is a *compressive* force, as shown in the figure. For kinematics, only the speed sign convention is shown on the diagram—the position and the acceleration also have the same positive sense as the speed. If this positive assumption is incorrect, then the values that come from the solution will be negative.

The *rotational spring* data is shown in Fig. 4-25b. As in the translational spring, the pure rotational spring usually demonstrates a nonlinear increasing stiff-

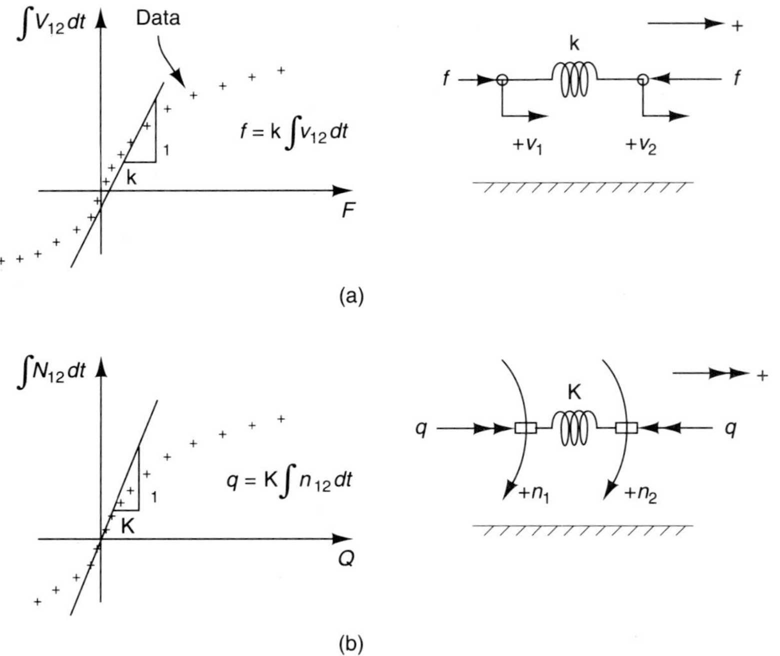

Figure 4-25. The Ideal Spring. (a) Translational spring, (b) rotational spring

ness as the displacement increases, and the element is idealized about the zero deflection point. The idealized model is

$$q = K \int n_{12}\, dt, \tag{4-39}$$

which has the English units

$$\text{ft-lb}_f = (\text{ft-lb}_f/\text{rad})\ (\text{rad/s})\ (\text{s}),$$

and the SI units

$$n\text{-}m = (n\text{-}m/\text{rad})\ (\text{rad}/s)\ (s).$$

The symbol for the rotational spring is shown at the right of Fig. 4-25b. The double-headed arrows are again used for the torque in order to show the equal and opposite external torques which *must* act at opposite ends of the massless spring. Also, the spring operates in such a way that its ends may take on different angular speeds, n.

All springs *store* energy by virtue of the amount of *deflection* that they experience.

Friction and the Ideal Damper. Friction occurs in two main types: coulomb friction and viscous friction.

Coulomb friction is also called "dry" friction or "contact" friction and is composed of the static ($V = 0$) and sliding ($V > 0$) types, as shown in Fig. 4-26a. Note

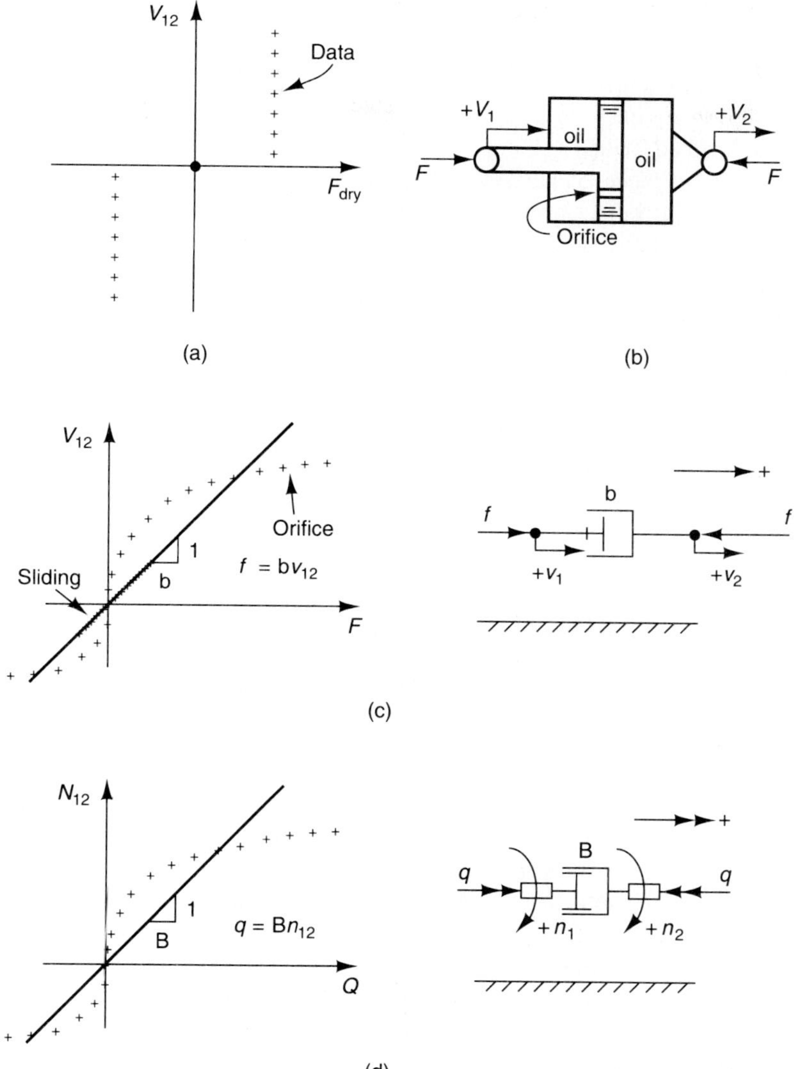

Figure 4-26. The Ideal Damper. (a) Coulomb friction, (b) real damper, (c) translational damper, (d) rotational damper

that this is a *nonlinearizable* type of friction due to the discontinuity that it demonstrates around $V = 0$. If coulomb friction is a significant part of the mechanics under analysis, then you have no choice but to use a nonlinear friction model.

If, however, some form of viscous lubricant is used to reduce contact friction, then the friction response looks more like that shown in Figs. 4-26c and d, and a linear model works quite well in some region near $F = 0$. This type of friction is called *viscous friction* and it often demonstrates a square-law relationship of the form

$$F = b' (V_{12})^2. \tag{4-40}$$

Viscous friction occurs in two primary forms: as a resistance to motion between lubricated sliding surfaces, or as a calibrated resistance to motion coming from a precision-manufactured orifice.

A common piece of hardware known as a *damper* is often used in structural components to introduce precise viscous friction. A cutaway view of a damper is shown in Fig. 4-26b. Note that the damper is designed to force a viscous fluid (oil in this case) back and forth through a calibrated orifice to achieve its friction effect. Dampers are used by designers to absorb the excess energy in a structure. Automobile shock absorbers are a type of damper used to reduce suspension oscillations. In this text, it is assumed that the designer has the necessary data to model this effect without having to resort to a fluid-mechanical analysis to determine it from theory (see also Sec. 4-8 on fluid resistance).

For well-lubricated *sliding surfaces*, the modeler uses an ideal Newtonian fluid model,

$$\tau = \mu V_{12}/d, \tag{4-40a}$$

where τ is the shear force in the fluid, lb_f/ft^2,
μ is the fluid absolute viscosity, $lb_f\text{-}s/ft^2$,
V_{12} is the relative speed of the two surfaces,
and d is the separation of the two surfaces, ft.

Rearranging of the Newtonian model gives the viscous friction model,

$$F = (\mu A/d) V_{12}, \tag{4-40b}$$

where A is the area over which the friction force, F, acts. Some values for μ are given in Table 4-4. Note that these values are temperature dependent. (The values given are useful near 100°F (±20°F).) Note again, this model of friction assumes *no significant contact* of the sliding surfaces. If significant contact friction is expected, then a nonlinear coulomb-friction model should be used ($F = \mu_e N$, where $\mu_e \sim 0.8$-static, 0.2-sliding, and N is the normal force on the contact area).

The data for *translational friction* is shown in Fig. 4-26c. This data may come from either sliding-friction measurements or a damper, but the same ideal model is used for both cases. Thus, the idealized translation damper model becomes,

Table 4-4. Absolute Viscosities (at 100°F)

Fluid	μ, $lb_f\text{-}s/ft^2$
Castor oil	5.0×10^{-3}
Crude oil	1.155×10^{-4}
Hydraulic fluid	2.19×10^{-4}
Water	1.358×10^{-5}
Gasoline	5.28×10^{-5}
Air or oxygen	4.5×10^{-7}

$$f = bv_{12}. \tag{4-40d}$$

The units of the translational damper constitutive relationship (Eq. 4-40d) are

$$\text{lb}_f = (\text{lb}_f\text{-s/ft})\ (\text{ft/s}),$$

with the SI units of

$$\text{n} = (\text{n-s/m})\ (\text{m/s}).$$

Again, notice that the ideal translational damper must be *massless*—it cannot tolerate any unbalanced forces. The reasoning for the positive sign conventions which are shown on the idealized damper diagram are essentially the same as for the spring under compression.

The rotational damper data is shown in Fig. 4-26d. The idealized model of this damper is

$$q = Bn_{12}. \tag{4-41}$$

The English units of this equation are

$$\text{ft-lb}_f = (\text{ft-lb}_f\text{-s/rad})\ (\text{rad/s}),$$

and the SI units are

$$\text{n-m} = (\text{n-m-s/rad})\ (\text{rad/s}).$$

The idealized symbol at the right of the figure is slightly different from the translational damper, correctly suggesting that the friction is taking place around the sides of a rotating cup. Again, the torques are indicated by the double arrows, in the right-handed sense. The torques must always balance each other since this, too, is a massless element. The rotational speeds are indicated by the curved arrows in the figure.

Dampers of any type are always used to represent the *dissipation of energy* through viscous friction losses. An ideal damper has no energy storage whatever, it merely models the frictional losses in part of a structural component.

HOMEWORK

4-18. A diving board designer is considering the problem shown in Fig. 4-27. Based on previous designs, the designer has tentatively selected a board spring rate of 0.6 lb_f/in per inch of board length, and a board density of 0.0027 slugs/in^3. The preliminary dimensions of the board were selected as 8 ft × 1 ft × 2 in. For this design configuration, complete the following:

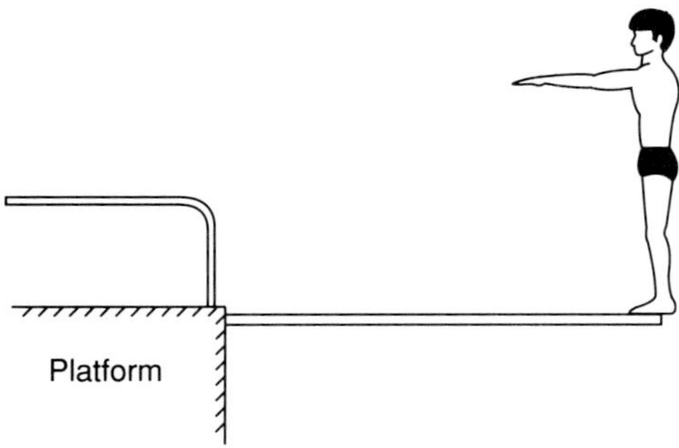

Figure 4-27. Springboard Geometry for Prob. 4-18

a) Draw a multiport diagram of the board, showing its connections to the diver and the platform. Neglect air friction resistance to board motions.

b) Draw a one-lump model of the system of (a), (hint: see Fig. 4-3),

c) Compute the values of the parameters of the model for (b),

d) Redraw the model of (b), adding in a one-lump viscous friction model of the air resistance to board motions.

4-19. A car weighs 3,220 lb_f, what is its mass?

4-20. A designer is developing a new method of engraving metal records ($W = 6$ oz $= 6/16\ lb_f$, $r = 5$ in.) as shown in Fig. 4-28. In this system, the motor turns a drive axle by using a flexible drive belt. This axle turns the metal record with no slipping, as the oiled cutting stylus engraves the desired information. Your job is to develop a preliminary model of this system.

You should assume that the motor is a torque source, that the drive belt is flexible enough to model its spring rate, that the record inertia dominates other inertias, and that the stylus exerts a viscous friction on the record.

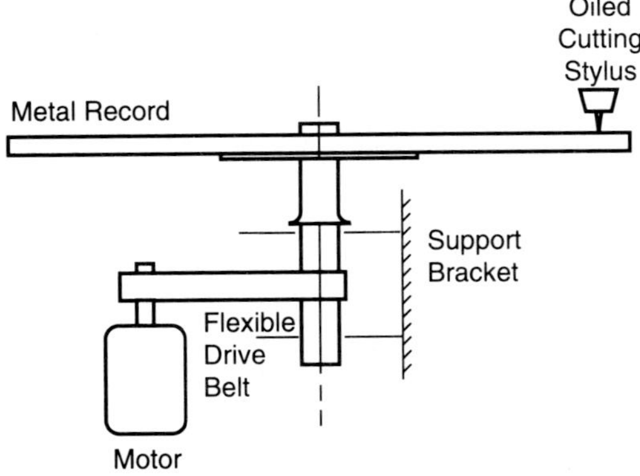

Figure 4-28. Engraver for Prob. 4-20

Do the following:

a) Draw a multiport diagram of the system, from the motor to the stylus.

b) Draw the ideal physical model.

c) Compute the system (e.g., the record) inertia.

d) Write out the equation which governs the record dynamics using a free-body diagram of the record.

4-21. A forensic engineer (one who analyzes for legal proceedings) is studying an auto accident. The conditions of the wreck are that an auto is sliding across a rain-slick roadway which provides some friction to motion. Twisting motions are negligible compared to the translational motions. Sketch a first model for the translational motion of the auto.

4-22. A marine engineer is interested in the performance of the rotating drive shaft connected between a ship's engine and the ship's propeller. The shaft is massive and flexible. Sketch a preliminary model of the shaft.

4-6 COMMON STRUCTURAL COMPONENTS

Mechanical designers often use structural *sources*, *transducers*, and *amplifiers* to create their structural systems. These products/components are combined with structural elements (masses, dampers, and springs) to accomplish various useful engineering tasks. While these devices are actually components, they are usually treated as elements for preliminary physical analysis purposes. This is possible since they are made in such a way that they may be readily idealized as elements within a preliminary system design.

The Structural Power Sources. The origin of structural power comes from hydraulic, wind, gravity, nuclear, and/or chemical resources, and various types of *prime-mover machinery* have been created to take advantage of each of these primary resources. Thus, when systems engineers speak of a "structural power source" they really mean a prime-mover machine which is designed to furnish mechanical power to a structural component, and drawing this power from one or more of the prime resources. However, for this text, only those sources which are powered by gravity and/or chemical resources are considered in order to introduce this type of modeling.

Note that a machine which converts electrical power into mechanical power (e.g., an electric motor) is really a *transducer*, not a prime mover. Transducers are discussed in a later part of this section.

There are four types of structural power sources with which you must be concerned. These produce force, torque, translational speed, or rotational speed.

The *force sources* are designed to output a force in a fashion which is more-or-less independent of translational speed at the output port. Some common types of this source are jet and rocket engines, which are both prime movers operating on chemical resources. The force sources also use various field forces, such as electromagnetics and gravity.

By way of example, consider the modeling of gravity acting on a mass as shown in Fig. 4-29. This multiport shows that the gravity force, F_g, acts on the mass which responds with some speed, V. The ideal source element which may be used to model this effect is shown in Fig. 4-29b. Note the similarity of this source representation with that of the current source presented earlier in Sec. 4-3. This is not an accidental selection—it has to do with the analogy between current and force which will be clarified later in this chapter.

Of course, there often is some kinematic constraint which restricts the allowed one-dimensional motion of the mass so that the full force of gravity does not act in the direction of motion. Such a case, and the use of the idealized-gravity source, is shown in Fig. 4-29c. Here, the mass element has some elevation, h (ft), and some speed, v, along a straight, frictionless, inclined path. Note that the ideal-force source representing gravity is directly connected to the earth, and is shown acting in the direction of the one-dimensional motion. The strength of this source, f_g, is dependent on the angle, β, that the path makes with the horizontal and the height of the mass above the earth.

The strength of the gravity model comes from the law of mutual attraction of two masses

$$F_g = GM_1 M_2/R^2, \tag{4-42}$$

where F_g is the force of mutual attraction between the two masses whose centers are separated by the distance R. The constant of proportionality is

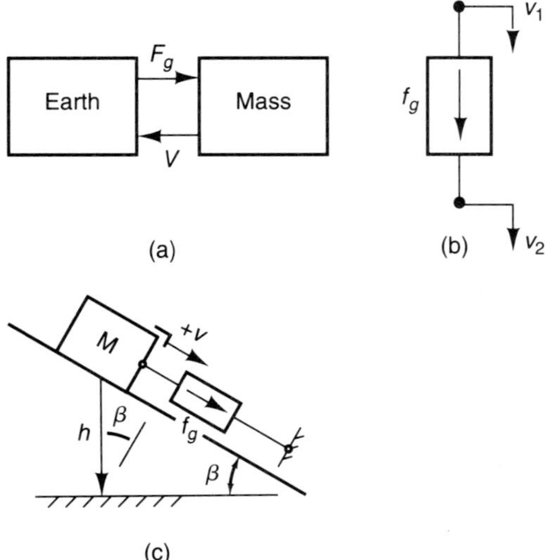

Figure 4-29. Gravity Modeling. (a) Multiport, (b) ideal symbol, (c) example

$$G = 6.67 \times 10^{-11} \text{ m}^3 \text{ kg}^{-1} \text{ s}^{-2}.$$

If motions are restricted to those near the surface of the earth then

$$R \approx R_{earth}.$$

Also,

$$M_2 = M_{earth}.$$

The gravitational equation thus reduces to the familiar *idealized gravitational model*

$$f_g = W \sin\beta = Mg \sin\beta. \tag{4-43}$$

In this text,

$$g = 32.2 \text{ ft/s}^2,$$

or

$$= 9.81 \; m/s^2.$$

Note that you could linearize (idealize) F_g about any R. And, as long as height changes (R changes) were small, then the ideal model would work quite well. If very large changes in R are expected to occur, then you may have to use a nonlinear model of gravity to achieve satisfactory accuracy in predictions.

Torque sources are similar to force sources in that they are designed to supply torque as independently as possible from angular speed at the output port. Unfortunately, the mechanical design problem for these sources is extremely complex, and this class of sources is very nonideal. Also, unfortunately, this type of structural source is very common; in fact, most *prime movers* are regarded to act as torque sources.

We will focus on internal combustion engines as the main interest in this study of torque sources. However, the general torque-speed characteristics of nuclear, steam, and jet engines are all very similar, needing only slight changes to model preliminary system designs.

The modeling of a torque source (e.g., a prime-mover engine) is shown in Fig. 4-30. Here the engine takes in chemical fuel flowrate, Z_f (sl/s), and converts it to output torque, Q (ft-lb$_f$), as a function of input load speed, N (rad/s).

The performance curves shown in Fig. 4-30a are created by varying the load on the source while holding the fuel flowrate constant. When the load and engine come to a steady condition, then a data point is taken. Thus, Fig. 4-30 shows only *steady operating point loci*. Notice that none of the fuel loci go through the origin of the graph—this occurs since the engine has start-up transients, none of which are steady conditions. Further, if too low a load speed is attempted for the chosen fuel flowrate, then the engine stalls and the output torque goes to zero, again in a transient fashion.

However, the origin of the idealized model for the engine must be chosen so that it is at a reachable point among a field of steady operating conditions. Further-

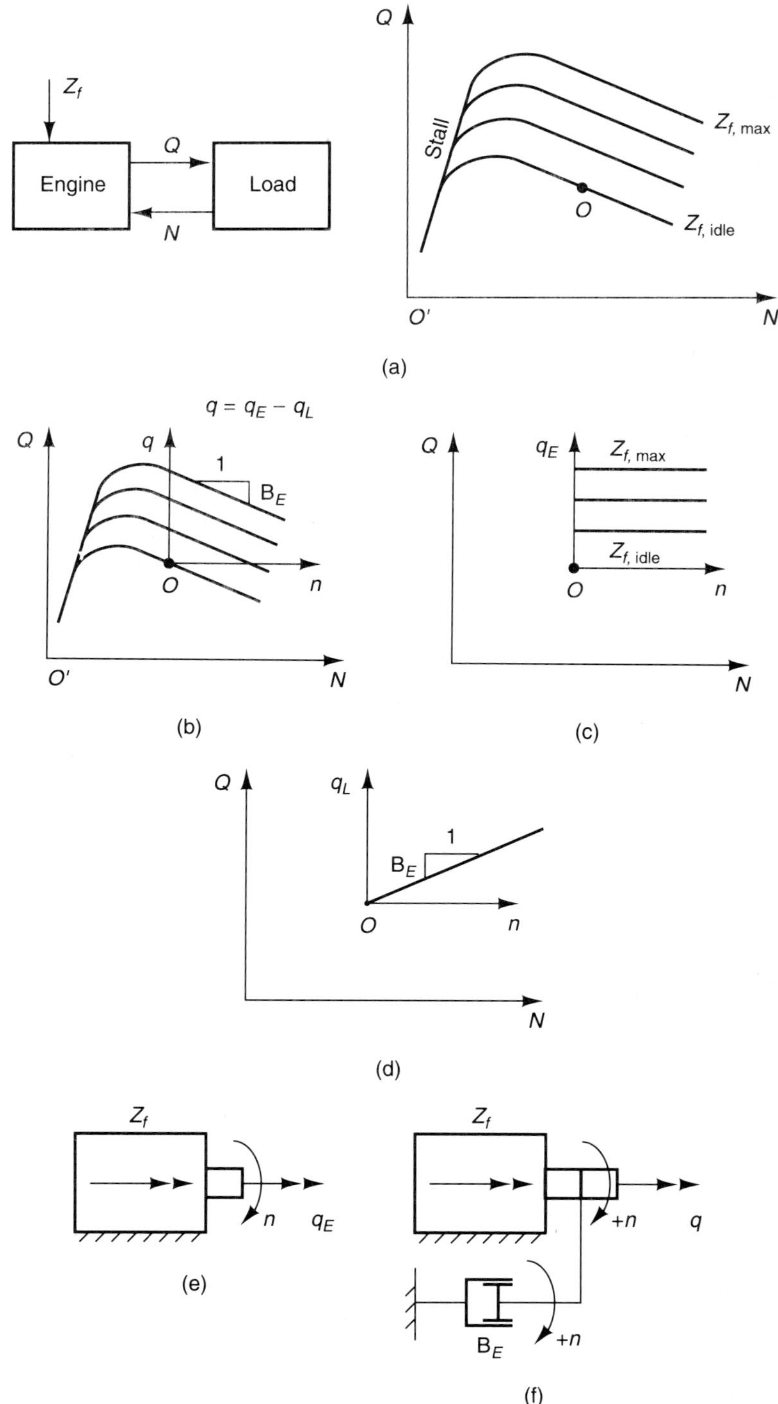

Figure 4-30. The Torque Source. (a) Multiport and performance, (b) linearization origin, (c) ideal engine, (d) droop damper, (e) ideal engine symbol, (f) idealized engine model

more, the origin must be chosen so that it is also a steady operating point for the load; that is, the origin for the idealization is a steady operating point for the entire connected system. You want to linearize the engine performance for some local region about this point. Suppose O is such a point on the "idle" curve. The linearization about O is called an *offset linearization*, and it amounts to constructing the q-n axes as shown in Fig. 4-30b. Note that each of the fuel loci are assumed to have the same slope, $-B_E$, in the region of O.

In Figs. 4-30c and d, the offset linearization is idealized into two contributing effects: an ideal source, q_E, and a damper-like droop term, q_L, such that,

$$q = q_E - q_L,$$

and

$$= q_E - B_E\, n.$$

The *ideal engine* characteristic is shown in Fig. 4-30c, note that it has no droop. That is, for a given fuel flowrate, the ideal engine produces torque independent of load speed. The characteristic for the droop-modeling damper is shown in Fig. 4-30d. Note the similarity in this modeling to that used for real batteries discussed earlier in Sec. 4-3.

The symbol for an ideal engine is shown in Fig. 4-30e. Note the "Z_f" above the rectangle, and the double-headed arrow inside to indicate a torque source dependent on fuel flowrate. The source is shown here connected to "ground" but this depends on the problem of interest. For example, an auto engine is a prime mover, but it is connected to a movable frame.

The model for an *idealized* engine with droop is shown in Fig. 4-30f. Here, the damper is used to draw off some of the ideal engine torque to model the torque droop in q, as observed in Fig. 4-30b.

Example 4-4

Consider the three-element structural component of Fig. 4-31. This component is comprised of an engine connected to a flywheel, which is connected to a water-filled dynamometer. The performance of each of the elements is shown in the figure. The problem is to determine the steady operating points for this system to prepare for linearization, and to sketch out the idealized model.

Solution

The first thing to do is to find a steady operating point for the system so that linearizations may be computed.

To find the steady operating points, you can resort to Newton's Law for the flywheel. Since the wheel is in steady motion,

$$Q_E = Q_D.$$

This realization means that you can plot the two data sets on the same axes, as in Fig. 4-32. The overlay plot of steady data (Fig 4-32a) shows that idealization axes may be selected at the those points where the two data sets intersect (e.g., at O_A,

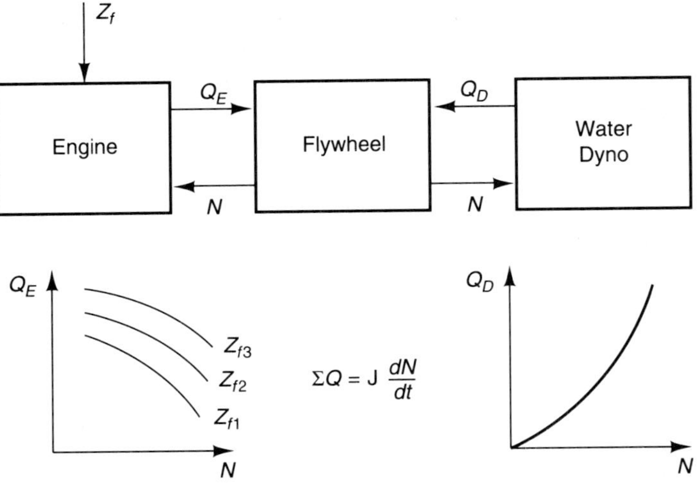

Figure 4-31. Example Structural Component

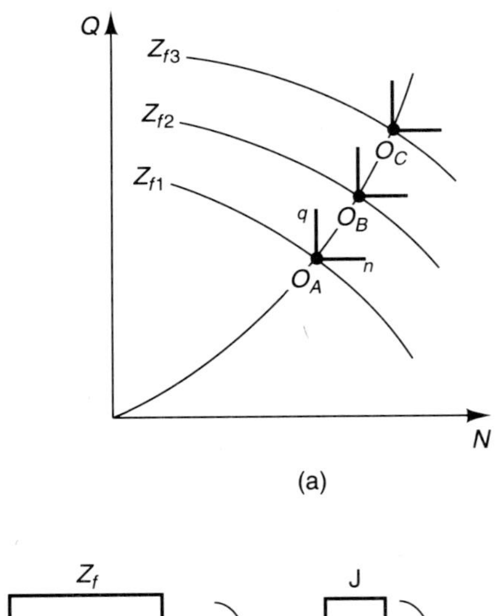

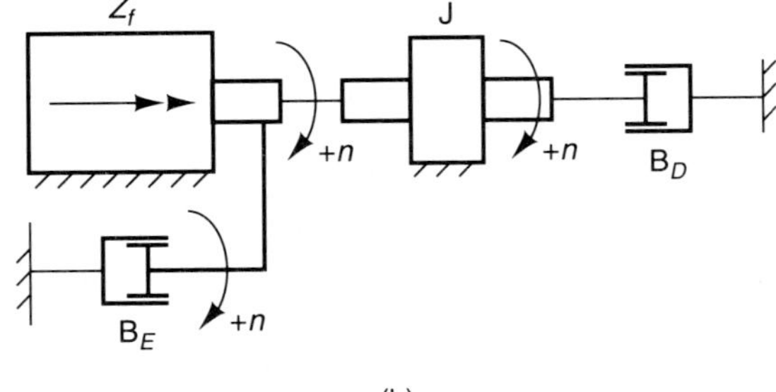

Figure 4-32. Example Structural Idealization. (a) Idealization points, (b) idealized model

O_B, and O_C). These are the steady operating points for this system. For this example, the origin is chosen at O_A. In general, you should choose an idealization origin which is closest to the steady conditions of interest to the designer. Thus, you can calculate all the necessary linearizations in the region of this new origin.

To find a proper ideal model of the engine means that you have to account for the torque droop (remember, the ideal source has constant torque for all input speeds, at a given fuel setting). In this case, you may use the approach discussed earlier and shown in Fig. 4-30f, where the engine torque is modeled as a combination of an ideal-engine torque and an idealized damper.

The component model is shown in Fig. 4-32d. Here all the ideal elements are shown, including the positive conventions and distinct labeling of all the speeds. Note that the internal torques are not shown on the model diagram. The calculations for the parameters in this model (B_E, J, B_D) are addressed in a homework problem.

Comments

This approach to finding the steady system operating point is called *input/output matching* or *impedance matching*. When you do this, you *must* match *all* the component input/output characteristics for the system. You must find a *system* steady operating point for the linearization.

In this problem a graphical method is used to perform the impedance matching. In a later chapter you will study how to convert the steady performance data into equations which are used to do mathematical input/output matching, and you will learn how to use steady data in the modeling of dynamics. But for now, this graphical procedure offers a convenient means of finding the steady operating points.

The *kinematic sources* for translational and angular speed appear in many forms. For example, under the right conditions, a roadway surface is a kinematic input to a tire which is passing over it. But there are times when the same roadway may be very soft to the tire—it thus becomes an impure kinematic input, offering some force input as well. A rotating cam is a kinematic input which creates a precise, scheduled, time-varying position for a cam follower in an engine. In any case, a general symbology and theory is not developed for this class of sources, they are simply indicated on the idealized structural models as a known inputs.

The Structural Transducers. These components/elements work to change power from one type to another. An element which changes rotational-structural power to DC-electrical power is the *DC generator*. An element which changes DC-electric to rotational-structural power is the *DC motor*. Also, rotational-structural power is converted back and forth between translational power in rack-and-pinion-gears, cams-and-followers, and in crankshaft-and-connecting-rods. Each of these transducers is studied in this section.

The operation of a DC motor is shown in Fig. 4-33. This type of motor is always driven by a DC current source, and it always has a torque output to the load, as shown in Fig. 4-33a. The motor acts as a device which takes an electrical current input into its uninsulated wire windings. Some of these windings are about an output

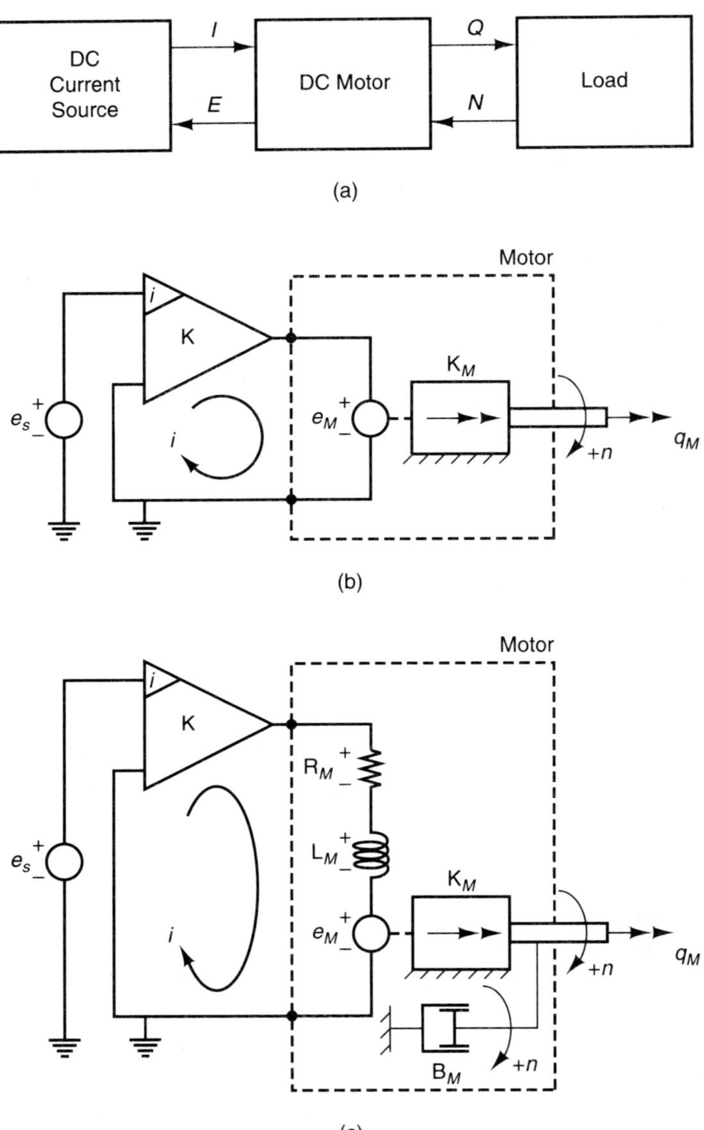

Figure 4-33. The DC Motor. (a) Multiport, (b) lossless model, (c) lossy model

shaft, and some are about an opposing housing. The motor then produces an inductive, electromagnetic force (emf) between the winding sets to create torque about the output shaft. The load is thus driven by the current input to the motor.

However, just as the load windings are moved by the electromotive torque exerted on them, so also does the load induce a voltage back in the source windings. This "back emf" is proportional to the speed of the load.

In Fig. 4-33b a motor model is shown connected to an electronic drive circuit. The motor model contained in the dashed box is called the *ideal-lossless motor model*. It neglects the losses and dynamics which occur in the induction windings of a real machine. Note that the motor back emf is shown as a voltage source, e_m, which *opposes* the direction of the input current flow, i, thus reflecting the load back to the driving current. In other words, the current source must work *against* the back emf to drive the load.

Note the similarity of the torque output symbol to that for the prime mover (a torque source), but in this case with "K_M" above the symbol. The interdependence between the back emf and the torque output is represented by the dashed line between the two source symbols.

In the physical analysis of the ideal motor, it is assumed that the electrical power, $\mathcal{P}$, into the machine is equal to the structural power output. That is,

$$\mathcal{P} = Ce_M i = q_M n, \tag{4-44}$$

where C is a units conversion factor, C = 0.7376 [ft-lb$_f$/w-s]. Note that this idealized model ignores power losses.

The two output equations for the ideal motor are extracted from Eq. 4-44,

$$q_M = Ci\text{K}_M, \tag{4-45}$$

$$e_M = n\text{K}_M, \tag{4-46}$$

where the units on K_M are v-s.

If a more realistic model is desired, then that of Fig. 4-33c should be used. In the realistic case, the motor losses and dynamics are more thoroughly accounted for, hence it is called an *ideal-lossy motor model*. The figure shows that winding resistance and inductance dynamics have been included on the electric side, and an ideal damper has been added to account for output torque droop on the structural side. An ideal inertia could also be added to account for internal motor inertia, but that inertia is usually negligible compared to the rest of the load structural dynamics.

A *DC generator* works like a motor with reverse causality, as shown in Fig. 4-34. Now the prime mover is controlled to keep the flywheel speed, N, *constant*. This keeps the output *voltage* constant for an ideal generator. The ideal generator thus has a voltage output which is directly proportional to its speed according to

$$e_G = n\text{K}_G. \tag{4-47}$$

Note that this equation is the same as Eq. 4-46, except for the constant, K_G. This is because the ideal generator must conserve power just as the ideal motor does.

Similarly, the torque output from the ideal generator is thus

$$q_G = Ci\text{K}_G. \tag{4-48}$$

The units for the generator equations are exactly the same as for the motor.

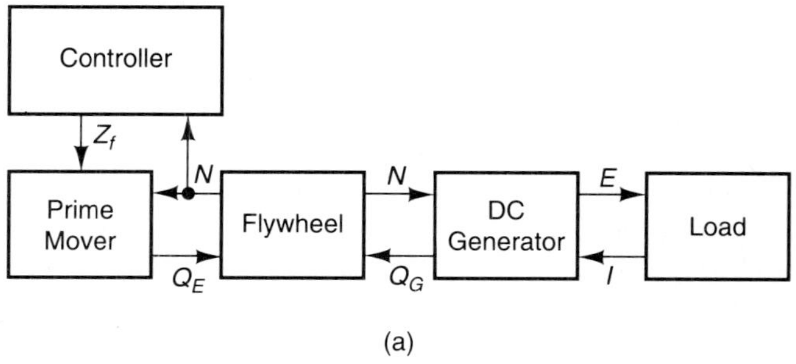

(a)

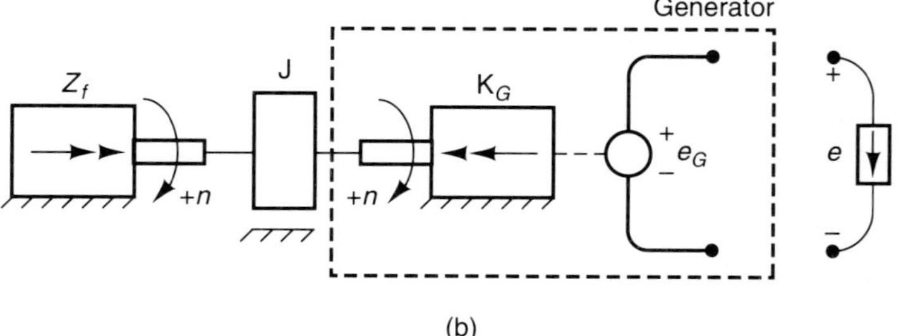

(b)

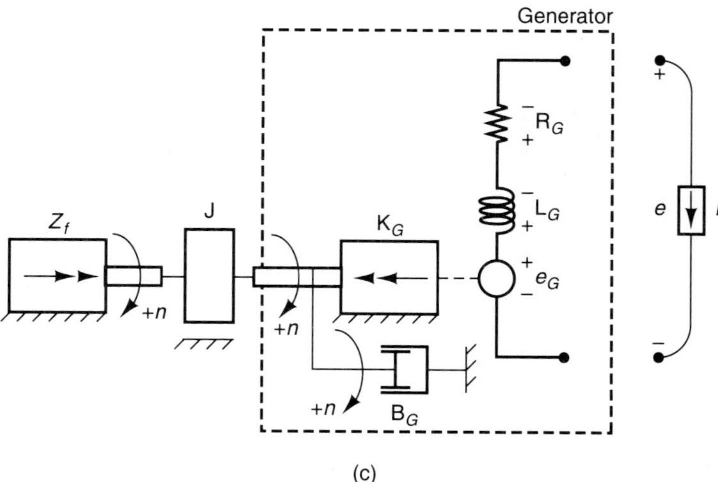

(c)

Figure 4-34. The DC Generator. (a) Multiport, (b) lossless model, (c) lossy model

The *ideal-lossless model* for the DC generator is shown in Fig. 4-34b in the dashed box. Notice that the torque source is here labeled with the generator constant, K_G. Notice also that the generator voltage and the port voltage are usually assumed to be in opposition, with a positive current coming out of the "+" side of the generator. Thus, the generator drives the load.

The *ideal-lossy* model of the DC generator is shown in Fig. 4-34c. The winding dynamics and the input torque droop are once again added to achieve this model.

The remaining structural transducers all work to change rotational power into translational power, or vice versa. They are shown in Fig. 4-35.

The rack-and-pinion (Fig. 4-35a) is very commonly used to translate oscillating motions, as in a sewing machine. Belts and pulleys (Fig. 4-35b) are used both as transducers (in power sanders, moving walkways, etc.) and as a type of gearing (in

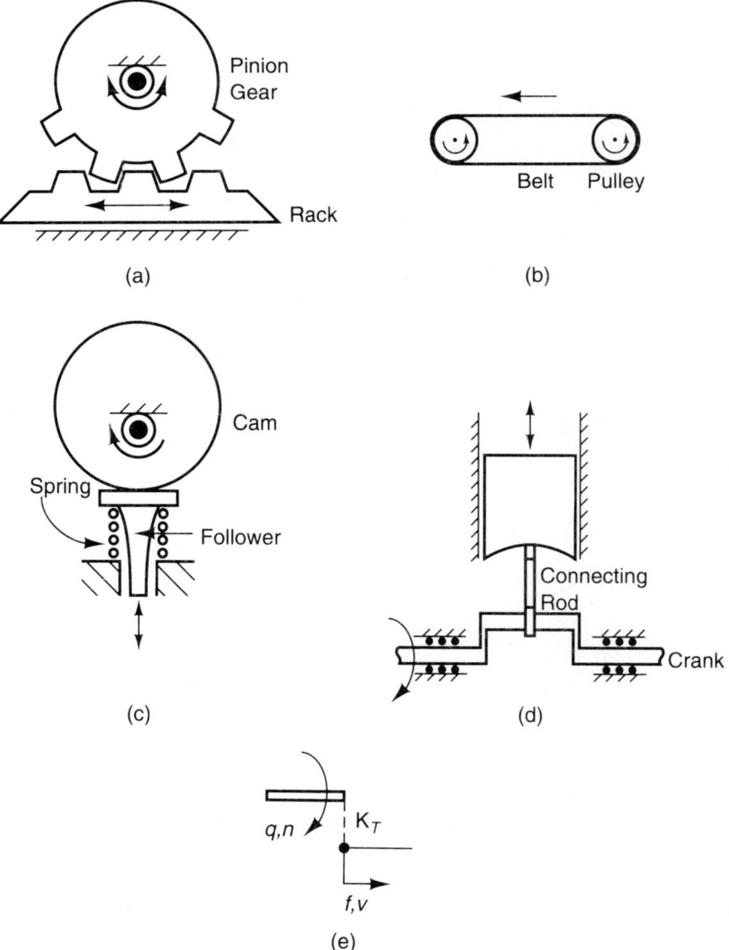

Figure 4-35. Structural Transducers

automobile engines). The cam-and-follower (Fig. 4-35c) is used in engines of all types to control the timing of valve openings and closings. The crank-and-connecting-rod (Fig. 4-35d) is used in pumps and engines of all types.

The general symbology for an ideal rotational/translational transducer is shown in Fig. 4-35e. The *gain* of the transducer, K_T, is computed from the kinematic relationship between input and output. For example, the rack and pinion gear has,

$$v_{RACK} = r\, n_{GEAR} = K_T\, n_{GEAR}.$$

Thus, $K_T = r$.

The Structural Amplifiers. In order to match a structural power source with its structural load, mechanical designers must often use a structural amplifier. These are recognized as *levers* and *gears*. They can be used to laterally transfer power, but they are more often used to *passively* amplify one of the port variables for load matching.

The operation of the *lever* is shown in Fig. 4-36. Note that the torque resultant about the pivot, P, must always be zero since this is a *massless* lever. Any nonzero torque thus results in an infinite angular acceleration, so the figure must be drawn in equilibrium. A *rigid* and *frictionless* lever is also assumed to give the idealized model.

The physical analysis of the ideal lever begins with establishing the notation as shown in Fig. 4-36a. Note that the positive x-y sign convention must be used at *both* ends of the lever. The analysis begins with summing the torques about P,

$$\Sigma_P Q = 0 = f_1 \ell_1 \cos\beta - f_2 \ell_2 \cos\beta. \qquad (4\text{-}49)$$

Or,

$$f_2/f_1 = \ell_1/\ell_2. \qquad (4\text{-}50)$$

Further, the kinematics gives

$$d(\ell_2\beta)/dt \cos\beta = -v_2, \qquad (4\text{-}51)$$

and

$$d(\ell_1\beta)/dt \cos\beta = v_1, \qquad (4\text{-}52)$$

which can be combined to give

$$v_2/v_1 = -\ell_2/\ell_1 = -1/K_L. \qquad (4\text{-}53)$$

To see how these equations work, consider $K_L = 1/3 = \ell_1/\ell_2$ (roughly the case of Fig. 4-36a), and assume that 1 is the input port with 2 the output port. In this case, v_2 gets bigger than v_1 and f_2 gets smaller than f_1. That is, this lever *amplifies* v_1 at the expense of f_1.

In terms of the multiport diagrams of Fig. 4-36b, the two possible cases of causality for the lever are shown. Either causality may be true depending on the nature of the connected elements. The equations above are true for both causalities, thus you may chose either causality to model the same lever geometry, depending on the intent of the design.

The symbol for the idealized lever is shown in Fig. 4-36c, along with the governing equations. Again, note that the sign convention regarding positive speed and

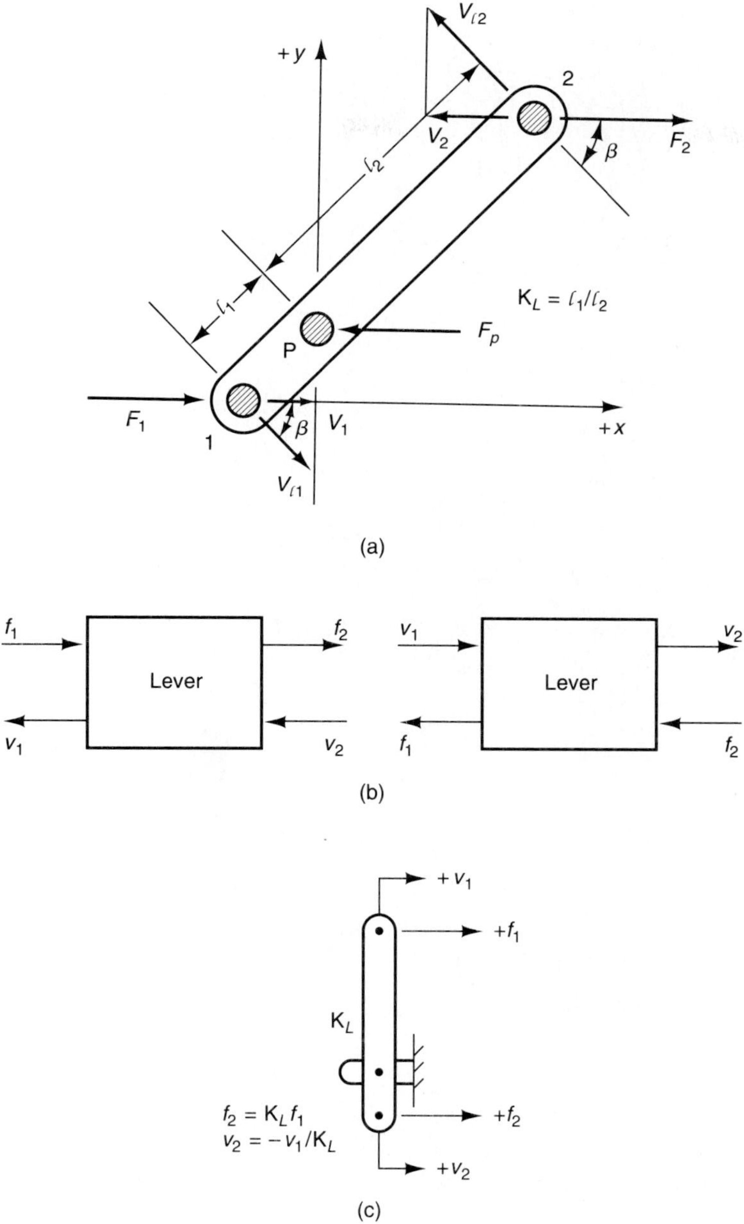

Figure 4-36. The Lever. (a) Notation, (b) multiports, (c) ideal symbol

force directions applies at *both* ends of the lever. Thus, the lever acts to reverse the sign of the speed. You must be careful to maintain your positive convention because you will eventually combine equations from both sides of the lever during a component analysis.

A *gear train* is shown in Fig. 4-37a. While only two gears are shown on this

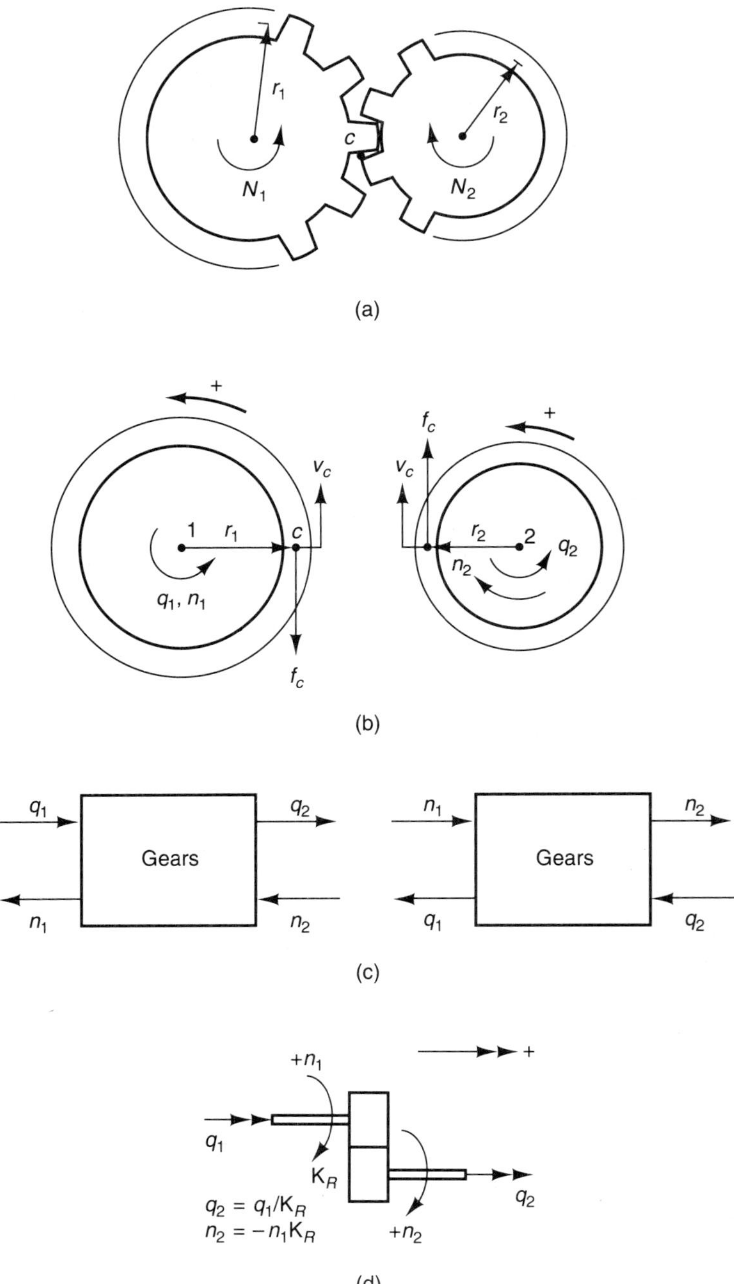

Figure 4-37. The Gears. (a) Geometry, (b) free bodies, (c) multiports, (d) ideal symbol

figure, the procedure for analysis can be repeatedly applied for as many gears as given in the design.

The gears have been separated into free bodies in Fig. 4-37b to begin the physical analysis. Note that the analysis depends on your model of what is happening at the contact point, C. At that point, each gear is subjected to an equal and opposite contact force, f_C. Also at that point, both gears must have the same tangential speed, v_C. Once more, a *massless*, *frictionless*, and *rigid* gear is ideally assumed. Thus, summing torques about point 1 and 2 gives

$$\Sigma_1 Q = 0 = q_1 - f_C r_1, \tag{4-54}$$

and

$$\Sigma_2 Q = 0 = q_2 - f_C r_2. \tag{4-55}$$

These can be combined to give

$$q_2 = q_1/\mathrm{K}_R, \tag{4-56}$$

with the definition

$$\mathrm{K}_R = \mathrm{r}_1/\mathrm{r}_2. \tag{4-57}$$

Also, since the tangential velocities must be the same,

$$v_C = r_1 n_1, \tag{4-58}$$

and

$$v_C = -r_2 n_2. \tag{4-59}$$

These also can be combined to give

$$n_2 = -n_1 \mathrm{K}_R. \tag{4-60}$$

Again one port variable is amplified at the expense of another by using the gear train. The two possible cases of causality are shown in Fig. 4-37c; the case selected depends upon the connected elements.

The ideal symbol for the two-gear train is shown in Fig. 4-37d, along with the governing equations. Gears are usually massive and will often have associated inertia elements. Note that the positive sign convention for speed must be preserved at both ends of the gears, just as for the lever.

HOMEWORK

4-23. A systems engineer wants to design the system shown in Fig. 4-38. The current source feeds a DC motor as shown. The design engineer has chosen a gear pair with $\mathrm{K}_R = 0.1$ to match the motor output to the torque rod shown on the right. The torque rod acts as an ideal spring with rate K for the length ℓ. At the midpoint of the rod a lever arm of length r is attached which is to be used as a transducer for small translational motions. For this component, answer the following:

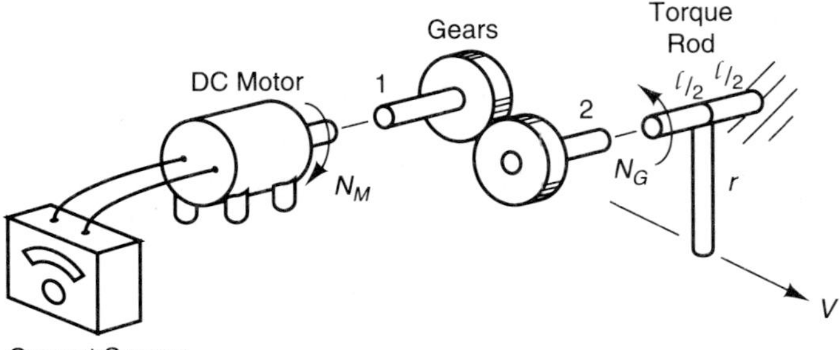

Figure 4-38. Rotational System for Prob. 4-23

a) What is the transducer gain, K_T, which relates the torque rod speed, N_G, to the translational speed, V?

b) Since r_2 is so much bigger than r_1 for the gear set, the modeler wants to model the r_2 mass, but not the r_1 mass. Sketch a lossless component model for this component.

c) Sketch the lossy component model which corresponds to (b) above.

4-24. A mass is resting on a spring, which is resting on the earth. Sketch an ideal model of this simple system showing gravity as a source.

4-25. An engine-dynamometer component which is under study is shown in Fig. 4-39. For this component, complete the following:

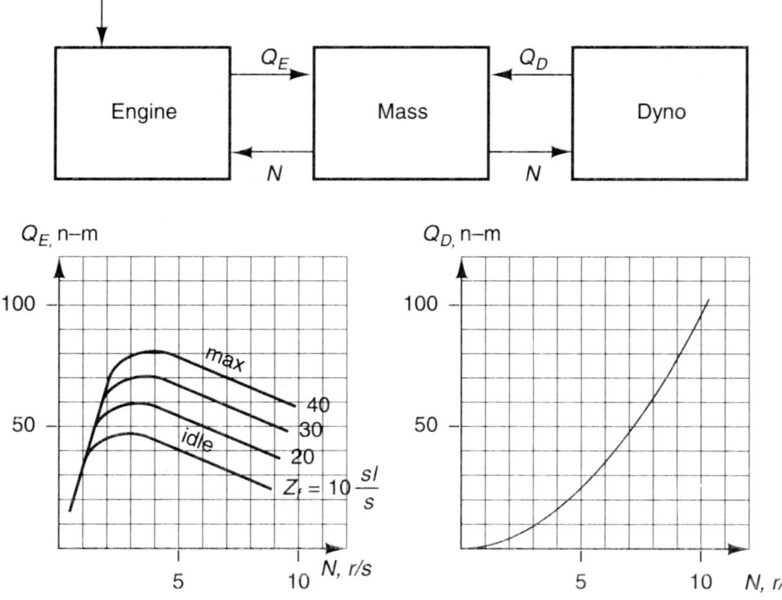

Figure 4-39 Engine System for Prob. 4-25

a) Use the figures to graphically approximate the idealization point, O, which lies on the idle curve.

b) Build an ideal model of Q_E using an ideal source q_E, and an ideal rotating damper, B_E (Hint: assume that all the fuel curves have the *same* negative slope.) Draw the q_E versus n plot for your ideal source, and compute the value of B_E.

c) Sketch your idealized structural source model.

4-26. Answer the following for the DC motor:

a) What is "back emf?"

b) Why is it necessary to include a resistance and an inductance to model a lossy motor?

c) Why does the motor produce torque, not speed?

4-27. Why is a controller necessary in the DC generator system?

4-28. A rack and 3 in. pinion gear are to be used in a machine. What is K_T for this component?

4-29. A certain lever has $K_L = 2$, and is 3 inches long. The pivot is fixed. Answer the following,

a) Sketch the geometry of the lever.

b) If the end of the short arm is moving tangentially at 3 m/s around the fixed pivot, how fast is the end of the long arm moving?

c) If you push on the end of the long arm at 2 newtons, what is the force exerted at the end of the short arm?

4-7 THE STRUCTURAL-ANALOG MODEL

Given an idealized model of structural physics coming from the methods of the past two sections, the next concern is to build the structural mathematical model. The two fundamental approaches which are generally used to do this are *free-body diagramming* or *analogies*. However, whichever method you use to find the equations of structural dynamics, there will only be *one* set of these equations. Thus, since free-body diagrams are thoroughly covered in other texts, analogies are discussed in some detail in this text.

The analysis plan using analogies is in three steps. First, an *electrical-network analogy* of the structural model is created. This model is called the *structural-analog model*. Second, the powerful analysis methods of Kirchhoff are used to find the corresponding output equations for the analog model. And third, the electrical analog variables are translated into structural variables in order to find the structural output equations.

In this section, the electrical analogies for the ideal structural elements (the mass, damper, and the spring) are first defined. This is followed by examining the analogies for the structural sources, transducers, and amplifiers. The creation and solution of network analogs is then discussed. The section is completed with a discussion of reversing the analogy to obtain the output equations for the structural component.

The Ideal Element Analogies. The fundamental basis for the analogies in this text comes from a comparison of the *dissipative elements* in each of the two types, electrical and structural. Recall that these are the elements whose primary function is to model the dissipation of energy in their respective components. They are the *resistor* and the *damper*. These elements are shown, along with the other ideal elements, in Fig. 4-40. Other types of analogies are possible, and even common. Still, the analogy used here is based in the element energy behavior and it thus has far-reaching consequences in terms of consistency and interpretation of results.

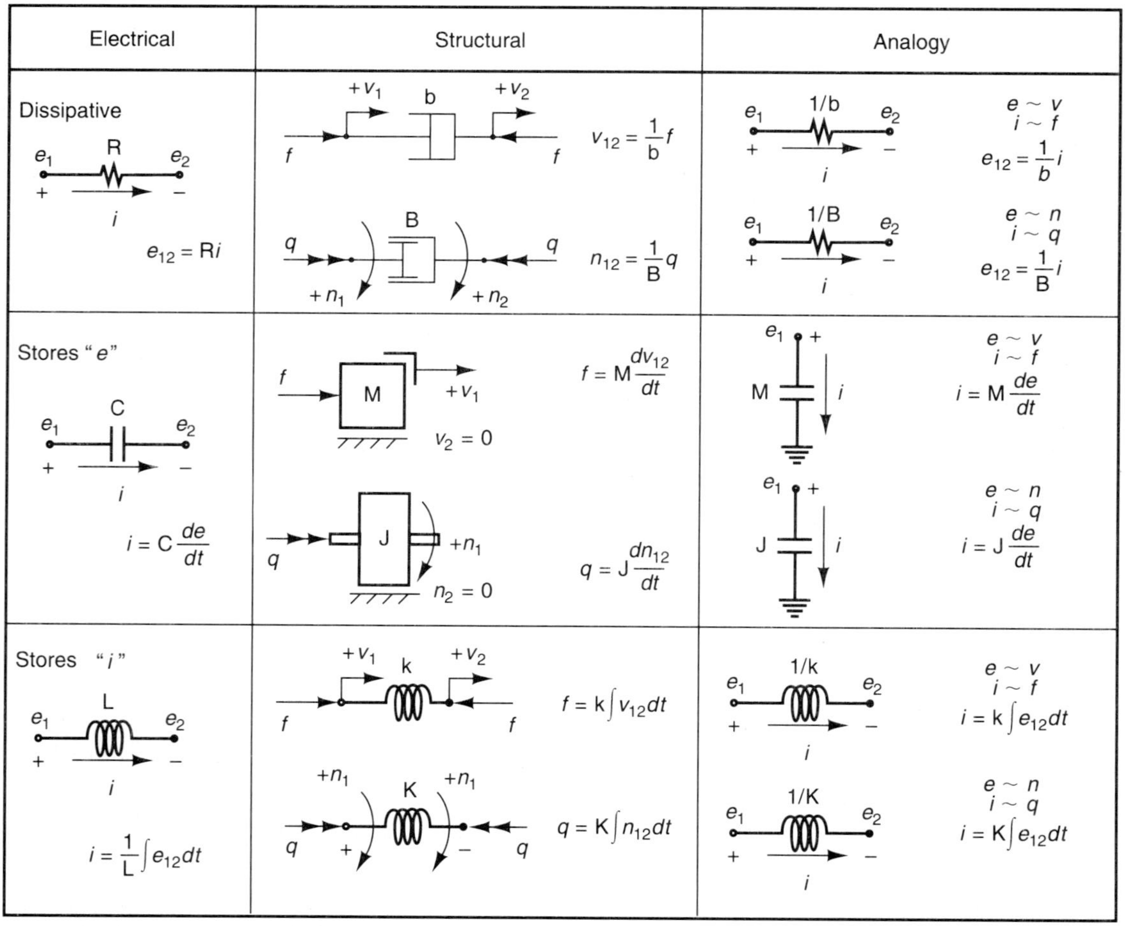

Electrical	Structural	Analogy
Dissipative R, e_1, e_2, +, −, i $e_{12} = Ri$	$+v_1$, b, $+v_2$, f, f: $v_{12} = \frac{1}{b}f$ B, q, q, $+n_1$, $+n_2$: $n_{12} = \frac{1}{B}q$	1/b, e_1, e_2, +, −, i: $e \sim v$, $i \sim f$, $e_{12} = \frac{1}{b}i$ 1/B, e_1, e_2, +, −, i: $e \sim n$, $i \sim q$, $e_{12} = \frac{1}{B}i$
Stores "e" C, e_1, e_2, +, −, i $i = C\frac{de}{dt}$	f, M, $+v_1$, $v_2 = 0$: $f = M\frac{dv_{12}}{dt}$ q, J, $+n_1$, $n_2 = 0$: $q = J\frac{dn_{12}}{dt}$	e_1, +, M, i: $e \sim v$, $i \sim f$, $i = M\frac{de}{dt}$ e_1, +, J, i: $e \sim n$, $i \sim q$, $i = J\frac{de}{dt}$
Stores "i" L, e_1, e_2, +, −, i $i = \frac{1}{L}\int e_{12}dt$	$+v_1$, k, $+v_2$, f, f: $f = k\int v_{12}dt$ $+n_1$, K, $+n_1$, q, +, −, q: $q = K\int n_{12}dt$	1/k, e_1, e_2, +, −, i: $e \sim v$, $i \sim f$, $i = k\int e_{12}dt$ 1/K, e_1, e_2, +, −, i: $e \sim n$, $i \sim q$, $i = K\int e_{12}dt$

Figure 4-40 The Ideal Element Structural Analogies.

The ideal electrical resistance symbol is shown at the top left of Fig. 4-40, along with its constitutive relationship

$$e_{12} = \mathrm{R}i. \tag{4-61}$$

The structural dampers are shown in the top center, along with their constitutive relationships

$$v_{12} = (1/\mathrm{b})f, \tag{4-62}$$

and

$$n_{12} = (1/\mathrm{B})q. \tag{4-63}$$

The obvious similarities between these equations gives rise to the *analogous variables* for these dissipative elements.

Note first that the terms on the left side of the last three equations are differences measured "across" the elements—they are therefore called *across variables* by systems engineers. Note also that the variables on the right side of the equations pass unchanged "through" the elements—they are therefore called *through variables* by systems engineers. The systems analogy is thus

across variables: $v_{12} \sim n_{12} \sim e_{12}$,

through variables: $f \sim q \sim i$.

So, the structural-analog, *analogous dissipative element* is adopted as shown in the top right of Fig. 4-40 whenever structural dampers are seen. Note that the analog equation is obtained by using the electrical equation with the structural parameter. Also, the electrical resistance R is replaced by the structural parameter (1/b) or (1/B) in the network diagram. This shows at a glance whether you have an electrical network or a structural-analog network.

The ideal "capacitor" is shown in the second row of Fig. 4-40. Recall that the electronic capacitor has the constitutive relationship

$$i = \mathrm{C}\ de_{12}/dt. \tag{4-64}$$

The analogous structural elements are shown in the center column, with their constitutive relationships

$$f = \mathrm{M}\ dv_{12}/dt, \tag{4-65}$$

and

$$q = \mathrm{J}\ dn_{12}/dt. \qquad \text{e(4-66)}$$

These elements are analogous since they all *store energy* by virtue of their *across variables*. Also, note the similarity in the equations. The through variables all appear on the left side and the across variables are all differentiated on the right side.

Recall that the idealized masses are inertial, with their speed measured with respect to a non-accelerating reference. This idea is shown in the structural element symbols using the ground reference indication. Further, this idea carries over into the electrical analogy since you *always* show the mass analogy *connected to electronic "ground,"* as at the far right of the second row of Fig. 4-40. Again, the equation form is taken from the electrical element, using the parameter from the structural element, and the electrical network symbol is labeled with the structural parameter.

The bottom row of Fig. 4-40 is for the "inductor" elements. All of these elements *store energy* by virtue of their *through variable*. Note the similarities of the element constitutive relationships:

$$i = (1/\mathrm{L}) \int e_{12}\, dt, \tag{4-67}$$

$$f = k \int v_{12}\, dt, \tag{4-68}$$

and

$$q = \mathrm{K} \int n_{12}\, dt. \tag{4-69}$$

This time though, all the through variables appear on the left side, and the right side has the integral of the across variables.

The Source Analogies. The source analogies are shown in Fig. 4-41. The *e type* sources are based on the analogy between across variables. This compares voltage sources to kinematic sources as shown in the figure. The kinematic source analogy thus appears as a voltage source labeled with the kinematic variable. Note that these analogous kinematic sources must always be connected to "ground" in order to show the role of the reference datum.

The *i type* sources are based on the analogy between through variables. Thus, the *force source* analogy is represented with an ideal current-source symbol, and labeled with the force magnitude.

Care must be exercised, however, to ensure that the direction of the analogous current-source reflects the true physics of the situation. That is, if the force is acting in the + *direction,* then the current source should be shown feeding *into* the analogous node. Note in Fig. 4-41 that the force is acting *opposite* to the positive speed, and the analogous source is pointed *away* from the node. Note also that the force source is here connected to "ground," allowing for an inertial reference. But the force source need not always be connected to ground.

The analogous *torque source* is also represented with an ideal current-source symbol since it too is a through-variable source. Again, if the torque is acting in the positive speed direction, then the current-source direction in the analogy is into the node (as indicated by Fig. 4-41).

Fig. 4-41 also shows an idealized model of a prime mover with some output droop, and its corresponding analogy. Note the similarity between the analogy and the ideal model of a idealized current-source studied earlier. This is just the sort of confirmation that you may expect with a fundamental analogy.

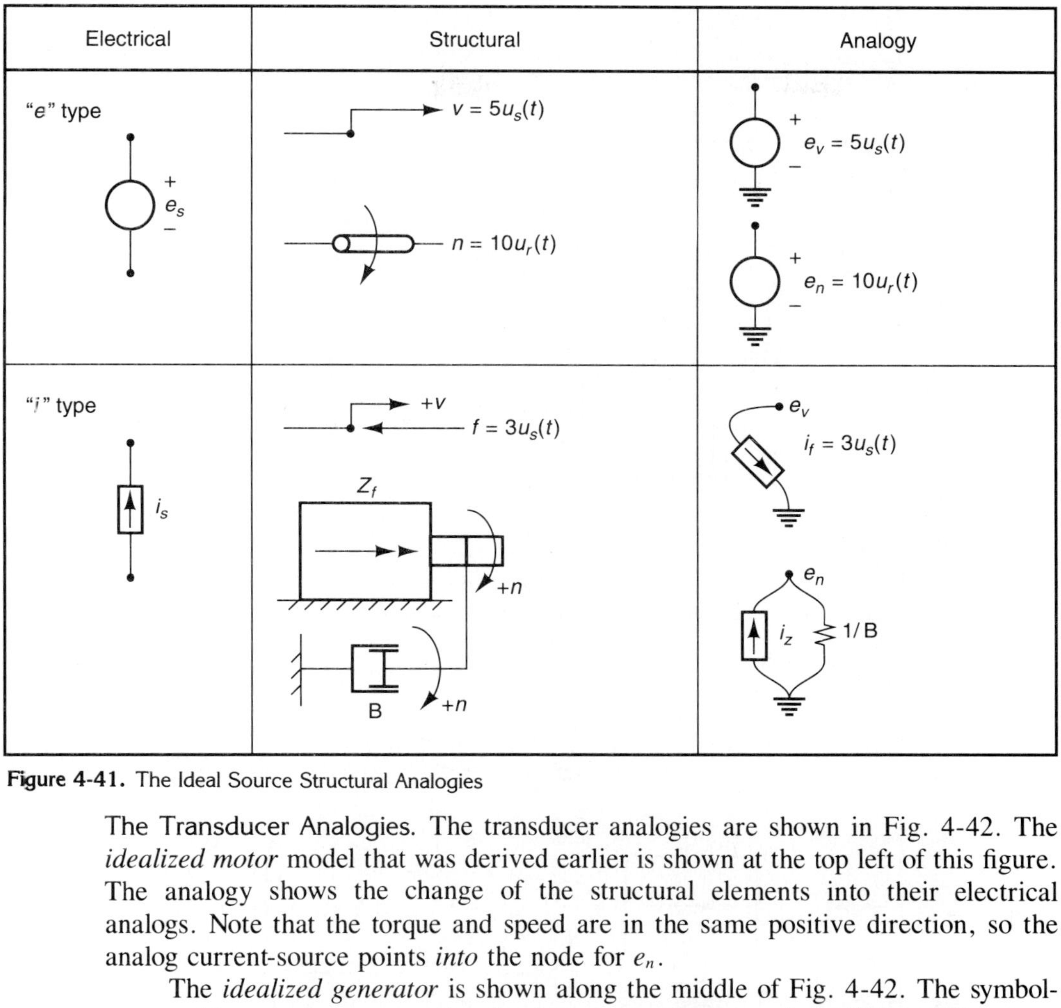

Electrical	Structural	Analogy
"e" type + e_s −	$v = 5u_s(t)$ $n = 10u_r(t)$	+ $e_v = 5u_s(t)$ − + $e_n = 10u_r(t)$ −
"i" type i_s	$+v$ $f = 3u_s(t)$ Z_f $+n$ B $+n$	e_v $i_f = 3u_s(t)$ e_n i_z 1/B

Figure 4-41. The Ideal Source Structural Analogies

The Transducer Analogies. The transducer analogies are shown in Fig. 4-42. The *idealized motor* model that was derived earlier is shown at the top left of this figure. The analogy shows the change of the structural elements into their electrical analogs. Note that the torque and speed are in the same positive direction, so the analog current-source points *into* the node for e_n.

The *idealized generator* is shown along the middle of Fig. 4-42. The symbology used here is the same as that for the motor, with several significant differences. First, notice that the torque source, K_M, is now applied opposing the input torque. This is because the torque source here represents the electrical load which the input torque must overcome. Thus, when the generator analogy is drawn, the analogous current source always points *away* from the node for e_n, as in the middle figure on the right.

The other difference with the motor analogy is that the generator operates in a reverse fashion to the motor; consequently, the structural analogy appears on the left rather than on the right as shown for the motor.

The idealized structural *translational/rotational transducer* is shown at the bottom of Fig. 4-42. However, since there are so many different types of this transducer, no specifics are given here other than noting that these analogs all appear as some form of coupled analogous sources of like type.

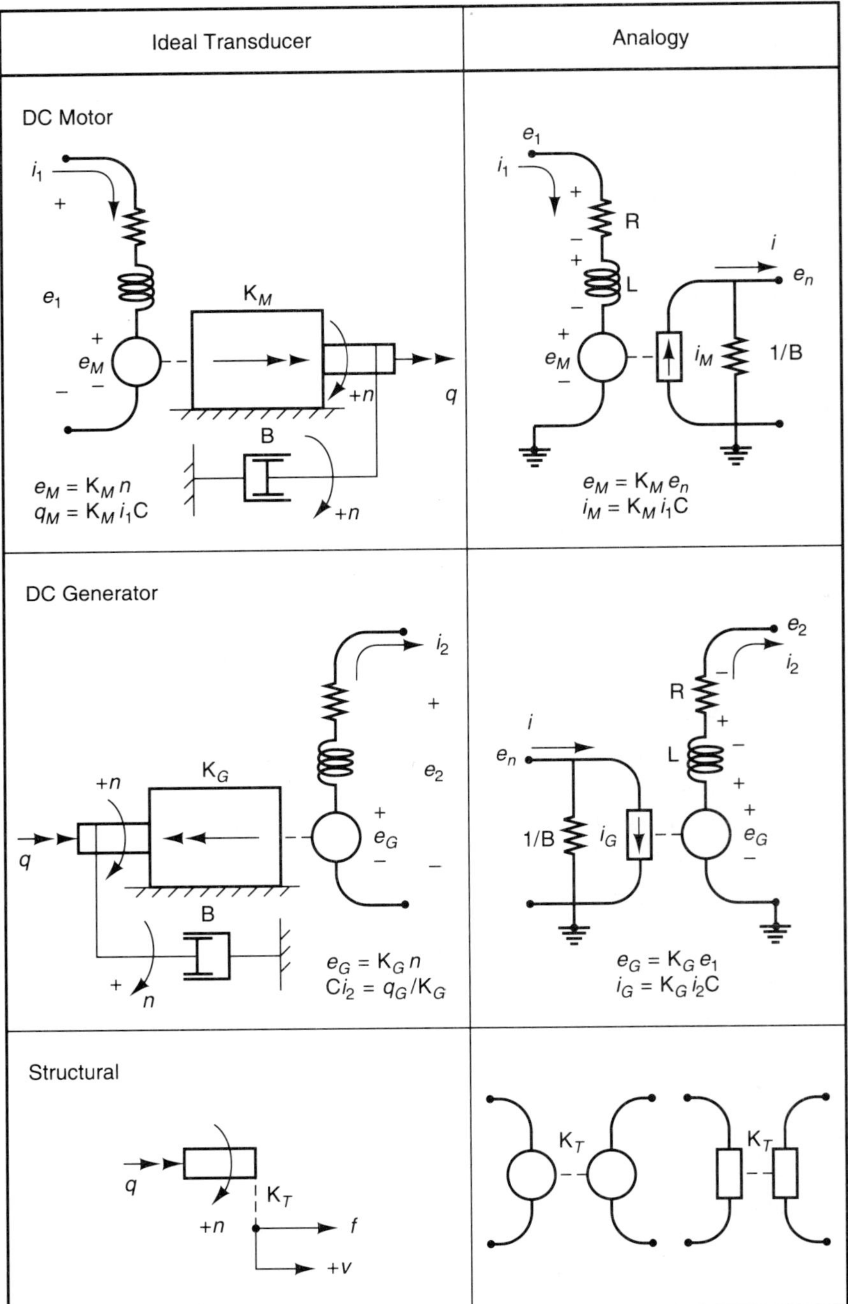

Figure 4-42. The Transducer Structural Analogies

The Amplifier Analogies. The two structural amplifiers and their analogies are shown in Fig. 4-43. Recall that the ideal electrical amplifiers are all *isolation* amplifiers, as shown at the left of Fig. 4-43. This allows you to separate two independent networks in the analysis. In the structural amplifiers, you also find amplification taking place, but along with it occurs a *mutual dependence* between the two "networks."

Thus, you find that you must always return to the multiport diagram in order to correctly interpret and apply the structural amplifier equations and find the electrical analog.

The *lever* is a translational amplifier, as shown in the top center of Fig. 4-43. Note that the equations have a positive direction associated with them (towards the right). Also note that there are only two possible cases of causality, as shown in the multiport diagrams. Thus, neither f_1 *and* f_2 nor v_1 *and* v_2 can be inputs since this causality cannot occur for the lever.

So, if f_1 and v_2 are the lever inputs, then you see the causality of the upper multiport. In this case, a *through* input is caused in the right circuit ($f_2 = f_1 K_L$), and an *across* input is caused in the left circuit ($v_1 = -v_2 K_L$). Thus, the analogy has a current source in the right circuit, and a voltage source in the left circuit.

Note that great care must be taken to ensure that the signs and variable assignments are consistent in the amplifier analogy. Consistent sets of variables are shown in Fig. 4-43. Note also that the sign reversals through the lever appear in the orientation of the two analog sources.

If v_1 and f_2 are the lever inputs then the lower multiport diagram of the lever has the proper causality. The adjacent analogy shows the consistent interpretation of this causality.

For the *gears*, there are again only two possible cases of causality as shown in Fig. 4-43. The inputs once more dictate the way the gears function, and the way they should be consistently shown on the analogous networks.

Thus, analogous electrical elements exist for each of the structural elements, and they may be used to formulate an analogous network. Given this network, you may correctly expect to use the methods of Kirchhoff to find the output equation(s), as discussed in Sec. 4-4 earlier.

Structural Network Analysis. Before the analog analysis is begun however, it is essential to examine the applicability of Kirchhoff's Laws to the analogous network.

Kirchhoff's Voltage Law (KVL) and Kirchhoff's Current Law (KCL) are used to create mathematical models for *electrical* idealizations. In general, both of these laws are needed to define the independent equations which describe the component output dynamics. Thus, if the analogy we are using is true, these laws must yield equations which could be found by another method, an *analogous method*.

In the case of KVL, you are summing voltages around a loop. According to the structural analogy, this would amount to summing speeds around a "loop" in a structure. Of course, you know from the study of dynamics that the speeds of various parts of a structure must meet certain kinematic compatibilites to be allowable dynamic motions. Thus, KVL and speed kinematic considerations are called *compatibility relationships* by systems engineers.

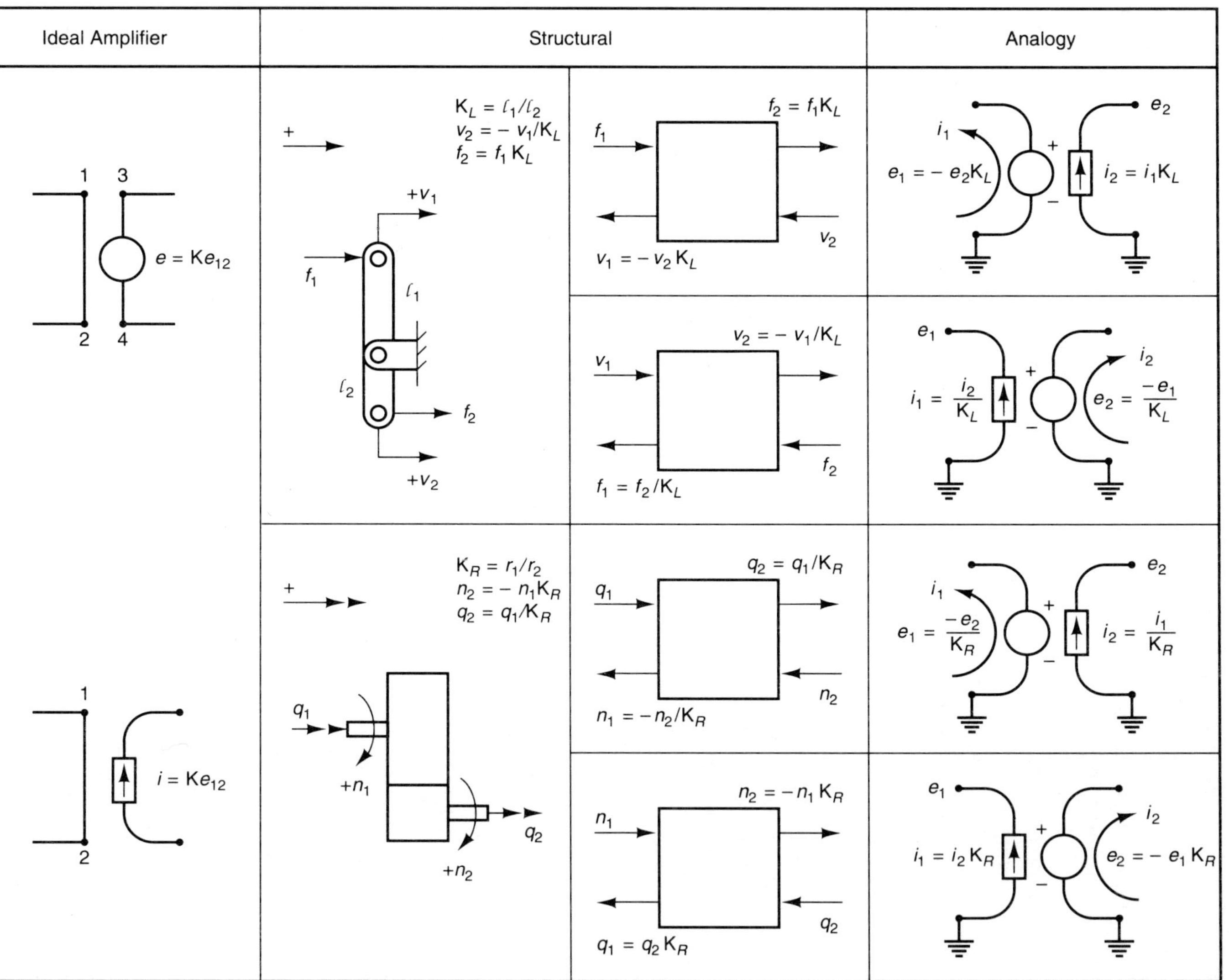

Figure 4-43. The Amplifier Structural Analogies

In the case of KCL, you are summing currents flowing into a node as you seek to maintain the principle of continuity of charge at that node. The structural analogy of this is summing forces into a structural "node." This is another view of the concept of load path discussed in Chap. 2. Systems engineers call these *continuity relationships* since they come from worrying about continuity of charge (or force).

Thus you find that the analogy holds for general and specific relationships. It holds on an element-by-element basis, and it allows you to use KVL and KCL to find the network equation set.

The final step to the method of analysis using analogy is to reverse the analogy of variables to find the structural equations of motion,

$$i \rightarrow f \quad \text{or} \quad q,$$

and

$$e \rightarrow v \quad \text{or} \quad n.$$

Example 4-5

Find the equation of motion for the idealized component of Fig. 4-44a using the method of analogy. The motion starts from rest against the compressed spring. Find the structural output equations. You may ignore the output f_R by assuming a rigid foundation at R.

Solution

The analogous network is shown at the right of Fig. 4-44a. Note that the current source is shown into node a since the force, f, acts in the positive v direction. Also note that gravity is not shown in the analogy since the motion is measured from the position where gravity and the spring offset each other. You are now ready to start the five-step analysis using Kirchhoff's laws.

Step 1: The notation is assumed as in Fig. 4-44b.

Step 2: KVL is written for the two meshes;
Mesh 1, there is a current source, don't write KVL.
Mesh 2, $e_m - e_k = 0$.
(Note that this last equation really says that the speeds of the mass and the spring must be the same for their motion to be *compatible*.)

Step 3: KCL is written for node a,

$$i_1 - i_m - i_2 = 0.$$

(This *continuity* equation says that the net force on the mass is equal to the difference between the force input and the restraining force of the spring.)

Step 4: There are only two element relations,

$$i_m = \mathrm{M}\, de_m/dt,$$

and

$$i_2 = \mathrm{k} \int e_k \, dt.$$

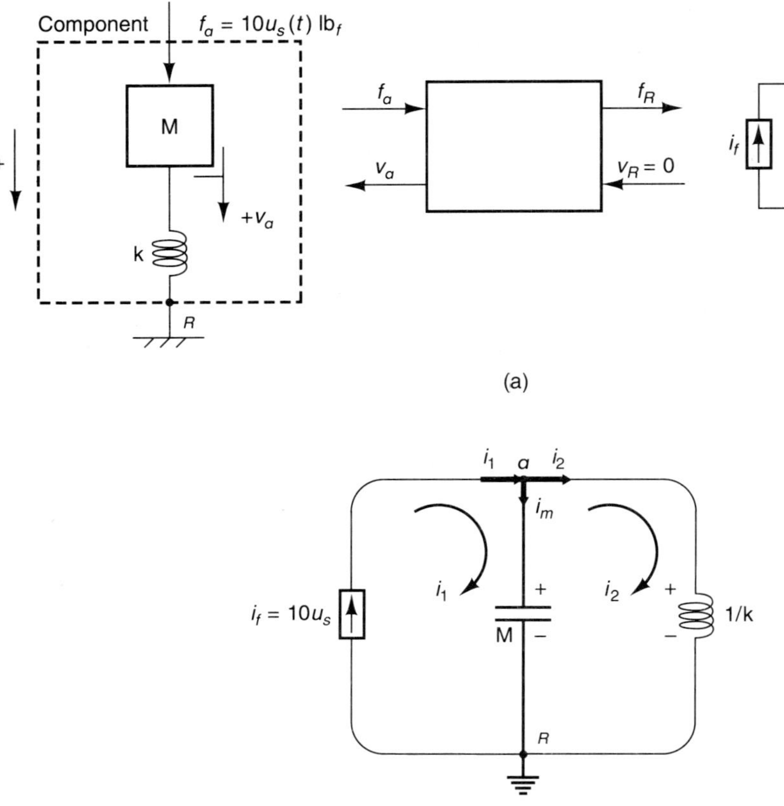

Figure 4-44 Example SISO Analysis. (a) Multiport and analog network, (b) notation.

Step 5: Equation manipulations. Since there is a current source, you should start with KCL at node a

$$i_1 - i_m - i_2 = 0,$$

and substitute the element relations

$$i_f - \mathrm{M}\, de_m/dt - \mathrm{k}\int e_k dt = 0.$$

Now you can use the compatibility relationship

$$e_m = e_k = e_a,$$

and substitute for the source

$$10u_s(t) - \mathrm{M}\, de_a/dt - \mathrm{k}\int e_a dt = 0.$$

Clearing the integral by differentiation gives

$$10u_i(t) - \mathrm{M}\, d^2e_a/dt^2 - \mathrm{k}e_a = 0.$$

(Note the change from step to impulse source from the differentiation.) Rearranging this gives

$$\mathrm{M}\, d^2e_a/dt^2 + \mathrm{k}e_a = 10u_i(t).$$

The final step is to reverse the analogy

$$\mathbf{M}\, \boldsymbol{d^2v_a/dt^2} + \mathbf{k}\boldsymbol{v_a} = \mathbf{10}\boldsymbol{u_i(t)}.$$

Comments

For this simple problem, you can see almost by inspection that the free-body diagram methods of dynamics would give the same equation of motion. However, for more complicated structures, especially those involving rotational dynamics, the analogy method is quite powerful.

Example 4-6

For a second and more complicated example, consider the structural component shown in Fig. 4-45a. Note that the structure has three inputs and three outputs. It has two kinematic inputs: v_o, at 0, and $v_a = 0$ at a; and it has an external kinetic input, f, at 3. Find the structural output equations. You may ignore the reaction, f_a, under the assumption of a rigid foundation.

Solution

The place where most analyses get confused in a structure like this is in the treatment of the lever. However, the analog method handles this very directly.

The causality of the lever is shown in Fig. 4-45b. Note that the *mass inputs must be forces* according to Newton's Second Law of Motion. Also note that the force at 3 must pass through the spring to the lever at 2. Both situations say that the causality of Fig. 4-45b is correct. Given the lever causality, you are in a position to sketch the analog model and begin the five-step analysis.

Step 1: Assign the notation of Fig. 4-45c. Note the precise agreement with Fig. 4-43 for the lever variables.

Step 2: Write KVL for the meshes;

Mesh 0, $e_0 - e_{01} - e_1 = 0$.

Mesh 1, no KVL due to current source.

Mesh 2, $e_2 - e_{23} - e_3 = 0$.

Mesh 3, no KVL due to the current source.

Step 3: Write KCL for the nodes;

Node 1, $i_0 - i_m + i_1 = 0$.

Node 3, $i_2 - i_b - i_3 = 0$.

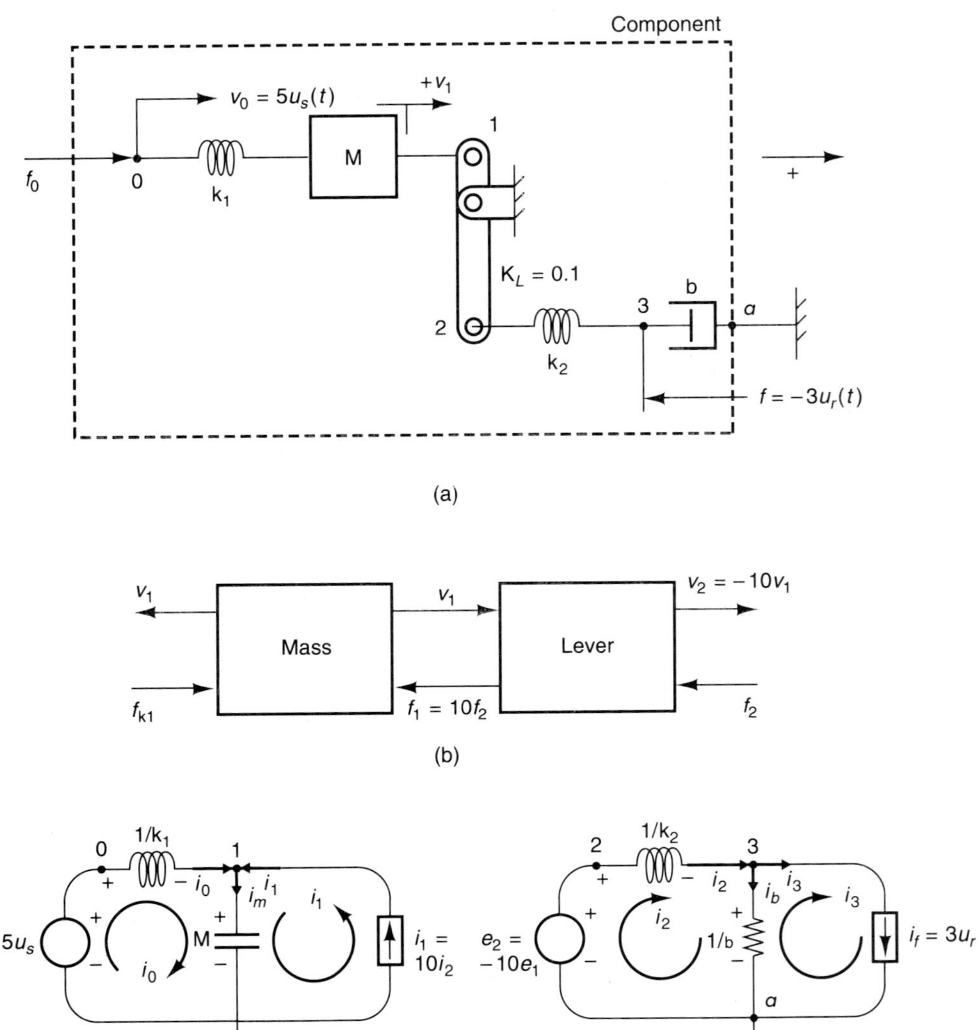

Figure 4-45. Example MIMO Analysis. (a) Idealized model, (b) lever causality, (c) analog network

Step 4: Write out the elemental relations. Since there are current sources, you will need the current on the left side of the element equations,

$$i_0 = k_1 \int e_{01} dt.$$

$$i_m = M\, de_1/dt.$$

$$i_2 = k_2 \int e_{23} dt.$$

$$i_b = be_3.$$

Step 5: Manipulations to get the output equations. Since there are current sources attached to the nodes, start with KCL at *node 1,*

$$i_0 - i_m + i_1 = 0,$$

and substitute the element relations,

$$\text{k}_1 \int e_{01} dt - \text{M}\, de_1/dt + 10 i_2 = 0.$$

The last term can be clarified from the KCL relationship for node 3,

$$\text{k}_1 \int e_{01} dt - \text{M}\, de_1/dt + 10(i_b + i_3) = 0.$$

Again, the last term can be clarified,

$$\text{k}_1 \int e_{01} dt - \text{M}\, de_1/dt + 10(\text{b}e_3 + 3u_r(t) = 0.$$

Clearing out the integral and rearranging gives,

$$\mathbf{M}\, d^2e_1/dt^2 + \mathbf{k}_1 e_1 = (5\mathbf{k}_1 + 30)\, u_s(t) + 10\mathbf{b}\, de_3/dt.$$

While this output equation may be reduced further, it is left in this form to clarify the final result.

For node 3, you again start with the KCL relation

$$i_2 - i_b - i_3 = 0,$$

and substitute the element relations

$$\text{k}_2 \int e_{23} dt - \text{b}e_3 - \text{i}_f = 0.$$

Clearing out the integral gives

$$\text{k}_2 e_{23} - \text{b}\, de_3/dt - di_f/dt = 0.$$

The first term of this can be clarified using KVL for mesh 2 to give

$$\text{k}_2(-10e_1 - e_3) - \text{b}\, de_3/dt - di_f/dt = 0.$$

Rearranging,

$$\mathbf{b}\, de_3/dt + \mathbf{k}_2 e_3 = -3u_s(t) - 10\mathbf{k}_2 e_1.$$

Again, this is as far as this output will be reduced.

Note that the simultaneous solution of the two output equations above leads to the solution for any variable in the network analog. The equations may be expressed in terms of the analog outputs by realizing that the output f_0 corresponds to i_0, the mesh 0 "current." Thus, the mesh 0 "inductor" may be used to find the output from,

$$di_0/dt = k_1 e_{01},$$

or
$$\boldsymbol{di_0/dt = k_1(e_0 - e_1).}$$

The other analog output is e_3, which is already represented in the two earlier output equations.

The final step is to reverse the analogy to find the structural output equations,

$$\mathbf{M}\, d^2\mathbf{v}_1/dt^2 + \mathbf{k}_1\mathbf{v}_1 = (5\mathbf{k}_1 + 10)u_s(t) + \mathbf{b}\, d\mathbf{v}_3/dt.$$

and
$$\mathbf{b}\, d\mathbf{v}_3/dt + \mathbf{k}_2\mathbf{v}_3 = -3u_s(t) - 10\mathbf{k}_2\mathbf{v}_1.$$

These two equations must be solved simultaneously to give v_1 and v_3, the last of which is one of the outputs. The other output comes from reversing the spring 1 analogy

$$df_0/dt = 5\mathbf{k}_1 u_s(t) - \mathbf{k}_1\mathbf{v}_1.$$

Comments

The presence of levers and gears always results in sets of coupled equations which must be solved simultaneously. If these are coupled differential equations then all you can do is to list them, as done in the example above. However, if the output equations are algebraic then you can find the output(s) explicitly.

Note that this component has three energy storage elements in it, these being two springs and a mass. This agrees with the order of the coupled differential equations, they likewise being of the order three in combination.

HOMEWORK

4-30. Use the analogy method to find the equation of motion for the component shown in Fig. 4-46.

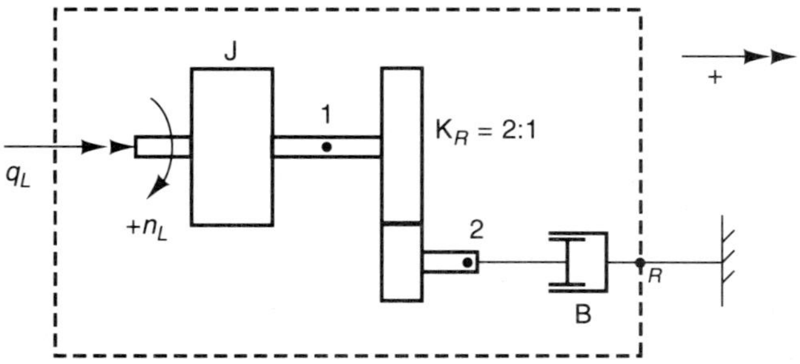

Figure 4-46. Structure for Prob. 4-30

4-31. What are the through and across variables in an ideal spring?

4-32. Given the analog equation,

$$(1/\mathrm{b})\, de_{12}/dt + (1/\mathrm{k})e_{12} = i.$$

a) What sort of analog is this?

b) Reverse the analogy to find the equation of motion for the component.

4-33. Why is the electrical analog of a structural amplifier always connected to ground?

4-34. Find the idealized equations of motion using the analogy method for the MIMO component of Fig. 4-47. Use a lossless motor model, and assume that the torque input, Q, acts on the spring to decrease the output speed, N.

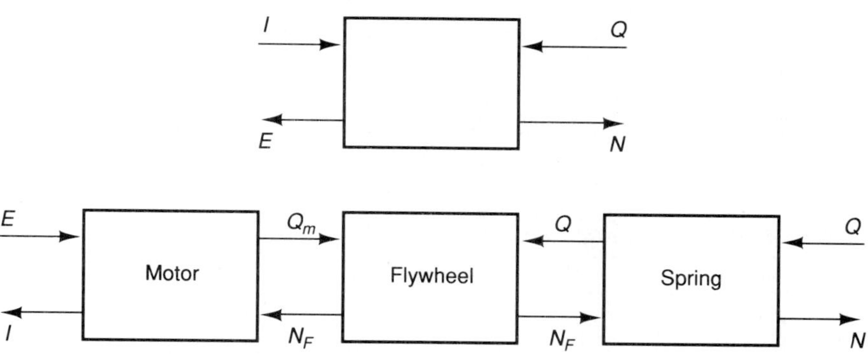

Figure 4-47. Multiports for Prob. 4-34

4-8 THE IDEAL FLUID ELEMENTS

Fluid components also have fundamental physical effects which are well-modeled using ideal system elements. These effects include the friction of internal fluid flows, the storage capacity of fluid volumes, and the inductance-like inertia of a flowing fluid. In this section, these fundamental fluid phenomena are examined to create a menu of idealized fluid-flow elements for modeling.

The Fluid Model. The first issue in modeling a fluid component is achieving a macroscopic understanding of the fluid which is flowing. This begins with categorizing the flow as either thermodynamic or non-thermodynamic.

The *thermodynamic flows* are characterized by a significant temperature change in the fluid. This may happen by design, as in a heating, ventilation, and cooling system; or it may happen incidentally, as in the heat loss from a hot fluid flowing in a thin copper pipe which goes through a cold environment. Thus, the presence of a *heat exchanger* in a design is a positive indication of a thermodynamic flow. The difficulty in these flows is that they have both fluid dynamics and convective heat transfer, so their analysis requires detailed thermodynamic modeling which is not in keeping with the spirit of preliminary analysis. These flows are thus treated in preliminary analysis using the input/output modeling methods of Chap. 7, Sec. 7-6.

The *non-thermodynamic* flows may be either liquid or gas, but system modelers idealize both types by assuming an *incompressible* fluid model. This assumption means that the density (ρ) is assumed to be constant everywhere in the flow. For liquid flows, this is usually an excellent model. For gas flows, this assumption may be checked after preliminary analysis by computing the sustained upstream-over-downstream absolute-pressure ratio across each constriction in the fluid-flow area. Experiments show that if the greatest of these pressure ratios is greater than 1.7, then non-ideal, compressible-gas physics is occurring, and the modeler must again turn to the methods of Chap. 7.

In this section, only *non-thermodynamic* flows are studied to develop a menu of idealized fluid-flow elements.

So, given a non-thermodynamic flow, the two properties which are used to characterize the fluid are the fluid *absolute viscosity*, μ, (lb_f-s/ft^2, n-s/m^2) and the fluid *density*, ρ (sl/ft^3, kg/m^3). Some typical values for fluid density are given in Table 4-5; values for viscosity were given earlier in Table 4-4, Sec. 4-5.

Table 4-5. Fluid Densities (at 100°F, 14.7 psia)

Fluid	ρ *(sl/ft^3)*
Air or oxygen	2.05×10^{-4}
Gasoline	1.32
Oil, crude	1.65
Hydraulic fluid	1.8
Water	1.94

A fluid that is highly viscous (high viscosity) flows very slowly, like cool honey. A fluid that is not very viscous (low viscosity) flows very easily, like water. Note that both the properties of the fluid are sensitive to the *temperature* of the fluid, thus the fluid temperature must be known or estimated *before* modeling is begun. The temperature is then used to anticipate the physical properties of the fluid and to compute system parameter values.

For the non-thermodynamic flows, the system variables are the *pressure*, P (lb_f/in^2 or psi, n/m^2), and the fluid *volume flowrate*, Z (ft^3/s, m^3/s).

The fluid *pressure* is a measure of how much constraining force is necessary to contain the fluid. In liquid flows, the pressure is called "gage" pressure (psig) because this is what is measured on a pressure gage after drilling a small hole in the wall of a straight pipe with a liquid flowing in it and measuring the pressure exerted by the liguid on the inside pipe wall. Thus, 0 psig is equal to an atmospheric pressure of 14.7 psi absolute (psia). In gas flows, the pressure variable is the absolute pressure. This is done in keeping with the equation of state for the gas ($P/\rho = R\Theta$, with R tabulated in Table 7-2, requires the absolute pressure to be used).

In all pipe flows, the pressure is assumed to be the same at any point on a plane perpendicular to the axis of the conduit containing the flow.

The fluid *volume flowrate* is a measure of the net volume of fluid which crosses a plane perpendicular to the axis of its conduit, per unit time.

Note that fluid pressure is very much like electrical voltage—both create a *potential* for flow. Similarly, the flow of fluid particles is like the flow of electrical charge. In fact, the strength of these similarities leads the system modeler to create the fluid physical model directly in electrical-analog form, thus skipping the intermediate idealized physical model like that built for structural components.

The Ideal Fluid Resistance. An ideal fluid resistance element models the amount of fluid energy lost due to fluid *friction*. This friction occurs mainly at two places in the flow: along the walls of the conduit, or at constrictions in the cross-sectional flow area.

Consider first the fluid friction exerted on the *inside wall* by the flowing fluid in a long, straight conduit. A "long" conduit is one whose length is more than 20 diameters; a "straight" conduit is one that may have gradual, gentle turns (turn radius greater than 10 diameters), but no sharp turns. A conduit of length less than 20 diameters usually has negligible wall friction.

But, in order to compute the wall friction in the conduit, the flow must first be sorted based on the *Reynolds number*,

$$Re = \left(\frac{4\rho}{\pi d\mu}\right) Z, \tag{4-70}$$

where ρ = fluid density (Table 4-5),
d = inside diameter of conduit (ft, m),
μ = absolute viscosity of fluid. (Table 4-5).

If the inside cross-sectional shape of the conduit is not a circle, then you must calculate and use the *hydraulic diameter* in Eq. 4-70. This diameter comes from equating the conduit cross-sectional area to an equivalent circular area; the equivalent circle diameter is the hydraulic diameter.

The test for flowrate is:

$$Re \leq 2{,}000 \text{ (low flowrate)}, \tag{4-70a}$$

$$Re > 2{,}000 \text{ (high flowrate)}. \tag{4-70b}$$

These two conduit flow conditions correspond to the cases where the fluid-flow structure is *laminar* (low-flow case) or *turbulent* (high-flow case). A laminar flow is one where the fluid particles move in parallel lines. The fluid is thus composed of orderly layers, or *laminates* of fluid. A turbulent flow is a random, chaotic tumbling of fluid particles through the conduit, characterized by much mixing and audible noisiness in the flow.

Note that laminar flow may exist above a Reynolds number of 2,000, but that would be under special conditions. The normal dividing line between the two types of flow is near a Reynolds number of 2,000.

The importance of this is that, if a high flowrate (turbulent) condition exists, then the friction is *less* than that for a low flowrate. Consequently, a low-flowrate model is either accurate or conservative, and, therefore, is often used in a preliminary-design model.

Low flowrate experimental measurements of wall friction are observed to follow a linear Hagen-Poiselle flow model,

$$P_{12} = \frac{128\mu\ell}{\pi d^4} Z, \tag{4-71}$$

where μ = absolute viscosity (Table 4-5), (n-s/m^2, lb$_f$-s/in^2),
ℓ = length of the conduit (m, in),
d = hydraulic diameter (m, in).

Thus, a conservative ideal model of the lumped wall friction for a length of straight conduit between ends 1 and 2 may be constructed using the low-flowrate, linear, Hagen-Poiselle physics. This *idealized* wall-friction model is

$$p_{12} = R_f z, \tag{4-72}$$

with

$$R_f = 128\,\mu\ell/(\pi d^4). \tag{4-73}$$

Eq. 4-72 has the English units,

$$\text{lb}_f/\text{ft}^2 = (\text{lb}_f\text{-s/ft}^5)\,(\text{ft}^3/\text{s}),$$

and the SI units,

$$\text{kg/m}^2 = (\text{kg-s/m}^5)\,(\text{m}^3/\text{s}).$$

The plots of typical turbulent and laminar flow data are shown in Fig. 4-48a, along with the ideal model of wall friction. Note the ideal analog symbol for fluid friction to the right of this figure. The analogous flow variables are shown in parenthesis in the figure, but this is not done on the fluid-analog network when electronic-analog variables are used.

Also shown in Fig. 4-48a is data from a "fixed orifice" which has been drilled in a flat plate and mounted in the pipe perpendicular to the flow. This gives rise to a type of fluid friction that is due to a *constriction* in the flow cross-sectional area. This area reduction may be due to incidental or controlling-area constrictions.

Incidental conduit constrictions are modeled using added conduit lengths and the wall-friction model in Eqs. 4-72 and 4-73. Typical added lengths for various fittings are shown in Table 4-6.

Table 4-6. Added Lengths for Fittings

	ℓ/d
Gate valve, fully open	13
90°, standard elbow	20-30
90°, street elbow	50
Standard tee,	
Straight-through	20
Flow to/from branch	60
Butterfly valve, fully open	20

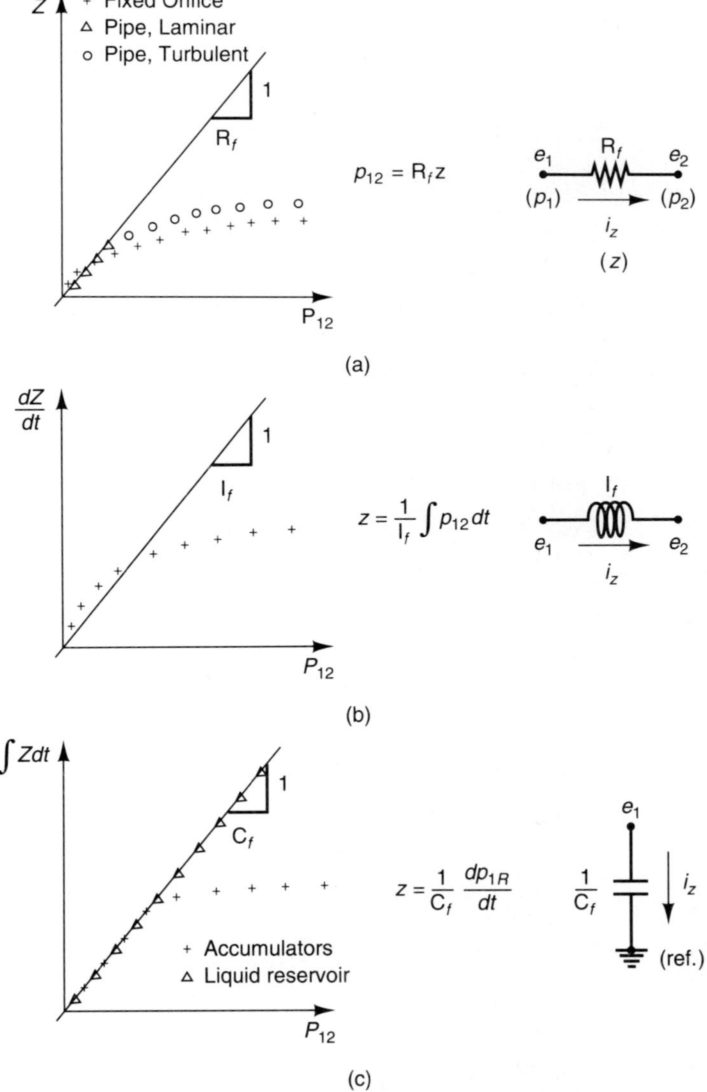

Figure 4-48. The Ideal Fluid Elements. (a) Fluid resistance, (b) fluid inductance, (c) fluid capacitance

Controlling-area constrictions are created by designers who use variable-area *valves* and fixed-area *orifices* to cause severe flow restrictions in order to modulate the flow. For modeling purposes, the valve flow resistance is estimated by comparing the constricted flow to the flow through a flat-plate orifice of the same size opening. The geometry of this is shown for various valves in Fig. 4-49.

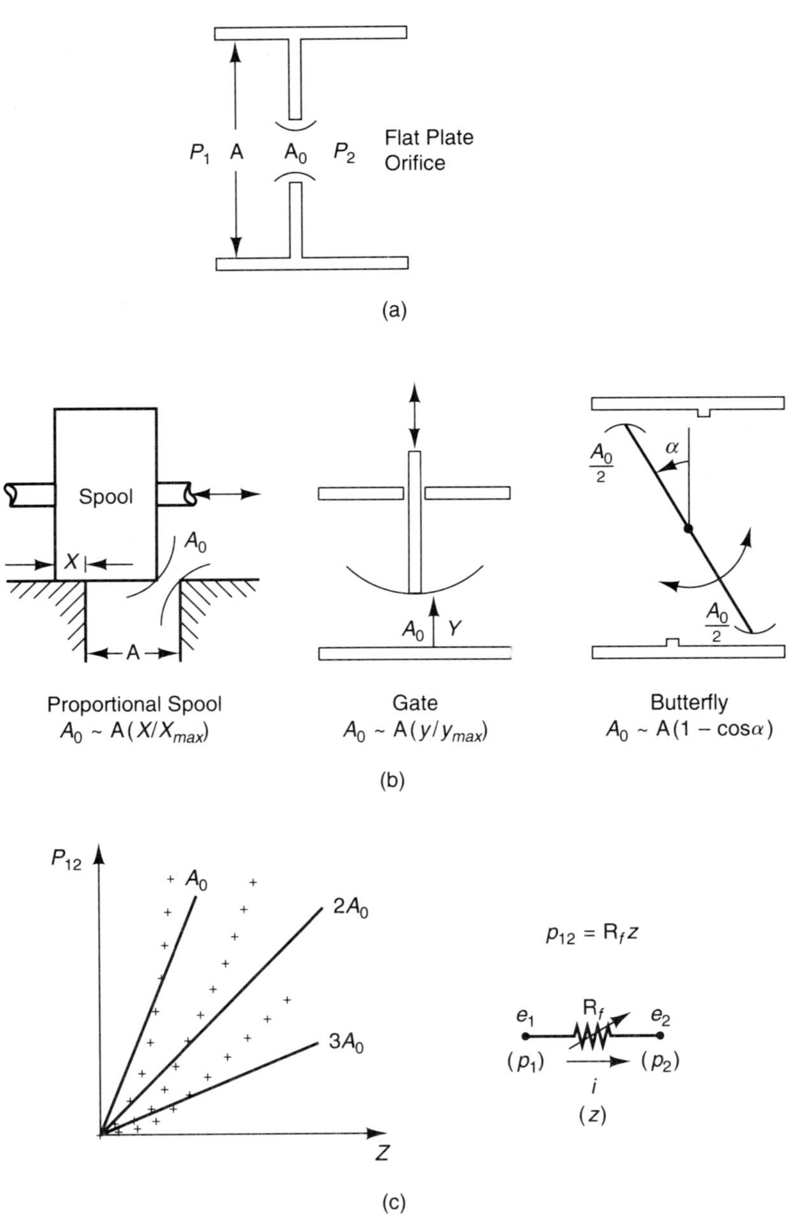

Figure 4-49. Fluid Resistance Due to Constriction. (a) Flat-plate orifice, (b) control valves, (c) model

The *flat-plate orifice* is made by creating a precise circular hole in a flat plate, and mounting it in the flow as in Fig. 4-49a. The pressure-flow *empirical* relationship for such an orifice is observed to be

$$Z^2 = (2/\rho)\, A_0^2\, C_D^2\, \beta P_{12}. \tag{4-74}$$

where P_{12} = pressure drop across the orifice, lb_f/ft^2,
A_0 = orifice area, ft^2,
A = conduit area, ft^2,
C_D = orifice discharge coefficient, 0.6 (unitless),
$\beta = [1 - (A_0/A)^2]^{-1}$,
ρ = fluid density, sl/ft^3,
and Z = volume flowrate, ft^3/s.

Thus, the pressure data is seen to fit a quadratic in the volume flowrate, Z.

For the *valves* of Fig. 4-49b, a variable area occurs depending upon a modulating variable (x, y, or α in the figure). *Spool* valves are used with either gaseous or liquid flows, *gate* valves are used mainly with liquid flows, and *butterfly* valves are used with gaseous flows. So, when one of these area functions is inserted into Eq. 4-74, a very complicated pressure-flow relationship is seen. Fortunately, however, the expression can be simplified by considering small openings only, since that is what normally occurs during flow modulation. These "small" openings are defined as, $(A_0/A) \leq 0.2$. For these, Eq. 4-74 becomes,

$$Z^2 = (2/\rho)A_0^2C_D^2\, P_{12}. \tag{4-75}$$

For the *proportional spool valve,* the area function is $A_0 = A_{max}\,(X/X_{max})$. So, these valves are called "proportional" because the controlling area is proportional to the spool position, X. The customary idealization is accomplished by fitting the valve pressure-flow data to a *bi-linear* model form (see Chap. 7 for further details on how to do this). This gives, for a linearization near A_0 and $x = 0$,

$$p_{12} = R_V\,(1 - x/X_{max})\, z. \tag{4-76}$$

This model is linear if either x or z is regarded as constant, but is nonlinear if both are allowed to vary simultaneously. Similar idealizations for the other valves shown in Fig. 4-49b may also be found.

Some typical experimental data for the controlling valves, and a fitted-bilinear model is shown in Fig. 4-49c. Note that as the valve area, gets larger, there is less pressure drop associated with the valve at a given flowrate due to less flow resistance—thus less slope in the linearization.

The idealized fluid-analog element for the valve is shown at the far right of Fig. 4-49c. Note the arrow through the "resistor" element to indicate the variable-area fluid resistance. Again, the fluid variables are shown here in parentheses simply to show the fluid analogy.

Of all the valves, the spool valve is such an important element in a fluid system that it deserves further modeling attention—a cut-away view of this valve is shown in Fig. 4-50.

The "spool" in the spool valve has two solid disks mounted on a push rod. The disks are positioned in the valve to control the flow by an electric positioning device as shown in Fig. 4-50a (note that $X = x = 0$ is shown). The disks direct the high-pressure fluid entering at *S* (for supply), into either a path towards 1 or a path towards 2. Whichever path is chosen for the supply flow, the remaining path is automatically connected to *R* (for return). The *R* path is necessary to capture the depressurized fluid and complete a fluid circuit.

The modeling of this valve is according to the nature of the positioning device. That is, the positioner is either a motor or a solenoid. If a *motor*, then the position of the spool is according to $x = K_v i$, and an orifice-like area constriction results. This gives rise to the controlling-area performance which is diagrammed in Fig. 4-49 and is modeled using the fluid network diagrams shown in Fig. 4-50b. Note the two cases of connections and the variable-area indicated by the dashed connection between the arrow heads in the model network. Also note that the *same area* occurs in both flow paths at the same time—this area has the variable fluid resistance R_V.

The fluid flow through a proportional spool valve thus experiences *two* pressure drops, for *two* cases of x:

for $x > 0$
$$p_{S2} = R_V (1 - x/X_{max})\, z_S, \tag{4-76a}$$
$$p_{1R} = R_V (1 - x/X_{max})\, z_R. \tag{4-76b}$$

And,

for $x < 0$
$$p_{S1} = R_V (1 + x/X_{max})\, z_S, \tag{4-76c}$$
$$p_{2R} = R_V (1 + x/X_{max})\, z_R. \tag{4-76d}$$

When $x = 0$, the flow through the valve may be zero for some valve designs, but not for others. The valve design details will sort this out for you.

If the spool-positioning device is a *solenoid*, then the valve is either fully opened or fully closed (the latter case is drawn in Fig. 4-50a). For the fully-opened valve, the fluid network connections depend on which way it is opened, as shown in Fig. 4-50c. Often, the solenoid spool valve is drawn in a fluid-system design in the manner of Fig. 4-50d, which is a much more concise diagram than that of Fig. 4-50c.

The Ideal Fluid Inductance. Reasoning from the electrical analogy, this element can be expected to store energy by virtue of the *flow* through it, just as the electrical inductor stores energy by virtue of the *current* through it. Thus, any liquid flow must generally demonstrate fluid inductance. A gas flow in a conduit may demonstrate this phenomenon if the mass flowrate is large enough, though gas inertia effects are usually ignored. Since this phenomenon is mainly due to the inertia of the flow, this property is also sometimes called fluid *inertance*.

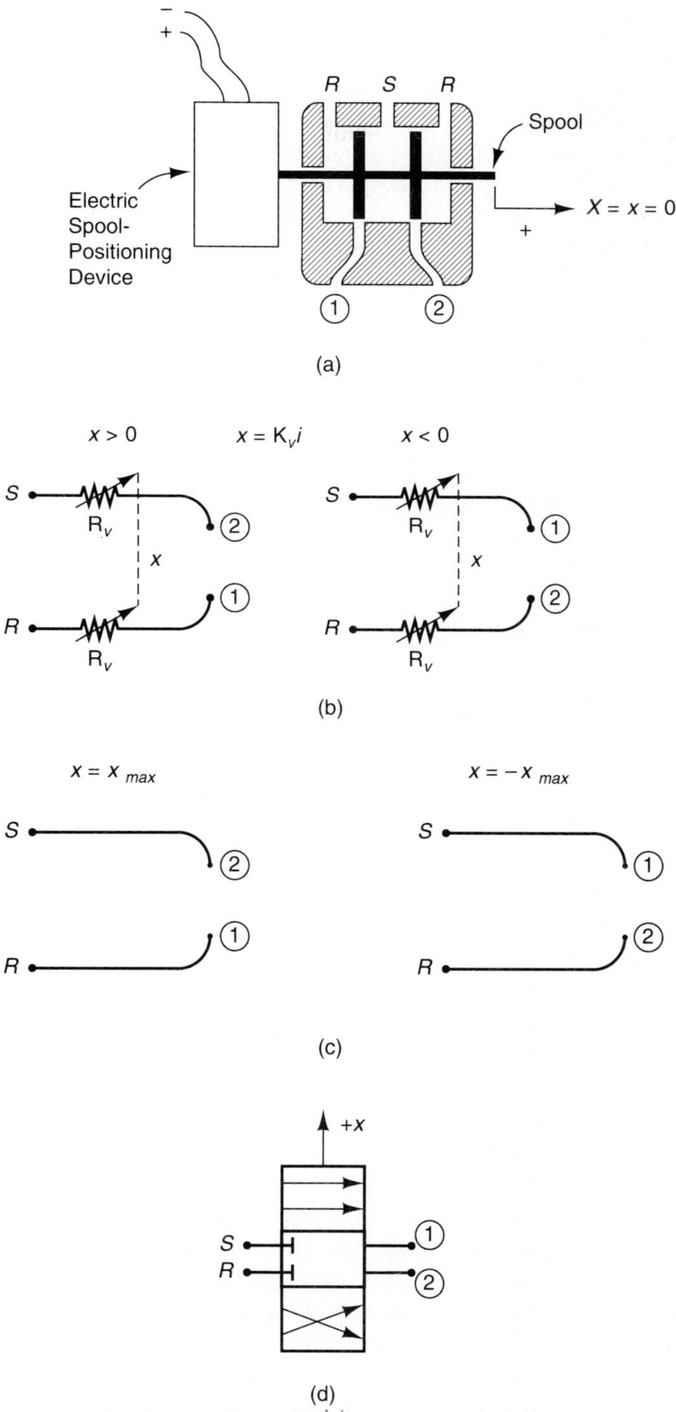

Figure 4-50. The Spool Valve. (a) Physical arrangement, (b) motor-driven, (c) solenoid-driven, (d) solenoid circuit symbol

A Newtonian fluid-momentum analysis is usually used to compute the fluid inductance. An *idealized* version of this analysis is based upon a "plug" of incompressible fluid moving through a pipe of constant area, A. This plug is analytically subjected to a pressure difference, and the acceleration of the plug mass is computed. (Note the similarity to using a voltage to overcome a current reluctance in an inductor). The velocity distribution of the fluid particles in the plug is ideally assumed to be the same at any cross-section in the plug, and every cross-section is assumed to have the same average speed, V_{AVG}. Thus Newton's Second Law for the plug is

$$\Sigma F = d(\mathrm{M}V_{AVG})/dt. \tag{4-77}$$

Or, since the plug is ℓ in length, and of uniform density ρ,

$$\mathrm{A}\ (P_1 - P_2) = (\rho \mathrm{A} \ell)\ dV_{AVG}/dt. \tag{4-78}$$

Further, since $Z = \mathrm{A}V_{AVG}$, linearization gives

$$p_{12} = (\rho \ell/\mathrm{A})\ dz/dt = \mathrm{I}_f\ dz/dt. \tag{4-79}$$

This model has the English units

$$\mathrm{lb}_f/\mathrm{ft}^2 = (\mathrm{sl}/\mathrm{ft}^4)\ (\mathrm{ft}^3/\mathrm{sec}^2),$$

and the SI units

$$\mathrm{n/m}^2 = (\mathrm{kg/m}^4)\ (\mathrm{m}^3/\mathrm{sec}^2).$$

The plot of the actual data is shown in Fig. 4-48b. The upward departure of the actual data from the linear model of Eq. 4-78 is partly due to intra-plug frictional losses and partly due to other nonideal effects. The ideal fluid inductance symbol is shown at the right side of the figure.

The Pure Fluid Capacitance. Finally, the fluid analogy requires that fluid capacitance occur whenever energy is stored by virtue of fluid *pressure*, which is the fluid across variable. Figure 4-51 shows four of the common types of fluid capacitors.

The capacitors in Figs. 4-51a-c are used in liquid flows, and that in Fig. 4-51d is used in gaseous flows. Note that the rigid tanks in Figs. 4-51b and d use gas *compressibility* as a means of storing fluid power. You must be careful when you see this, however, since you are limited to *non-thermodynamic* flows only. That is, you must know or assume that all gas compressions/decompressions occur at, or near, constant temperature. Fortunately, this is the way these devices are usually designed to be used, employing small-amplitude pressurizations.

Figure 4-51a shows a *small liquid reservoir*. This reservoir is "small" in the sense that the net flowrate in or out, Z, has a significant effect on the liquid volume

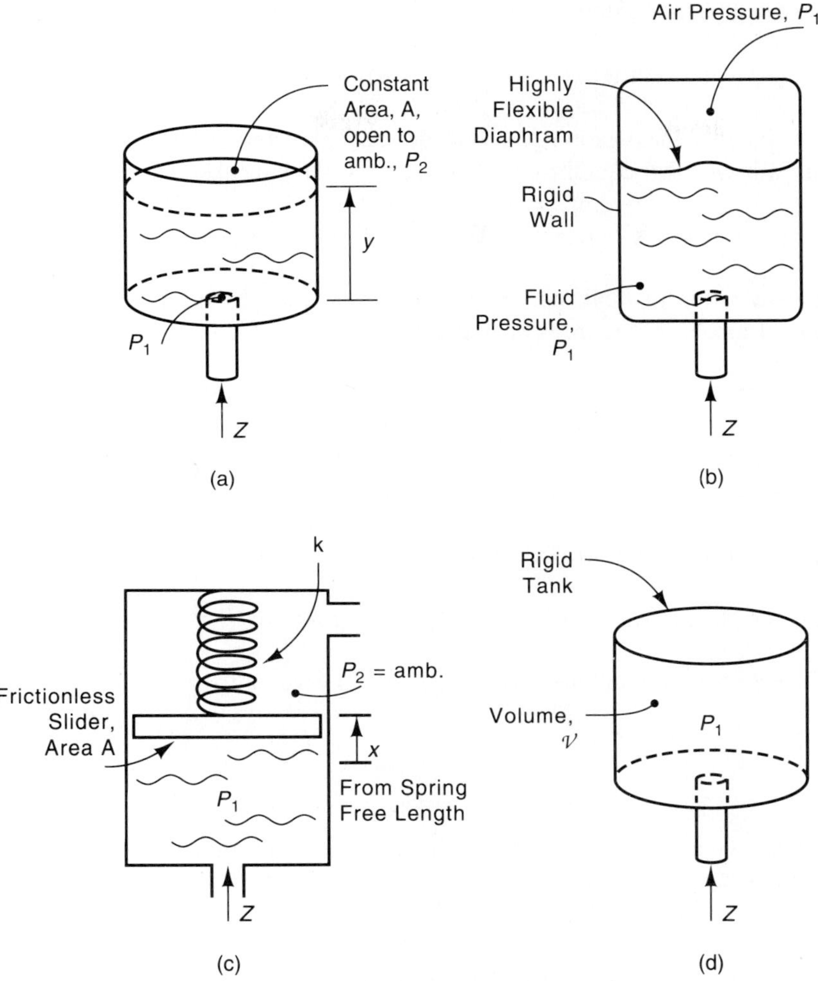

Figure 4-51. Fluid Capacitance. (a) The small reservoir, (b) diaphragm accumulator, (c) spring accumulator, (d) tank accumulator

stored in the capacitance. If Z does not have a significant effect on the volume, then the large reservoir model of the next section should be used.

The physical analysis of the small reservoir capacitance, and fluid capacitors in general, always begins with the application of *conservation of mass* for the element. This law states

$$\text{(mass flowrate in)} - \text{(mass flowrate out)} = d/dt\text{(mass inside)}. \tag{4-80}$$

In this case,

$$\rho Z - 0 = d/dt\,(\rho A y). \tag{4-81}$$

Since liquids are ideally regarded as incompressible, the above relation idealizes to

$$z = \mathrm{A}\ dy/dt. \tag{4-82}$$

Further idealization is accomplished by assuming that the liguid in the reservoir is in vertical equilibrium, or close to it. This means

$$\Sigma f = d(MV)/dt \approx 0,$$

where V is the vertical speed of the center of mass of the liquid in the reservoir. The applied forces on the mass of liquid are

$$\Sigma f = (\text{pressure}) \times (\text{area}) - \text{weight} \approx 0,$$

or

$$(p_1 - p_2)\,\mathrm{A} - \rho g \mathrm{A}\, y \approx 0, \qquad ((4\text{-}83))$$

and

$$p_{12} = \rho g y. \tag{4-84}$$

Taking a differential of both sides gives

$$dp_{12}/dt = \rho g\ dy/dt. \tag{4-85}$$

Eq. 4-85 can be combined with Eq. 4-82 to give

$$p_{12} = (\rho g/\mathrm{A}) \int z dt. \tag{4-86}$$

This last equation is recognized to be in the form of an ideal fluid capacitance, with the capacitance value of

$$\mathrm{C}_f = \mathrm{A}/(\rho g). \tag{4-87}$$

The model for the ideal fluid capacitor is thus

$$p_{12} = (1/\mathrm{C}_f) \int z dt, \tag{4-88}$$

which has English units

$$\mathrm{lb}_f/\mathrm{ft}^2 = (\mathrm{lb}_f/\mathrm{ft}^5)\ (\mathrm{ft}^3/\mathrm{s})\ \mathrm{s},$$

and SI units

$$\mathrm{n/m}^2 = (\mathrm{n/m}^5)\ (\mathrm{m}^3/\mathrm{s})\ \mathrm{s}.$$

The data for a fluid capacitor is shown in Fig. 4-48c. Note the linear response of the small liquid reservoir over the entire range of interest.

Also note that the ideal fluid capacitor symbol at the right of Fig. 4-48c is always connected to "ground," thus showing the need for a reference ambient pressure to give meaning to the pressure-energy which stored in the fluid capacitor.

Figures 4-51b and c show two types of *accumulators* which are used to store pressure energy from a liquid flow. In these, the air-charged, diaphragm accumulator stores energy in the compressed air, and the spring accumulator stores energy in the spring deflection. The coefficients of fluid capacitance for these two elements will be found in the homeworks.

Figure 4-51d shows a small, rigid tank for storing pressure energy from a gas flow. To understand the physics of this element, you begin again with the law of mass conservation

$$\rho Z = d/dt\,(\rho \mathcal{V}) = \mathcal{V}\, d\rho/dt, \tag{4-89}$$

where $\mathcal{V}$ is the volume of the rigid tank.
Thus,

$$Z = (\mathcal{V}/\rho)\, d\rho/dt. \tag{4-89a}$$

For ideal gas flows, you can write the equation of state in the form

$$d\rho/\rho = dp/p. \tag{4-90}$$

This relationship comes from taking the natural logarithm of the equation of state for the ideal gas ($P/\rho = R\Theta$), recognizing that Θ is constant in a non-thermodynamic flow, and differentiating both sides. Further, this can be idealized to

$$d\rho/\rho = (1/p_{1avg})\, dp, \tag{4-91}$$

since the pressure inside the tank does not vary much from the constant pressure, p_{1avg}. If the last equation is substituted into Eq. 4-89, you find,

$$z = (\mathcal{V}/p_{1avg})\, dp/dt. \tag{4-92}$$

and integration gives,

$$\int z dt = (\mathcal{V}/p_{1avg})\, p_{12}. \tag{4-93}$$

Thus, the fluid capacitance for the rigid gas tank is,

$$C_f = \mathcal{V}/p_{1avg}. \tag{4-94}$$

The units for this capacitance must be the same as those for the capacitance of the liquid reservoir discussed earlier.

Note that it is usual for the average pressure in a gas flow to be near atmospheric pressure. Thus, $p_{1avg} \approx p_{amb}$, which is another common idealization for the tank capacitance.

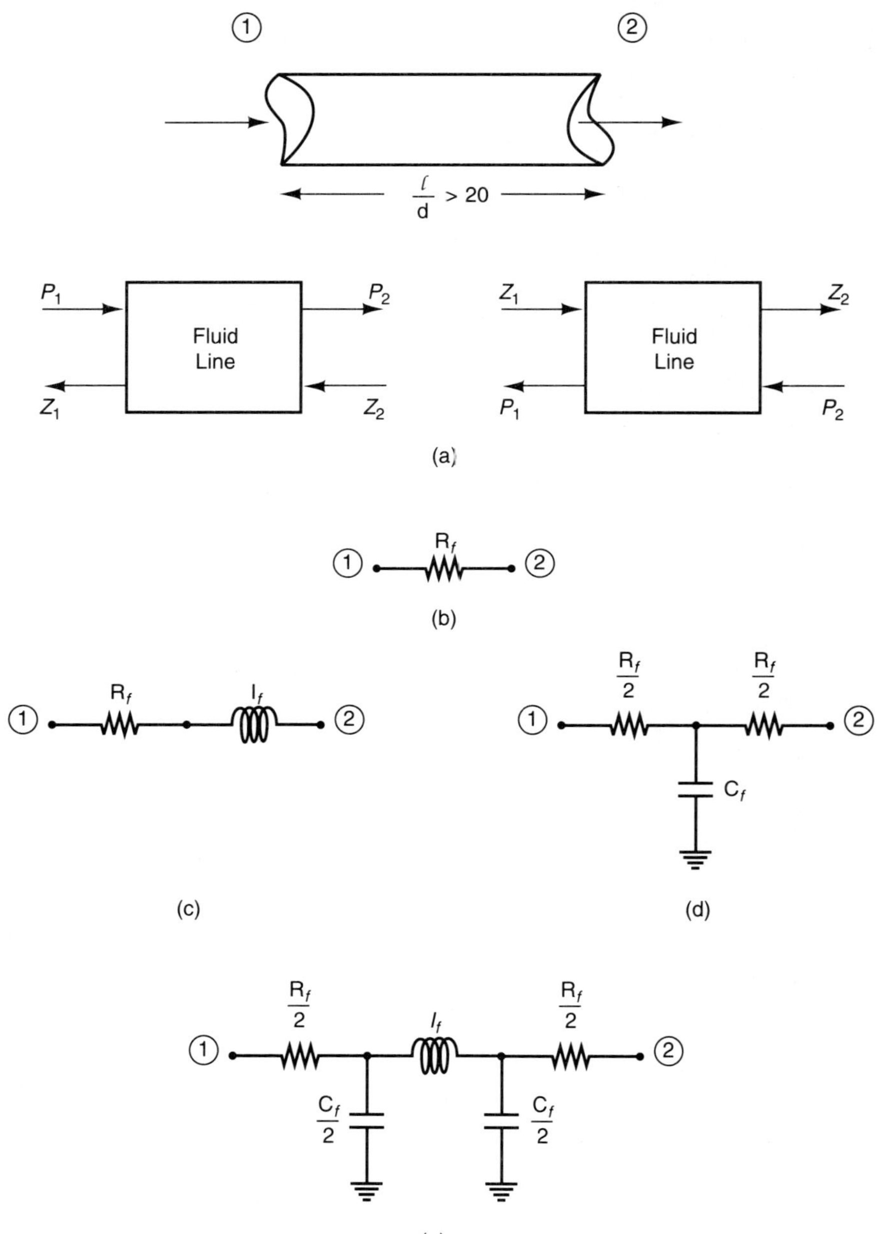

Figure 4-52. Fluid Line Models. (a) Geometry and causality, (b) simplest, (c) liquid flow, (d) gas flow "T" model, (e) "π" model

Finally, returning again to Fig. 4-48c, the non-ideal departure of the tank capacitor data from the linear ideal curve is the same as that measured in the accumulators, it is due to the *increased stiffening* of the capacitance as mass is input to the element.

Fluid Lines. Armed with this introductory understanding of fluid resistance, inductance, and capacitance, it is time to consider a crucial aspect of fluid system modeling — the modeling of the fluid flow in the connecting lines. These line dynamics can be extremely complex, yet your simple ideal models can be very effectively used to evaluate most preliminary-system designs. Several effective modeling approaches are outlined in Fig. 4-52.

Usually, you need not model the lines at all if ℓ/d for them is less than about 20. For greater ratios, you need to consider their function and dominant physics. Lines are put into fluid systems to function as either pressure conduits or as flowrate conduits, as indicated by the nature of the connected elements, and which should be determined from a multiport analysis, such as shown in Fig. 4-52a.

The simplest model of a fluid flow is that of a lone fluid resistance, as shown in Fig. 4-52b. This model would apply to a liquid flow with a small flowrate, or a gas flow which has a small pressure difference between the conduit ends.

For liquid flows with higher flowrates, the inertia of the flow must also be modeled. Thus, the network must include an inertance, as shown in Fig. 4-52c.

For gas flows with a high pressure-difference between the ends of the conduit, the "T" model of Fig. 4-52d is often used. The capacitance here is the same as that of the tank capacitance computed earlier for Fig. 4-51d.

Sometimes, the compressibility in a liquid flow or the inertance in a gas flow must be modeled. When these situations arise, the "π" model of Fig. 4-52e is used. However, the compressibility of a liquid flow is a very complicated and difficult evaluation, and it will only be mentioned here in passing.

The most significant effect that these simple models ignore is that of the time delay between the start of an effect at one end of the conduit and its perceived response at the other end. This effect can be very significant for gas-flow conduits with very large ℓ/d ratios. Should the designer require you to model the delay in the fluid lines, you must resort to the methods of Chap. 7, using input/output modeling.

HOMEWORK

4-35. Use the accumulator sketch shown in Fig. 4-51b to complete the following modeling steps. You should assume that the diaphragm adds no spring rate to the air compressibility. Also, assume that the air and liquid are at the same pressure and temperature.

a) Find the fluid capacitance for this accumulator.

b) How could you increase this capacitive value? (Hint: how would you change the spring rate?)

4-36. Hydraulic oil at 100°F is flowing in a 1 inch pipe (inside diameter). The volume flowrate in the pipe is $Z = 0.1$ ft^3/s.

a) What is the average speed of the oil in the pipe?

b) What is the Reynolds number for this flow?

c) Will the Hagen-Poiselle fluid resistance model be a good preliminary model in this case? (Why?)

4-37. Given a pipe with a liquid flow in it at 100°F. The dimensions of the pipe are, 1/4 inch inside diameter (ID), and 5 ft. long. The designer must choose between a flow of water and a flow of hydraulic fluid. Which of the two flows has:

a) a higher fluid resistance?

b) a higher fluid inductance (inertance)?

4-9 COMMON FLUID COMPONENTS

Complete fluid subsystems usually incorporate some sort of fluid power *source*, as well as various *transducers* and/or *amplifiers* for effectiveness. The physics of each of these classes of common fluid components is examined in this section.

Once again, this section presents the idealized modeling of real components in order to treat them as idealized elements within a larger system model. Should the designer indicate a desire for a non-ideal model, then you should use the methods of Chap. 7.

Fluid Power Sources. In liquid components, fluid power may be developed from *gravity* or *pumps*. In gas systems, fluid power usually comes from a *compressor*.

Manufactured liquid "pumps" typically include a motor or engine to provide structural power, and a mechanical pump to convert the structural power into fluid power. In this text, such a "pump" is modeled as a single fluid-power source element, including its motor or engine.

Similarly, manufactured gas "compressors" include a structural power source and a compressor. Unfortunately, these products are far too complicated to develop a fundamental analysis from a limited background. They are only mentioned in this section in passing, and again later in Chap. 7 on input/output modeling.

Gravity often provides a useful pressure to move liquids; such a case is shown in Fig. 4-53. Here the *large liquid reservoir* is connected by a flexible hose to the gate valve at c. The designer wishes to study an ideal model of this component which neglects fluid inertia, predicts the fluid pressures at a, b, and c, and treats the reservoir as an infinite fluid capacitance by assuming that the liquid level remains approximately constant.

The ideal model of this component is shown in Fig. 4-53b. This model is based on the hydro-static observation that the pressure at any point in a non-moving

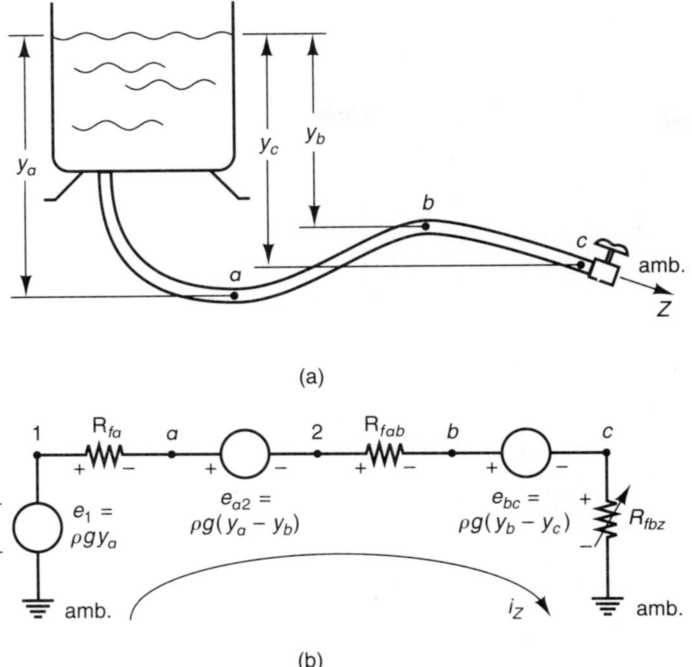

Figure 4-53. Gravity-driven Flow. (a) Schematic, (b) idealized model

body of liquid is $p = \rho g \ \Delta y$, where Δy is the depth below the free surface. So, the gravity-driven pressure potential for causing flowrate in liquid components may be modeled using a voltage-analog source of the form $e_i = \rho g \ \Delta y_i$.

Note that as the fluid in Fig. 4-53b flows from the reservoir to point a, it experiences a pressure rise due to the hydro-static effect and a pressure drop due to fluid friction. Thus, the ideal network shows a pressure source for hydro-statics connected to a mythical point 1, which is attached to a lumped resistance connected to a to model the friction effect. So, with zero flowrate in the network, the pressure at 1 is the same as the pressure at a. As flowrate occurs, these pressures drop.

Note also that the relative-height effects of various points in the hose are also modeled in Fig. 4-53b. The connection between a and b, for example, shows another mythical point 2 which is used to connect an analog source and a friction model for that length of hose. However, the voltage source here is shown *opposing* the first source in sign due to the rise, and the subscripts must be chosen as shown.

The last section of hose from b to c is either too short for a fluid resistance model, or the resistance for that section may be lumped into the gate valve model.

Often, fluid systems require a more accurate and portable liquid-power source than gravity to provide fluid power. To meet this need, a variety of *constant-speed pumps* are manufactured, as shown in Fig. 4-54. This figure shows the two common types of pumps, those for liquid flowrate and those for liquid pressure. But the general function of these pumps is very similar.

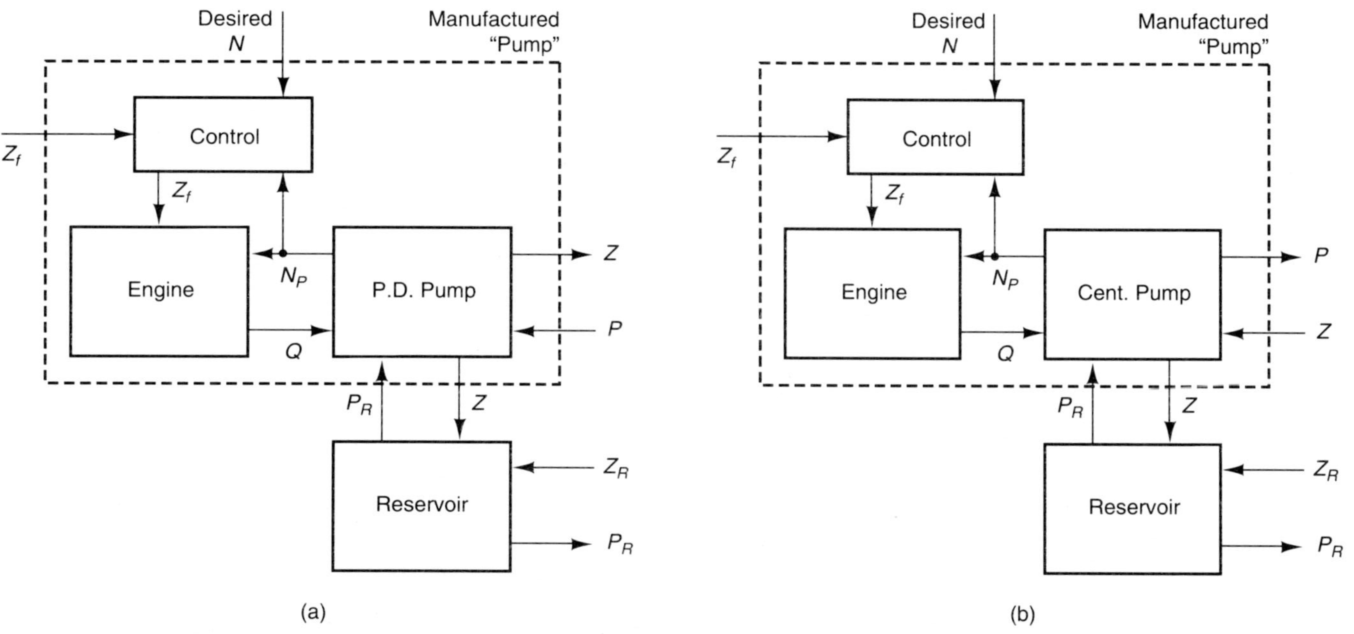

Figure 4-54. Constant-speed Liquid Pump Multiport Diagrams. (a) Positive displacement, (b) centrifugal

The pump multiport diagrams in the figure show a "desired" setting being input to the engine control. In Fig. 4-54a, this selection is used to control the pump speed, N_P, by changing the fuel flowrate to the engine, Z_f, until it gives the desired speed at some nominal system operating condition. Once chosen, the desired setting is left constant, and the speed is held constant for various loads by the controller. For example, if a heavy fluid load gives a low pump speed at a given torque, Q, then the controller gives more fuel to the engine, thus increasing the torque and the speed. The pump in Fig. 4-54b has a similar function for system pressure.

Also shown in Fig. 4-54 is the function of the liquid *reservoir* which is part of every liquid system. This reservoir acts as a collecting point for the depressurized liquid, and as a supply for the system pump. It is a device which has the physics of a liquid capacitance, collecting flowrate inputs and supplying liquid at pressure P_R to the pump.

Figure 4-54a shows the concept of operation of positive displacement pumps ("p.d. pump" in the figure). These pumps are designed to displace, or pump, a certain fixed-volume of liquid on each stroke by the action of pistons, vanes, or gears. They are, therefore, called "positive" displacement pumps. Their output is the volume flowrate, Z.

Figure 4-54b shows the concept of operation for centrifugal pumps ("cent. pump" in the figure). These pumps are designed to increase the pressure of the liquid using centrifugal forces from specially designed blades and impellers. This type of pump is usually regarded as a pressure source by fluid-power modelers.

Cutaway views of the three most common types of *positive displacement pumps* are shown in Fig. 4-55a. Each of these rotary pumps displaces a small amount of liquid as each piston, gear, or vane travels around its rotation. Typical steady operating characteristics are shown in Fig. 4-55b. The slight droop in these curves is mainly due to leakage and friction, which can be significant problems in this type of pump.

Note that the model of the droop in the pump output is created using a combination of two ideal elements: an ideal flowrate source, z_p, and a constant resistor-like element, R_f. Also note the similarity of this ideal model to that for the current sources discussed in Sec. 4-3. These both produce a *through* variable, they therefore have a similar network model. The network model for this type of pump is shown in Fig. 4-55c.

A *centrifugal pump* works by spinning a fluid outward from an axis of rotation, then recovering the kinetic energy of the fluid as *pressure*. The design of the blades used to send the fluid outward from the axis are thus very important to the performance of the pump. The geometry and performance of several types of liquid centrifugal pumps are shown in Fig. 4-56a, note the wide range of efficiencies and speeds which are available.

Typical input/output data for the centrifugal pump is shown in Fig. 4-56b. The large droop that occurs in the data comes mainly from the off-ideal mismatch of rotational speed and output flowrate. The idealized network model of a centrifugal pump is shown in Fig. 4-56c. Again, the function of this pump at a given speed is very similar to that of a battery since both are sources of an *across* variable.

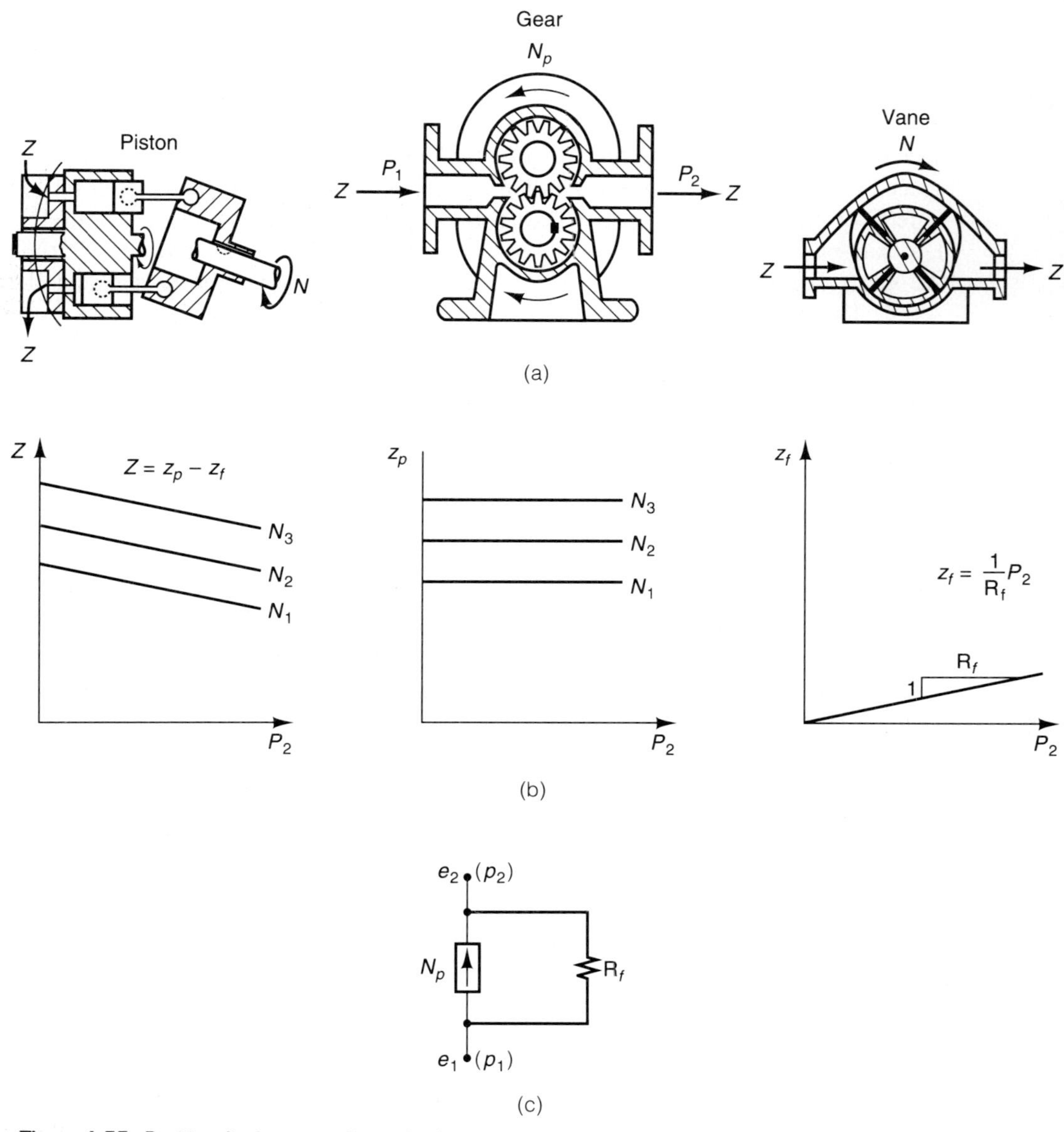

Figure 4-55. Positive-displacement Pump Performance. (a) Cutaway views [redrawn from Baumeister and Marks, *Standard Handbook for Mechanical Engineers,* McGraw-Hill Book Co., 1967 with permission], (b) performance, (c) idealized model

Many common *gas compressors* are constructed very much like the centrifugal liquid pumps in that they use centrifugal forces to create a gas pressure source. The gas may also be pressurized using an axial compressor, which operates like an air fan in a large pipe. However, the large temperature variations occurring in typical gas compression indicate that you must use a thermodynamic gas model, and you must use the input/output compressor modeling methods of Chap. 7.

Unfortunately, the basic compressibility of gases leads to a situation where there is *no flowrate source for gas components*.

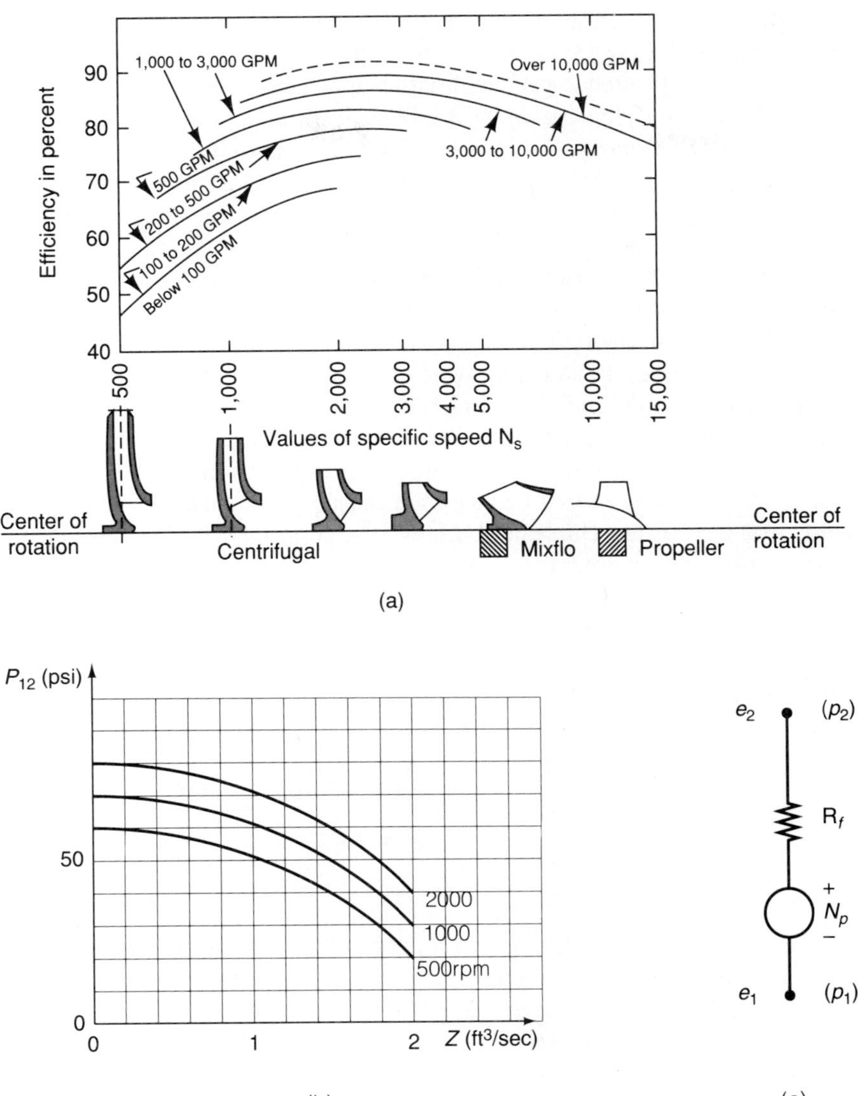

Figure 4-56. Centrifugal Pump Performance. (a) Impeller efficiences [redrawn from Baumeister and Marks, *Standard Handbook for Mechanical Engineers,* McGraw-Hill Book Co., 1967 with permission], (b) performance, (c) idealized model

Fluid Transducers. Pure fluid transducers are used to transform between fluid power and some other type of power. By far, the most common are the fluid/structural transducers. These include the pumps and compressors which convert rotary structural power into fluid power, as you just studied. These also include the hydraulic motors, actuators, and turbines which convert fluid power into structural power. Hydraulic motors and liquid turbines are not covered in this text due to their complexity. However, the methods of Chap. 7 on input/output modeling will work quite well for them, as well as for the thermodynamic gas flows which occur in gas turbines. In this section the *actuators* are studied as a type of fluid-to-structural transducer, and which often appear with either liquid or gaseous components.

The actuator physical analysis begins with the assumption that the fluid is not only *non-thermodynamic*, but it is *incompressible* as well. This may seem like an unreasonable assumption for gaseous flows, but it turns out to be a good model of most actuator flow cases. Compressibility may be an important fluid feature in the study of pressure lines, but it is usually not too important at the actuators. Further, gas-based components are usually designed to deliver low power levels to the load, and thus operate at lower pressure levels than liquid-based systems. All things considered, the incompressible fluid model is a good first model for all actuator flows.

The two types of actuators which are commonly used are those which deliver translational motion, and those which deliver rotational motion.

A schematic view of the *translational actuator* is shown in Fig. 4-57a. This device converts fluid power (P, Z) into structural power (F, V). When you model this element, you must first decide what causality is being used in the fluid component; that is, you must first decide how it is connected to the other elements. Two cases of this are possible: it is either flowrate forced, or it is pressure forced.

The multiport diagram of Fig. 4-57b shows the case of flowrate forcing, with flowrate as the input for chamber 1. Note that speed, V, *must* be the output if flowrate is an input. This comes from the law of conservation of mass for the incompressible fluid, applied to either chamber

$$\rho Z = d(\rho \mathrm{A} X)/dt, \tag{4-95}$$

or

$$Z = \mathrm{A}V. \tag{4-96}$$

This last equation must *always* be true, regardless of the type of forcing of the actuator. Thus, if flowrate is an input, then speed is an output. Conversely, if speed is an input, then flowrate must be an output. No other case is possible.

The ideal fluid actuator is a passive element without power loss. Thus, it must have the same power output as is input. That is,

$$\text{power in} = \text{power out},$$

or

$$zp_{12} = fv. \tag{4-97}$$

Note the lower case variables for the *idealized* transducer. Some of the things that are usually ignored here are leakage between the two chambers, frictional losses, and the mass and springiness associated with the actuator material.

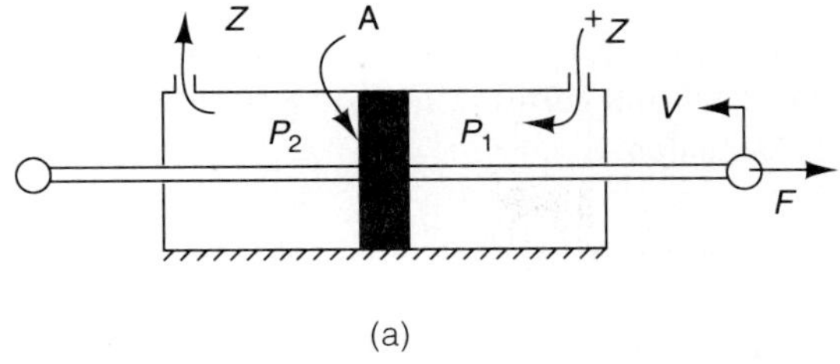

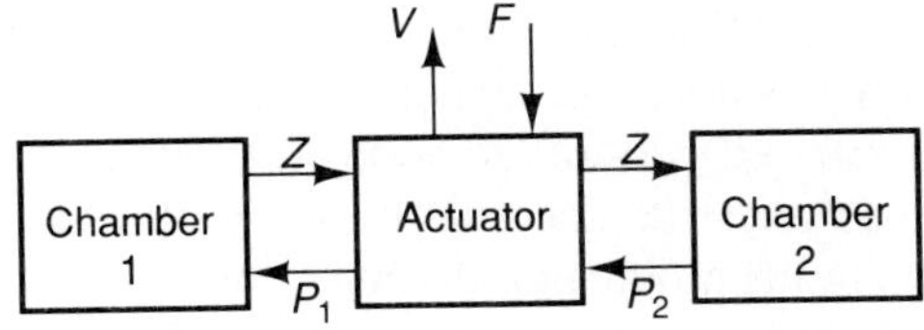

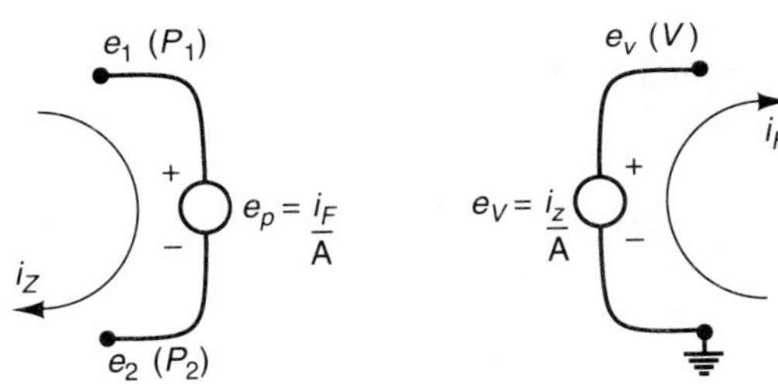

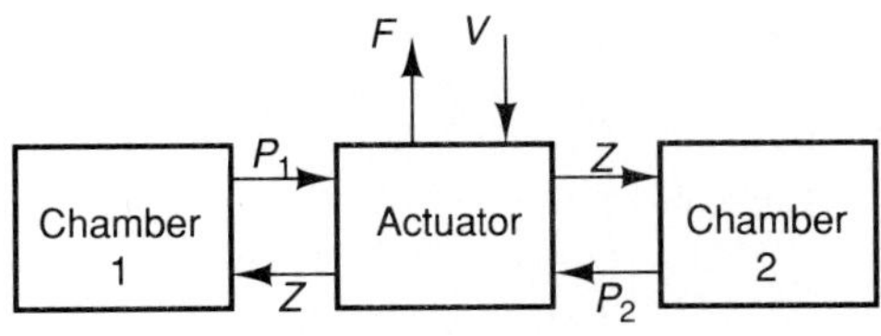

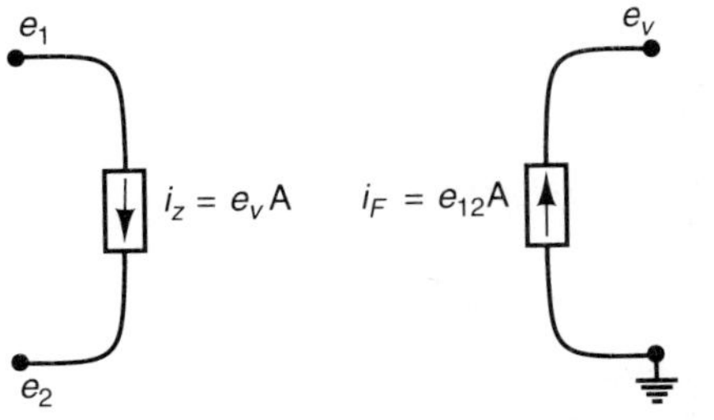

Figure 4-57. The Translational Actuator. (a) Schematic, (b) flowrate-forced actuation, (c) pressure-forced actuation

Also note that it is the pressure *drop* across the element which determines the power transfer. Rearranging Eq. 4-97 and substituting Eq. 4-96 gives the governing relationship for the translational actuator

$$z/v = f/p_{12} = A. \tag{4-98}$$

The idealized analog model of the translational actuator is shown in Figs. 4-57b and c. Again, the causality determines which model to use for the actuator.

Notice at the bottom of Fig. 4-57b that the actuator operates to input the pressure drop to the fluid side, and it inputs a voltage analog of speed into the structural side. That is, a pressure source is needed *between* p_1 and p_2 as shown in the figure. Also, since the speed, v, is input to the structural circuit by the flowrate, an *across* source connected to ground in the structural analog to model this interaction.

Figure 4-57c for pressure forcing shows two through-sources due to the fact that the actuator operates to input the through variables to the two analog circuits. Otherwise, the symbology and polarities are the same as in Fig. 4-57b.

The physical analysis of the *rotary actuator* is exactly parallel to that discussed above. It is reserved for the homework.

Fluid Amplifiers. Pure fluid amplifiers are passive devices which operate in much the same way as gears and levers: they amplify one fluid variable at the expense of another.

A schematic of a fluid amplifier is shown in Fig. 4-58a. A device very similar to this is used to lift cars in service stations. In the configuration shown, it amplifies pressure ($P_2 > P_1$) at the expense of flowrate ($Z_2 < Z_1$). The center area of the amplifier is open to ambient conditions, and it draws in or pushes out a flowrate as the center piece moves back and forth.

In order to analyze the fluid amplifier, begin with the observation that the rigid center-piece must provide *compatible motions*, and mass must be conserved in the element. That is,

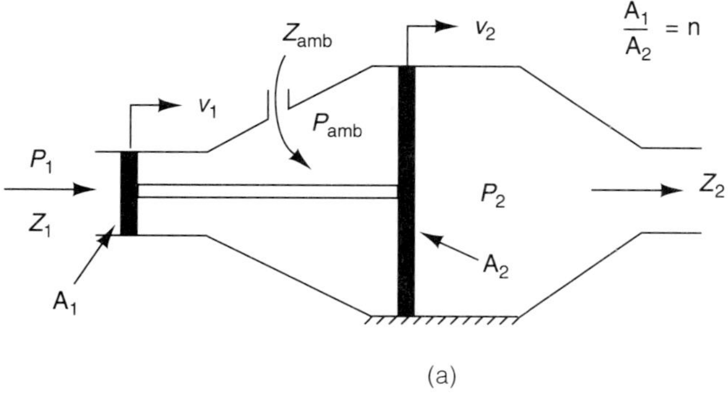

Figure 4-58. The Fluid Amplifier. (a) Schematic, (b) pressure-forced amplification, (c) flowrate-forced amplifciation

Pressure-Forced

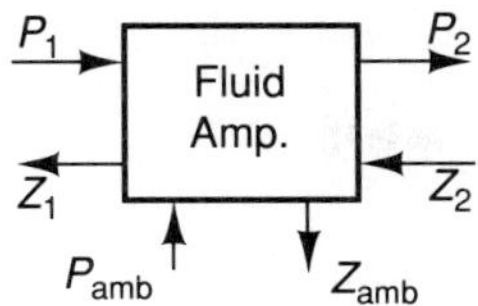

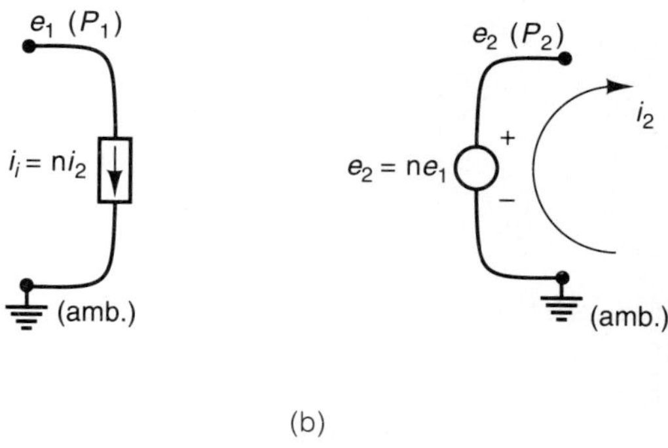

(b)

Flowrate-Forced

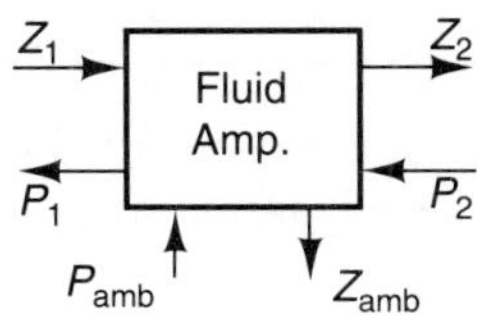

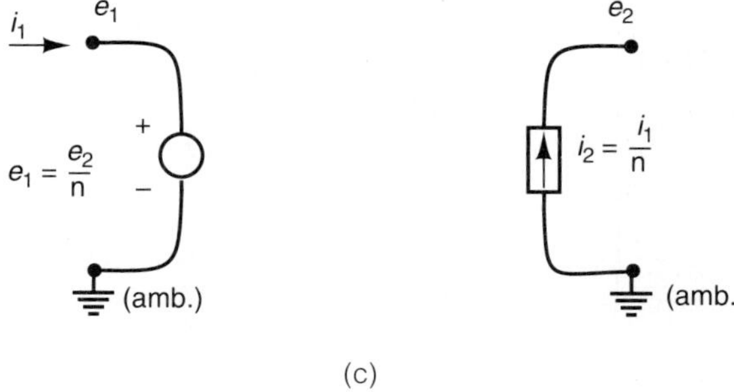

(c)

Figure 4-58. *(cont.)*

$$V_1 = V_2 = V = Z_1/A_1 = Z_2/A_2. \tag{4-99}$$

And

$$Z_1/Z_2 = A_1/A_2 = n. \tag{4-100}$$

Further, the *ideal* amplifier must satisfy

$$\text{power in} = \text{power out},$$

which is

$$p_1 z_1 = p_2 z_2, \tag{4-101}$$

by neglecting friction, compressibility, and mass effects. Combining equations gives the fluid amplifier model

$$p_1/p_2 = z_2/z_1 = 1/n. \tag{4-102}$$

Equation 4-102 is used after the *causality* of the amplifier is selected. Again, only two cases are possible, as shown in Figs. 4-58b and c. In Fig. 4-58b the element is seen as a device to transmit *pressure*. In Fig. 4-58c it is seen as a device to transmit *flowrate*. No other cases of causality are possible.

Thus, the two idealized fluid analogs of Figs. 4-58b and c may be drawn. Once more, care must be excersized to ensure that the positive sign conventions are preserved in the analog. Note that both circuits are connected to "ground," showing the role of the ambient pressure in the mechanics.

Example 4-8

An example of a fluid component is shown in Fig. 4-59. This component is proposed as a preliminary design to position the aerodynamic surface shown at the right of the

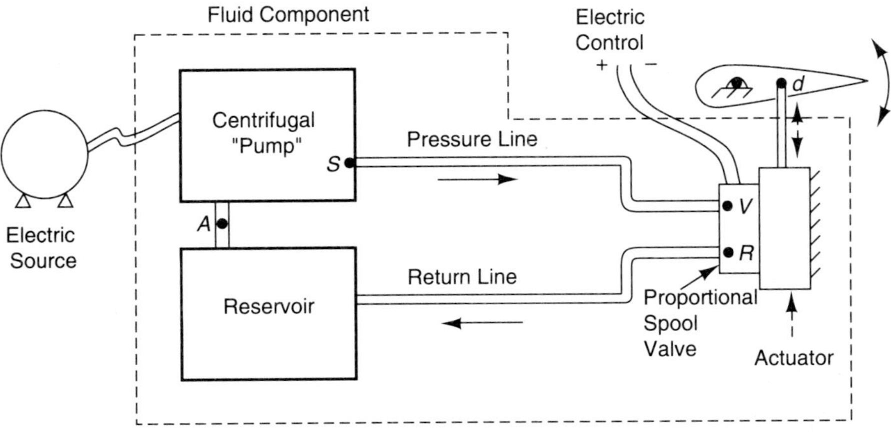

Figure 4-59. An Aerospace Positioning Fluid Component

figure. The component is powered with hydraulic oil, coming from a centrifugal constant-speed pump giving 1,500 psi pressure.

An idealized fluid-analog model is desired of this system. The electric source may be lumped with the pump, which may be idealized without output droop. The compressibility of the oil may be neglected, but the resistance and inertance of the oil must be modeled. The valve motor current, i_m, and the translational speed of point d are the system inputs. The system output is the force at d.

Solution

The multiport diagram for this component is shown in Fig. 4-60a. Note that the centrifugal pump is modeled as an ideal pressure source for the liquid. This amounts to assuming that the components of the "pump" and the electric supply are also ideal and the pump thus works perfectly. Also, the reservoir is assumed to act to supply the necessary oil that the pump flow demands, without producing any back-pressure to the return oil-line.

Given a centrifugal pump, the lines are therefore seen as conduits for pressure. The "pressure line" model is assumed to contain both inertia and friction, and the "return line" is assumed to have the same dynamics. The "reservoir" is modeled as a physical place for the depressurized fluid to collect, and need not be shown on the multiport diagram.

The proportional valve orifices are modulated by the valve opening, x, so this must be shown on the multiport. However, the analysis neglects the matching port variable (force) which goes with the x under the assumption that the force that is necessary to position the spool is negligible and may be ignored. The model is built assuming that $i_m > 0$ means $x > 0$ and the supply is connected to chamber 1 of the actuator.

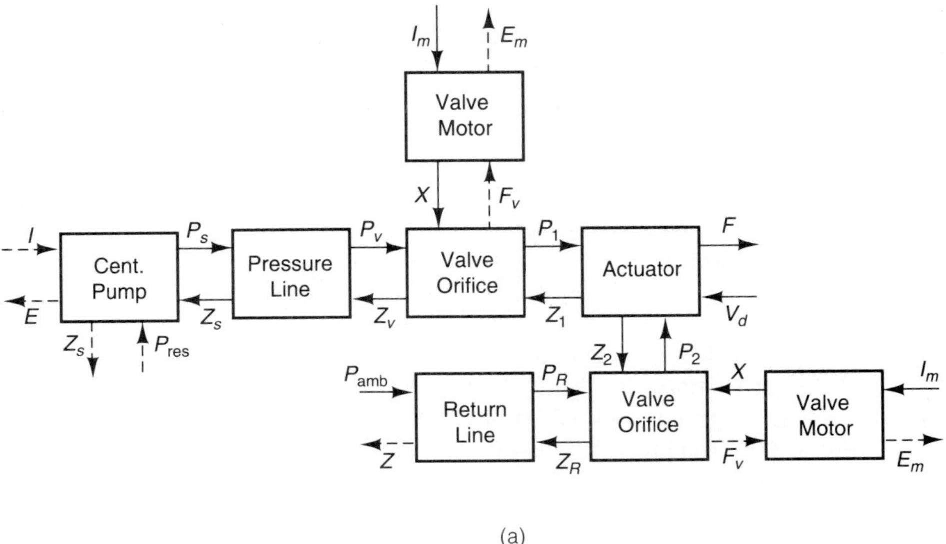

Figure 4-60. Modeling the Positioning Component. (a) Multiport diagram, (b) sign consistency, (c) fluid-analog model

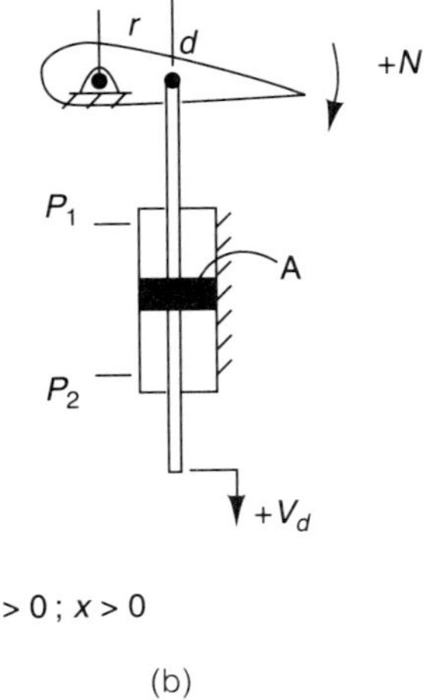

(b)

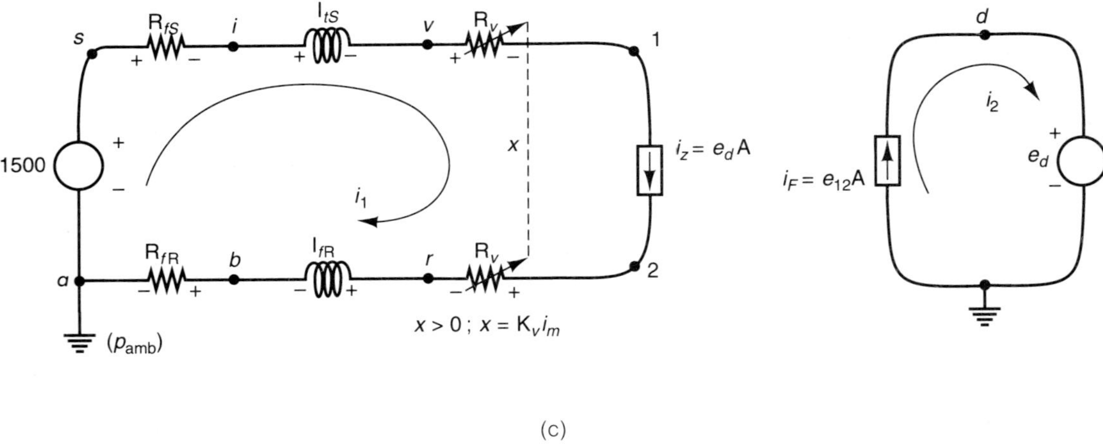

(c)

Figure 4-60. *(cont.)*

Thus, the component inputs are the valve motor current i_m, and the input speed of the aero surface, V_d. The system output is the force exerted on the aero surface, F.

The sign conventions for the analysis of the valve and actuator are shown in Fig. 4-60b. Note that the pressure difference and the translation must be consistently chosen in terms of positive directions.

The fluid-analog network is thus as shown in Fig. 4-60c. Note that the actuator analog must be forced with a voltage-analog velocity since that is one system input. The solution of this network is considered in the next section of the text.

HOMEWORK

4-38. The pump of Fig 4-56b is to be used in a design with the expected volume flowrate less than 1.4 ft^3/s. If the speed of the pump is to be 1,000 rpm (rev/min), sketch the idealized analog model of the pump and label it with the proper parameter values determined from the data.

4-39. This problem concerns the water system shown in Fig. 4-53 for a no-flow condition.

a) What is the pressure difference (in psi) between b and c if y_{bc} = 3ft. and the temperature of the water is 100°F?

b) What is the pressure in the liquid stream as it exits the valve at the hose end?

4-40. In this problem you are to develop an idealized model of a rotary actuator. In the parts of this problem you will follow the approach taken for the translation actuator of Fig. 4-57. You should assume that the fluid variables are Z and P_{12} and the structural variables are Q and N. Also, the actuator is characterized by the relationship

$$Z = DN,$$

where D is the amount of fluid displaced by the actuator per revolution. To complete the model, you should complete the following:

a) Draw the multiport diagrams for the two cases of forcing,

b) Solve for the idealized actuator model equation, and

c) Sketch the idealized analog model.

4-41. The fluid amplifier shown in Fig. 4-58 is usually a leaky device. Sketch an analog model of this lossy component assuming pressure forcing.

4-10 FLUID NETWORK ANALYSIS

So idealized fluid network models are created by combining ideal fluid elements during a physical modeling process. This task is made somewhat easier due to the fact that the fluid elements are modeled directly as electrical analogs. After the fluid network model is assembled, Kirchhoff's Laws are used to find the analog output equations for that component. In the final step, the component math model is found by reversing the analogy to the true fluid variables.

However, notice that this modeling procedure rests on applying Kirchhoff's Laws to a fluid network analogy. If the analogy is proper, then Kirchhoff's Laws must be an analogous expression of fundamental fluid physics. If this is an improper analogy, then Kirchoff's Laws may not be used.

So, recall that the fluid analogy is

$$p_{12} \sim e_{12} \text{ and } z \sim i.$$

This analogy is based on the comparison of the *energy dissipative elements* in the two systems, these being the fluid friction and the electrical resistance. The variables measured *across* these elements are seen to be pressure and voltage, and the variables measured *through* them are the flowrate and current.

Kirchhoff's Current Law (KCL) states that the sum of all currents into a node, n, must be zero. Or,

$$\Sigma_n i = 0.$$

The fluid analogy to this is that the net *incompressible* fluid flow into any rigid volume (a node) must also be zero. Or,

$$\Sigma_n z = 0,$$

which is recognized as a statement of the law of *conservation of mass*. These expressions thus represent analogous forms of the fundamental physics of *continuity*, whether for charge or mass.

Kirchhoff's Voltage Law (KVL) states that the sum of the voltage drops around any closed circuit (e.g., a mesh, m) must be zero. Or,

$$\Sigma_m e = 0.$$

The fluid analogy to this equation is

$$\Sigma_m p = 0.$$

This last equation for the pressure is seen in any fluid circuit. That is, no matter where you start in a fluid circuit, as you travel around it adding pressure rises and subtracting falls as you pass through various elements, you must finish with the pressure where you started because it is a *circuit*. Thus, KVL represents a type of *compatibility* which is analogous to that required in fluid circuits.

Consequently, both KVL and KCL may be used to analyze fluid networks since they are analogous requirements to fundamental fluid-physical relationships.

Example 4-9

A preliminary design for a pneumatic gate-raiser system is shown in Fig. 4-61a. The designer wants an idealized model to study the motion of the gate *raising* for different cases of friction at the gate rails.

The gate is raised when an operator throws a switch (not shown) which sends current to energize the solenoid and push the valve into the "raise" position. This connects the compressor to chamber 2, and exerts the lifting force needed for the mass, M, of the gate. The designer wants to study line dynamics (there is a long pressure line), but is willing to consider the compressor as an ideal, constant-pressure source. The fluid resistance offered by the vent line to chamber 1 may be ignored. Find the equations of motion for this system.

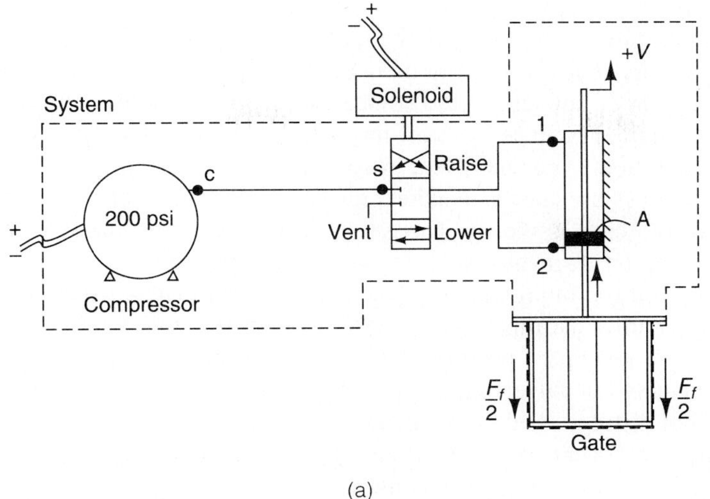

(a)

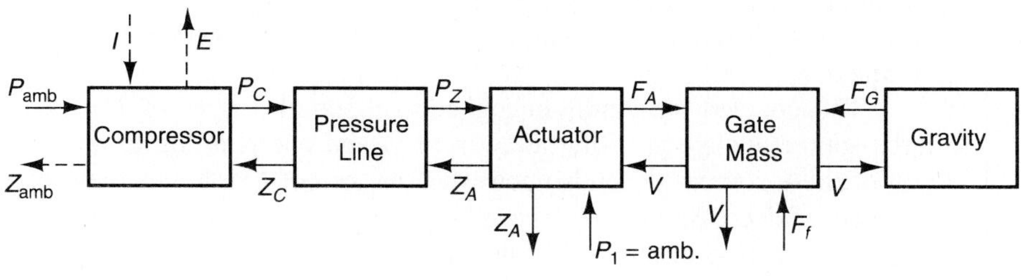

Raising Function

(b)

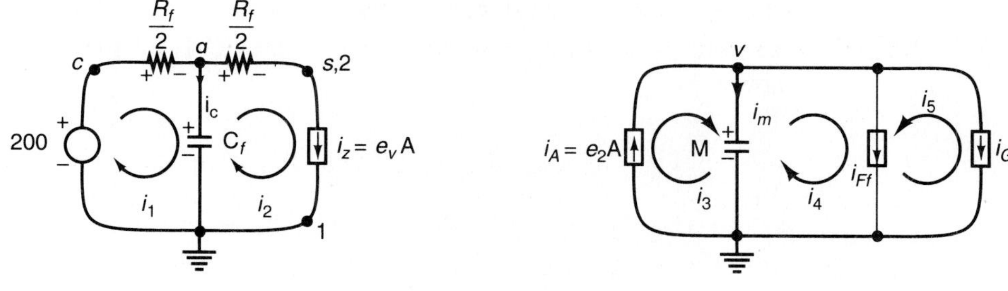

Raising Dynamics

(c)

Figure 4-61. Pneumatic Gate-raising System. (a) Schematic, (b) raising multiport, (c) fluid/structrual analog

Solution

The first thing to do is to draw the multiport diagram as in Fig. 4-61b, this gives you the causality that you need to model the actuator. Note that the gate-raising gas circuit goes from ambient to the compressor, to the pressure line, to the actuator, and back to ambient. Since the vent line offers no fluid resistance, it need not be shown in the multiport, nor modeled. Also note that the compressor is a pressure source, thus the lines are seen as conduits for pressure. Moreover, the actuator must thus be pressure-forced, as shown. The gate is seen to be a pure mass, with gravity and friction acting on it; it thus has force inputs and speed outputs.

Given the multiport diagram, the electrical analog is drawn, as in Fig. 4-61c. Note carefully that this figure is only valid for the time after the solenoid moves the valve into the "raise" position. Other positions of the valve require other figures due to the change in valve connections. Also note that the two schematic points of s and 2 are not distinguished in the analog figure, they are the same node for this analysis. Further, the "vent" and 1 nodes are not distinguished in the figure, which amounts to neglecting the vent-line dynamics.

The fluid-analog network on the left of Fig. 4-61c shows a fairly detailed "T" model of the long pressure line. In this case, resistance and capacitance are included, but fluid inertance is neglected because this is a flow of air. This assumption may have to be checked after an estimate of the computed flowrate in the pressure line is obtained.

The coupled structural-analog network on the right of Fig. 4-61c has two force-source analogs pointing away from v, and one pointing towards v. The direction of these sources depends upon whether the positive forces they represent act to increase or decrease positive speed, v. The positive sense on v is chosen to agree with the direction of the raising flowrate in the fluid circuit. You are thus in a position to begin the Kirchhoff analysis,

Step 1: The analog notation is shown in Fig. 4-61c. Note that several of the mesh currents have been chosen opposite to a given current source, they should therefore come out negative in the final solution.

Step 2: There are five meshes, as shown in the figure. KVL for these meshes is as follows:

$$\text{Mesh 1, } 200 - e_{ca} - e_{ag} = 0.$$

$$\text{Meshes } 2 - 5\text{, no KVL, a current source is present.}$$

Step 3: There are two nodes which have KCL relationships:

$$\text{Node } a,\ i_1 - i_2 - i_c = 0,$$

$$\text{Node } v,\ - i_G + i_A - i_M - i_{Ff} = 0.$$

Step 4: The element relationships are:

$$e_{ca} = (R_f/2)\, i_1 .$$

$$e_{ag} = (1/C_f) \int i_c dt .$$

$$e_2 = e_a - (R_f/2)\, i_2 .$$

$$i_2 = i_Z = e_v A .$$

$$i_3 = i_A = e_2 A .$$

$$i_M = M\, de_v/dt$$

$$i_5 = -i_G = -Mg .$$

Step 5: Manipulations can begin with *loop 1*,

$$200 - e_{ca} - e_{ag} = 0.$$

Substituting element relations

$$200 - (R_f/2)i_1 - (1/C_f) \int i_c dt = 0.$$

Also, using the node currents gives the first equation

$$\mathbf{(R_f C_f/2)\, di_1/dt + i_1 = A e_v .}$$

The second equation comes from starting at node v,

$$-i_G + i_A - i_M - i_{Ff} = 0,$$

and substituting the element relations,

$$-Mg + e_2 A - M\, de_v/dt - i_{Ff} = 0.$$

But,

$$e_2 = e_a - i_2 (R_f/2) = e_a - e_v (AR_f/2).$$

Further,

$$e_a = 200 - i_1 (R_f/2),$$

so $$e_2 = 200 - i_1 (R_f/2) - e_v (AR_f/2).$$

Combining equations gives the second governing equation,

$$\mathbf{M\, de_v/dt + (A^2 R_f/2)\, e_v = -Mg + 200A - (R_f/2)\, i_1 - i_{Ff} .}$$

The final step in the analysis is to reverse the analogy. This results in the two equations of motion for the gate-raising system

$$\mathbf{(R_f C_f/2)\, dz_1/dt + z_1 = Av ,}$$

$$\mathbf{M\, dv/dt + (A^2 R_f/2)\, v = -Mg + 200A - (R_f/2)\, z_1 - F_f .}$$

Comments

Again, as in previous examples, the last two bold equations must be solved *simultaneously* in order to compute the response of this system. These equations can be solved if the *constant* design parameters A, R_f, C_f, M, and g are all known, and if the friction model, F_f, is specified.

The checks on the equations support this solution formulation. The two first-order differential equations agree with the two energy storage elements, one in each network (mass and fluid capacitance). And the equations show higher differentials of the output than that of the inputs.

Example 4-10

The network model for this system was shown in the previous section, in Ex. 4-8, Fig. 4-60. The problem here is to convert the analog network found earlier into a math model using the network analysis approach.

Solution

The coupled fluid and structural networks are as shown in the analog of Fig. 4-60c, and you are ready to begin the analysis.

Step 1: The notation is shown in Fig. 4-60c.

Step 2: No KVL equations are possible in these networks since each mesh has a current source in it.

Step 3: No KCL equations are necessary since there are no connected meshes here.

Step 4: The element relations are the most important part of this analysis.

$$e_{si} = R_{fS} i_1 = R_{fS} i_Z = R_{fS} A e_d.$$

$$e_{iv} = I_{fS}\, di_1/dt = I_{fS}\, di_Z/dt = I_{fS} A\, de_d/dt.$$

$$e_{v1} = R_V(1 - x/X_{max}) i_1 = R_V(1 - x/X_{max})\, i_Z$$

$$= R_V(1 - x/X_{max}) A e_d.$$

$$e_{2r} = R_V(1 - x/X_{max}) A e_d.$$

$$x = K_v i_m.$$

$$e_{rb} = I_{fR}\, di_1/dt = I_{fR}\, di_Z/dt = I_{fR} A\, de_d/dt.$$

$$e_{ba} = R_{fR} i_1 = R_{fR} i_Z = R_{fR} A e_d.$$

$$i_F = A e_{12}.$$

Step 5: Equation manipulations.

The key to this problem is to recognize that the last equation is the starting point, and the difficulty is in finding e_{12}. This is true since i_F is the analog of the output F, and e_{12} may be found directly from the element relations.

To find e_{12}, you must form each voltage in the difference starting from the known conditions at a. That is, you must build up e_1 starting from the known 1,500 psi source at the left of the analog figure. So,

$$e_1 = 1{,}500 - e_{si} - e_{iv} - e_{v1},$$
$$= 1{,}500 - R_{fS}Ae_d - I_{fS}A\, de_d/dt - R_V(1 - K_v i_m/X_{max})Ae_d.$$

Similarly, working the other way from a gives

$$e_2 = e_{ba} + e_{rb} + e_{2r},$$
$$= R_{fR}A\, e_d + I_{fR}A\, de_d/dt + R_V(1 - K_v i_m/X_{max})Ae_d.$$

This allows you to compute

$$e_{12} = 1{,}500 - (R_{fS} + R_{fR})Ae_d - (I_{fS} + I_{fR})\, de_d/dt$$
$$- 2R_V(1 - K_v i_m/X_{max})Ae_d.$$

Substitution of this last equation into the i_F equation gives

$$i_F = 1{,}500A - [R_{fS} + R_{fR} + 2R_V(1 - K_v i_m/X_{max})]A^2e_d$$
$$- (I_{fS} + I_{fR})A\, de_d/dt.$$

Reversing the analogy gives the fluid/structural idealized equation of motion

$$\boldsymbol{F = 1{,}500A - [R_{fS} + R_{fR} + 2R_v(1 - K_v i_m/x_{max})]A^2v_d}$$
$$\boldsymbol{- (I_{fS} + I_{fR})A\, dv_d/dt.}$$

The two energy storage elements in the fluid mesh work together as a net-single inductance, thus there is only one first-order differential equation in the model. The differentiated input on the right side of the equation may be a concern, until you recognize that the speed and flowrate *must* be as shown in the multiport diagram of Fig. 4-60a. All the parameters of the network appear in the model equation, thus the model fully checks out.

Comments

Note that the pressure drop across the flow source, from 1 to 2, *is not zero*. In fact, you need to compute this critical pressure drop in this problem term-by-term due to the presence of the flow source in the actuator model.

Also note that the governing equation for this component is a *nonlinear* differential equation. It is nonlinear due to the product of i_m and speed, v_d, which are two component inputs. This has occurred because the idealized proportional valve is in a bi-linear form, which is really a nonlinear form of idealization.

HOMEWORK

4-42. A fluid-power designer wants to use an idealized model to investigate the response of the hydraulic oil supply component of Fig. 4-62. As you model this component, you should neglect the inertance of the oil. Assume an ideal

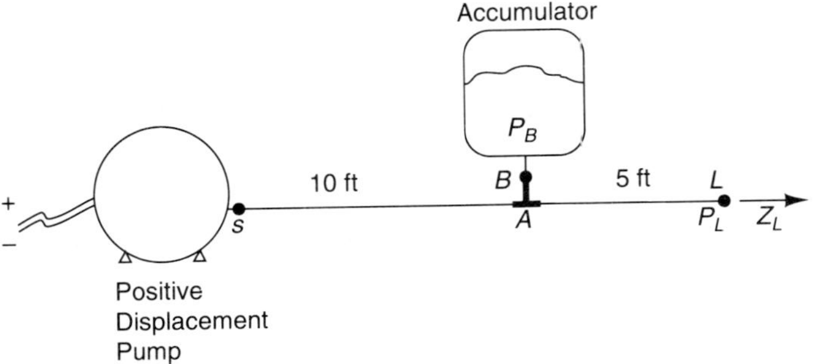

Figure 4-62. Liquid Supply Design for Prob. 4-42

pump model, with no output droop. Don't forget the resistance of the "T" joint at A (ignore the through-flow resistance compared to the line lengths). Do not try to compute the parameter values for R, C, etc.

Complete the following:

a) Find the output equation, in analog form, for the case where the *pressure* at L is the *output*.

b) Starting again from an analog figure, find the analog output equation when *flowrate* at L is the *output*.

c) Is there an easier way to find the output equation for the case of reversed forcing than starting over from first principles? What is it?

4-43. Let Z_c be the output of the fluid component shown in Fig. 4-53b. Use the network analysis method to find the output equation for this component.

4-44. A schematic for a shop compressed-air supply system is shown in Fig. 4-63. The designer is interested in the component's pressure response to a step input in flowrate demand (say, at $t = 0$ sec.), such as might happen in the case of the sudden opening of a valve in the system. You should model the compressor droop, and the capacitance and resistance of the lines. Use the analog method to find the output equation.

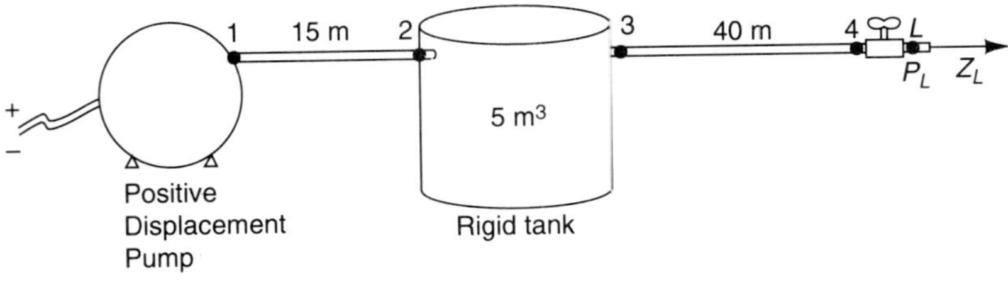

Figure 4-63. Air Supply Design for Prob. 4-44

4-11 THE THERMAL-ANALOG MODEL

Fundamental, thermal physical effects usually occur within hybrid components. These hybrids are seen as thermo-electric, thermo-structural, and thermo-fluid components. A typical thermo-electric component is the electric heater which operates as an electric resistance in an electrical circuit, and as a heat source in a thermal "circuit." Thermo-structural effects are seen in thermal sensors (low-power transducers such as the thermometer), and in structural designs with important expansion and contraction. Thermo-fluid effects are the basis for almost all engine design and all heating, ventilation and air conditioning (HVAC)—they involve heat transfer to and from a fluid (called a *thermo-fluid*) which does the work of the component.

However, in this section, the physics of heat transfer is isolated so that a menu of *pure,* fundamental thermal elements may be defined. In addition, the common types of thermal sources and reservoirs are discussed, including the thermo-electric heater. But the thermo-structural components are not discussed in this text as this is generally regarded as a detail problem by designers, and is of lesser importance in preliminary design. Also, the thermo-fluid components are deferred to Chap. 7 on input/output modeling since the fundamental physics of these components is generally beyond the assumed background of the reader.

So, when the common engineering systems are inventoried, there are few pure-thermal components, but many hybrid thermal components. Again, as in structural and fluid systems, you must work to identify and isolate the pure thermal elements in the system. And again, you use the electric analogy to build and analyze effective system models.

The Thermal Variables. The variables of the thermal physical analysis are the heat flowrate, H (Btu/s, kg-cal/s), and the temperature, Θ (°F, °C).

The *temperature* is a thermodynamic variable used to describe the state of all solids and fluids along with the pressure and volume. It is a measure of the molecular activity of a substance, and it is always measured relative to some reference temperature, of which there are several.

The first common reference is the temperature at which water freezes at 14.7 psia. This is usually matched to the temperature at which water boils at 14.7 psia to create a temperature *scale*. The SI (*Celsius*, °C) temperature scale assigns freezing water the temperature of 0 °C, and boiling 100 °C, with uniform divisions between. The English (*Farenheit*, °F) scale is created by letting freezing be 32 °F, and boiling be 212 °F, with uniform divisions between. Thus the conversion between the two scales is

$$°C = 5/9\,(°F - 32). \qquad (4\text{-}103)$$

Often, it is useful to measure the temperature from a reference condition at which the theoretical molecular activity has substantially ceased. These two points are 0 °K (SI, *Kelvin* scale), and 0 °R (English, *Rankine* scale). These two scales are related to the water scales as follows,

$$°K = °C + 273.2° \text{ (SI)}, \tag{4-104}$$

$$°R = °F + 491.6° \text{ (English)}. \tag{4-105}$$

So, when two bodies of different temperatures are brought into contact their temperatures are seen to converge as time increases. This temperature change is attributed to an exchange of *heat*, Q, from the hotter body to the colder body. The units of heat are kilogram-calories (kg-cal, SI) or British-thermal-units (Btu, English). One Btu is the amount of heat which must be added to 1/32.2 slug of water to raise its temperature 1 °F. One kg-cal is the amount of heat which must be added to raise the temperature of 1 kg of water 1 °C. These observations lead to the physical conclusion:

Wherever there is temperature difference, there is heat flow.

The rate at which heat is exchanged is called the *heat flowrate*, H. Thus,

$$H = dQ/dt. \tag{4-106}$$

The units of heat flowrate, H, are kg-cal/s or Btu/s.

There are three fundamental ways that heat flows through substances—these being conduction, convection, and radiation. Each of these demonstrates its own type of resistance to heat flow. And, as the heat flows in these ways, it is stored in the substance, thus demonstrating heat capacitance. However, common heat-flow situations do not demonstrate the phenomenon of inductance, which is the reluctance of heat flow to change once established. So, it is possible to build ideal thermal-analog modeling elements for heat-flow resistance and heat-flow capacitance.

We begin with heat-flow resistance, also called *thermal resistance*.

Thermal Resistance. All physical matter demonstrates the property of resistance to heat flow. This resistance must be analyzed for each of the three ways that heat can flow, by conduction, convection, or by radiation.

Conduction operates to transfer molecular heat energy by direct physical contact of molecules. An example of this is shown at the left of Fig. 4-64a for a block of solid material. The left-side surface area of the block is at a uniform temperature, Θ_1. The right-side surface area is at a uniform temperature, Θ_2. The top and bottom of the block are insulated to stop heat flow out of those surfaces. This situation leads to a uniform linear temperature change through the solid material as shown in the figure. The temperature difference between the two surfaces thus forms a driving potential for the heat flow which is *always* directed from high to low temperatures.

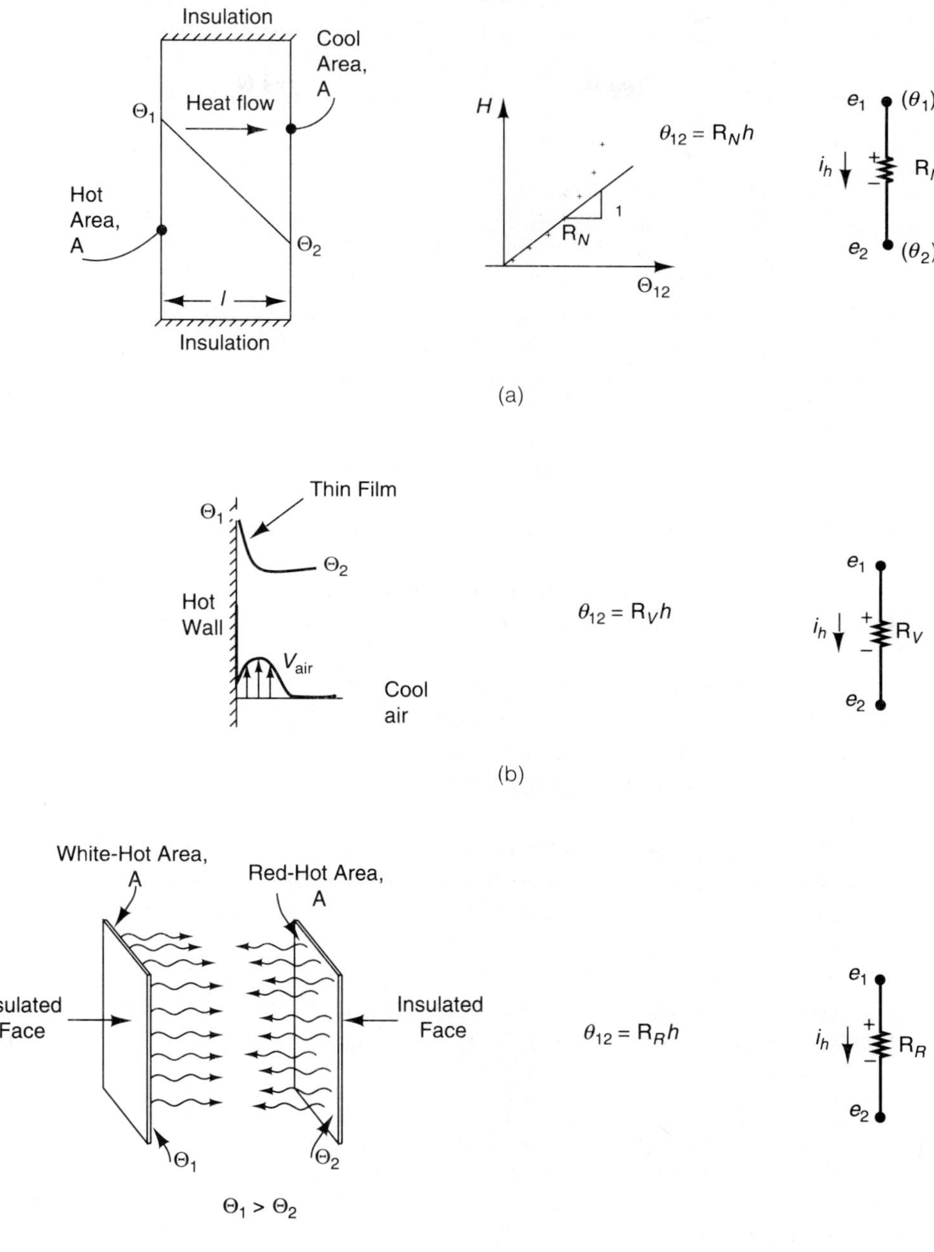

Figure 4-64. Heat-flow Resistance. (a) Conduction, (b) natural convection, (c) radiation

When H versus Θ_{12} data is taken for heat conduction through a particular material, the plot is similar to that shown in the center of Fig. 4-64a. The nonlinear behavior of the material arises, in part, because the higher temperatures eventually lead to a change of material phase due to melting or vaporization. When the conduction response is linearized, the idealized response may be modeled as:

$$\theta_{12} = \mathrm{R}_N\, h. \tag{4-107}$$

Your first insight to the thermal analogy for modeling becomes apparent when this last equation is compared to that for the electrical resistor:

$$e_{12} = \mathrm{R}i. \tag{4-108}$$

Thus, the *temperature* must be the *across* variable, while *heat flowrate* must be the *through* variable. Consequently, the thermal-analog element symbol for conduction is as shown at the right of Fig. 4-64a. It embodies the thermal analogy

$$h \sim i,$$

and

$$\theta_{12} \sim e_{12}.$$

Note that the thermal-analog model is formed directly, just as it was for the fluid-analog elements.

The computation of the heat-conduction resistance, R_N, comes from Fourier's Law for heat flow, which is in itself an idealization of conduction,

$$h = \mathrm{k}_C\, \mathrm{A}\, \theta_{12}/\ell, \tag{4-109}$$

where k_C = thermal conductivity (kg-cal/s-m-°C, Btu/s-ft-°F),
A = crossectional area through which the heat flows (m^2, ft^2),
ℓ = length of the heat path (m, ft).

Some representative values of k_C are given in Table 4-7. Note the small value of conductance for still air. This implies its excellent ability as an insulator, and shows why walls are designed and built with "dead air" spaces to improve insulation qualities. Fourier's Law thus gives

Table 4-7. Thermal Conductivities

	(k_C, *Btu/s-ft-°F*)
Air, still	0.39×10^{-5}
Glass wool, blanket	0.61×10^{-5}
Wood, pine	0.181×10^{-4}
Sand, dry	0.52×10^{-4}
Brick	0.75×10^{-4}
Water, still	0.103×10^{-3}
Concrete	0.114×10^{-3}
Glass	0.164×10^{-3}
Earth, wet	0.172×10^{-3}
Steel	0.73×10^{-2}
Aluminum	0.036
Copper	0.064

$$\theta_{12} = \ell/(k_C A)\, h. \qquad (4\text{-}110)$$

$$= R_N\, h,$$

and

$$R_N = \ell/(k_C A). \qquad (4\text{-}111)$$

There are two types of heat transfer by *convection* with which modelers are concerned: those coming from *natural* convection and those coming from *forced* convection. Natural convection occurs whenever a fluid is heated. This heating creates pockets of warm fluid which are less dense than the surrounding cool fluid. These warm pockets are thus more buoyant and tend to rise, thus "convecting" the heat away. The heated fluid is replaced by cooler fluid which, in turn, is heated. Thus, natural convection tends to *decrease* the resistance to heat flow in a fluid compared to conduction alone. Forced convection works in a similar fashion, except the fluid motion is forced through the use of a pump or fan.

However, both kinds of heat convection in fluids are much too complicated to approach with a fundamental analysis given the assumed background of the reader. Nevertheless, a *conceptual* understanding of a convection process is often achieved using idealized elements in evaluating preliminary designs. The difficulty is that the resistance parameter in the lumped convection model is almost impossible to estimate.

Consider, for example, Fig. 4-64b (left) which shows the upward velocity of heated air near a hot, vertical wall. The figure also shows the temperature profile of the air near the wall. Note that the temperature change mostly takes place in a thin film near the wall. In fact, system modelers often use a *thin-film heat transfer coefficient* to model the convection resistance, which works very much like the k_c used in the conduction heat transfer discussed earlier. Using this approach, an ideal convection resistive element may be defined,

$$\theta_{12} = R_V\, h, \qquad (4\text{-}112)$$

with the idealized analog element as shown at the right of Fig. 4-64b.

The danger in such a simple convection model is that it completely decouples the heat circuit from the fluid circuit. At the very least, you might intuitively expect that the convection heat resistance, R_V, is dependent on the type of fluid and its speed over the heated surface. These thermal-fluid coupling issues are best dealt with through an experimentally based input/output model, using the methods described in Chap. 7. The simple, idealized model above is presented merely to show one kind of common *preliminary* thermal model. Its utility, however, is severely undermined by the difficulties in the calculation of R_V for most practical situations in preliminary design. In other words, the idealized convection model (Eq. 4-112) works well for conceptual evaluation problems, but is ineffective for quantitative estimation problems.

The final form of heat transfer is *radiation*. Once again, a situation is found where the fundamental approach to compute thermal resistance has theoretical difficulties which are severe. However, some useful idealization is possible in this case.

Consider the pure-radiation case shown in Fig. 4-64c (left). Here, two plates of area, A, are parallel and facing each other across a vacuum. The facing sides of the plates are exposed and heated, while the opposite sides of the plates are insulated. Each of the uninsulated surfaces absorbs and emits electromagnetic radiation, which is perceived at the surfaces in the form of temperature. The plates are closely spaced, so they "see" little besides the opposing surface. In this way, the background is neglected which is also exchanging radiation with the two exposed surfaces. Surface 1 is at temperature 1, and surface 2 is at temperature 2, such that,

$$\Theta_1 > \Theta_2.$$

This means that heat flows from surface 1 to surface 2 in the form of electromagnetic radiation, and the resistance to this heat flow may be calculated.

The physical analysis is begun with an energy balance written for the receiving surface, surface 2,

$$\text{(heat flow in)} = \text{(radiated power in)} - \text{(radiated power out)}. \qquad (4\text{-}113)$$

Ideally, if surface 2 is a hemisphere painted dull black, then the Stefan-Boltzman Law would apply,

$$\text{radiated power out of 2} = A\sigma\, \Theta_2^4, \qquad (4\text{-}114)$$

where A = area of the emitting surface (ft^2, m^2),
σ = the Stefan-Boltzman constant,
= 4.76×10^{-13} Btu/s-ft^2-°R^4,
= 1.37×10^{-5} kg-cal/s-m^2-°K^4,
= 5.67×10^{-8} watts/m^2-°K^4.

This approach is used to compute the so-called *black-body radiation*. It forms the standard of comparison which is used to estimate the *actual* radiated heat for other surface shapes and types.

Similarly, if surface 1 is also a hemispherical black body, then it would emit the radiation

$$\text{radiated power out of 1} = A\sigma\, \Theta_1^4. \qquad (4\text{-}115)$$

This situation is idealized by assuming that the radiated power out of 1 is equal to the radiated power into 2. Thus,

$$\text{radiated power into 2} = A\sigma\, \Theta_1^4. \qquad (4\text{-}115a)$$

So, the net black-body heat transfer received at surface 2 is

$$H = A\sigma\, (\Theta_1^4 - \Theta_2^4). \qquad (4\text{-}116)$$

This is the same heat flowrate emitted by surface 1. And, because both surfaces are assumed to be hemispherical black bodies, the radiation heat transfer only depends upon their surface temperatures and areas.

Practically though, you must account for non-ideal geometry and surfaces. Every surface is not spherical, nor are they all colored dull black. To accomodate these variables, the dimensionless parameters of *emittance*, ε, and *absorptance*, α, are used to characterize surfaces in conjunction with the basic formula (Eq. 4-116). Further, these parameters are defined for the special case of *plane surfaces*, when they are referred to as the *emissivity* and the *absorptivity*. For this special case, the basic formula becomes

$$H = A\sigma\,(\alpha_{12}\,\Theta_1^{\,4} - \varepsilon_2\,\Theta_2^{\,4}). \tag{4-117}$$

Here, the absorptivity, α_{12}, characterizes surface 2, the receiving surface, as well as the nature of surface 1, the emitting surface. The emissivity, ε_2, characterizes the surface 2 only. Some representative values for these radiation parameters are given in Table 4-8.

Table 4-8. Planar Surface Radiation Properties. (All Dimensionless)

Material	α[1] *(Solar source)*	α *(70°F source)*	ε *(70°F)*
Flat-black lacquer	0.95	0.95	0.95
Rough red brick	0.55	0.92	0.95
Rubber, soft gray	0.65	0.86	0.85
White enamel	0.40	0.86	0.95
Polished aluminum	0.30	0.08	0.07

[1] Values based on a solar source at 10,600 °R, with area of 1.638×10^{19} ft^2, at a distance of 4.91×10^{11} ft.

The ideal radiation heat-flow resistance is obtained by linearizing Eq. 4-117 about some operating point, Θ_{1A}, Θ_{2A}, and assuming $\alpha_{12}\,\Theta_{1A}^3 \approx \varepsilon_2\,\Theta_{2A}^3$. Thus,

$$\theta_{12} = R_R\,h, \tag{4-118}$$

where

$$R_R = 1/[4A\sigma\,\alpha_{12}\,\Theta_{1A}^3]. \tag{4-119}$$

This resistance remains accurate when the absolute temperatures do not vary much from the operating point. Note that this *offset linearization* is very similar to that for the prime movers discussed earlier in the section of the text on structural sources. Again, the linearization leads to a resistor-like equation, and the ideal symbol is as shown in Fig. 4-64c (right).

In summary, the three kinds of fundamental heat flow are conduction, convection, and radiation. In each case, an ideal, fundamental temperature-flowrate model is developed which has the same *form* as that for an electrical resistor. The ideal thermal elements rest on this similarity of form.

However, the thermal resistance *does not dissipate energy* as does the electrical resistance, and the other resistances. In the thermal case, the through variable is the heat flowrate, H, which is also the heat power (energy rate). Thus, since the heat-in is typically the same as the heat-out for the thermal resistor, no energy is lost to dissipation.

The Pure Thermal Inductance. A thermal inductor would be an element which would store heat energy by virtue of the thermal through variable, H. That is, it would show a reluctance to change its heat flowrate once the heat flow is established. However, a search for cases where this fundamental property occurs reveals that there are no practical situations which demonstrate thermal inductance.

The Pure Thermal Capacitance. The ability to store heat energy by virtue of temperature, the thermal across variable, is a property demonstrated by all matter. This concept is shown in Fig. 4-65.

In Fig. 4-65a, a substance is shown which is thermally isolated to measure its heat-holding capacity. The substance is assumed to have a uniform temperature, Θ_1, measured relative to some reference temperature (say, the freezing of water). The insulation restrains loss of heat from the volume.

When heat is added to the substance the temperature rises, as in Fig. 4-65b. The nonlinearity of the data is due to the inter-molecular frictional losses as the temperature rises and, possibly, to the start of a change of state. For idealization purposes, the thermal capacitance is linearized as shown, and the electrical analogy of Fig 4-65c is used in a thermal network.

In order to evaluate the coefficient of thermal capacitance, C_T, the substance in the capacitance volume must be considered.

For solids, the thermal capacitance is the same as the *specific heat*, and the governing physics follow the relationship

$$Q = MC_M\Theta_1, \tag{4-120}$$

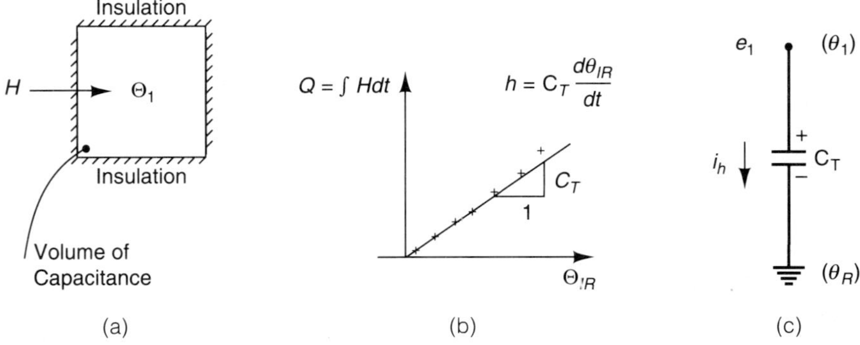

Figure 4-65. Thermal Capacitance. (a) Schematic, (b) linearization, (c) analogy

where Q = heat (Btu, cal),
M = mass of the solid (sl, kg),
C_M = specific heat of the solid (Btu/sl-°F, kg-cal/kg-°C).

Note that the last equation is in the form

$$\int H dt = MC_M \Theta_1. \tag{4-121a}$$

And, under the assumption of a constant specific heat, the idealized thermal capacitance form is obtained,

$$h = C_T \, d\theta_1/dt, \tag{4-121b}$$

such that

$$C_T = MC_M. \tag{4-122a}$$

Again, like the fluid capacitive elements, the heat capacitance of a substance is *always* relative to a reference temperature, so the thermal capacitance must *always* be connected to "ground" in the thermal-analog network. This idea is shown in Fig. 6-65c.

For liquids and gases, the *manner* in which the heat is added is important to the heat capacitance of the fluid. The usual approach is to add heat at either constant pressure or at constant volume. This text focuses on the *constant-pressure* approach, since that is most common for the widest variety of thermal systems.

The constant-volume approach of adding heat to fluids is mostly found in combustion designs, where the volume of fluid is trapped in a chamber, as in a car engine. This approach thus usually leads to a coupled thermo-fluid analysis which may be best treated using the methods in Chap. 7 on input/output modeling.

For the constant-pressure addition of heat to liquids and gases, the constant-pressure specific heat, C_P, is used for the lumped capacitance,

$$C_T = MC_P. \tag{4-122b}$$

Some representative values for C_M and C_P are given in Table 4-9 (note that ρC is tabulated since $MC = \mathcal{V}\rho C$).

Thermal Power Sources. The thermal power sources may occur in two types: those which are temperature sources (the thermal across-variable) and those which are heat flowrate sources (the through-variable). Several common forms of the thermal sources are shown in Fig. 4-66.

The ideal *across-source* models are used when the temperature remains constant, regardless of heat flow—this is true whenever a substance acts as an infinite heat capacitance. Such capacitances are called heat *reservoirs*. The atmosphere, for

Table 4-9. Volume Heat Capacity of Substances

(Btu/ft³-°F at 68°F, 14.7 psia)

Solids, ρC_M		*Liquids, ρC_P*		*Gases, ρC_P.*	
Aluminum	38.0	Alcohol	29.0	Air	0.1817
Asbestos	30.6	Gasoline	22.2	Helium	0.01303
Brick	33.0	Machine Oil	22.7	Hydrogen	0.01786
Concrete	20.3	Sea Water	60.2	Methane	0.0247
Glass	32.2	Water	62.4	Nat. gas	0.0284
Granite	40.1				
Gypsum	41.2				
Steel	53.3				
Wood	20.1				

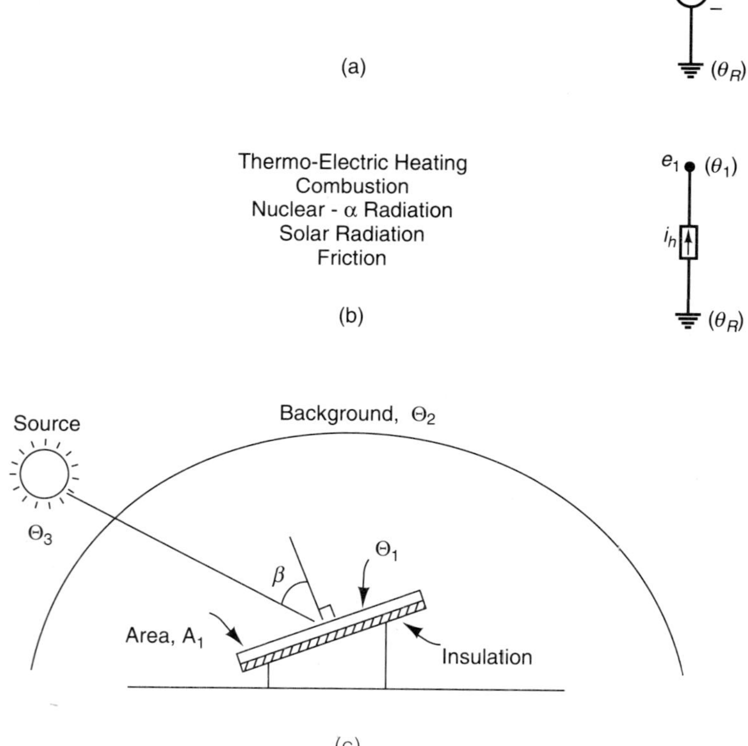

Figure 4-66. Thermal Sources. (a) Across, (b) through, (c) radiation geometry

example, acts like a constant-temperature heat reservoir. This is also true of space. In addition, this is true whenever a substance changes its physical phase. For example, when heat is removed from water, it eventually passes through a solidifying phase as it changes to ice. As it passes from all-water to all-ice, or vice versa, the temperature remains essentially constant at 32 °F (0 °C). Similar behavior occurs when water becomes steam at 212 °F (100 °C). In fact, any substance undergoing a *phase change* behaves as a very nice constant-temperature source.

The ideal *through-source* models are used whenever the heat flowrate is constant, regardless of temperature. However, the naturally occurring forms of heat-flowrate sources are usually less ideal than the temperature sources since all substances possess thermal resistance. Thus, a change of temperature *must* accompany heat flowrate. Still, this is often a useful idealization in the cases listed in Fig. 4-66b.

For practical examples, consider the case of solar radiation which is used in designs to cause heat flow on both small and large scales. On a small scale, solar heating is accomplished by painting the inside of an open box with a flat-black color, and enclosing the open side with clear glass. The glass side is then directed to the sun so that the visible light falls on the black interior. When the "hot box" interior is exposed to the electro-magnetic sunlight, it warms up and emits heat in the form of infrared waves, which find the glass to be opaque. Heat is thus trapped inside the box. This scheme has been used successfully to cook meat, to boil water, and to heat air for house warming.

On the large scale of solar heating, a vast array of mirrors is used to concentrate the sunlight energy so that enough heat is generated to run a medium-sized electric power plant. In fact, many forms of large-scale heat-flow designs use a radiant mechanism to transmit or receive heat, so it is well worth the effort to examine this further.

The geometry of a radiant-heat receptor is shown in Fig. 4-66c. Note that this is different from the earlier radiation analysis in that here the source only covers a small part of the receptor's horizon, and there is a substantial background with which to exchange radiation. Further, the receptor is now allowed to misalign with the incoming rays. The receptor is assumed to be a plane area, A_1, at temperature Θ_1. The background is assumed to be at temperature Θ_2, and the source at temperature Θ_3.

This time, the energy balance for the receptor area is,

$$H_{IN} = \sigma[A_1(\alpha_{21}\,\Theta_2^4 - \varepsilon_1\Theta_1^4) + (A_1\cos\beta)\,\alpha_{31}\,(\Theta_3)^4\,(A_{3/1})/(\pi R^2)]. \tag{4-123}$$

The first two terms in this equation are the same as before (Eq. 4-117), but they now represent the exchange of radiant energy with the *background*. The last term is the added term for the energy received from the source. It comes from the black-body radiator equation (Eq. 4-114), which is modified to account for distance of the source area ($A_{3/1}/\pi R^2$), and the angle the rays hit the receptor (β). Thus,

α_{21} = the absorptivity of plane surface 1 due to a background at temperature Θ_2 (dimensionless),

α_{31} = the absorptivity of plane surface 1 due to a black body source at temperature Θ_3 (10,600 °R for the sun),

$A_1 \cos\beta$ = the receptor area normal to the incoming source radiation,

$A_{3/1}$ = the apparent (visible) area of the source when viewed from the direction of the receptor (e.g., a circle for the sun = 1.638×10^{19} ft^2),

R = the distance from the source to the receptor (sun to earth = 4.91×10^{11} ft).

Given this model, it is possible to compute the heat-flowrate source strength for a radiatiant-heat source, whether it be the sun or some other thermal energy source.

Thermal Amplifiers and Transducers. The process of thermal amplification is unknown, so there are no thermal amplifiers.

Thermal transducers are only found in low-power designs (e.g., as sensors). These are most common in designs using thermo-electric resistance changes, and thermo-structural shape changes due to expansion and contraction. However, sensors are not studied in this text.

So, the menu of ideal thermal elements is complete. These elements may be used to idealize the physics in the vast majority of thermal-system designs. And a thermal network model may thus be assembled for mathematical analysis.

Thermal Network Analysis. As in the other network types, it would be very useful if the thermal-analog model could be analyzed using Kirchhoff's Laws. Before this is done, however, you must be certain that the electrical laws have fundamental meaning in terms of the thermal physics.

Kirchhoff's Voltage Law for an electrical mesh is

$$\Sigma_m e = 0,$$

which has the thermal analog

$$\Sigma_m \theta = 0.$$

This is recognized as a statement of the fundamental *compatibility* of the temperatures around a mesh or circuit. It is a statement that, no matter where you start in the circuit, when you complete a tour around that circuit adding and subtracting temperature changes as you go, you must arrive back where you started, temperature-wise.

Kirchhoff's Current Law for a node is

$$\Sigma_n i = 0,$$

which has the thermal analog,

$$\Sigma_n h = 0.$$

This is recognized as a statement of the fundamental *continuity* of heat flow. That is, the net heat flowing into a node must equal the net heat flowing out of that node in a pure-thermal component. This is a restatement of the principle of conservation of energy, that heat (energy) may not be created or destroyed at a node.

Consequently, both of Kirchhoff's Laws for electrical networks may be applied to a thermal analog network with complete fundamental accuracy.

HOMEWORK

4-45. In the experiment of Fig. 4-67, a shelter designer wants to evaluate the effectiveness of a red brick wall of 1 ft thickness to insulate the interior of a $8 \times 8 \times 10$ft room from a nearby heat blast.

The input is to be the ignition of a spherical black-body radiant source at 5,000°F, 20 ft from the wall, which starts at $t = 0$ sec. when everything is at 70°F. An atmospheric reservoir at 70°F surrounds the experiment. The designer is interested in the temperature in the center of the room (at 4 in the figure). You should ignore convection everywhere, and radiation inside the room. Choose your model nodes at the points shown in the figure (e.g., 1, 2, 3, 4, and 5), and complete the following:

a) Build an idealized analog model of the heat transfer, modeling the radiant source as a temperature source. Don't forget to model the aluminum back wall.

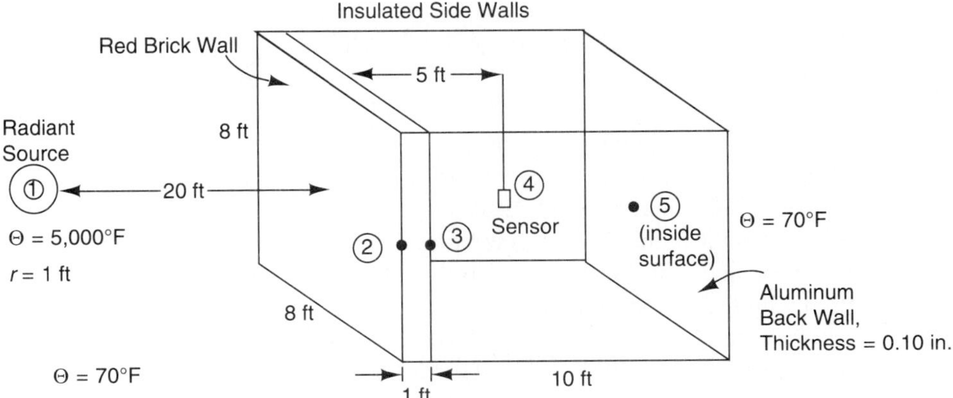

Figure 4-67. Geometry for Prob. 4-45

b) Build another idealized analog model, modeling the radiant source as a heat source feeding directly into the brick wall. Since aluminum is such a good heat conductor, ignore its presence in this model and treat the wall as if it were at a constant temperature of 70 °F.

c) What is the output equation for (b)?

d) Compute the values of all resistances, sources, and capacitances for the models of (a) and (b) above. Linearize around 5000°F, assume constant α and ε, and treat the source/wall as in Eq. 4-119.

4-46. Compute the heat flowrate into the solar collector of Fig. 4-66c. You may assume that the atmosphere has no effect on the incoming radiation. The background and the collector are at a temperature of 70 °F. The collector is inclined at an angle of $\beta = 30°$, and the collector area is 10 sq ft. The collector interior is painted with flat black lacquer.

4-47. An HVAC engineer is interested in modeling the temperature in an insulated room. Heat flows into the room from a forced air heating system, and heat flows out through a large picture window. The room is 10 × 10 × 10 ft, and the window is a single pane of glass which is 4 × 4ft and 0.25 inches thick.

You should model the room as a perfectly insulated volume, which is a single, lumped capacitance at a uniform temperature Θ_1. The heat input to the room can be modeled as a heat source. Ignore convection and radiation, and the heat capacity of the glass (it is a small volume). Treat the outside of the picture window as an infinite heat reservoir at a uniform temperature of 10 °F.

Complete the following:

a) Sketch the analog circuit model.

b) Compute the heat capacitance of the room and the heat resistance of the glass.

c) What is the equation for the system output, the room temperature?

4-12 SUMMARY, BIBLIOGRAPHY, AND REVIEW PROBLEMS

This chapter presented fundamental modeling methods for those components designed to operate using time-domain inputs (impulse, step, ramp, etc.). The objective was to define a menu of idealized elements to be used to model the fundamental physics found in an array of common engineering designs. The fundamental physics that was analyzed and modeled in this chapter was in the form of electrical, structural, fluid, and thermal effects.

Three types of element idealizations were found to occur again and again in the physical analyses—these being the resistance, the inductance, and the capacitance. These element types were developed into a rigorous analogy based upon the comparison of the energy-dissipative elements in the electrical, structural, and fluid physics. This was done to develop your understanding of the underlying similarity of the basic physical processes in all engineering systems. Further, this type of analogous reasoning is very characteristic of systems thinking, and can be a powerful ally—if properly applied.

The analogies presented here were expressed using an across-variable and a through-variable for each ideal element. In this, the across-variable was measured "across" the element (voltage, velocity, and pressure), while the through-variable was measured "through" the element (current, force, and fluid flowrate). In addition, the equation form for the thermal elements was found to be such that they can also be included in the analogies even though there was no dissipative element among the thermal physics. For the thermal elements, the across-variable was the temperature, and the through-variable was the heat flowrate. Thus, the analogies were defined as summarized in Fig. 4-68.

Recall that the idealized physical elements summarized in Fig. 4-68 are lumped, pure, and linear. Further, the lumped elements shown here strictly model only one-dimensional physics. For higher dimensions in modeling, other techniques should be used (e.g., input/output modeling, free-body diagrams, or some other method to handle the inter-dimensional physics).

The idealizations shown in Fig. 4-68 are also pure. That is, they are based on experimentation which is designed to isolate the pure, fundamental, physical response which they model. It is recommended that another modeling method be found for those situations where the element physical types are too closely coupled to be accurately separated into pure effects (i.e., thermo-fluid convective components are best treated with input/output modeling).

Finally, the fundamental physics of Fig. 4-68 are idealized by linearization to allow for quantitative understanding of the cause-and-effect of preliminary designs without the confusing aspect of nonlinearity.

Thus, the fundamental, idealized elements may be assembled to create an electrical-analog network model for preliminary design studies. Given this network model, Kirchhoff's Laws for electrical circuits are used to develop mathematical models of the designs. In each case of electrical, structural, fluid, and thermal components, these laws have some fundamental analogy to the actual physics of the component under study. Thus, they may be used successfully to find mathematical models for quantitative design studies.

During all of your physical analysis and modeling, you are well-advised to remember the Fundamental Truth of Modeling—that all physical effects are nonlinear and distributed. Therefore, *every* model is an approximation and idealization, not only those presented in this chapter. This leads you to the *First Law of Modeling:*

> Keep the model simple.

When you model, it is better to err on the side of being too simple than on the side of being unnecessarily complicated. Do not use two lumps when one will do; do not use a nonlinear model when a linear one is essentially correct. You seek to cap-

Fundamental Element		Analogy	
Electrical Resistor	$e_{12} = (R)i$		
Structural Translational Damper	$v_{12} = (1/b)f$	i_h	
Structural Rotational Damper	$r_{12} = (1/B)q$	e_1 + − e_2	$e_{12} = (\ )^*i$
Fluid Friction	$p_{12} = (R_f)z$	$(\ )^*$	
Thermal Resistor	$\theta_{12} = (R_T)h$		
Electrical Inductor	$i = (1/L)\int e_{12}dt$	i_h	
Structural Translational Spring	$f = (k)\int \upsilon_{12}dt$		$i = (\ )\int e_{12}dt$
Structural Rotational Spring	$q = (K)\int n_{12}dt$	e_1 + − e_2 $1/(\ )$	
Fluid Inertance	$z = (1/I_f)\int e_{12}dt$		
Electrical Capacitor	$i = (C)de_{12}/dt$	e_1	
Structural Mass	$f = (M)d\upsilon_{IR}/dt$	+ i	
Structural Inertia	$q = (J)dn_{IR}/dt$	$(\ )$ −	$i = (\ )\dfrac{de_{12}}{dt}$
Fluid Capacitor	$z = (C_f)dp_{IR}/dt$	e_2	
Thermal Capacitor	$h = (C_T)d\theta_{IR}/dt$		
Voltage Source	e_1	Current Source	
Translational Speed Source	+	Force Source	
Rotational Speed Source		Torque Source	i
Fluid Pressure Source	−	Fluid Flowrate Source	
Temperature Source	e_2	Heat Flowrate Source	
Amplifiers		Multiport first, then analog sources	
Transducers			

*( ) = constant from element equation

Figure 4-68. The Idealized-element Analogies

ture the essential physical function of the component in the simplest model possible. In this way, you create a model which presents the *important* cause-and-effect dynamics for design studies and improvement.

BIBLIOGRAPHY

Allocca, J.A. and A. Stuart, *Transducers, Theory and Application*, Reston, Virginia: Reston Publ. Co., 1984.

Baumeister, T. and L.S. Marks (eds.), *Standard Handbook for Mechanical Engineers*, New York, N.Y.: McGraw-Hill Book Co., 1967.

Cannon, R.H. jr., *Dynamics of Physical Systems*, New York, N.Y.: McGraw-Hill Book Co., 1967. This classic textbook has an excellent discussion of the usefulness and limits of analogies. It also has further reading on two- and three-dimensional mechanics problems in a systems context.

Close, C.M. and D.K. Frederick, *Modeling and Analysis of Dynamic Systems*, Boston, Mass.: Hougton Mifflin Co., 1978.

Karnopp, D.C., Margolis, D.L., and R.C. Rosenberg, *System Dynamics: A Unified Approach*, New York, N.Y.: J. Wiley & Sons, Inc., 1990. Karnopp, et. al., have led the way with power bond graphs as a means for system analysis. These graphs are a more detailed method than analog network modeling. There is a good deal of theory behind the method, so it is more powerful. It makes a good follow-on study to the analog method.

Kay, J.M. and R.M. Nedderman, *Fluid Mechanics and Transfer Processes*, New York, N.Y.: Cambridge University Press, 1985.

Kiehul, H.A. (ed.), *Battery Technology Handbook*, New York, N.Y.: Marcel Dekker, Inc., 1989.

Nilsson, J.W., *Electric Circuits*, Reading, Mass.: Addison-Wesley Publ. Co., 1990.

Ogata, K., *Modern Control Engineering*, Englewood Cliffs, N.J.: Prentice-Hall, Inc., 1970. This is a classic text, with the first few chapters devoted to system modeling. Analogies and a method of signal flow graphs are used to formulate the math models. This book is mainly SISO systems.

Reid, J.G., *Linear System Fundamentals*, New York, N.Y.: McGraw Hill Book Co., 1983. This is a thorough treatment of SISO systems.

Say, M.G. and E.O. Taylor, *Direct Current Machines*, New York, N.Y.: J. Wiley & Sons, 1980.

Shearer, J.L., Murphy, A.T., and H.H. Richardson, *Introduction to System Dynamics*, Reading, Mass.: Addison-Wesley Publ. Co., 1967. This is perhaps the classic text in the area of modeling engineering systems. The book uses linear graphs and SISO block diagrams to formulate component models.

Shepherd, D.G., *Elements of Fluid Mechanics*, New York, N.Y.: Harcourt, Brace & World, Inc., 1965.

Soo, S.L., *Direct Energy Conversion*, Englewood Cliffs, N.J.: Prentice-Hall, Inc., 1968. This book has a nice discussion of solar energy use.

REVIEW HOMEWORK

4-48. What are the three characteristics of any ideal element?

4-49. If a spring has the nonlinear constitutive relationship

$$F = 2\ \mathrm{k}X^2,$$

show that superposition fails for this element.

4-50. Classify the following expressions for linearity. Indicate the reason(s) for non-linearity as appropriate:

a) $(\sin t)\ dy/dt + y^2 = 0.$

b) $u\ dy/dt + (\cos t)y = t^3.$

c) $3\ dy/dt + uy = \log(t + 1) + 2\ du/dt.$

4-51. An electrical designer wants a model for the electrical component in Fig. 4-69. You should ignore the resistance of the capacitor and the connecting leads, and complete the following:

a) Sketch the ideal model of the component,

b) Calculate the values of the element constants.

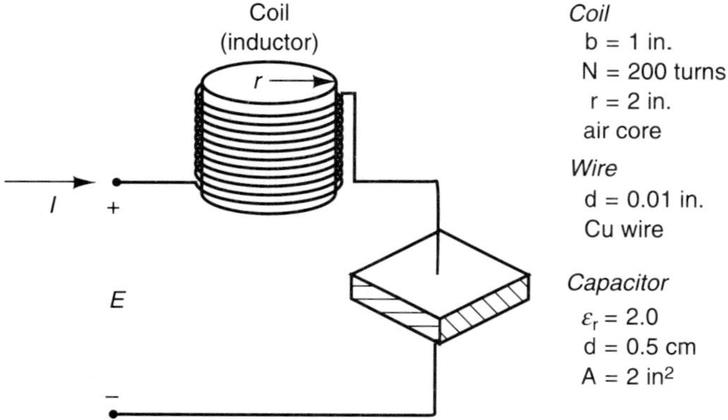

Figure 4-69. Electronics for Prob. 4-51

4-52. Given the battery data in Fig. 4-70:

a) What value of internal battery "resistance" does this data indicate?

b) Sketch the idealized model for this battery, and quantitatively assign the element constants.

4-53. For the components shown in Fig. 4-71

a) What element could you put between the computer and the motor which would:

i. Isolate and protect the computer from the motor?

ii. Supply the current needed by the motor?

iii. Provide the high power level needed by the motor, yet controlled by the computer?

b) Sketch the schematic for the circuit element you chose.

4-54. Find the output equations for the component in Fig. 4-72.

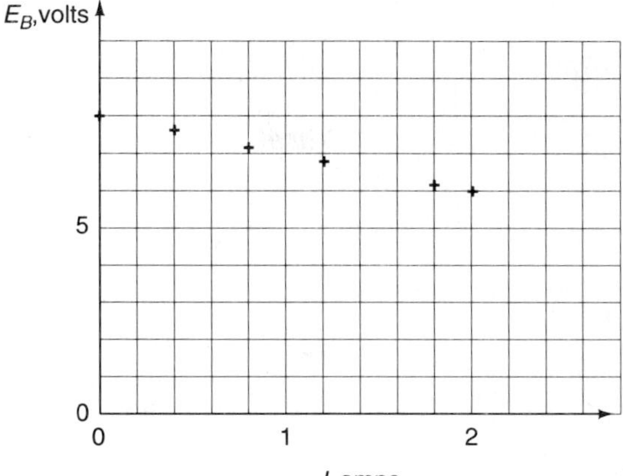

Figure 4-70. Battery Data for Prob. 4-52

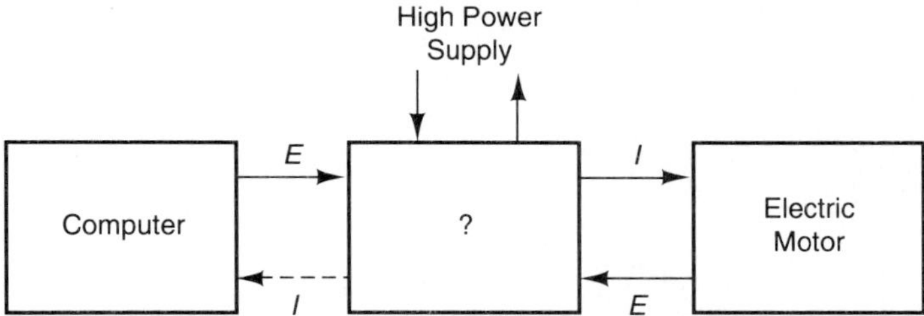

Figure 4-71. Multiport for Prob. 4-53

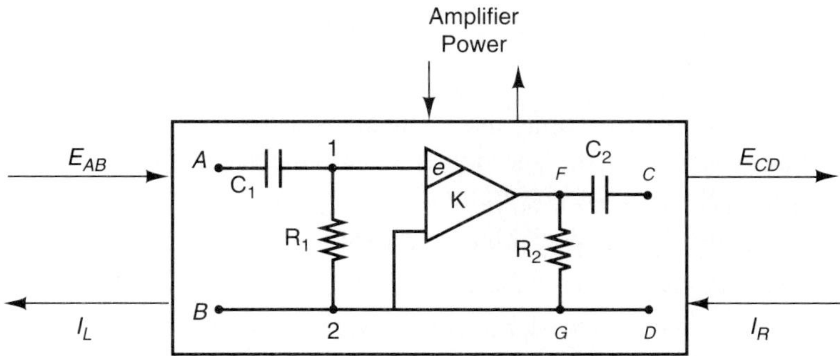

Figure 4-72. Multiport for Prob. 4-54

4-55. Find the output equation(s) for the component in Fig. 4-73.

4-56. A mechanical designer is interested in the performance of the speedometer cable shown in Fig. 4-74. The cable is 10 ft long, 1/8 in in radius, and is made of steel (density = 0.00875 sl/cu-in). In modeling the cable, you should include its inertia, spring-rate, and the viscous friction it has with the fluid in the housing. Treat the dial as if it exerts a negligible torque on the cable, and do the following:

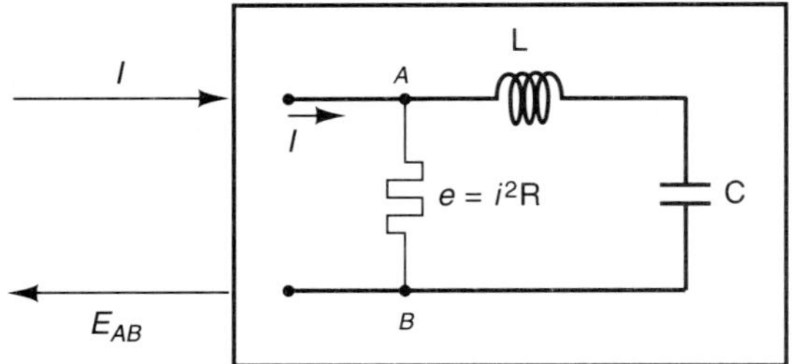

Figure 4-73. Multiport for Prob. 4-55

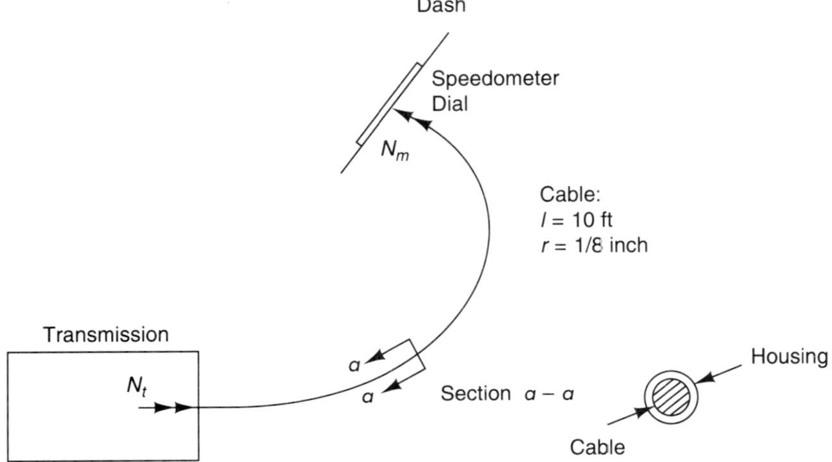

Figure 4-74. Speedometer Design for Prob. 4-56

a) Sketch a multiport diagram showing how the cable works between the transmission and the speedometer dial, including single-lump models of spring-rate, inertia, and friction.

b) Sketch the ideal model which goes with (a).

c) Compute the inertia of the cable in the proper units for this model.

4-57. An automotive engine designer is interested in the dynamics of piston motion, as sketched in Fig. 4-75. A few measurements of the compressive force of the piston air are also shown in the figure.

Build a model using this data as follows:

a) Sketch the ideal-analog model for the motion of the piston. Don't forget gravity.

b) Compute the spring-rate that you would recommend for the air compression near $X = 0$.

4-58. An engineer/restaurant-owner has designed the system of Fig. 4-76 for advertising. The constant-thrust rocket is designed to burn for about 3 minutes. As the rocket burns, it turns the disk on which it is mounted, and the disk turns

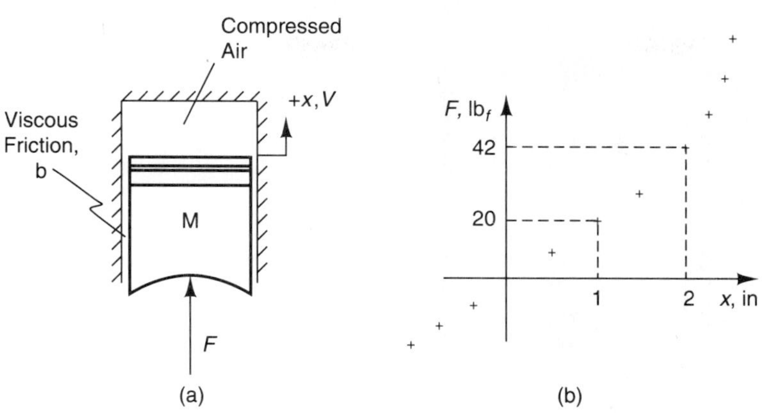

Figure 4-75. Piston for Prob. 4-57

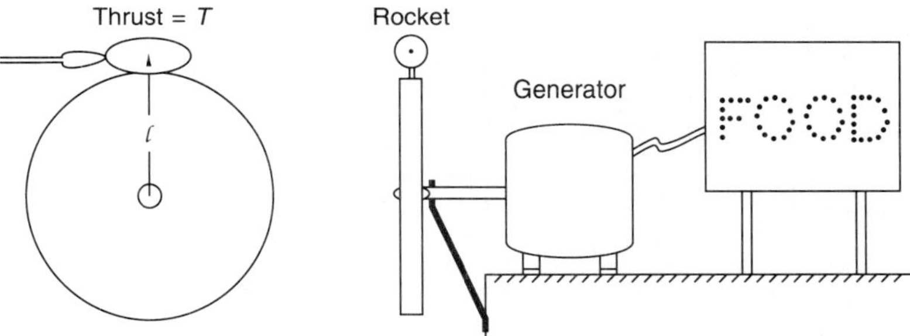

Figure 4-76. Rocket System for Prob. 4-58

the generator which is on the same axle. The generator thus produces power to light the advertising sign as the rocket burns. After burnout, the lights slowly dim as friction in the mechanical parts slows the disk and generator.

For this problem you should:

a) Sketch the lossless ideal model. Treat the lights as a pure-resistive load on the generator and ignore all structural ports to the reactive supports. This model will be useful in initially sizing the generator.

b) Sketch the lossy ideal model. This model will be useful in refining the computations of (a) and in computing the length of time that the lights will be visible.

4-59. Two springs are attached to a lever as shown in Fig. 4-77. Both springs have the same free length of 2 inches from which the displacement, X, is measured. Find the steady position, $X_1 = X_2 = X_P > 0$, where the lever is in equilibrium.

4-60. An electrical engineer is interested in a battery recharging system, as shown in Fig. 4-78. The engineer has asked you to build a mathematical model of the system with Z_f as the input and I as the output. Use the ideal engine and

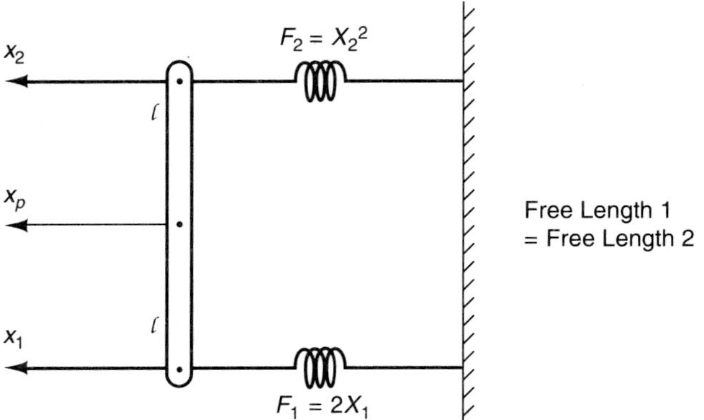

Figure 4-77. Springs for Prob. 4-59

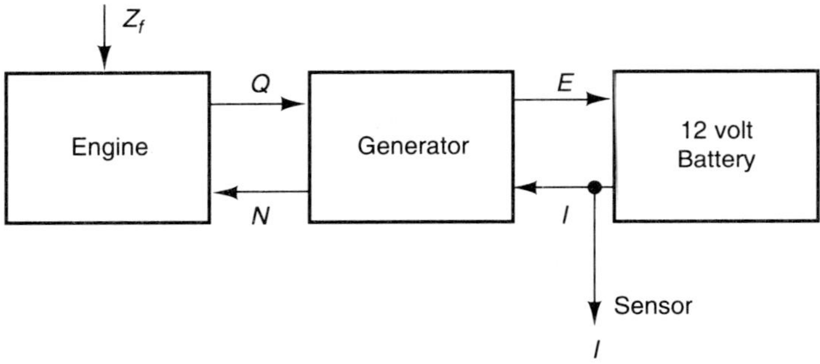

Figure 4-78. Multiport for Prob. 4-60

battery models (e.g., no droop in either element), and use a lossy generator model. Use the analog approach and build a math model of the system.

4-61. A mechanical designer wants you to find the output equations for the component in Fig. 4-79. Use the method of analogy.

4-62. For the fluid capacitance shown earlier in Fig. 4-51c, complete the following:

a) Find the fluid capacitance, C_f.

b) Describe how you would make this into a variable capacitor. (Hint: How would you change the spring rate?)

4-63. Given that a component has a 0.25 inch inside diameter pipe which is 5 ft long, and has a laminar flow of air in it at 100 °F. Compute the following for this flow:

a) The fluid resistance.

b) The fluid inertance.

c) The fluid capacitance near atmospheric pressure.

4-64. A hydraulic designer wants to get an estimate of the dynamics of the supply component shown in Fig. 4-80. Three-inch inside-diameter pipe is used

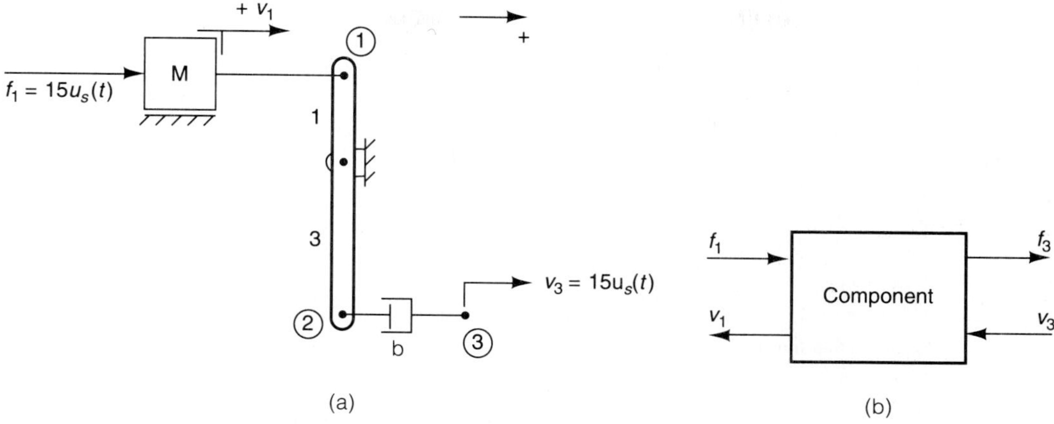

Figure 4-79. Mechanical Component for Prob. 4-61

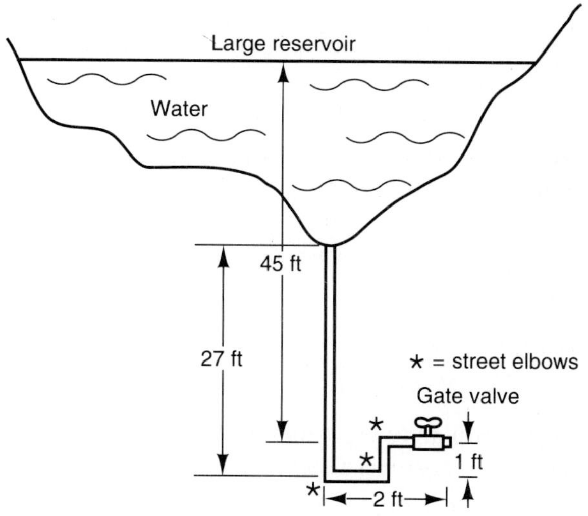

Figure 4-80. Reservoir System for Prob. 4-64.

throughout. Ignore the capacitance of the pipes and do the following:

a) Sketch an idealized analog model of the component above.

b) Compute the constants for the model of (a).

4-65. A hydraulic engineer has designed the lifting system shown in Fig. 4-81. You are to build a fluid-analog model to study the dynamics of this system. You should ignore pipe capacitance, leakage, and nonideal pump behavior. You should include fluid friction where appropriate, along with fluid inertance, and a finite-sized reservoir.

Use the analog method to sketch a fluid-analog model of the system from the top of the open reservoir to the load weight, *F*.

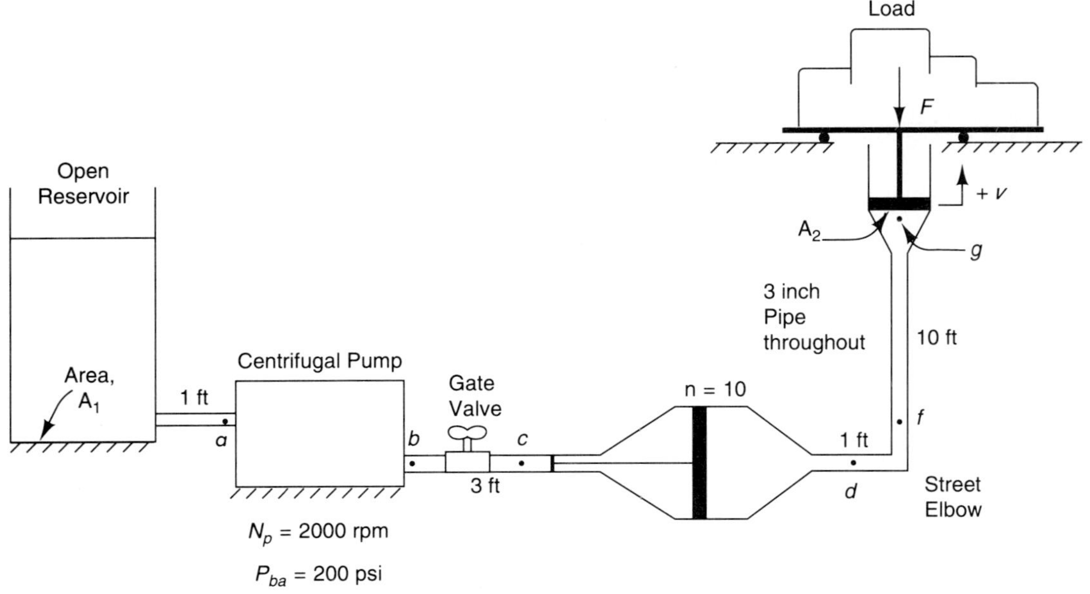

Figure 4-81. Lifting System for Prob. 4-65

4-66. A fluid power designer has designed the tank propulsion system shown in Fig. 4-82. The tank goes forward when $x > 0$, and the fluid power supply at a is connected to port 1 in the motor.

The hydraulic motor in this system works as a rotary actuator in the manner of Fig. 4-82b, with the equations

$$q = K_M z,$$

$$p_{12} = K_M n,$$

where p_{12} is the pressure drop across the inlet to the motor.

Also, since the motor is a positive-displacement actuator,

$$z = D_M n,$$

where D_M is the displaced fluid per radian of motor motion.

The motor drives the tank inertia as seen at the drive sprocket, J, and the friction, whose model is as in Fig. 4-82c. Ignore line dynamics and non-ideal behavior from the 1,500 psi centrifugal pump. What is the output equation for an input of x?

4-67. Find the output equation for the system of Prob. 4-65. Your inputs are the valve opening, and the load weight, F. Your output is the load speed, v.

4-68. A satellite engineer has created the polished aluminum, cubic vehicle of Fig. 4-83 (2 ft edge length) for space exploration and is now interested in its ther-

Tank Top View

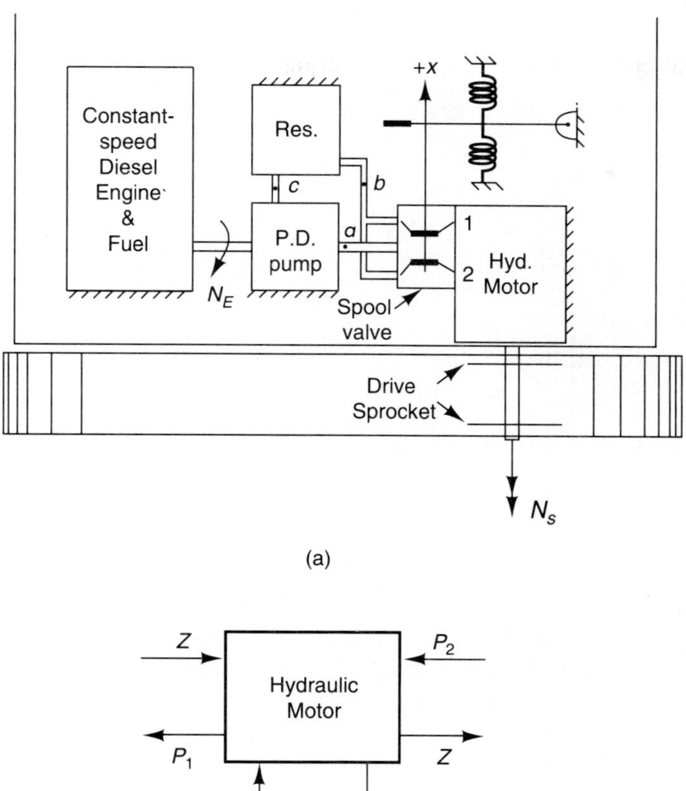

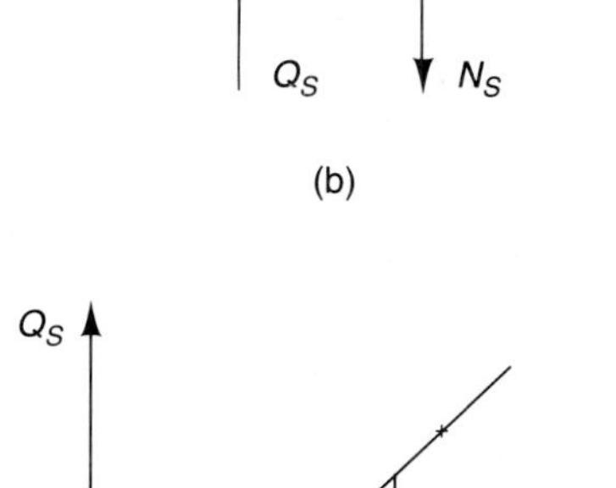

(b)

Q_S

B

1

N_S

(c)

Figure 4-82. Tank Propulsion for Prob. 4-68

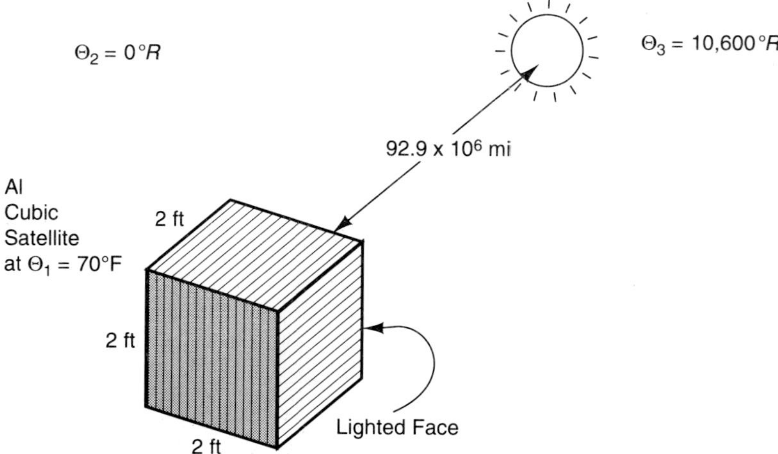

Figure 4-83. Satellite for Prob. 4-68

mal dynamics. When the vehicle is deployed, it will be at 92.9 million miles from the sun, and at a uniform temperature of 70 °F. Calculate the net heat gain when one face is directly facing the sun, and all the rest are in shadow.

4-69. An automotive engineer is interested in the heating dynamics of the brakes shown in Fig. 4-84. When the force, F, is applied to the brake pads, they cre-

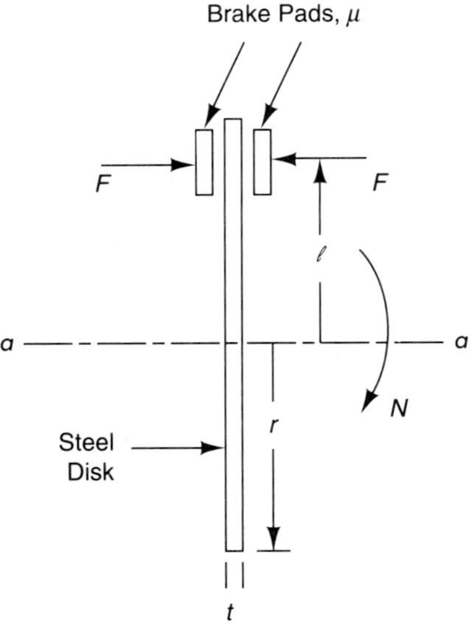

Figure 4-84. Brake Geometry for P?rob. 4-69

ate a slowing torque from the Coulomb force (μF) about the a-a axis as shown in the figure. This slowing torque dissipates energy through heating of the brakes.

In this problem you should:

a) Compute the power consumed by the brakes when the axle changes speed from N to zero,

b) Sketch an analog model of this thermal component, labeling the source strength. Ignore convection and radiation heat transfer.

c) What is the output equation for the temperature at the axle.

Chapter 5

Mathematical Analysis: Simple Differential Equations as Time-Domain Models

5-0 INTRODUCTION

Even simple differential equations are very powerful tools for capturing the essential response of a component. Such equations are "simple" in the sense that they are linear, of low order, and are subjected to idealized forcing functions. In this chapter, the properties of simple differential-equation models are examined in order to better understand the design message in this form of time-domain math model.

To give the discussion some substance, consider the SISO electrical component shown in Fig. 5-1. Please note that the electrical component in this figure only represents a *type* of component. The results apply equally well to any other type of component, whether structural, fluid, or thermal. Moreover, the methods of this chapter apply equally well to any output equation, even if it comes from a MIMO component. This is true since a component output equation expresses the dynamics of only one output at a time. Further, the number of inputs in the output equation is

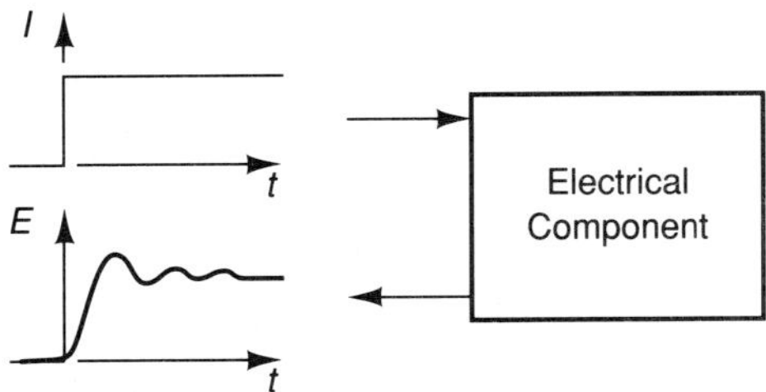

Figure 5-1. A SISO Electronic Component

irrelevant since they are all regarded as knowns, while the output is the single unknown in the equation. However, for the sake of clarity, only SISO models are discussed in this chapter.

In general, a time-domain input in an output equation may appear as an idealized singularity input (impulse, step, and/or ramp), or it may appear as some other less-orderly input-time history. However, in this chapter *all* the inputs are set to unit-step inputs so that correlations can be drawn between the output-side model form and the output response predicted by the equation. This eliminates input variations as a factor in comparing various responses, and creates a *standardized response* of the model.

Further, since these standardized responses are shaped by the attributes of model order and the model coefficients *alone,* the modelers and designers usually focus on these two attributes to verify and improve their designs. Such responses are thus studied at length here for various simple differential equations.

Recall, for example, that the *order* of an output equation is the same as the number of differentiations in the highest output derivative in the differential equation. Similarly, the order of a component is the sum of all the orders of the output equations. So, a zero-order component must be an algebraic component—a component with no output derivatives. For the component of Fig. 5-1, an algebraic model would be of the form

$$E = f'(I). \tag{5-1}$$

Note that the algebraic model always demonstrates an *instantaneous* change in E as soon as I is changed. But, of course, all real, physical components must have some finite time lag between the input stimulation and the completed output response. An algebraic model thus shows an insignificant input/output time lag based upon the physical idealizations of the modeler.

However, if you have found that

$$\tau_c \, dE/dt + E = f'(I). \tag{5-2}$$

Then the theory of differential equations says that you can expect an exponential response in the time history of E when f' is stepped. Further, physical observation tells you that this is more realistic than the algebraic model since nothing can change instantaneously.

Thus, Eq. 5-1 is really a special case of Eq. 5-2, without the rate term. The additional physical realism of Eq. 5-2 occurs because the model has an added energy-storage element which increases the order of the differential-equation model from 0 to 1, and this results in an additional response-shaping factor, τ_c, which is called the *time constant*.

Each unit increase in order adds another time constant to the model.

So, too low an order in the differential equation implies neglected physics. And too high an order in the output differential equation implies that the component might be overmodelled. Very high-order component models are simply not appropriate to preliminary design. The modeler is thus concerned with finding an acceptable-order model.

Remember the First Law of Modeling: you want to keep the preliminary model as simple as possible, yet capture the essence of the component input/output response. In this chapter, you begin to study this modeling tradeoff by understanding the nature of the differential equation(s) in the model. Later, this understanding may be used to anticipate the model output response, and the adequacy of the model and/or the design.

Thus, the order and time constants of the differential equation(s) are studied here as they shape the component response predictions. In this, a familiarity with ordinary differential equations is assumed to avoid lengthy review. For completeness, nonlinearity is included as a complicating factor and methods are developed to deal with it when it occurs. In a later chapter, the use of the response predictions to make design decisions is discussed.

5-1 THE FIRST-ORDER MODEL—LAG COMPONENTS

Perhaps the simplest form of rate model is the linear, first-order, ordinary differential equation. If such a model were to occur for the component of Fig. 5-1, its form would be

$$dE/dt + E/\tau_c = f(I). \tag{5-3}$$

Notice that only one output variable, E, appears on the left side of the equation, with the input variable, I, appearing on the right. Further, the equation is written with a *unity* coefficient for the derivative term. Thus, τ_c is a constant parameter which is determined by the modeler's choice of elements and is a grouping of those element values. The model is thus posed as a non-homogeneous, first-order, ordinary differential equation for the output E.

Physical examples of this type of first-order component are shown in Fig. 5-2. Note that each component is composed of a capacitance in parallel with a resistance element. Other arrangements of elements to give first-order components are possible—these are merely representative.

Several simplifying assumptions about the right-side of the model are first made in order to generalize conclusions about the left-side of the equation. For clarity, the simplest nonzero forcing function is taken for study

$$f(I) = \mathrm{D}_0 I, \tag{5-3a}$$

and a step-input model for I is assumed

$$I = \mathrm{I}_m u_s(t). \tag{5-3b}$$

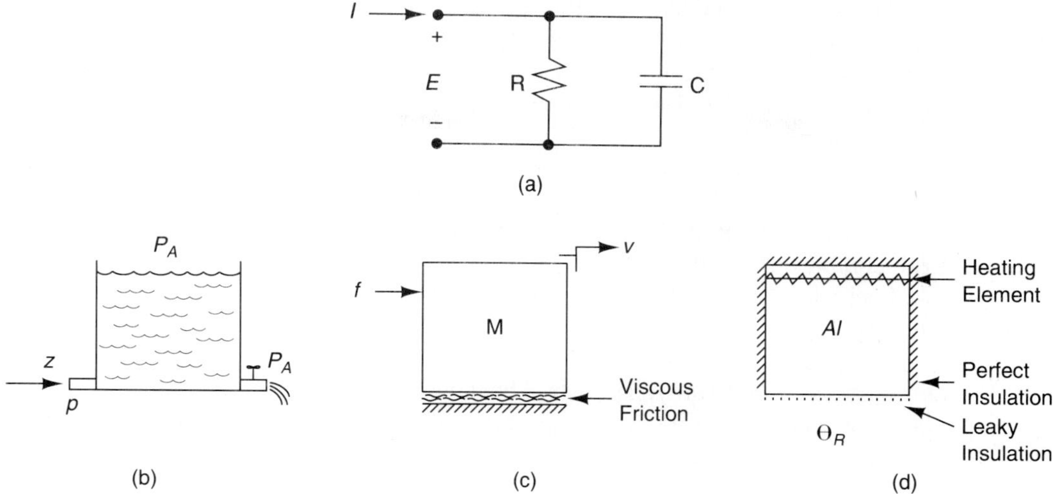

Figure 5-2. First-order Components. (a) Electronic, (b) fluid, (c) structural, (d) thermal

These assumptions create an *idealized* math model in the form

$$de/dt + e/\tau_c = \mathrm{D_0 I}_m u_s(t). \tag{5-3c}$$

Notice that the solution to this last equation can be used to find the solution to every linear, first-order equation, with any singularity forcing function. This is true since the differentiation of both sides of Eq. 5-3c, and the substitution $e' = de/dt$, yields the impulse-input response of the component. Similarly, the integration of both sides of the equation yields the ramp-input response. Further, superposition of these responses may be used to synthesize solutions to more complicated forcing functions. Thus, you need only solve for the step response, all other idealized linear responses may be computed from this one.

The final type of idealization that is typically done is to study these components as they respond from *rest*. This means that the idealized output, e, and all its derivatives are assumed to be zero at $t = 0$ sec. These "resting" initial conditions on the output are encountered on virtually every idealized problem, and are very useful as a basis for generalizations.

Initial conditions are not zero on E in general, however, and you are cautioned to account for the actual initial conditions in other modeling situations.

A mathematician will recognize that the solution to Eq. 5-3c has two parts: a homogeneous (*characteristic*) solution, e_h, and a *particular* solution, e_p. Further, since the equation is linear, these two solutions can be solved separately and added together (*superposed*) to find the general solution for e:

$$e = e_h + e_p. \tag{5-4}$$

The system modeler recognizes that component models are always nonhomogeneous since there are always inputs to all components that are truly part of a larger system. Further, the modeler speaks in terms of the input(s) *forcing* the response at the output(s). The particular response due to the input(s) is thus called the *forced response*.

In a similar way, the homogeneous response is characteristic and *natural* to the dynamics of the physical parts of the component—it doesn't depend on the input(s) at all. Consequently, the homogeneous response is also called the *natural response* of the component.

Each of these two responses is next studied to see how the component model is working.

Homogeneous (Natural) Response. The homogeneous response of the first-order component is found from the homogeneous differential equation

$$de_h/dt + e_h/\tau_c = 0. \tag{5-5}$$

This equation is solved by assuming a solution in the form

$$e_h = C_2 e^{\lambda t} = C_2 \exp(\lambda t), \tag{5-6}$$

and substituting into Eq. 5-5,

$$\lambda\ C_2 \exp(\lambda t) + (C_2/\tau_c) \exp(\lambda t) = 0. \tag{5-7}$$

This last equation can be reduced to a more useful form by dividing out the C_2 $\exp(\lambda t)$ factor,

$$\lambda + 1/\tau_c = 0. \tag{5-8}$$

This is possible since $C_2 = 0$ is a trivial solution, and $\exp(\lambda t)$ is never zero for finite t.

Equation 5-8 is the *first-order characteristic equation,* it tells you what the value of λ must be to solve the homogeneous equation. That is,

$$\lambda = -1/\tau_c, \tag{5-9}$$

and

$$e_h = C_2 \exp(-t/\tau_c). \tag{5-10}$$

The constant parameter, τ_c, is called the *time constant* of the equation/model. It has the units of seconds. Notice that τ_c must be positive so that the homogeneous solution does not become infinite as time gets large—this is the single *stability condition* for the first-order model. This requires that λ be negative for stability.

The unknown constant, C_2, cannot be evaluated until the other part of the solution (the forced, non-homogenous part) is solved and the two parts are added together (superposed). After the forced part of the solution is superposed with the homogeneous solution above, an *initial condition* on e is used to find the unknown C_2.

In fact, this pattern is true regardless of the order of the differential equation. The nth-order homogeneous equation similarly yields a characteristic equation, which has n time constants, plus n constants like C_2 which must be evaluated by n initial conditions.

Particular (Forced) Response. The idealized particular solution comes from letting $e = e_p$ in Eq. 5-3c,

$$de_p/dt + e_p/\tau_c = D_0 I_m u_s(t). \tag{5-11}$$

The method of undetermined coefficients may be used to solve this equation and find

$$e_p = \tau_c D_0 I_m u_s(t). \tag{5-12}$$

General Solution. The final step is to find the general solution to the differential equation by superposing the two previous solutions and applying the initial condition. Superposing gives

$$e = e_h + e_p, \tag{5-4}$$

$$= C_2 \exp(-t/\tau_c) + \tau_c D_0 I_m u_s(t). \tag{5-13}$$

The initial condition is applied by evaluating Eq. 5-13 at time 0^+:

$$e(0^+) = C_2 + \tau_c D_0 I_m = 0. \tag{5-14}$$

Notice that this initial condition needs to be applied at time $= 0^+$ sec. since there is a discontinuity precisely at time $= 0.0$ due to the unit step. That is, 0^+ is a time very slightly larger than 0.0, which begins the time-interval of interest. This evaluation results in an equation which can be solved for the remaining unknown, C_2,

$$C_2 = -\tau_c D_0 I_m. \tag{5-15}$$

This result can be inserted back into Eq. 5-13, and the terms rearranged to give the final form of the solution:

$$e = \tau_c \mathrm{D}_0 \mathrm{I}_m \left[1 - \exp(-t/\tau_c) \right], \ t \geq 0^+. \tag{5-16}$$

The general solution (Eq. 5-16) thus shows two terms. The first term is due to the input. The second term is due to the natural solution—it is substantial near $t = 0.0$, but it fades out exponentially as time increases.

The exponential decay which is shown in the general solution is *characteristic* of the first-order model. The sequence of events is started by a step input, which causes the exponential change of the output to a new value. This time *lag* between input and the final output gives rise to the first-order component's name as a *lag component*.

Figure 5-3 shows two step responses in e, both of which start at $e(0^+) = 0.0$ and have $\mathrm{D}_0 = 10.0$. One response has been plotted (from Eq. 5-16) for a larger time constant (τ_{cL}), and one for a smaller time constant (τ_{cS}). Notice that both components start at an initially steady value of zero voltage, they pass through a transient, and they asymptotically approach the same steady final value of $e = 10$v. (This final value can be found from setting the derivative term in the differential equation to zero and solving the resulting algebraic equation)

Notice next that both components in the figure reached 63.21% of their final values when the time elapsed was equal to one time constant. Following this initial period, for each subsequent time interval of one time-constant, 63.21% of the *remaining* change to final value is completed.

Thus, a component with a small time constant responds faster than one with a large time constant.

Standard Form. In order to generalize these results, the first-order model discussed earlier must be converted to a more general form. This is done by letting $E = y$ in Eq. 5-3,

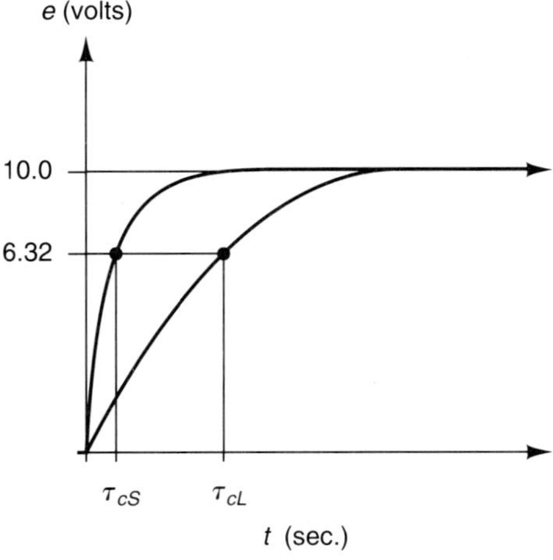

Figure 5-3. Time Constant Comparison

$$dy/dt + y/\tau_c = f(t). \tag{5-17}$$

The *standard response* to this equation is *defined* as that to the scaled unit-step input starting at time $t = 0.0$ sec,

$$f(t) = (1/\tau_c)\ u_s(t),$$

such that all initial conditions on y are equal to zero.

Again, this situation of zero initial conditions has to do with starting the component at rest when the step is applied. This approach is very important in order to standardize the effect of initial conditions on the response. This is especially true when talking about those component models which have higher-order differential equations, which are soon to be studied.

Given these conditions, the standard response is

$$y(t) = (1 - e^{-t/\tau_c}),\ \text{for } t > 0.0. \tag{5-18}$$

Example 5-1

A modeler has developed the following first-order model of a component

$$dE/dt + E/2 = 2.5I. \tag{5-19}$$

Plot the response of this model to the input $I = \mathrm{I}_m u_s(t)$, with the conditions

$$E(0^+) = 14.7 \text{ volts},$$

$$\mathrm{I}_m = 5 \text{ amps},$$

$$\mathrm{D}_0 = 5.0 \text{ v/a-s}.$$

Do this in such a way that the response can be plotted against time in a *linear* plot.

Solution

The linear plot can be achieved through a change of variable which smooths out the characteristic exponential response of the first-order model. This requires the following algebra, starting with Eq. 5-14 developed earlier,

$$E = E(0^+)e^{-t/\tau_c} + \tau_c \mathrm{D}_0 \mathrm{I}_m(1 - e^{-t/\tau_c}),\ t \geq 0^+, \tag{5-14}$$

$$E - \tau_c \mathrm{D}_0 \mathrm{I}_m = [E(0^+) - \tau_c \mathrm{D}_0 \mathrm{I}_m]\ e^{-t/\tau_c}, \tag{5-20}$$

$$(E - \tau_c \mathrm{D}_0 \mathrm{I}_m)/[E(0^+) - \tau_c \mathrm{D}_0 \mathrm{I}_m] = e^{-t/\tau_c}. \tag{5-21}$$

Taking the natural log of both sides gives

$$ln\ \{(E - \tau_c \mathrm{D}_0 \mathrm{I}_m)/[E(0^+) - \tau_c \mathrm{D}_0 \mathrm{I}_m]\} = -t/\tau_c. \tag{5-22}$$

A change of variable at this point accomplishes the linearization. For example, the substitutions

$$\alpha = ln\left\{(E - \tau_c \mathrm{D}_0 \mathrm{I}_m)\Big/\left[E(0^+) - \tau_c \mathrm{D}_0 \mathrm{I}_m\right]\right\}, \tag{5-23}$$

and $$m = -1/\tau_c, \tag{5-24}$$

give an equation of the form $\alpha = mt$ when they are substituted into Eq. 5-22. The equation for α in terms of the data given is

$$\alpha = ln\,[(25.0 - E)/10.3]. \tag{5-25}$$

When this equation is used to compute the α values corresponding to voltage response predicted by Eq. 5-16, the (t, α) graph of Fig. 5-4 is the result.

Comments

Notice that the slope of the α line in Fig. 5-4 $(-1/2)$ is equal to the negative reciprocal of the time constant of the given model (2 sec.). Thus you can expect that this type of plot may be very useful in the identification of an unknown time constant given the response data, a topic which will arise again in Chap. 7.

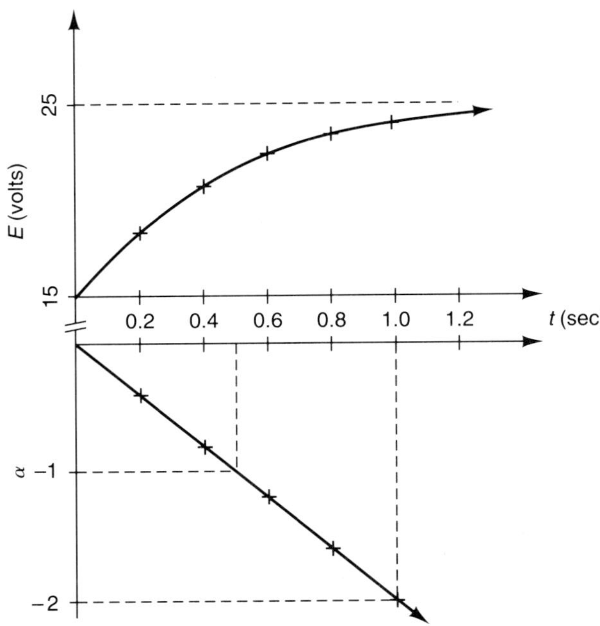

Figure 5-4. First-order Response Linearization

HOMEWORK

5-1. Given Eq. 5-19 in the example problem:

a) What is the standard form of this equation?

b) What is y in terms of E?

c) What is $f(t)$ in terms of I?

5-2. Find τ_c and D_0 for the causality of the component of Fig. 5-1, and the network of Fig. 5-2a.

5-3. How close to its final value (in %) is a step response in an ideal first-order component when $t = 4\tau_c$?

5-4. What output time history would you expect for the fluid component which has the following model:

$$3\ dP/dt + 6P = 2Z,$$

if $P(0^+) = 2$, and $Z = 10u_s(t)$?

5-2 LAG COMPLICATIONS

So far, this chapter has dealt with the linear, first-order, ordinary differential equation as a model for the lag component. Complications in such a model often occur in two forms: first, there may be delay modeled in the equation; and second, there may be nonideal behavior in the form of nonlinearities and/or time-varying coefficients.

Delay. A component with delay has the response shown in Fig. 5-5, such that the delay (τ_d, sec.) is observed in the lack of change between the start of the input and the beginning of the component response. The component response then proceeds as if the input had started at the end of the delay.

A linear, first-order component model with delay appears in the following standard mathematical form:

$$\tau_c\ dy/dt + y = u_s(t - \tau_d). \tag{5-26}$$

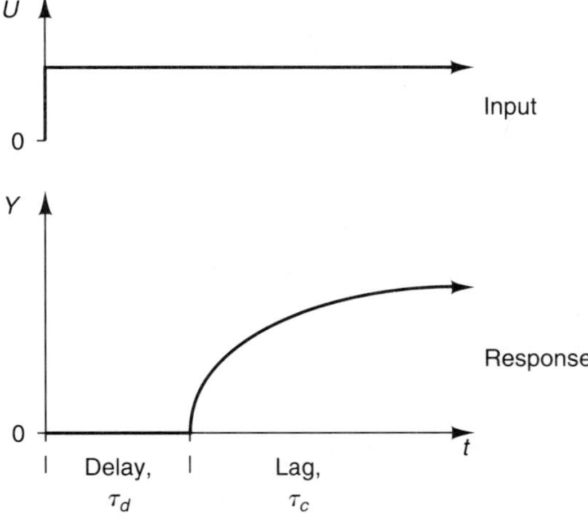

Figure 5-5. Lag Response with Delay

Note that the delay of the *output* is accomplished by mathematically delaying the *input*.

Further, although it is quite arbitrary when $t = 0$, it still is necessary to synchronize the input/output responses of all the components. Thus, $t = 0$ represents the start of central *system* time, and the component delays work on the inputs to create the external outputs in much the same way as a stream flowing down a series of small waterfalls. The *net* delay from the top of the falls to the bottom is composed of a series of smaller delays for each waterfall. So, inputs feed to delayed outputs, which feed to delayed outputs, and so on until the net system effect of all the components delays is realized.

Physically, delay is seen in every real component. However, it is most dominant in pneumatics, between the time when a pressure step is input at one end of a long tube, and the measurement of that step at the other end. This form of delay is also called *transportation lag*. Such delays in the operation of systems can be very troublesome in designing their operations. Delay is also sometimes seen in certain sensors and controller actions, where it is called *dead time*.

Mathematically, the delay response of a linear, first-order component (e.g., the solution of Eq. 5-26) can be easily computed from the non-delayed model studied in the last section of the text. That is, the first-order response *without* delay,

$$y = (1 - e^{-t/\tau_c})\, u_s(t), \tag{5-27}$$

is transformed into the first-order response *with* delay,

$$y = [1 - e^{-(t-\tau_d)/\tau_c}]\, u_s(t - \tau_d), \tag{5-28}$$

by substituting the time-shift variable, $(t - \tau_d)$, for t in Eq. 5-27.

Nonideal Components. Two approaches to solving the first-order, nonideal, differential equation are possible. First, you may be able to find a theoretical solution. For the purposes of this text, however, this is assumed to be beyond the skill of the reader. The second approach is that you may numerically integrate the output equation on a digital computer. Analog computers are still sometimes used to do this sort of computation, but the performance, cost, and availability of micro-computers has led to their steadily taking over this task.

Numerical Integration. There are many possible schemes of numerical integration. A thorough review of all these numerical methods is not attempted here as there are many large texts which do that quite adequately. Instead, the rectangular integration method presented in Chap. 3 is applied here as well.

A first-order, nonlinear equation form is assumed

$$g(dY/dt, Y, U) = f(U, U^*), \tag{5-29}$$

where f and g are some nonlinear functions. Thus, given an initial condition on Y, and the time histories of the inputs, the problem is to determine the time history for Y.

The approach to the numerical integration of a differential equation (linear or nonlinear) is to first discretize the equation. The graphical interpretation of such a *uniform time-step* discretization is shown in Fig. 5-6. This figure shows a plot of $Y(t)$ which is broken down into N uniform time-intervals according to the modeler's selection of the constant time-step size, Δt (sec.). In other words, you are given the value of Y at time $t = k\Delta t$, and you are trying to predict the value of Y at time $t = (k + 1)\Delta t$ using the differential equation.

Recall from Chap. 3 that the numerical approximation for the derivative of Y is

$$dY/dt \cong \Delta Y/\Delta t. \tag{5-30}$$

This discrete approximation of the slope is shown in Fig. 5-6 as it is applied using the discrete data near k,

$$(\Delta Y/\Delta t)_k = (Y_{k+1} - Y_k)/\Delta t. \tag{5-31}$$

The approximation given by Eq. 5-31 is called a *forward-difference* formula, since the derivative is evaluated between the forward point at $(k + 1)$ and the point at k. Formulas also exist for central-difference and backward-difference versions of the derivative—these are discussed later in the text.

Even a cursory examination of the actual derivative evaluated at k in Fig 5-6 reveals that it is reasonably close to the finite approximation given by Eq. 5-31. Further, you can rightly expect that the derivative approximation gets better as the step,

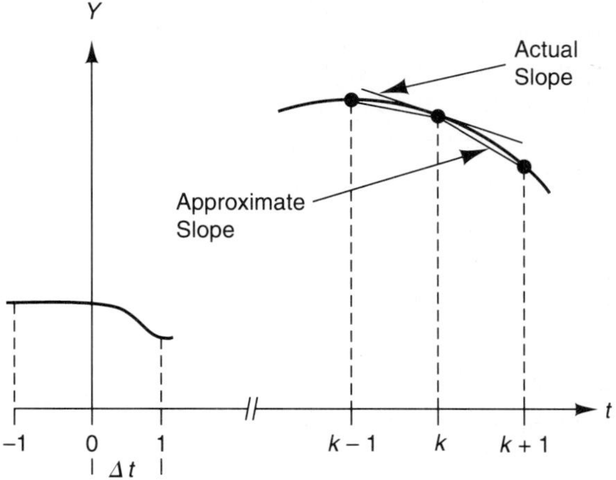

Figure 5-6. Numerical Approximation of First-derivative

Δt, is made smaller—within certain limits. Substitution of the derivative approximation back into the differential equation thus allows you to develop a numerical equation for the value of Y_{k+1} at the next-time step. One problem remains: the selection of a proper time step for the computations.

The *time-step selection* comes from a consideration of the limiting errors in the numerical computations. That is, if Δt is chosen too large, then the derivative approximation does a poor job of fitting the true response. If Δt is chosen too small, then excessive computing leads to computational-error buildup. The key to this tradeoff rests in using the time constant from the model to guide your time-step selection. System modelers often use the rule-of-thumb for integration

$$\tau_c/1{,}000 \leq \Delta t \leq \tau_c/10. \tag{5-32}$$

For nonlinear integrations, this rule is applied to the time constant from the *linearized* form of the output equation. This linearization is done according to the three simplified rules of *Time-step Linearization:*

1. First, simply erase exponents, and replace nonlinear functions with linear forms (e.g., $\sin U$ becomes u, $\cos U$ becomes 1, etc.).
2. Second, given a product of input and output (or their derivatives), erase the input parts.
3. Third, time-average any time-varying coefficients.

These rules are used to find an *initial estimate* for Δt, which is then used to compute the component response. This is followed by computing the response again with $\Delta t/10$. If there is no significant change in the response, then you may have some confidence that the simulation is correct (perhaps to be verified by experiment). If the response changes significantly in the second computation, then the time step is reduced again and the computation rerun. This process is continued until successive responses show acceptable convergence between their results.

Example 5-2

A modeler has developed a nonlinear model of a component in the form

$$(dY/dt)^2 + Y^2 = \sin U, \tag{5-33}$$

subject to the knowledge that Y is never negative. Find Y as predicted by this model subject to the step input

$$U = (\pi/2)u_s(t),$$

and starting from the initial condition $Y(0^+) = 0.0$. The time interval of interest is $0.0 < t \leq 2.0$ sec.

Solution

The first thing to do is to transform the differential equation into its approximate finite-difference form,

$$[(\Delta Y/\Delta t)_k]^2 + (Y_k)^2 = \sin(\pi/2) = 1.0. \tag{5-34}$$

The first term is then expanded according to Eq. 5-31,

$$[(Y_{k+1} - Y_k)/\Delta t]^2 + (Y_k)^2 = 1.0. \tag{5-35}$$

Solving this equation for Y_{k+1} gives the desired *recursion formula* (this is a recursion formula since the expression for Y_{k+1} keeps "recurring" on every time step k),

$$Y_{k+1} = Y_k + \Delta t[1.0 - (Y_k)^2]^{1/2}. \tag{5-36}$$

Note that the positive root is used since Y can never be negative.

A *simulation* of the response of Y is started by letting $k = 0$ in the above formula

$$Y_1 = Y_0 + \Delta t[1.0 - (Y_0)^2]^{1/2}.$$

The Y value on the right-hand side comes from the initial condition,

$$Y_0 = 0.0.$$

The time-step linearization gives a linearized time-constant estimate of about 1.0 sec. Thus, an initial estimate of Δt is

$$\Delta t = 0.01 \text{ sec.}$$

Substituting this into the recursion formula, you calculate

$$Y_1 = 0.01.$$

You keep applying the recursion formula (Eq. 5-36), and building the output history file on Y_k until the value at the final time ($t = 2.0$ sec.) is computed. In doing this, you accumulate a data file containing 201 evenly spaced Y data points which go from Y_0 through Y_{200}.

The simulation then is rerun with $\Delta t = 0.001$ sec. to verify the results.

HOMEWORK

5-5. Given the following component model with a delayed flowrate input:

$$R_f C_f\, dp/dt + p = R_f z(t - 5.0).$$

And the following input model:

$$z(t) = u_r(t) - u_r(t - 5.0).$$

a) Compute a table of $z(t - 5.0)$ at one-second intervals and compare it to $z(t)$ at the same times for the time interval $0 \le t \le 11.0$ sec.

b) If $z(t) = u_s(t)$, and $p(0^+) = 0$, what is $p(t)$ for the above model with delay?

5-6. Given $dy/dt + y = u_s(t)$, with the initial condition $y(0^+) = 0$.

In this problem you will conduct a time-step effectiveness study for the given linear equation. To do this, you must find the forward-difference recursion formula for the given equation using Eq. 5-31. You must then encode the formula onto a digital computer for solution.

Compare the exact solution (Eq. 5-27) to five different numerical integrations using five different time steps, $\Delta t = 0.01, 0.1, 0.5, 1.0, 2.0$ sec. That is, to do this problem, you need to solve the differential equation five different times, once for each of the time steps.

Compare your results at the times $t = 0.0, 1.0, 2.0, 3.0, 4.0, 5.0, 6.0$ sec. Tabulate your results at the times of interest, and comment on the results with respect to size of the time step and the component time constant.

5-7. Given the model

$$(dY/dt)^2 + UY = \sin U.$$

a) What maximum initial estimate of Δt would you recommend for rectangular integration of this model?

b) How would you mathematically check the integration accuracy?

5-8. Explain the difference between a lag in output and a delay in output in 2 or 3 sentences.

5-3 THE SECOND-ORDER MODEL, OSCILLATING COMPONENTS

The next level of model complexity is the second-order, linear, ordinary differential equation with constant coefficients. In terms of the component in Fig. 5-1, the correct general form of this model is

$$d^2E/dt^2 + C_1\, dE/dt + C_0 E = f(I). \qquad (5\text{-}37)$$

In general, the forcing function, f, may be a differential equation in I, such as

$$f(I) = D_1\, dI/dt + D_0 I. \qquad (5\text{-}38)$$

Again, as in the first-derivative equation, Eqs. 5-37 and 5-38 are written with a unity coefficient for the highest derivative of the output and the coefficients are subscripted according to derivative order.

However, in order to gain focus for the discussion, the simplest nonzero input-forcing expression is assumed,

$$f(I) = D_0 I, \qquad (5\text{-}39)$$

with a step-input model for I,

$$I = I_m u_s(t). \quad (5\text{-}40)$$

As in the first-order model, these simplifying input assumptions are the basis for assembling the response of the linear second-order model through the use of superposition. These assumptions create an idealized equation of the form

$$d^2e/dt^2 + C_1\, de/dt + C_0 e = D_0 I_m u_s(t), \quad (5\text{-}41)$$

which requires two initial conditions for its solution.

Again, as for the first-order model, resting initial conditions are assumed for the solution, and the solution is composed of superposed particular (forced) and homogeneous (natural) parts,

$$e = e_p + e_h. \quad (5\text{-}42)$$

Forced Response. The particular (forced) solution is found using the method of undetermined coefficients. This method gives

$$e_p = (D_0 I_m/C_0)\, u_s(t). \quad (5\text{-}43)$$

Natural Response. The natural solution comes from the second-order homogeneous equation

$$d^2e_h/dt^2 + C_1\, de_h/dt + C_0 e_h = 0.0. \quad (5\text{-}44)$$

The solution to Eq. 5-44 is assumed to be of the form

$$e_h = C_4 \exp(\lambda t), \quad (5\text{-}45)$$

which is substituted into the equation to find

$$\lambda^2 C_4 \exp(\lambda t) + C_1 \lambda C_4 \exp(\lambda t) + C_0 C_4 \exp(\lambda t) = 0.0. \quad (5\text{-}46)$$

Now, since $C_4 \exp(\lambda t) \neq 0.0$ (this would give a trivial solution), this term is cancelled out of the above expression to give

$$\lambda^2 + C_1 \lambda + C_0 = 0.0, \quad (5\text{-}47)$$

which is called the *second-order characteristic equation*. The solution to this equation is

$$\lambda = -C_1/2 \pm \left[(C_1/2)^2 - C_0\right]^{1/2} \tag{5-48}$$

So, four cases of homogeneous solutions are possible depending on the relative size of the two output-side coefficients,

Case 1: $C_1 = 0.0$, two equal, imaginary λ's,
Case 2: $(C_1)^2 < 4C_0$, two unequal, complex λ's,
Case 3: $(C_1)^2 = 4C_0$, two equal, real λ's,
Case 4: $(C_1)^2 > 4C_0$, two unequal, real λ's.

If you remember back to the last chapter, the first-derivative term in a second-order output equation is due to friction or energy-loss elements. So, C_1 is a measure of the friction, or energy-loss, or *damping* in the component. Note, therefore, that the four cases of second-order coefficients are arranged in a fashion of increasing damping, while holding C_0 constant. This section of the text presents the first two cases where damping is small, the next section presents the last two cases and a generalization of the linear second-order model.

Case 1—Undamped Solutions. When $C_1 = 0$, the roots of the characteristic equation are

$$\lambda = \pm j\,(C_0)^{1/2}, \tag{5-49}$$

where $j = (-1)^{1/2}$.

The substitution of these roots into Eq. 5-45 gives two solutions to the homogeneous equation. Thanks to linearity, these two solutions may be superposed to give

$$e_h = C_5 \exp[+j(C_0)^{1/2}t] + C_6 \exp[-j(C_0)^{1/2}t]. \tag{5-50}$$

Notice here that two constants (C_5 and C_6) come out of the second-order homogeneous solution. These two constants are evaluated using two resting initial conditions (say, one on e, and one on de/dt) *after* the particular solution is superposed onto the homogeneous solution.

Equation 5-50 is translated into something more sensible using Euler's Equations,

$$\exp(+j\beta) = \cos\beta + j\sin\beta, \tag{5-51}$$

$$\exp(-j\beta) = \cos\beta - j\sin\beta, \tag{5-52}$$

plus the substitution

$$\beta = (C_0)^{1/2}t = \omega_n t. \tag{5-53}$$

When these are substituted into the homogeneous solution, the result is

$$e_h = C_5(\cos\omega_n t + j\sin\omega_n t) + C_6(\cos\omega_n t - j\sin\omega_n t). \tag{5-54}$$

$$= C_7\cos\omega_n t + C_8\sin\omega_n t. \tag{5-55}$$

The general step-response of the Case 1, second-order system is obtained by superposing the natural and forced responses

$$e = D_0 I_m/C_0 + C_7\cos\omega_n t + C_8\sin\omega_n t. \tag{5-56}$$

An equivalent form of this solution is

$$e = D_0 I_m/C_0 + C_9\sin(\omega_n t + \phi). \tag{5-57}$$

The two unknown constants (C_7 and C_8, *or* C_9 and ϕ) are determined using the two resting initial conditions.

The impact of C_1 being equal to zero is now seen in the general solution above; namely, that there are here two pure-sine terms which oscillate at a constant amplitude and frequency. The frequency of these oscillations, ω_n (rad/s), is called the *undamped,* or *natural frequency* of the component.

Example 5-3

A modeler has developed a model in the following form for a design

$$2\,d^2e/dt^2 + 8e = 6i. \tag{5-58}$$

The designer wants to know the *natural frequency* and the *maximum magnitude* of the step response in order to design a connecting component.

The designer specifies

$$i = 40u_s(t) \text{ amp},$$

and the component starts from rest.

Solution

Comparison of the general form of the component model (Eq. 5-41) with that above reveals the following:

$$C_0 = 4.0,$$

$$C_1 = 0.0,$$

$$D_0 = 3.0,$$

$$I_m = 40.0.$$

Thus, since $C_1 = 0.0$, you know the solution is of the Case 1 general solution form,

$$e = D_0 I_m / C_0 + C_9 \sin(\omega_n t + \phi). \tag{5-57}$$

Such that

$$\omega_n = (C_0)^{1/2},$$

and

$$\boldsymbol{\omega_n = 2.0 \text{ rad/s.}}$$

Substituting the constants into the general solution gives

$$e = (3.0)(40.0)/4.0 + C_9 \sin(2t + \phi), \tag{5-59}$$

or

$$e = 30.0 + C_9 \sin(2t + \phi). \tag{5-60}$$

The derivative of this equation is next computed in order to apply the resting initial condition

$$de/dt = 2C_9 \cos(2t + \phi). \tag{5-61}$$

Application of the initial conditions yields

$$e(0^+) = 30 + C_9 \sin(\phi) = 0.0,$$

$$de/dt(0^+) = 2C_9 \cos(\phi) = 0.0.$$

Solving the last equation gives

$$\phi = \pm\pi/2.$$

$\phi = +\pi/2$ is arbitrarily chosen and this choice gives

$$C_9 = -30.0,$$

and results in the general solution

$$\boldsymbol{e = 30.0 - 30.0 \sin(2t + \pi/2) \text{ volts.}} \tag{5-62}$$

The result in the last equation is plotted in Fig. 5-7. Notice that the slope and magnitude of the voltage at $t = 0$ sec. is in agreement with the resting initial conditions. Also notice that the amplitude of the sine wave is 30.0 volts, with the phase shift of $\pi/2$ rad., as discussed in Chap. 3. The figure shows that the *maximum magnitude* of the wave is *60.0 volts*.

Case 2—Under-damped Solutions. When the damping term C_1 is increased a little, into the range $(C_1)^2 < 4C_0$, the roots of the characteristic equation become complex and of the form

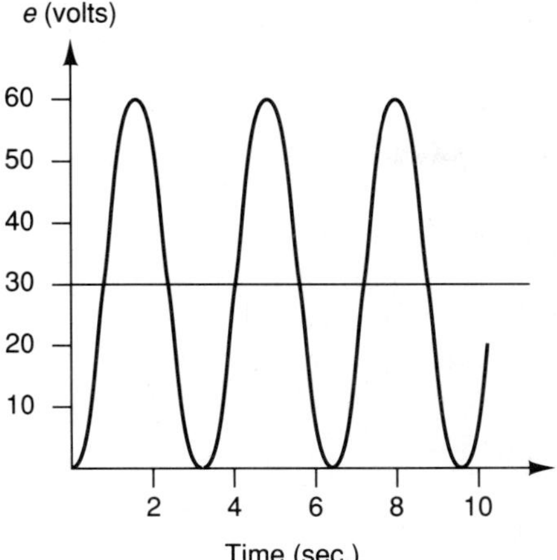

Figure 5-7. Case 1 Response (Pure Sinusoid)

$$\lambda_1 = -C_1/2 + j\left[C_0 - (C_1/2)^2\right]^{1/2}, \tag{5-63}$$

$$\lambda_2 = -C_1/2 - j\left[C_0 - (C_1/2)^2\right]^{1/2}, \tag{5-64}$$

which are also of the form $\lambda_{1,2} = -C_1/2 \pm j\omega_d$. These two roots thus lead to two solutions which can be once again superposed to find

$$e_h = C_5 \exp(\lambda_1 t) + C_6 \exp(\lambda_2 t). \tag{5-65}$$

Furthermore, the imaginary part of the roots now contains a term, ω_d, which is very much like the natural frequency for the Case 1 component. Now, however, the frequency term (in brackets in 5-63, 5-64) is smaller due to the presence of the nonzero damping coefficient, C_1. For this reason, this part of the roots is called the *damped natural frequency*

$$\omega_d = \left[C_0 - (C_1/2)^2\right]^{1/2} \tag{5-66}$$

Using this definition, some algebra gives a more usable expression for the homogeneous solution

$$e_h = C_5 \exp(-C_1 t/2 + j\omega_d t) + C_6 \exp(-C_1 t/2 - j\omega_d t), \tag{5-67}$$

$$= \exp(-C_1 t/2)\left[C_5 \exp(j\omega_d t) + C_6 \exp(-j\omega_d t)\right] \tag{5-68}$$

As before, Euler's Equations (5-51 and 5-52) can be used to simplify the two imaginary exponentials in the last expression

$$e_h = \exp(-C_1 t/2)\left[C_7 \cos(\omega_d t) + C_8 \sin(\omega_d t)\right], \tag{5-69}$$

$$e_h = \exp(-C_1 t/2)\ C_9 \sin(\omega_d t + \phi). \tag{5-70}$$

Once again, the homogeneous solution shows that the response of the Case 2 component is oscillatory in nature due to the presence of sine terms. Consequently, oscillation occurs in both the Case 1 and Case 2 components *regardless of the nature of forcing function*. This is true since the oscillation comes from the *homogeneous* equation—it is *characteristic* of Case 1 and 2 (lightly damped) second-order models.

The general Case 2 solution comes from superposing the homogeneous solution with the forced solution (Eq. 5-43),

$$e = D_0 I_m / C_0 + \exp(-C_1 t/2)\ C_9 \sin(\omega_d t + \phi). \tag{5-71}$$

The final step in the solution is to apply two initial conditions to Eq. 5-71 in order to evaluate the two unknown constants, C_9 and ϕ.

Example 5-4

Consider the Case 1 model in Ex. 5-3, and add in a little damping through a derivative term

$$2\, d^2e/dt^2 + 2\, de/dt + 8e = 6i. \tag{5-72}$$

The remainder of the problem statement is kept the same. An evaluation of the new *frequency of oscillation* and the *time history* of the output will show the effect of this added damping on the Case 1 response computed earlier.

Solution

Comparison of Eq. 5-72 with the general model (Eq. 5-41) shows the following:

$$C_0 = 4.0,$$

$$C_1 = 1.0,$$

$$D_0 = 3.0.$$

These values are used to compute the frequency of oscillation, which is the damped natural frequency

$$\omega_d = [C_0 - (C_1/2)^2]^{1/2},$$

$$= [4.0 - (1/2)^2]^{1/2},$$

and

$$\boldsymbol{\omega_d = 1.94\ \mathrm{rad/sec}.}$$

Notice that the frequency of oscillation with damping is less than the natural frequency of this component without damping (ω_n = 2.0 rad/sec.). This is always true for linear components.

As before, the forcing step is of magnitude

$$I_m = 40.0.$$

So, substituting the constants into the general solution for the step response of the Case 2 component (Eq. 5-71) yields

$$e = (3)(40)/(4) + \exp(-t/2)\, C_9 \sin(1.94t + \phi)$$

or

$$e = 30.0 + \exp(-t/2)\, C_9 \sin(1.94t + \phi) \text{ volts.} \tag{5-73}$$

In order to apply the initial conditions, you must first compute

$$de/dt = (-1/2) \exp(-t/2)\, C_9 \sin(1.94t + \phi)$$
$$+ (1.94) \exp(-t/2)\, C_9 \cos(1.94t + \phi). \tag{5-74}$$

The initial conditions now give

$$e(0^+) = 30.0 + C_9 \sin(\phi) = 0.0,$$

and

$$de/dt(0^+) = (-1/2)\, C_9 \sin(\phi) + (1.94)\, C_9 \cos(\phi) = 0.0. \tag{5-75}$$

Simultaneous solution of these two equations gives

$$\phi = 1.32 \text{ rad.} = 75.5 \text{ deg.},$$

$$C_9 = -31.0$$

Substitution of these values back into Eq. 5-73 gives the *time history* of the output,

$$\mathbf{e = 30.0 - \exp(-t/2)\, 31.0 \sin(1.94t + 1.32) \text{ volts.}} \tag{5-76}$$

Equation 5-76 is plotted in Fig. 5-8, which also shows the previous Case 1 solution with the same initial conditions. Note that the damping increase has led to a slight

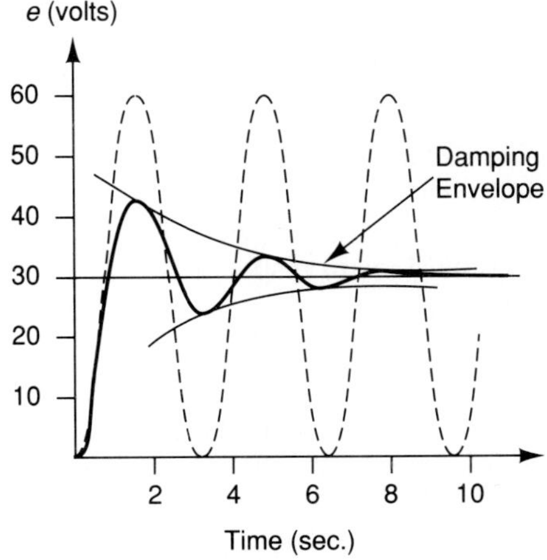

Figure 5-8. Case 2 Response (Damped Sinusoid)

decrease in the phase angle, a slight increase in the sine amplitude, and an exponential *damping envelope* ($\exp[-t/2]$) for the sinusoidal oscillations.

Comments

Notice first that the oscillation of the response is now settling to a final steady value of 30 volts. In doing this, the frequency of the oscillations remains constant while the amplitude decreases. Also, the frequency here is somewhat smaller (larger period) than the same model without damping. In fact, if you were to slightly increase the damping again, you would see the period get longer still. At some value of increased damping, the oscillations would not occur and you would observe a response very similar to that from a first-order model. The damping value at which the oscillations start or stop is called *critical damping*, it is discussed further in the next section.

Also notice that the amplitude of the oscillations is now dominated by the exponential decay term, $\exp(-t/2)$. Referring back to Eq. 5-71, the decay time-constant is $(2/C_1)$. That is, a larger C_1 means a smaller time constant and a quicker decay of the sine amplitude to the steady value.

Thus, the effect of the increased damping in the model partly results in the exponential decaying amplitude of the sine wave, and partly in the decreased frequency of oscillation, ω_d.

The effect of damping on this response leads to its name of the *damped-sinusoid response*. Further, with the Case 2 damping less than the critical value needed to inhibit oscillations, the component is described as *under-damped*.

In summary, for values of C_1 in a small range near zero, the characteristic equation yields complex roots. In turn, the complex roots lead to sine terms in the output response. Further, a small C_1 corresponds physically to a small-friction case. Thus, small friction may lead to oscillating second-order responses, which are predicted by the complex roots in the characteristic equation.

Physically, you see these oscillating components as analog mass-spring combinations. A large bell, for example, or an electronic-tank circuit (a parallel inductor-capacitor connection with negligible resistance), or as water hammer in a fluid component.

In this, you must be very careful to distinguish between the time-domain problem here versus the frequency-domain problem which is in the next chapter. In the time-domain problem, a time-domain input gives an *unforced* oscillating response, as in the ringing in a large bell. Such a time-domain response is called a *free vibration*. An electronic speaker may create the same vibrations in the air as the bell, but it must be continuously *forced* with a frequency-domain electronic input. Thus, the speaker performs in the frequency domain to create a *forced vibration*. The vibration of the bell is inherent in its physics, but the speaker must be forced. You have just studied the former case; you will study the latter case in the next chapter.

HOMEWORK

5-9. A modeler has found the following model for an electrical component:

$$C\, d^2e/dt^2 + (1/L)\, e = i.$$

a) At what frequency will this component oscillate if forced with a unit step input, $i = u_s(t)$?

b) What maximum magnitude will the oscillations have if the component is initially at rest?

c) How much will the response be offset from $e = 0$?

5-10. A property of linear models is that the unit-impulse response for a linear component may be obtained by differentiating the unit-step response for that component. Use this idea to compute the unit-impulse response for the component in the last problem.

5-11. Given the following electrical component model:

$$0.2\, d^2e/dt^2 + de/dt + 5e = 0.2i.$$

a) At what frequency will this component oscillate?

b) What will the output be in response to a unit-step input, $i = u_s(t)$, if the component starts from rest?

5-12. If you double the amplitude of the input forcing function in the last problem, how much does this change the *frequency* of the response?

5-4 THE SECOND-ORDER MODEL, NON-OSCILLATING COMPONENTS

In the previous section, the effect of very small damping on the second-order response was investigated (Case 1 and Case 2 components, C_1 small). Those *underdamped* cases were seen to result in oscillations in the response to the general second-order model

$$d^2e/dt^2 + C_1\, de/dt + C_0 e = D_0 I_m u_s(t). \tag{5-41}$$

In this section C_1 is increased further with respect to C_0 and the oscillations are observed to vanish. When the damping is equal to the critical level at which oscillations stop, the Case 3 components are observed. Further increases in damping give ever-slower, non-oscillating responses from the Case 4 components. Both of these cases of components are discussed in this section. At the conclusion of the section, the modeling results of all four cases of second-order responses are generalized for the linear second-order model.

Case 3—Critically Damped Responses. When the damping is exactly $(C_1)^2 = 4C_0$, the model is said to be *critcally damped*. In this case, the roots of the second-order characteristic equation (Eq. 5-47) are real and repeated,

$$\lambda_1 = -C_1/2,$$
$$\lambda_2 = -C_1/2,$$

and a special form of the homogeneous solution is used,

$$e_h = C_4 \exp(-C_1 t/2) + t C_5 \exp(-C_1 t/2). \tag{5-77}$$

Note the presence of t (time) in front of the second term of Eq. 5-77.

The general solution is found by superposing the forced-step solution with the homogeneous solution,

$$e = D_0 I_m/C_0 + C_4 \exp(-C_1 t/2) + C_5 t \exp(-C_1 t/2). \tag{5-78}$$

As before, a specific general solution is obtained using the two initial resting conditions to find the constants, C_4 and C_5.

Example 5-5

Consider again the model studied in the examples of the previous section. Again let $C_0 = 4$, but this time make

$$(C_1)^2 = 4C_0,$$
$$= 4(4) = 16.$$

So that $$C_1 = 4.0.$$

This gives a critically damped model.

Thus, the model now becomes

$$2\, d^2e/dt^2 + 8\, de/dt + 8e = 6i. \tag{5-79}$$

Subject to the same conditions as before,

$$I = 40\, u_s(t) \text{ amp}.$$

Find the solution to this model.

Solution

Substitution of the model constants into Eq. 5-78 gives

$$e = 30.0 + C_4 \exp(-2t) + C_5 t \exp(-2t). \tag{5-80}$$

The derivative expression must next be computed in order to evaluate the initial conditions,

$$de/dt = -2C_4 \exp(-2t) + C_5 \exp(-2t) - 2C_5 t \exp(-2t). \tag{5-81}$$

Evaluation of the initial conditions gives

$$e(0^+) = 30.0 + C_4 = 0.0,$$

$$de/dt(0^+) = -2C_4 + C_5 = 0.0.$$

From these two equations,

$$C_4 = -30.0,$$

$$C_5 = -60.0.$$

The numerical solution to this problem thus becomes

$$\mathbf{e = 30.0 - 30.0 \exp(-2t) - 60.0t \exp(-2t) \text{ v.}} \tag{5-82}$$

This solution is plotted in Fig 5-9. Also plotted in Fig. 5-9 is the previous underdamped solution.

Comments

Notice first that only one constant in the model was slightly changed (C_1, from 1 to 4), but the *character* of the response has changed greatly. Now there are no oscillations, only exponentials in the solution. Further, these exponentials work together to give the fastest possible non-oscillating transient to the final steady value. Any further increase in C_1 means more damping (more friction), which could only further slow down the response.

Also notice the shape of the initial segment of the response in Fig. 5-9. This initial response distinguishes the second-order component from the first-order com-

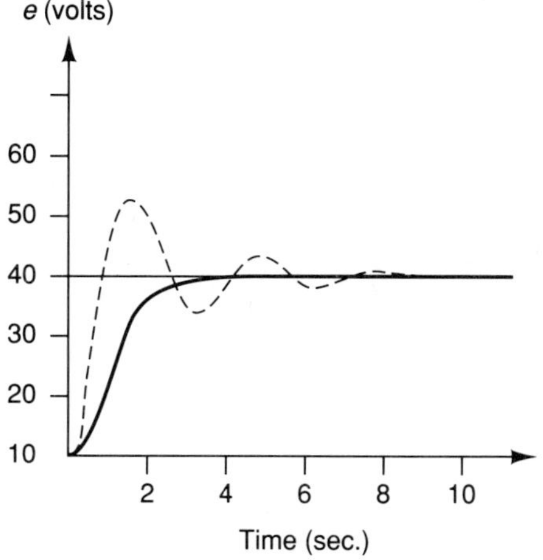

Figure 5-9. Case 3 Response (Fastest Exponential)

ponent. That is, the second-order step response with zero initial conditions has zero slope at $t = 0^+$, while the first-order step response has a sharp corner at the initial condition (e.g., see Figs. 5-3 and 5-4)

Case 4—Over-damped Responses. In the final case of second-order models, the damping is such that $(C_1)^2 > 4C_0$. Now, since the damping is greater than the critical value, the equation model (and the component) is described as *over-damped.* In this case, the roots of the characteristic equation (Eq. 5-47) are real and unequal, according to

$$\lambda_1 = -C_1/2 + [(C_1/2)^2 - C_0]^{1/2}, \quad (5\text{-}63)$$

$$\lambda_2 = -C_1/2 - [(C_1/2)^2 - C_0]^{1/2}. \quad (5\text{-}64)$$

Once more, each of these roots gives a solution to the homogeneous equation. Superposition of the solutions gives the homogeneous solution,

$$e_h = C_4 \exp(\lambda_1 t) + C_5 \exp(\lambda_2 t). \quad (5\text{-}83)$$

Since both of the roots are real, no further manipulation of the terms in Eq. 5-83 is necessary.

Superposition with the forced-step response gives the general solution,

$$e = D_0 I_m / C_0 + C_4 \exp(\lambda_1 t) + C_5 \exp(\lambda_2 t). \quad (5\text{-}84)$$

Example 5-6

For this last case of linear second-order component, return to the same model as the last example and increase the damping a little further (change C_1 from 4 to 6). The rest of the problem statement remains the same.

In this case, the model is

$$2\, d^2e/dt^2 + 12\, de/dt + 8e = 6i. \quad (5\text{-}85)$$

Find the solution to this over-damped model.

Solution

The roots to the characteristic equation are

$$\lambda_1 = -6/2 + [(6/2)^2 - 4]^{1/2},$$
$$= -3 + 2.24,$$
$$= -0.76,$$

and

$$\lambda_2 = -6/2 - [(6/2)^2 - 4]^{1/2},$$
$$= -5.24.$$

Notice that now there is one very fast-acting exponential (−5.24), and one slow-acting exponential (−0.76). So, the additional damping has acted to *separate the roots*. If yet more damping is added, the roots will be further separated. That is, more damping means that the fast root gets faster, and the slow root gets slower.

The step response in this case is

$$e = D_0I_m/C_0 + C_4 \exp(-0.76t) + C_5 \exp(-5.24t). \tag{5-86}$$

When the first term is evaluated and the initial conditions imposed, the solution becomes

$$\mathbf{e = 30.0 - 35.1 \exp(-0.76t) + 5.1 \exp(-5.24t)\ v.} \tag{5-87}$$

Comments

When the over-damped response is plotted, as in Fig. 5-10, the fast-acting root (−5.24) is only discernible near $t = 0.0$ sec. since it dies out so quickly. This conclusion is emphasized by also plotting a first-order response meeting the same initial and final values, and having the same time constant as the slow root (−0.76). Figure 5-10 shows that such a first-order model is not too bad a representation of the response of the over-damped second-order component. In fact, as the roots separate further with more damping, you may decide that the additional complexity of the extra dynamics is not worth the small improvement in accuracy associated with the extra root.

Standard Form. A standard form for second-order models exists in much the same manner as that for first-order models. That form is

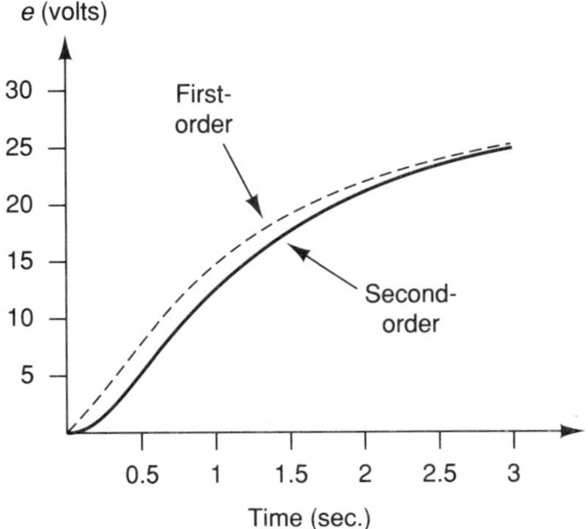

Figure 5-10. Case 4 Response. First-order is $e = 30[1 - \exp(-0.76t)]$. Second-order is $e = 30[1 - 1.17\exp(-0.76t) + 0.17\exp(-5.24t)]$

$$d^2y/dt^2 + (2\zeta\omega_n)\, dy/dt + (\omega_n)^2 y = f. \tag{5-88}$$

When the roots are extracted from the homogeneous (natural) form of this standard equation, they are found to be

$$\lambda_1 = -\zeta\omega_n + j\omega_n\, (\zeta^2 - 1)^{1/2}, \tag{5-89}$$

$$\lambda_2 = -\zeta\omega_n - j\omega_n\, (\zeta^2 - 1)^{1/2}. \tag{5-90}$$

Here, ζ is called the *damping ratio*. Of course, some similarity exists between C_1 in the earlier discussion with ζ here. However, in the earlier discussion, C_0 was held constant while C_1 was varied. Here C_1 is shown to have two contributions ($C_1 = 2\zeta\omega_n$).

When the standard model (Eq. 5-88) is compared to the idealized model (Eq. 5-41), the damping ratio is found to be $\zeta = C_1/[2(C_0)^{1/2}]$. Thus, the four cases of second-order responses may be correlated to the effect of ζ as follows

Case 1: $\zeta = 0$, undamped, $\omega = \omega_n$,
Case 2: $0 < \zeta < 1$, under-damped, $\omega = \omega_d = \omega_n(1 - \zeta^2)^{1/2}$,
Case 3: $\zeta = 1$, critically damped, no oscillations,
Case 4: $\zeta > 1$, over-damped, no oscillations.

So, in the standard model, the ζ *alone* determines the case of response in place of the previous comparisons between C_0 and C_1. In other words, here you are varying only ζ, that part of C_1 which is *pure* friction effect.

In the second-order component, a *standard unit-step response* is also defined as that to the unit step $f = (\omega_n)^2 u_s(t)$, with *zero initial conditions*. This leads to the four standard solutions, for $t > 0.0$ sec.:

Case 1: $y = 1 - \cos(\omega_n t)$, (5-91)
Case 2: $y = 1 - \exp(-\zeta\omega_n t)\, \{\cos(\omega_d t) + [\zeta \sin(\omega_d t)]/(1 - \zeta^2)^{1/2}\}$, (5-92)
Case 3: $y = 1 - (1 - \omega_n t), \exp(-\omega_n t)$ (5-93)
Case 4: $y = 1 + [\omega_n/(2\lambda_1\lambda_2)]\, [1/(\zeta^2 - 1)^{1/2}]\, [\lambda_2 e^{-\lambda_1 t} - \lambda_1 e^{-\lambda_2 t}]$. (5-94)

These four, standard, second-order responses are plotted in Fig. 5-11. Comparison of this figure with the results of the previous four examples (Figs. 5-7 through 5-10) shows that the effect of the damping ratio on these curves is very similar to that of the coefficient C_1 discussed in the examples.

The theoretical output responses given above show that oscillation occurs whenever $\zeta < 1$. However, a careful inspection of the curves in Fig. 5-11 reveals that the oscillations only become visually apparent in the output when $\zeta < 0.5$. In the range $0.5 > \zeta > 1.0$, The sine and cosine terms give an "oscillation" which appears more as an *overshoot* of the output past its final value, followed by a settling to the final value. That is, for a damping ratio slightly below 1, no more than about one sine-cycle is apparent in the response plot.

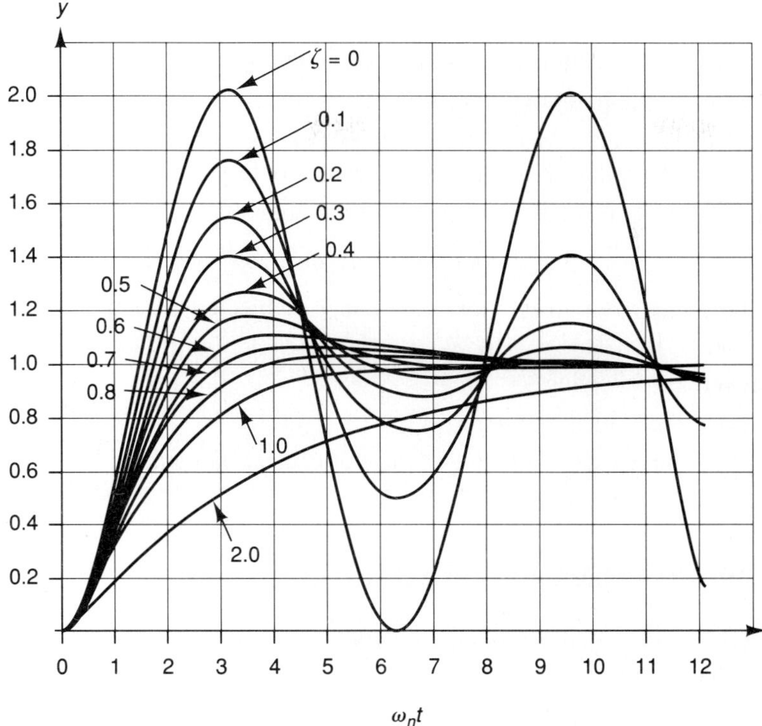

Figure 5-11. Standard Response of the Second-order Ideal Model. [after Katsuhiko Ogata, MODERN CONTROL ENGINEERING, 2/e, 1990, pg. 264. Reprinted by permission of Prentice Hall, Englewood Cliffs, New Jersey.]

The *stability conditon* for the second-order model rests on the idea that the *real* part of the charàcteristic roots (Eqs. 5-89 and 5-90) must remain nonpositive or the component response will grow to infinity as time gets large. Thus, stability requires that $\zeta\omega_n \geq 0$. This, in turn, requires that $\zeta \geq 0$ *and* $\omega_n \geq 0$. This condition can only be met if $C_1 \geq 0$ and $C_0 \geq 0$.

HOMEWORK

5-13. For the critically-damped model of Ex. 5-5,

$$2\, d^2e/dt^2 + 8\, de/dt + 8e = 6i,$$

$C_0 = C_1 = 4$. Convert this model to the standard form to answer the following:

a) What is the natural frequency of this component?

b) What is the damping ratio of the component?

c) Explain why this critically-damped component has a natural frequency?

5-14. For the over-damped model of Ex. 5-6,

$$2\, d^2e/dt^2 + 16\, de/dt + 8e = 6i,$$

$C_0 = 4$, $C_1 = 8$. Use the standard form of this model to answer the following:

a) What is the natural frequency of this component?

b) What is the damping ratio of the component?

c) If you wanted to make this component respond faster, what would you do to the damping ratio? Physically, what would you have to do?

5-15. Find the general solution to the structural component model

$$3\, d^2V/dt^2 + 12\, dV/dt + 6V = 6F,$$

if the component starts from rest and is subjected to a unit step in force, $F = u_s(t)$.

5-16. Answer the following:

a) Describe critical damping.

b) What numerical value does ζ have at critical damping?

5-5 SECOND-ORDER MODEL COMPLICATIONS

Linear second-order models contain several complications which may be treated very much like those in a first-order model. For example, multiple inputs pose no special difficulty to a second-order theoretical solution since superposition can be used to find the response. Moreover, if a model contains a differentiated input, then the correct theoretical response is the differentiation of the computed response to the *undifferentiated* input. For example, a unit impulse is a differentiated unit step. Thus, to find the theoretical response from an *ideal model* to a unit impulse, first find the response to a unit step—then differentiate it. This technique also works for integrated inputs (a ramp is an integrated step input).

Delayed-input responses may be analyzed by computing the response to an undelayed input, then time-shifting the response to account for the delay. Consequently, the computed step responses of the previous two sections can be well-used as "building blocks" to assemble a general, linear, second-order component response to a variety of input functions.

However, as in first-order models, the primary complication in a second-order model comes from nonideal terms in the output-side of the model equation. These appear as time-variable coefficients or as nonlinear terms. Nonlinearity is especially difficult since it causes the failure of superposition and thus makes the theoretical solution of the differential equation much more difficult, if not impossible.

The approach in this text to nonideal second-order models is to use numerical integration to find the component response.

Numerical Integration. Again, the method of rectangular integration is used, with the notation of interest as shown earlier in Fig. 5-6. Recall that the basis for this integration is a time axis which is divided into k equal segments, each of Δt seconds in width. You are at time $t = k\Delta t$, and you desire to predict the value of the output at the forward data point at time $(k + 1)\Delta t$. This leads to a numerical recursion formula which predicts the response of interest.

The analysis method is based on the fact that the highest derivative in the equation should be allowed to guide the numerical solution. This is accomplished using a combination of *forward-difference* and *backward-difference* numerics for the second derivative. That is, let the second derivative be

$$(d^2Y/dt^2)_k \cong \Delta(dY/dt)_k/\Delta t, \tag{5-95}$$

$$= [(dY/dt)_k - (dY/dt)_{k-1}]/\Delta t, \tag{5-96}$$

which is a backward-difference formula in terms of the *derivatives* at k and $(k - 1)$. When the right-side of Eq. 5-96 is expanded in terms of forward differences, the predictive term at $(k + 1)$ comes forth,

$$(d^2Y/dt^2)_k \cong [(Y_{k+1} - Y_k)/\Delta t - (Y_k - Y_{k-1})/\Delta t]/\Delta t, \tag{5-97}$$

and

$$(d^2Y/dt^2)_k \cong (Y_{k+1} - 2Y_k + Y_{k-1})/\Delta t^2. \tag{5-98}$$

For first-derivative terms in the model, a *backward-difference formula* must be used so that they do not contain the predictive term at $(k + 1)$,

$$(dY/dt)_k \cong (Y_k - Y_{k-1})/\Delta t. \tag{5-99}$$

Thus, Eqs. 5-98 and 5-99 are substituted into the nonideal model to find the predictive *recursion formula* for Y_{k+1}.

The *time-step criteria* for the selection of Δt comes from the time-step linearization of the second-order model (see Sec. 5-2). The criteria is now based upon the period of the linearized, undamped frequency, T_n (sec.). That is,

$$T_n = 2\pi/\omega_n, \tag{5-100}$$

and

$$T_n/1{,}000 \leq \Delta t \leq T_n/10. \tag{5-101}$$

In this criteria you are trying to set Δt small enough so that the numerical derivatives can properly approximate the curves in Y associated with an oscillating response, if that should occur. With the finite-difference method of integration, a good rule of thumb is 10 to 1,000 steps per undamped cycle. Remember, ω_n is the *fastest* frequency at which the idealized component can oscillate. Any form of damping only slows down the response.

As a final step, Δt should be decreased an order of magnitude and the solution rerun. Agreement between the two solutions using the two Δt's gives a strong indication of accuracy.

Example 5-7

Given the component model

$$3d^2Y/dt^2 + 15t\,(dY/dt)^2 + 108Y = U + dU/dt. \qquad (5\text{-}102)$$

Estimate the output, Y, for $0 < t \leq 5$ sec. if $U = u_s(t)$. The component starts from rest.

Solution

The model is nonlinear in the second term. So, the first step is to linearize Eq. 5-102 in order to estimate the time step for the numerical integration. This linearization is expressed in terms of y_L,

$$3d^2y_L/dt^2 + 37.5\,dy_L/dt + 108y_L = u + du/dt. \qquad (5\text{-}103)$$

This linear model has a natural frequency of

$$\omega_n = (108/3)^{1/2} = 6 \text{ rad/sec.}$$

Thus, a proper initial time step selection would be

$$\begin{aligned} \Delta t &= T_n/10 \\ &= 2\pi/(\omega_n 10) \\ &= 2\pi/[(6)(10)] \end{aligned}$$

so that

$$\Delta t \cong 0.1 \text{ sec.}$$

Treatment of the input in this problem requires some care. For example, since U is a step input, there is an impulse input to the component,

$$du_s/dt = u_i(t). \qquad (5\text{-}104a)$$

Consequently, the numerical problem is also one of approximating this unit impulse properly. However, now that the length of the time step is known, the impulse may be approximated as

$$[u_s(t) - u_s(t - \Delta t)]/\Delta t \simeq u_i(t). \qquad (5\text{-}104b)$$

Notice that Eq. 5-104b is really the numerical approximation of Eq. 5-104a. Furthermore, the formula given in Eq. 5-104 is quite general and can be used wherever needed. In the example, you would use

$$u_i(t) \simeq 10[u_s(t) - u_s(t - 0.1)].$$

Substituting the finite difference formulas into the nonlinear component model (Eq. 5-102) gives

$$3(Y_{k+1} - 2Y_k + Y_{k-1})/\Delta t^2 + 15(k\Delta t)\,[(Y_k - Y_{k-1})/\Delta t]^2 + 108Y_k$$
$$= 20u_s(k\Delta t) - 10u_s(k\Delta t - 0.1).$$

When this equation is solved for the next-time output, Y_{k+1}, the *recursion formula* is found,

$$\begin{aligned} Y_{k+1} &= 2Y_k - Y_{k-1} - 5.0(k\Delta t)(Y_k - Y_{k-1})^2 - \Delta t^2 36\, Y_k \\ &\quad + \Delta t^2 6.6667\, u_S(k\Delta t) - \Delta t^2 3.3333\, u_s(k\Delta t - 0.1), \\ &= 1.64Y_k - Y_{k-1} - 0.5k(Y_k - Y_{k-1})^2 \\ &\quad + 0.06667\, u_s(k\Delta t) - 0.03333\, u_s(k\Delta t - 0.1). \end{aligned}$$

This series starts with the component at rest, such that $Y_{-1} = Y_0 = 0.0$. The values at the first few time steps thus become

$$\begin{aligned} k = 0, \quad Y_1 &= 1.64(0.0) - 0.0 - 0.5(0)(0.0 - 0.0)^2 + 0.06667, \\ &= 0.06667. \\ k = 1, \quad Y_2 &= 1.64(0.06667) - 0.0 - 0.5(1)(0.06667 - 0.0)^2 + 0.03333, \\ &= 0.1404. \end{aligned}$$

The full simulated response of the component is plotted in Fig. 5-12.

Comments

Notice that the damping term in the model (e.g., the dY/dt term) has a coefficient that makes it grow with time. Thus, you would expect the initial response when the time is small to look something like an underdamped response of a linear second-order model. Figure 5-12 does indeed show such a response.

As time gets very large, you would expect the response to become overdamped and without oscillation due to the domination of the first-derivative term. The response seems to be headed in that direction at the end time. However, you must be careful about the final steady values in any nonlinear component.

The final steady output value can usually be found by setting all the derivatives

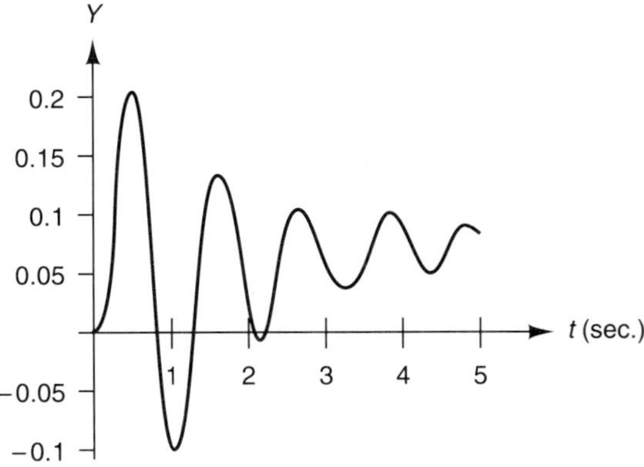

Figure 5-12. Nonlinear Second-order Component Response

equal to zero ($Y_\infty = 1/108$ in this example). But care must be exercised since, in general, the nonlinearities may be such that multiple final values may exist, or the final value may not be reachable from the initial conditions. This is one of the difficulties of nonlinear models—general rules are hard to find.

HOMEWORK

5-17. Use the rules of time-step linearization to idealize the following models over the time interval $0 < t \le 5$ sec. Estimate a proper initial time step for the rectangular integration of each.

a) $(2t + 5)\, d^2P/dt^2 + (dP/dt)^2 + 20P = 20Z$.

b) $10\, d^2E/dt^2 + t^2\, dE/dt + 10t\, E = 5I + dI/dt$.

c) $d^2V/dt^2 + dV/dt + 8FV = 16\, F^2$.

5-18. Find numerical recursion formulas for the integration of the three models of Prob. 5-17. Assume that each component is subjected to a unit step in its input.

5-19. Estimate the final steady value of the output to a unit step in the input for each of the three components of Prob. 5-17. Check your answers through a tabulation of the numerical integration, starting each of the components from rest.

5-6 THIRD-ORDER MODELS AND EIGENVALUES

Third-order models are almost too difficult when trying to determine a response using the classical methods of analytic solution. In this section, a brief look at the third-order classical method is followed in the next section by an examination of the modern state-space method. The modern method has a number of advantages which make it the method of choice for all components, third-order and higher. Yet the classical method offers several insights to the third-order component response.

Classical Solution. Consider now the ideal third-order component model

$$d^3e/dt^3 + C_2 d^2e/dt^2 + C_1\, de/dt + C_0 e = D_0 u_s(t). \qquad (5\text{-}105)$$

Again, a forced and homogeneous solution exists which must be superposed, and the component response from rest is to be studied.

The homogeneous response is assumed to be of the form

$$e_h = C_4 \exp(\lambda t), \qquad (5\text{-}106)$$

as in the previous sections. This leads to the *third-order characteristic equation*

$$\lambda^3 + C_2\lambda^2 + C_1\lambda + C_0 = 0.0. \qquad (5\text{-}107)$$

Obtaining the roots to Eq. 5-107 is by no means a trivial problem, and this forms the first obstacle to the classic method for higher-order differential equations. This, combined with the tedious algebra associated with the application of the initial conditions, makes a strong case for the use of numerical integration in the solution of all problems which are third-order and higher.

The roots of the characteristic equation are also called the *eigenvalues* of the equation, and they will henceforward be referred to as such. This comes from the German word "eigen" which means "singular." These values are singular in the sense that they uniquely dominate the solution. They give you great insight into the character of the component, just as they did in the previous studies of first- and second-order components.

The eigenvalues for the third-order component (Eq. 5-107) are all real, or they consist of one real root and a pair of complex roots.

In order to illustrate the properties of the third-order ideal model, consider the component which has the characteristic equation

$$g_1(\lambda) = \lambda^3 + 7.5\lambda^2 + 18\lambda + 11.5 = 0. \tag{5-108}$$

In the following discussion, this characteristic equation is slowly changed to observe the effect on the eigenvalues, and the component response, $y(t)$.

The eigenvalues of Eq. 5-108 are

$$\lambda = -1, -3.25 \pm 0.968j, \tag{5-109}$$

and the characteristic function $g_1(\lambda)$ is plotted in Fig. 5-13a (left).

Note that g_1 in the figure has a single crossing of the λ-axis which corresponds to the real eigenvalue at $\lambda = -1$ (at each crossing, $g = 0$, thus indicating a root). Such a plot is thus seen as a valuable tool to estimate at least one real eigenvalue since the third-order $g(\lambda)$ must cross the λ-axis at least once. Moreover, the real part of the complex pair is roughly indicated by the closest approach of $g(\lambda)$ to the λ-axis, as indicated by the circle drawn on the plot in Fig. 5-13a (left). Thus, the plot of $g_1(\lambda)$ in the figure indicates one real root and a complex pair. The output response which corresponds to these eigenvalues is shown in Fig. 5-13a (right)—yet the output does not show any oscillation as might be expected from the complex pair of eigenvalues.

The resolution of this difficulty lies in the value of the damping ratio of the complex roots. A little algebra on the characteristic equation shows that $\zeta = 0.959$ (this algebra is discussed further in Ex. 5-8 below). As mentioned in Sec. 5-4, you would expect to observe oscillations in the form of overshoot in the response when $0.5 > \zeta > 1.0$. However, this ζ is so close to 1 that the exponential-looking y_1 response shown in Fig. 5-13a (right) is the result.

Suppose now that the component is modified by redesign such that the last output-side coefficient is increased while the others remain the same (say by increasing a spring rate, or adding a little electrical inductance). This action creates the new characteristic equation

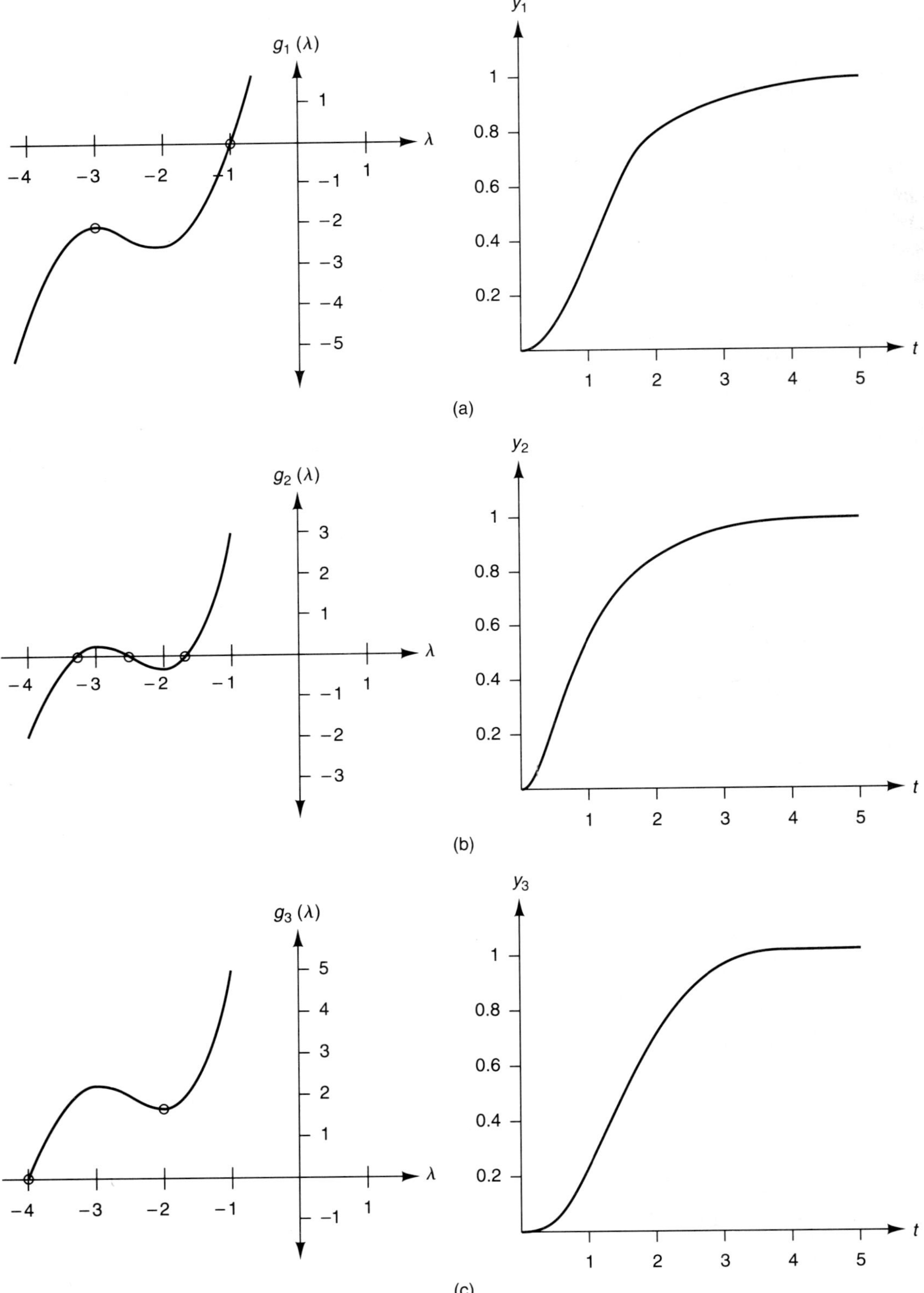

Figure 5-13. Third-order Ideal Responses. (a) Small real root, (b) all real roots, (c) large real root

$$g_2(\lambda) = \lambda^3 + 7.5\lambda^2 + 18\lambda + 13.7 = 0 \tag{5-110}$$

which is plotted in Fig. 5-13b (left). The three approximate eigenvalues are easily extracted from the λ-axis crossings in the figure,

$$\lambda \approx -1.6,\ -2.6,\ -3.3. \tag{5-111}$$

The superposed exponentials which correspond to these real eigenvalues create the pure-exponential response, y_2, as plotted in Fig. 5-13b (right).

As a final case, consider the further modification of the component to create the characteristic equation

$$g_3(\lambda) = \lambda^3 + 7.5\lambda^2 + 18\lambda + 16.0 = 0 \tag{5-112}$$

which again has one real eigenvalue and a complex pair as shown in Fig. 5-13c. Now the component has eigenvalues at

$$\lambda = -4,\ -1.75 \pm 0.968j. \tag{5-113}$$

Again, the presence of a complex pair would seem to indicate that the component should show an oscillatory response. However, the damping ratio is now $\zeta = 0.875$ which is still too large to show oscillations, only some overshoot. This behavior is demonstrated in the plot of y_3 in Fig. 5-13c (right).

Notice that all three cases of Fig. 5-13 show an exponential type of behavior in spite of the presence of complex eigenvalues. When this happens, the modeler may elect to reduce the order of the model in the spirit of economically characterizing the response of the component, and the designer may be able to simplify the design.

Remember, all real physical components are nonlinear and of infinite order, so all models are an approximation. The modeler desires to capture the essence of the response without undue complexity. Another way to look at this is that some of the less-important physical effects originally analyzed gave rise to the *higher-order mathematics*. The response shows the most important, first-order effects by inspection.

Further, all of the responses of Fig. 5-13 could be well reproduced by an overdamped second-order ideal model. It might even be possible to argue that a first-order model would be sufficient for preliminary design evaluations since oscillations are not observed.

So, the question arises, is it possible to anticipate when a third-order ideal model will show significant oscillation? In order to answer this, we must generalize the results so far. The tool to do this is the standard form for the third-order ideal model.

Standard Form. The standard, ideal third-order form is

$$d^3y/dt^3 + C_2\, d^2y/dt^2 + C_1\, dy/dt + C_0 y = D_0 u_s(t). \tag{5-114}$$

This model has the factored characteristic equation

$$(\lambda + \lambda_1)\,[\lambda^2 + 2\zeta\omega_n\lambda + (\omega_n)^2] = 0.0. \tag{5-115}$$

The case of three real roots ($\zeta \geq 1$) is ignored for now in favor of studying the oscillating component. Further, it is assumed that the real root, λ_1, may be obtained by some method (such as the graphical procedure discussed earlier) so that the factored form of Eq. 5-115 may be used.

From this, the standard third-order response starting from rest can be shown to be

$$y = 1 - C_5 e^{-\lambda_1 t} - e^{-\zeta\omega_n t}\left[C_6 \cos(\omega_d t) + C_7 \sin(\omega_d t)\right]. \tag{5-116}$$

Such that

$$\omega_d = \omega_n(1 - \zeta^2)^{1/2}. \tag{5-117}$$

The initial conditions may now be applied. First define,

$$\beta = -\lambda_1/\zeta\omega_n. \tag{5-118}$$

Notice that β is the ratio of the real root to the damping in the complex pair. The C's thus have the following values

$$C_5 = 1/[\beta\zeta^2(\beta - 2) + 1], \tag{5-119}$$

$$C_6 = C_5\beta\zeta^2(\beta - 2), \tag{5-120}$$

$$C_7 = C_5\beta\zeta[\zeta^2(\beta - 2) + 1]/(1 - \zeta^2)^{1/2}. \tag{5-121}$$

You are now ready to return to the question of oscillation in the third-order component. Start with a situation where $\beta = 1$ to keep the exponentials at the same relative speed (see Eqs. 5-116, 5-118). And take $\zeta = 0.169$ to give an underdamped component. Starting with these assumptions, you can completely derive the characteristic equation (Eq. 5-115) once you choose the lone real eigenvalue. In the case for $\lambda_1 = -1$, the characteristic equation is

$$g_4(\lambda) = \lambda^3 + 3\lambda^2 + 37\lambda + 35 = 0, \tag{5-122}$$

with eigenvalues

$$\lambda = -1, -1 \pm 5.92j. \tag{5-123}$$

The plot of the g_4 characteristic equation is shown in Fig. 5-14a (left). Note the absence of the clear maximum and minimum in the figure. This is due to the proximity of the real parts of all the eigenvalues (e.g., $\beta = 1$). It results in a component whose characteristic function cannot be simply shifted to generate three real eigenvalues in order to remove the oscillation. The low damping also gives the expected oscillating response, as shown in the plot of y_4.

Note that this type of oscillation is very different from that of a second-order model. Here the component is not oscillating about its steady value, as does a second-order model. Rather, it oscillates as it rises to that value. Model reduction is out

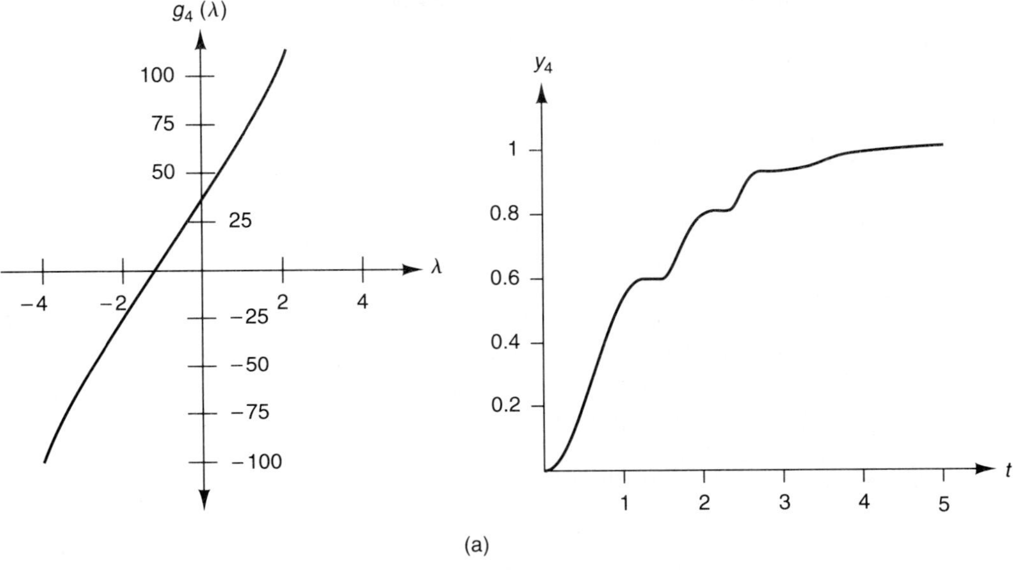

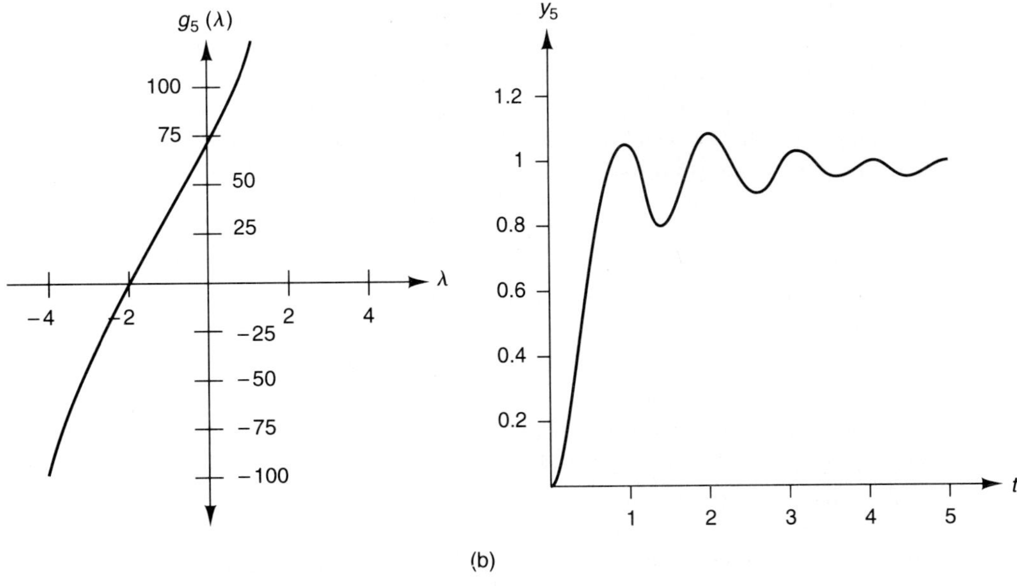

Figure 5-14. β in the Third-order Response. (a) $\beta = 1$, (b) $\beta = 4$

of the question here since it is impossible to capture this character of response in a lesser-order model.

If you modify this component as done earlier in this section, by increasing the C_0 parameter, you have

$$g_5(\lambda) = \lambda^3 + 3\lambda^2 + 37\lambda + 70 = 0, \tag{5-124}$$

with eigenvalues

$$\lambda = -2.0, -0.5 \pm 5.92j, \tag{5-125}$$

and you observe the response as indicated in Fig. 5-14b (right). Further, the more you shift the characteristic equation, the more you change β, and the more it looks like a second-order response.

The *stability conditions* for the third-order component require that all exponential terms in the solution (Eq. 5-115) go to zero as time gets large. Thus, $\lambda_1 < 0$, $\zeta \geq 0$, and $\omega_n \geq 0$ are all required for stability. A more general way to say this is that the *real parts* of the eigenvalues must all be *negative* in order for the response to reach some steady final value. Note that this was also true for the first- and second-order models, it is true for the third-order model, and it is true for *every* order of model.

Complications. Complications in third-order response estimation can arise from the left-hand side of the governing differential equation (i.e., the output side), or from the right-hand side (the input side).

On the output side, there may be nonideal terms in the form of nonlinearities and time-variable coefficients. Again, these can be treated with finite-difference numerical integration, as discussed in the next section of the text.

On the input side, there can be other input forms than the step, there can be multiple inputs, or there can be derivatives of inputs. However, inputs are always related to a particular solution to the governing differential equation. Each case must be considered separately. In the previous examples and homework, these complications have been considered for lesser-order models. The same approaches discussed there will work here as well.

The *general rule-of-thumb* is that numerical integration offers the best hope of always finding a solution to the higher-order differential equation. Further, you will soon find that numerical integration works best if you can transform the differential-equation model into an equivalent set of first-order differential equations. This transformation creates a set of *states* to describe the model in *state space*, and numerical integration is one of its most powerful applications.

Example 5-8

Given the component model

$$d^3y/dt^3 + 4\, d^2y/dt^2 + 10\, dy/dt + 12y = u. \tag{5-126}$$

The component designer claims that there are elements of the component which work together to give a lag factor in the response which has a time constant of 2.0 sec. The designer wants you to determine (a) if the component will oscillate and (b) the frequency of the oscillation.

Solution

You may begin by recognizing the linear form of the component model and the observation that superposition operates in this component. Thus, the homogeneous equation can be factored in such a way that the expected lag can be extracted from the characteristic equation. That is, the homogeneous equation is

$$\lambda^3 + 4\lambda^2 + 10\lambda + 12 = 0, \tag{5-127}$$

and the 2.0 sec. lag can be extracted from this equation by factoring

$$(\lambda + 2)(\lambda^2 + C_3\lambda + C_4) = 0. \tag{5-128}$$

Equating 5-127 and 5-128 gives the following

$$\lambda^2 + C_3\lambda + C_4 = (\lambda^3 + 4\lambda^2 + 10\lambda + 12)/(\lambda + 2). \tag{5-129}$$

This last equation may be evaluated with a technique called *synthetic division.* This type of division is just like the long division of algebra, except that it includes the symbols. In this case, you divide in the following manner

$$\begin{array}{rl} & \lambda^2 + 2\lambda + 6 \\ (\lambda + 2) & \overline{)\lambda^3 + 4\lambda^2 + 10\lambda + 12} \\ & -\underline{(\lambda^3 + 2\lambda^2)} \\ & 2\lambda^2 + 10\lambda \\ & -\underline{(2\lambda^2 + 4\lambda)} \\ & 6\lambda + 12 \\ & -\underline{(6\lambda + 12)} \\ & 0. \end{array}$$

Thus,

$$\lambda^2 + C_3\lambda + C_4 = \lambda^2 + 2\lambda + 6$$

in Eq. 5-128. When the right-side of this expression is factored with the binomial expression,

$$\lambda_{2,3} = -1 \pm 2.24j.$$

The complex roots show how to answer the designer's two questions. (a) The component has a small damping ratio ($\zeta = 0.408$, from Eqs. 5-89 and 5-90) which *will* show some slight oscillations. And (b) the frequency of oscillation will be *2.24 rad/sec*.

Comments

In applying the above method of synthetic division, it is important that the remainder of the division be zero. If this is not true, then you have not found a root. In the example, if there was some non-zero remainder at the end of the division, then the designers suspicion of a lag at 2 sec. would be incorrrect with this model. However, the zero remainder can be achieved by making slight adjustments in the lag root until the division is exact.

In general, the classical method of solving ideal differential equations requires that you find the roots of the characteristic equation (i.e., the eigenvalues). You may have to resort to numerical root-finding schemes to apply this method to higher-order equations, but for third-order models, a simple plot of the characteristic equation allows you to estimate at least one real root by inspection.

If the differential equation is nonideal, then you should turn to the state-space methods as described in the following section.

HOMEWORK

5-20. Given that the following models have an imbedded 3 sec. root in their characteristic equations. Find the remaining eigenvalues using synthetic division.

a) $d^2E/dt^2 + 5\ dE/dt + 6E = 7I^2$.

b) $d^3P/dt^3 + 6\ d^2P/dt^2 + 11\ dP/dt + 6P = 3t^2Z$.

c) $d^3V/dt^3 + 5\ d^2V/dt^2 + 15\ dV/dt + 27V = 10F$.

d) $d^3y/dt^3 + 11\ d^2y/dt^2 + 52\ dy/dt + 84y = 6\ du/dt$.

5-21. Plot each of the characteristic equations in Prob. 5-20 as a function of λ, the eigenvalue variable. Indicate the eigenvalues on your plots.

5-22. Given the plot of the characteristic function in Fig. 5-15 and the knowledge that $\beta = 2$, and $\zeta = 0.5$:

a) What is the standard solution if this third-order ideal component starts from rest and is subjected to a unit step input?

b) Would you expect the response to show much oscillation?

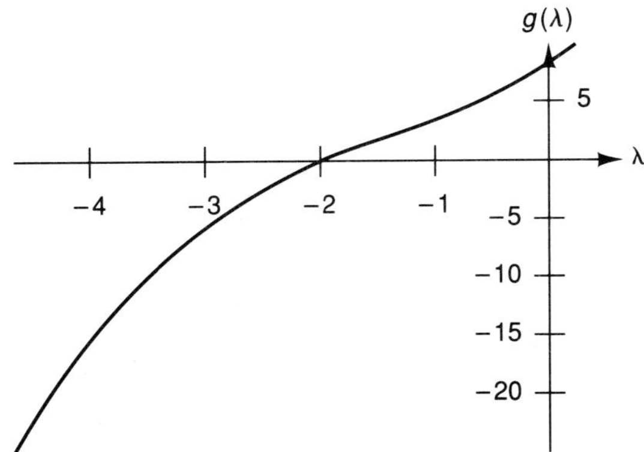

Figure 5-15. Characteristic Plot for Prob. 5-22

5-7 HIGHER-ORDER COMPONENTS AND STATE SPACE

Fourth- and higher-order ordinary differential equations are more easily solved using other methods than the classical method of analytic solution. One very common alternative is the state-space solution method. In this, a transformation is applied to the given time-domain model to change it into an equivalent vectorial representation called a *state model*. This is done since the state form of the model is very well-suited to numerical solutions on the digital computer, and is useful for solving ideal or nonideal models of any order. In this section of the text, the state method for the solution of differential equations is presented.

We begin with an nth-order, idealized-component SISO model which is given by the following equation

$$d^n y/dt^n + C_{n-1} d^{n-1} y/dt_{n-1} + \cdots + C_1 dy/dt + C_0 y = F(u). \quad (5\text{-}130)$$

Note the leading unity coefficient. In order to solve this equation for y, you need to know the time history of $F(u)$, as well as the n initial conditions $y(0)$, $dy/dt(0)$, $\ldots$, $d^{n-1}y/dt^{n-1}(0)$.

In transforming into state form, you perform a *change of variables* which assigns a *state variable* (X) for each of the output terms which need an initial condition. Thus, the nth-order component has n distinct and independent states, X, which collectively describe the component's *state* during its response to input(s). This is called state "space" since the states span the n independent derivatives which describe the component's output. That is, the states form a basis for the descriptive space of the component output.

The states are thus a set of independent variables which keep track of the output and its derivatives.

If the component is *nonlinear* and SISO, then the state-space transformation gives the following general vectorial form

$$d\boldsymbol{X}/dt = \boldsymbol{f}(\boldsymbol{X}, U), \quad (5\text{-}131)$$

and

$$Y = h(\boldsymbol{X}, U). \quad (5\text{-}132)$$

Note that $\boldsymbol{f}$ is a vector of *functions*—it has one or more nonlinear functions as entries. Also note the uppercase variables used here to indicate this nonideal model. However, even though the $\boldsymbol{f}(\ )$ functions are nonlinear, they are nevertheless associated with derivatives.

For the *linear* SISO component (e.g., Eq. 5-130), the state-space transformation gives two linear *vector* equations of the form

$$d\boldsymbol{x}/dt = \mathbf{A}\,\boldsymbol{x} + \mathbf{B}\,u, \quad (5\text{-}133)$$

and

$$y = \mathbf{C}\,\boldsymbol{x} + \mathrm{D}u. \quad (5\text{-}134)$$

The first of these equations is called the *state equation,* the second is called the *output equation*.

Note that when all the component inputs and outputs are unchanging, you must have a sustained condition of $d\mathbf{x}/dt = \mathbf{0}$. The component is then said to be in *steady-state*.

Two methods for the transformation of a time-domain model into state space are presented below. The first transformation is for a model *without* differentiated inputs. But, if the model does act to differentiate its inputs, then the first method must be modified to produce a state-space model in the form of Eqs. 5-133 and 5-134 (note the absence of input derivatives in the general form). The second method is the modified method.

Transformation Method 1 (No Input Derivatives). In this case, the forcing function is assumed to be

$$F(u) = \mathrm{b}u, \tag{5-135}$$

where b is a constant. This forcing function allows you to simply choose the states, X, to be the successive derivatives of the output. That is,

$$X_1 = Y, \tag{5-136}$$

$$X_2 = dY/dt, \tag{5-137}$$

$$\dots = \dots,$$

$$X_n = d^{n-1}Y/dt^{n-1}. \tag{5-138}$$

These n state selections are not mandatory, they may be shifted by a constant. Thus, $X_1 = Y + 2$ is acceptable in place of Eq. 5-136. Such shifting leads to another set of n distinct states which just as accurately represent the component's response.

Notice that successive rows of the above transformations are related in the following ways

$$dX_1/dt = X_2, \tag{5-139}$$

$$dX_2/dt = X_3, \tag{5-140}$$

$$\dots = \dots,$$

$$dX_{n-1}/dt = X_n. \tag{5-141}$$

which comprise $n - 1$ equations in the transformation. In order to complete the transformation of the original nth-order differential equation into n first-order differential equations, you need one more equation to complete the set. This nth equation comes from substituting the states back into the original differential equation. That is, you must algebraically solve the original equation (Eq. 5-130) for the nth-order derivative, you then substitute into that equation the states representing the lower-order output derivatives (e.g., Eqs. 5-139 through 5-141). This gives the nth differential equation

$$dX_n/dt = \mathrm{b}u - \mathrm{C}_{n-1}X_n - \mathrm{C}_{n-2}X_{n-1} - \cdots - \mathrm{C}_1X_2 - \mathrm{C}_0X_1. \quad (5\text{-}142)$$

This transformation works perfectly well even if the differential equation to be transformed is not linear, as long as there are no input derivatives present.

Transformation Method 2 (Input Derivatives Present). If input derivatives are present, then a well-known transformation can be used in place of the above procedure. This transformation, however, requires an ideal model with constant coefficients.

Suppose the component differential equation now is

$$d^ny/dt^n + \mathrm{C}_{n-1}d^{n-1}y/dt^{n-1} + \cdots + \mathrm{C}_1\,dy/dt + \mathrm{C}_0y$$
$$= \mathrm{b}_nd^nu/dt^n + \mathrm{b}_{n-1}\,d^{n-1}u/dt^{n-1} + \cdots + \mathrm{b}_1\,du/dt + \mathrm{b}_0u. \quad (5\text{-}143)$$

Then proper state selections are given by

$$X_1 = y - \beta_0u, \quad (5\text{-}144)$$

$$X_2 = dy/dt - \beta_0\,du/dt - \beta_1u = dX_1/dt - \beta_1u, \quad (5\text{-}145)$$

$$X_3 = d^2y/dt^2 - \beta_0\,d^2u/dt^2 - \beta_1\,du/dt - \beta_2u = dX_2/dt - \beta_2u, \quad (5\text{-}146)$$

$$\cdots = \cdots,$$

$$X_n = d^{n-1}X_{n-1}/dt^{n-1} - \beta_{n-1}u. \quad (5\text{-}147)$$

After some slight rearranging, Eqs. 5-144 through 5-147 comprise the first $(n - 1)$ differential equations.

The β's come from

$$\beta_0 = \mathrm{b}_n, \quad (5\text{-}148)$$

$$\beta_1 = \mathrm{b}_{n-1} - \mathrm{C}_{n-1}\beta_0, \quad (5\text{-}149)$$

$$\beta_2 = \mathrm{b}_{n-2} - \mathrm{C}_{n-1}\beta_1 - \mathrm{C}_{n-2}\beta_0, \quad (5\text{-}150)$$

$$\beta_3 = \mathrm{b}_{n-3} - \mathrm{C}_{n-1}\beta_2 - \mathrm{C}_{n-2}\beta_1 - \mathrm{C}_{n-3}\beta_0, \quad (5\text{-}151)$$

$$\cdots = \cdots,$$

$$\beta_n = \mathrm{b}_0 - \mathrm{C}_{n-1}\beta_{n-1} - \cdots - \mathrm{C}_1\beta_1 - \mathrm{C}_0\beta_0. \quad (5\text{-}152)$$

Again, as in the previous method, the nth first-order differential equation comes from solving the given differential equation for the nth derivative and substituting the transformations. This gives

$$dX_n/dt = \beta_n u - C_{n-1}X_n - C_{n-2}X_{n-1} - \cdots - C_1X_2 - C_0X_1. \quad (5\text{-}153)$$

If the original equation (Eq. 5-130) involves nonlinear terms, then this transformation will not work in general. Consequently, the transformation of higher-order, nonlinear, differential equation models with differentiated inputs is not considered in this introductory text. (See the Bibliography to this chapter for further study in this area.)

Given a transformed output equation, you may either seek an analytical solution or you may try for a numerical solution. Linear systems composed of linear components may be solved analytically in state space, but this topic is reserved for a later point in the text (Chap. 8), after the assembly of the component models into the system model is discussed. Nonlinear state-space models are almost always solved numerically, using the methods of numerical integration. Due to the simplicity of this method, it is often used for linear components as well.

Numerical Integration. One big advantage of the transformation into state-space is that it allows you to avoid the very messy problem of trying to find a finite-difference expression for an nth-order derivative as you did for the second-order time-domain model. Instead, you use the first-order, forward-difference expression studied in Sec. 5-2 on the n well-behaved, first-order state derivatives.

However, the related problem of correctly choosing the numerical-integration time step for the simultaneous integration of the n differential equations is much more difficult to address now since there are so many equations to be integrated. In fact, there are few, if any, general guidelines to help in this selection since the nth equation in the transformation may be nonlinear. However, by linearizing the nth equation (the others are already linear), you can find some help in the selection of the time step. To do this, you may use the time-step linearization procedure discussed in Sec. 5-2.

Thus you find a *linearized state-space* model in the following SISO form

$$d\mathbf{x}/dt = \mathbf{A}_L\,\mathbf{x} + \mathbf{B}_L\,u, \quad (5\text{-}154)$$

$$y = \mathbf{C}_L\,\mathbf{x} + D_L u. \quad (5\text{-}155)$$

In this notation, a lower-case x is used to represent the *linearized* value of the general state, X. If the original component model is linear, then $\mathbf{A}_L$ is the same as $\mathbf{A}$, $\mathbf{B}_L$ is the same as $\mathbf{B}$, and so on.

Now, a proper time step for the integration of the state-space model depends on the response of the model. This response is characterized by the eigenvalues hidden in Eqs. 5-131 and/or 5-133. Further, these eigenvalues may be initially estimated by using the matrix form of the *nth-order characteristic equation* on the linearized model of Eq. 5-154,

$$\det(\lambda\mathbf{I} - \mathbf{A}_L) = 0, \quad (5\text{-}156)$$

where **I** is the identity matrix. This determinant creates the nth-order characteristic polynomial in λ which must be solved to estimate the n eigenvalues. Again, all eigenvalues must have negative real parts for component stability.

Further, note that the most-negative of the real parts of the eigenvalues, λ_{RF}, gives the fastest exponential contribution. And the largest imaginary part, λ_{IF}, gives the largest frequency (shortest period). Thus, the time step you choose must be small enough to accurately integrate these fastest contributions to the response. You can meet these two conditions by choosing the *greatest absolute magnitude* of them, $|\lambda_G| = |\lambda_{RF}|$ or $|\lambda_{IF}|$, and using this number to estimate the smallest time constant, τ_s, in the following way

$$\tau_s = 1/|\lambda_G|. \tag{5-157}$$

So, to choose a proper initial estimate of the integration time interval, Δt, you may apply the following rule of thumb:

$$\tau_s/100 \le \Delta t \le \tau_s/10. \tag{5-158}$$

The time interval chosen in this way will work in the state-space integrations when the component model is linear. However, it is impossible to anticipate the nature of all nonlinearities. Consequently, there may be times when this rule of thumb fails for nonlinear components. In those cases, it is best to choose a time constant which is less than 1/100th of the time required for the most extreme dynamics of the system. That is, use the rule of thumb to get started in simulation, then rerun the simulation with a smaller time step to see the effect. Continue this process until the time-step change has negligible effect on the computed output(s).

Example 5-9

A modeler has found the component model

$$d^4y/dt^4 + 11d^3y/dt^3 + 41d^2y/dt^2 + 61dy/dt + 30y = 8u + 4du/dt. \tag{5-159}$$

Show how the modeler can simulate the response of this component using state-space variables.

Solution

The model is higher than second order and linear. The decision to use state variables is the correct one. Further, since there are derivatives of the input in the model, it is necessary to use Method 2 in the transformation to state space.

You may begin by comparing the given model to the general form in Eq. 5-130. The following coefficients may then be extracted according to Method 2,

$$n = 4$$

$$b_4 = 0$$

$$C_3 = 11 \qquad b_3 = 0$$

$$C_2 = 41 \qquad b_2 = 0$$
$$C_1 = 61 \qquad b_1 = 4$$
$$C_0 = 30 \qquad b_0 = 8$$

The coefficients are next used to compute the β terms according to Eqs. 5-148 to 5-152,

$$\beta_0 = b_4 = 0,$$
$$\beta_1 = 0 - (11)(0) = 0,$$
$$\beta_2 = 0 - (11)(0) - (41)(0) = 0,$$
$$\beta_3 = 4 - (11)(0) - (41)(0) - (61)(0) = 4,$$
$$\beta_4 = 8 - (11)(4) - (41)(0) - (61)(0) - (30)(0) = -36.$$

The β terms are then used in Eqs. 5-144 through 5-147 to find the state differential equations,

$$dx_1/dt = x_2, \tag{5-160}$$
$$dx_2/dt = x_3, \tag{5-161}$$
$$dx_3/dt = x_4 + 4u, \tag{5-162}$$

with the last equation coming from Eq. 5-159,

$$dx_4/dt = -36u - 11x_4 - 41x_3 - 61x_2 - 30x_1. \tag{5-163}$$

The state-space model can be constructed from the last four linear equations,

$$d\boldsymbol{x}/dt = \begin{bmatrix} dx_1/dt \\ dx_2/dt \\ dx_3/dt \\ dx_4/dt \end{bmatrix} = \begin{bmatrix} 0 & 1 & 0 & 0 \\ 0 & 0 & 1 & 0 \\ 0 & 0 & 0 & 1 \\ -30 & -61 & -41 & -11 \end{bmatrix} \boldsymbol{x} + \begin{bmatrix} 0 \\ 0 \\ 4 \\ -36 \end{bmatrix} u. \tag{5-164}$$

$$y = [1 \quad 0 \quad 0 \quad 0]\, \boldsymbol{x}, \tag{5-165}$$
$$= \mathbf{C}\, \boldsymbol{x}.$$

Using a forward-difference expression on these equations, the numerical-integration recursion equations for a computer are as follows

```
X1(k + 1) = X1(k) + Δt*X2(k).                          (5-166)

X2(k + 1) = X2(k) + Δt*X3(k).                          (5-167)

X3(k + 1) = X3(k) + Δt*[X4(k) + 4*U(k)].               (5-168)

X4(k + 1) = X4(k) + Δt*[-36*U(k) - 11*X4(k)
            -41*X3(k) - 61*X2(k) - 30*X1(k)]           (5-169)

     Y(k) = X1(k)                                      (5-170)
```

The eigenvalues of the **A** matrix are next used to estimate the time step, Δt, for the integrations. This is done using the matrix characteristic equation (Eq. 5-156). Evaluation of the characteristic equation for the present example gives

$$\lambda^4 + 11\lambda^3 + 41\lambda^2 + 61\lambda + 30 = 0. \tag{5-171}$$

In order to determine the eigenvalues precisely, you would have to find all the roots to Eq. 5-171. In this, you could use the *graphical eigenvalue procedure* of the previous section. The plot of Eq. 5-171 is shown in Fig. 5-16a. The four λ-axis crossings indicate that there are four real roots for this example problem. These roots are labeled as points a, b, c, and d in the figure (the rest of Fig. 5-16 is shown for the "Comments" discussion which follows).

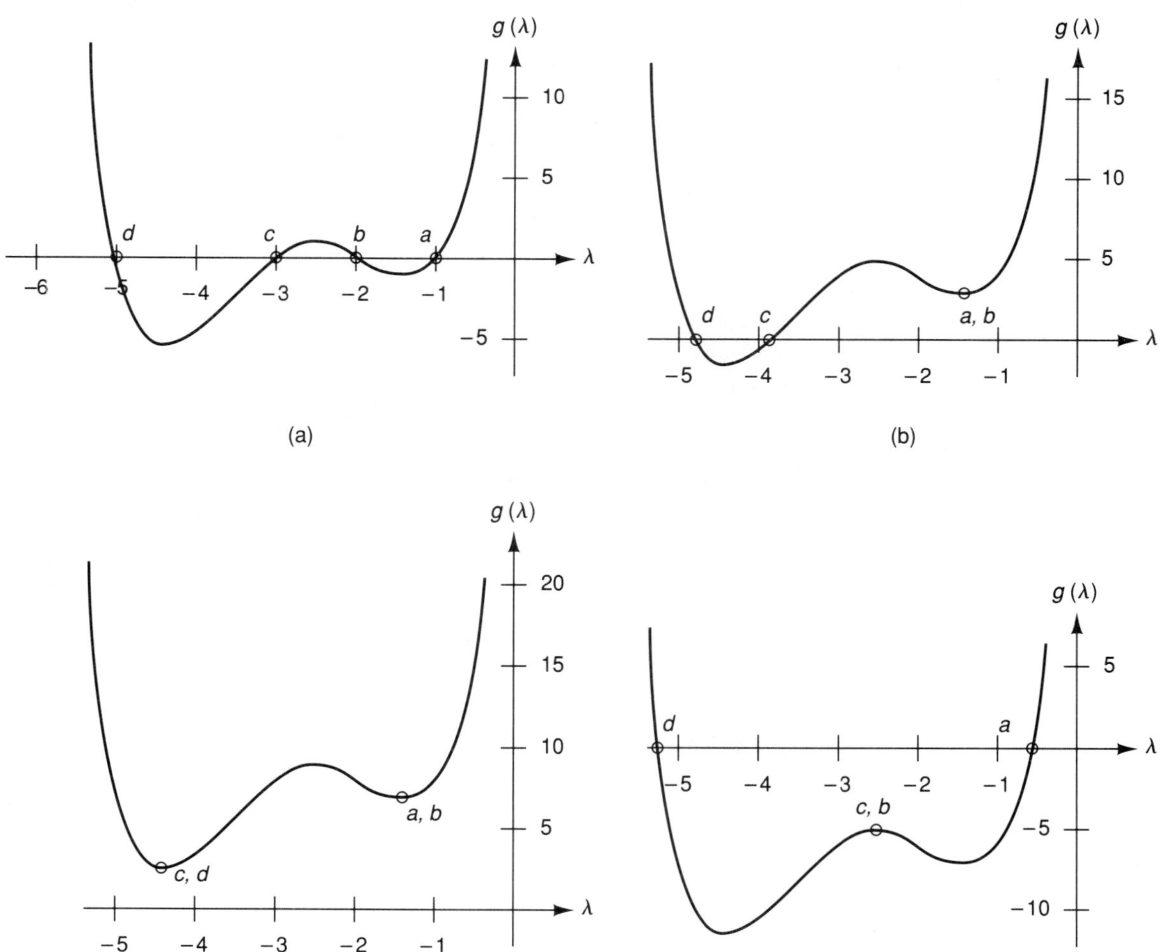

Figure 5-16. Fourth-order Characteristic Plots. (a) Four real roots, (b) two real roots, (c) four complex roots, (d) two real roots

Figure 5-16a indicates that the most-negative real eigenvalue part is at point d, and there are no imaginary eigenvalue parts. Thus, the root at d is used to compute the integration time step as follows:

$$\lambda_G \approx -5,$$

so

$$\tau_s \approx 0.2 \text{ sec}.$$

This gives you a conservative basis for choosing the time step for this component

$$\Delta t = 0.002 \text{ sec}.$$

Now you are ready to start the integrations from the initial conditions of the state variables (the output derivatives) as you seek to estimate the response.

Comments

It is very instructive to consider how the eigenvalues of this fourth-order component could change, depending on the values of the coefficients in the model.

The case just considered is shown in Fig. 5-16a. Notice first that all the real roots in this case are negative, hence this is a stable component. If there are any positive real eigenvalue parts, then there are positive exponents in the solution and the solution grows without bound as time gets large, which is an unstable condition.

In Fig. 5-16b, the last coefficient in the model is increased slightly, resulting in only two crossings of the λ-axis. These crossings correspond to the two real eigenvalues at c and d in the figure. However, since the equation is a fourth-order polynomial, you would expect two more roots. The other two roots are a complex pair which is indicated by the local minimum at point (a, b). So, the change in the model has changed the characteristic equation, and has resulted in the two real roots at a and b merging together. Further, the two real roots can be read directly from the plot. The complex pair can then be obtained through synthetic division as you did before (Ex. 5-8 in the previous section).

The plot of Fig. 5-16c results when the characteristic curve is shifted even further upward. In this case, two complex pairs of eigenvalues are created as indicated by the absence of λ-axis crossings. These complex pairs are roughly located by the two local minimums at (a, b) and (c, d). If this type of plot occurs, then the graphical method is useless since its purpose is to estimate the *real* roots.

Figure 5-16d comes about by *decreasing* the last coefficient in the original model. Here the two roots at b and c have merged into a complex pair, leaving two real roots.

Note that numerical methods to find the eigenvalues of a matrix exist, but this is regarded as an advanced topic—see the Bibliography to this chapter for further study in this area.

HOMEWORK

5-23. Given the following component models, find a state-space model of each (do not use shifted states). Show all your state definition equations in each case.

Express your answers in matrix form.

a) $dN/dt + 3N^2 = 5Q$.

b) $2\, d^2F/dt^2 + 2(dF/dt)^2 + 2F = 6V$.

c) $d^3P/dt^3 + (\sin t)\, d^2P/dt^2 + 7\, dP/dt + P = 6Z^2$.

d) $2\, d^3y/dt^3 + 6\, d^2y/dt^2 + 12\, dy/dt + 4y = 2\, du/dt + 4u$.

5-24. For each of the state space models of Prob. 5-23, find the equations to numerically integrate the model.

5-25. Given

$$d\boldsymbol{x}/dt = \begin{bmatrix} 0 & 1 & 0 \\ 0 & 0 & 1 \\ -2 & -6 & -3 \end{bmatrix} \boldsymbol{x} + \begin{bmatrix} 0 & 0 \\ -4 & 0 \\ 0 & -6 \end{bmatrix} \boldsymbol{u}.$$

a) How many inputs does this component have?

b) Estimate the eigenvalues of this component.

c) Recommend a maximum time step for the integration of this model.

5-8 SUMMARY, BIBLIOGRAPHY, AND REVIEW PROBLEMS

This chapter presented a study of the simple differential equation as a time-domain model. This study started with the first-order lag model, and continued with a gradually increasing model order to associate characteristic responses to ideal models.

In every case, the relative size of the damping in the model was seen to determine the character of the response through the eigenvalues of the model. Further, recall from the last chapter that the damping, friction, and energy-loss physics were the basis for the element analogs and the *form* of the system network models. In this chapter, friction and energy-loss were again seen to play a critical role—this time in shaping the *function* of the component response.

Nonideal effects were treated as complications to the classical model solution. These nonideal effects included nonlinearity, time-varying coefficients, and high-order models. As a remedy, nonlinear models and models without constant coefficients were treated through the use of numerical integration. Higher-order models were approached with state-space transformations and, again, numerical methods.

Two state-space transformation methods were discussed in this chapter. Method 1 handled linear or nonlinear models, but without input derivatives. Method 2 handled only linear models with constant coefficients, but it allowed input derivatives. The key idea in both of these transformations was to associate states with output derivatives. This naturally connected the order of a component (and system!) with the number of states.

So, at this point in the text, you should associate time-domain inputs with state-space models, both for design and analysis.

Finally, something must be said of algebraic components since these models violate physical observation. That is, when you observe real physical components, you see that outputs lag inputs. This is true because it always takes some finite amount of time for inputs to propagate to outputs. Therefore, every output equation must really be at least a first-order differential equation. But in an algebraic model, the outputs respond instantly to input changes. Of course, what the algebraic model is really telling you is that the time constant of the algebraic component is very small, *compared to the other components,* so that it is simpler to model this way.

Thus, much of the modeling problem is to find and characterize the dominant states. And much of the design problem is to create and control a proper set of dominant states.

BIBLIOGRAPHY

Arbenz, K. and A Wohlhauser, *Advanced Mathematics for Practicing Engineers,* Norwood, Mass.: Artech House, Inc., 1986.

Dodes, I. A., *Numerical Analysis for Computer Science,* New York, N.Y.: North-Holland, 1978.

Dorn, W. S. and D.D. McCracken, *Numerical Methods with FORTRAN IV Case Studies,* New York, N.Y.: J. Wiley & Sons, Inc., 1972. This is a classic in the field of numerical methods, with many FORTRAN programs.

Ferziger, J. H., *Numerical Methods for Engineering Application,* New York, N.Y.: J. Wiley & Sons, 1981.

Fox, L., *An Introduction to Numerical Linear Algebra,* London, Eng.: Oxford Univ. Press, 1964.

Kreyszig, E., *Advanced Engineering Mathematics,* New York, N.Y.: J. Wiley & Sons, Inc., 1988.

Ogata, K., *Modern Control Engineering,* Englewood Cliffs, N.J.: Prentice-Hall, Inc., 1990.

Slavadori, M.G. and M.L. Baron, *Numerical Methods in Engineering,* Englewood Cliffs, N.J.: Prentice-Hall, Inc., 1961.

Wait, R. A., *The Numerical Solution of Algebraic Equations,* New York, N.Y.: J. Wiley & Sons, 1979.

REVIEW HOMEWORK

5-26. Given the model for the component in Fig. 5-2b,

$$(A/\rho g)\, dP/dt + P/R_f = Z,$$

complete the following:

a) What is the time constant for this fluid component?

b) Check the units for your answer to (a) in English units.

c) Given that $P(0^+) = 14.7\ lb_f/in^2$, what is the theoretical pressure response to $Z = Z_m u_s(t)$?

5-27. Given the thermal model for Fig. 5-2d,

$$5\ d\theta/dt + \theta/2 = 4h,\ h = H_m u_s(t),$$

complete the following:

a) At what time will θ be at 63.21% of the final value of its response to a step input that starts at $t = 0$ sec.?

b) What will be the final value of θ when $t = \infty$ sec.?

c) What is the time constant for this component?

d) Plot θ-versus-time for $h = 10u_s(t)$ Btu/s, $\theta(0^+) = 0°$ F, and $0 < t \leq 50$ sec.

5-28. Find the numerical integration recursion formulas for $y(k + 1)$, for $t > 0$; use the forward-difference formula for the derivatives:

a) $dy/dt + y^2 = u_s(t)$.

b) $dy/dt + \sin y = u_s(t)$.

c) $y(dy/dt) + y^2 = u_s(t),\ y > 0$.

5-29. Given the two component models:

$$dy/dt + 4y^2 = u_s(t),$$

$$dy/dt + 2y = u_s(t).$$

a) Plot the two output responses starting from rest, for $0 < t \leq 5$ sec.

b) One of the two possible steady states for the nonlinear model is not reachable from the initial condition. Why?

5-30. A model for a water piping component is found to be

$$2\ d^2P/dt^2 + 18P = 10Z.$$

a) Will this component oscillate? If so, what is the frequency of oscillation?

b) What is the theoretical response of this component to a unit step input, if the component starts from rest?

5-31. A structural component is found to have the following model

$$M\ d^2V/dt^2 + B\ dV/dt + KV = 10\ dF/dt.$$

a) If this component oscillates, what will be the frequency of oscillation?

b) What will the output be if the component starts from rest and is given a unit step input in F?

5-32. Determine the nature of the damping (undamped, under- , critically, or over-damped), and indicate the type of output response to expect (sine, decaying sine, or exponential) for each of the following models:

a) $3d^2y/dt^2 + 6y = 5u_i(t)$.

b) $4d^2y/dt^2 + 2dy/dt + 8y = 7u_r(t)$.

c) $2d^2y/dt^2 + 4dy/dt + 8y = 10u_s(t)$.

d) $5d^2y/dt^2 + 20dy/dt + 5y = \sin(6t - 3)$.

e) $d^2P/dt^2 + 3dP/dt + 4P = 8Z$.

f) $4d^2V/dt^2 + 2dV/dt + 2V = 16dF/dt$.

5-33. Given the component model

$$d^2E/dt^2 + t(dE/dt) + 12E = I.$$

a) Compute and plot the component response which starts from rest and is subjected to the input

$$I = 10u_s(t) \text{ amps},$$

for the interval $0 < t \leq 5$ sec.

b) Comment on the approximate value of the damping ratio when $t = 0$ sec., $t = 2$ sec., and $t = 5$ sec. and how this is seen in the plot for (a) above.

5-34. Given the fluid component model

$$d^3P/dt^3 + 9\,d^2P/dt^2 + 26\,dP/dt + 24P = 12Z.$$

a) Estimate the eigenvalues for this component.

b) If the component starts from rest and is subjected to a unit step in Z, what is the steady value of the output?

c) What is the general solution to the model equation (don't solve for the coefficients from the initial conditions, but do show the necessary equations to do this).

5-35. Given the structural component model

$$d^3V/dt^3 + 2d^2V/dt^2 + 4dV/dt + 8V = 5F.$$

a) Given that one eigenvalue is a lag with a time constant of 2 sec., what are the other eigenvalues?

b) Will this component ever reach steady output for step input? Why or why not?

c) What would you recommend to a designer who wanted this component to reach steady value faster?

5-36. Transform the following models into state space, express your answers in matrix form:

a) $d^3y/dt^3 + 3d^2y/dt^2 + 6dy/dt + 8y = 7u_1 + u_2.$

b) $d^3y/dt^3 + 2d^2y/dt^2 + 4dy/dt + 12y = 2du/dt + 7u.$

5-37. Given

$$d\mathbf{x}/dt = \begin{bmatrix} 0 & 1 & 0 \\ 0 & 0 & 1 \\ -2 & -7 & -10 \end{bmatrix} \mathbf{x} + \begin{bmatrix} 0 \\ 0 \\ 1 \end{bmatrix} u.$$

$$y = x_1.$$

Show how you would find a proper time interval for the simultaneous integration of these three equations (you need not perform each step, just list the steps).

Chapter 6

Component Physical Analysis: The Fundamental Approach in the Frequency Domain

6-0 INTRODUCTION

Components designed to use periodic inputs are well-suited to a frequency-domain analysis since these components process frequency-rich inputs to create outputs. The usual preliminary modeling approach to these components relys heavily on creating an *ideal* model so that the property of superposition may be used in the mathematical analyses. So, just as in Chap. 4, this puts the modeler in conflict with the Fundamental Truth of Modeling—that all physical effects are nonlinear and distributed. This time, however, the nonideal physical effects can only be neglected through careful limits on the sinusoidal variables, and through restrictions on the modeled hardware.

Sinusoidal Variables. Mathematically, the variables of interest are expressed as

$$Y = \mathrm{Y}_A + y,$$

and

$$U = \mathrm{U}_A + u,$$

where the lower case now represents a *periodic* signal centered on the steady values, U_A and Y_A. An example of these variables is shown in Fig. 6-1.

Note that in the frequency-domain physics, the component never stays at a time-domain steady state. Instead, a "constant" frequency-domain variable has constant amplitude, frequency, and phase. This condition is called the *sinusoidal steady state* to distinguish it from the time-domain steady condition. Such constant periodic variables are shown in Figs. 6-1a and b. Thus, the transient change which occurs in the frequency-domain output between "steady" conditions is usually not of interest in a frequency-domain analysis. This is because the transient is inherently a time-domain effect. Further, the precise time when $t = 0$ is more of a system-model synchronization issue than an important question for frequency-domain analysis.

Recall from Chap. 4 that a steady Y-versus-U plot was used to assess component linearity. Now however, when these two time functions are plotted against each other, a *Lissajous figure* like that of Fig. 6-1c is the result. Note that the Lissajous figure is a parametric plot in the time, t, and is quite different from the steady-state input/output plots which were used in Chap. 4 for linearity. The modeler is thus faced with an initial problem in assessing the frequency-domain component linear-

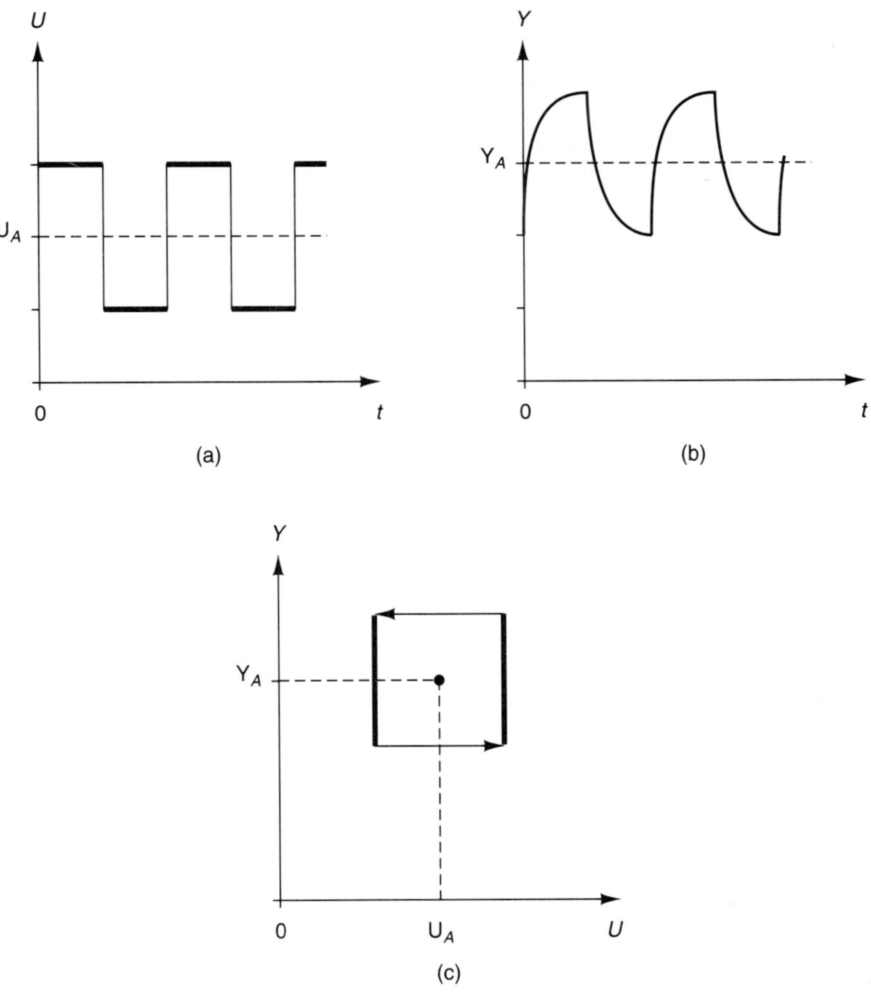

Figure 6-1. Sinusoidal Steady State. (a) Input, (b) output, (c) Lissajous figure

ity. This problem is solved in this chapter by restricting the allowable input(s) and hardware. It is solved in the next chapter (input/output modeling) using a more thorough understanding of frequency-domain linearity.

In a fundamental frequency-domain model, the amplitude of u is assumed to be small enough, *with respect to the component nonlinearities,* that the component behaves in a linear fashion. This assumption depends, of course, upon the design of the component, but this is often the case. If so, a linear model gives good results, and the mathematical property of *superposition* may be used by the modeler. This sort of linearization is called *small-amplitude linearization,* and it is very common in frequency-domain designs.

Recall from Chap. 3 that you modeled a periodic input, u, by superposing an infinite series of harmonic signals, u_n, each with distinct amplitude (M), frequency (ω), and phase (ϕ). Fourier analysis methods were used to find the superposition

$$u = \sum u_n = \sum M_n \sin(\omega_n t + \phi_n). \tag{6-1}$$

In this chapter you consider how each of these distinct signals, u_n, is transferred to a signal at the output, y_n. Then, thanks to linearity, you again use superposition to find the net output signal,

$$y = \sum y_n = \sum T_n u_n. \tag{6-2}$$

The focus of this chapter is the physical analysis to create T_n, the frequency-domain model in Eq. 6-2. You will create it in three forms for the ideal component. First, when a pure-sine input is physically transferred to an output, T becomes the *Transfer Ratio*. Next, when a more complicated periodic input is transferred to an output, T becomes the *Transfer Function*. Finally, when complicated periodic inputs are transferred to multiple outputs in a MIMO component, T becomes the *Transfer Matrix*. All of these models are easily found, if certain restrictions on the hardware are first assumed.

Hardware Restrictions. The most important hardware restriction is that the design must not include any elements with discontinuous input-output charcteristics. Examples of these are coulomb friction in structural components, and diode switches in electrical components. These are not allowed because the small-amplitude input does not linearize them—they are fundamentally nonlinear and must be treated as such. Advanced frequency-domain modeling methods for these nonlinear elements exist, but they will not be considered in this introductory study.

The remaining hardware restrictions are chosen to simplify and focus the analyses. For example, the two ways that all frequency-domain components physically create outputs are either by amplitude modulation (AM) or by frequency modulation (FM) of inputs. However, in order to limit this study, FM components are reserved for other texts.

Typical results from a physical analysis of AM effects are shown graphically in Fig. 6-2. Note that this electronic component is chosen for illustrative purposes only, the methods in this chapter also work quite well for the other types of frequency-domain components (thermal, fluid, and structural). The amplitudes of the periodic output harmonics from the component, $|E_n|$, are shown on a Bode diagram of the same type as in Chap. 3, along with a plot of the phase angle of the output harmonics, ϕ_n. Similarly, a Bode diagram will be used to give a graphical plot of the frequency-domain T model as a preparatory step to the computation of the output(s).

Sometimes a frequency-domain component in a preliminary system design has its hardware details hidden for patent reasons, or it is too complicated for a fundamental approach. In such cases, if frequency input/output data on the component is available, then you may resort to the input/output modeling method as discussed in Chap. 7.

The physical idealizations that you will study in this chapter are very similar to those in Chap. 4, with one exception—the network models in this chapter do not have any internal power-source elements. This restriction greatly simplifies the model, and the modeling task.

Thus, this study is limited to modeling the response of *passive* networks only. Source elements are allowed only at the ports of these networks to represent the forcing of the inputs. Further, even though some allowable AC components must be supplied by an external power source (as in the AC power amplifier), only those elements are allowed which appear as passive idealized elements in the network, without explicit supply representation. In any case, note that a component boundary may be redrawn to exclude a supply from a network, thus meeting this restriction.

Modeling Plan. The general modeling plan in this chapter follows closely with that used in Chap. 4, with the addition of a final step to create the frequency-domain T model.

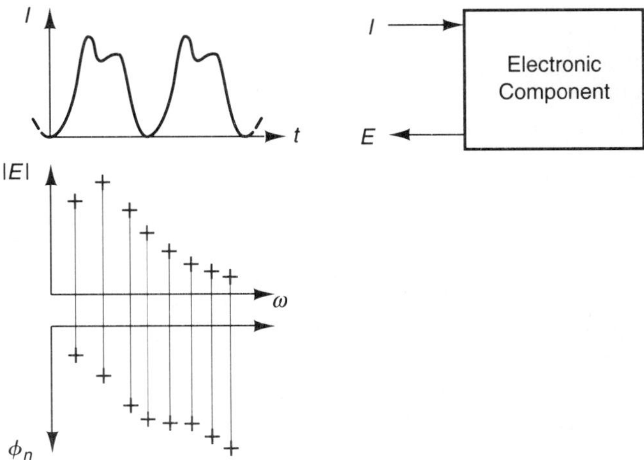

Figure 6-2. Frequency-domain Response

The first step is to conduct a fundamental physical analysis of the frequency-domain hardware. This is done to create an idealized physical-network model in terms of ideal-analog elements (R, L, and C analogs) and a few idealized components. These network models are formed from the fundamental effects of the electronic, structural, and fluid physics. Thermal networks are not considered in this introductory study since they are rarely designed to operate under sinusoidal forcing.

The second step is, given an idealized network, to use Kirchhoff's laws to find the time-domain, differential-equation model of the network.

The final step is to transform the time-domain model into the frequency-domain T model for further study and evaluation.

We begin with the study of electronic fundamentals, and the transfer-ratio model for pure-sine inputs.

6-1 ELECTRONIC ELEMENTS AND TRANSFER RATIO MODELING

Frequency-domain electronic idealizations arise from the same fundamental effects as studied in Chap. 4. Thus you find that frequency-domain elements are the same fundamental elements of resistance, capacitance, and inductance (R, C, and L). So again, as in Chap. 4, the conceptual physical hardware is first idealized using the modeling tools of purity, linearity, and lumping. And again, the First Law of Modeling is emphasized to encourage you to keep it simple.

If a pure-sine wave is input to an ideal electronic element, then the output is also a pure-sine. Since this is the simplest input in the frequency domain, it leads to the simplest model of these elements which is the ratio of the sine output over the sine input. This model is called the *Transfer Ratio, T.* We begin by computing this model for the fundamental electronic elements. This is followed by computing the transfer ratio for simple networks composed only of these elements.

Throughout this analysis you should bear in mind that these elements may also be analogs of structural and fluid effects. And these modeling methods may also be used to create transfer ratio models of those analog networks as well.

The Ideal Electronic Elements. The ideal electronic symbology of the frequency domain is shown in Fig. 6-3; note the similarity to the time-domain symbols. However, one distinction here is the small wave in the source symbols to distinguish them as periodic sources. Note also that the sources in the figure have a constant offset term, (E_o or I_o), which is shown for complete generality. But, in your frequency-domain analyses you neglect such time-domain contributions in favor of the small amplitude, pure-periodic waveform with zero offset, such as e_s or i_s.

The fundamental physics of the *ideal resistor* is

$$e = i\mathrm{R}. \tag{4-7}$$

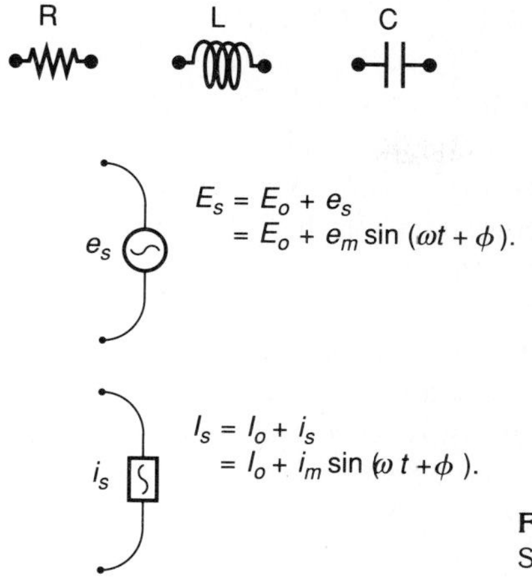

R L C

$E_s = E_o + e_s$
$= E_o + e_m \sin(\omega t + \phi).$

e_s

$I_s = I_o + i_s$
$= I_o + i_m \sin(\omega t + \phi).$

i_s

Figure 6-3. Frequency-domain Network Symbology

Note that this equation must be satisfied at every instant of time. Thus, if the input to the resistance is the pure-sine current,

$$i = \mathrm{i}_m \sin(\omega t), \tag{6-3}$$

then the pure-sine output is

$$e = \mathrm{Ri}_m \sin(\omega t), \tag{6-4}$$

and the transfer ratio is

$$T = \text{output/input} = e/i = \mathrm{R}. \tag{6-5}$$

The transfer ratio in the form of voltage-over-current *at the same port* is also called the *impedance,* T_I. This impedance may be computed regardless of the forcing at the port. Further, the ratio of a voltage at one port to a current at another port is *not* an impedance. Note that the impedance at an electrical port always has the units of ohms (Ω) since it is the ratio of voltage over current.

Further, in more general terms, the port impedance is defined as the transfer ratio of the across-over-through variables at that port. Thus, a port impedance may be computed for the other component types (fluid, thermal, and structural), and is expressed in consistent units.

If the input to the resistance happens to be the voltage

$$e = \mathrm{e}_m \sin(\omega t), \tag{6-6}$$

then the output is a sinusoidal current, and the transfer ratio is

$$T = \text{output/input} = i/e = 1/\text{R}. \qquad (6\text{-}7)$$

For this ratio of current-over-voltage, the transfer ratio is also called the *admittance*, T_A. It has the units of 1/ohms, or mhos, in the electrical component. Again, the admittance is only computed for variables at the same port, regardless of causality. Further, the admittance may also be computed for the other types of component ports, as the ratio of through-over-across variables at the port, and again in consistent units.

The situation for the *ideal inductor* is a little more complicated, and is shown in Fig. 6-4. Note that the inductor also has an impedance and an admittance. In fact, every network port has these two transfer ratios, and designers are usually very interested in their values, for reasons which will be discussed shortly.

The computed responses in Fig. 6-4c are based upon the fundamental, ideal-inductor relationship,

$$e = \text{L}\, di/dt. \qquad (4\text{-}13)$$

Impedance forcing is shown in the multiport diagram at the left of Fig. 6-4a. This is translated into the simple circuit at the center of the figure—note the use of a current source to model the current input shown in the multiport diagram. The transfer ratio in this case is shown at the right of the figure.

Also note that the frequency-domain circuit notation has an assumed positive-direction for current flow, and a corresponding voltage drop across the inductor. You must indicate these positive senses in order to use the laws of Kirchhoff to analyze the circuits. But you may choose these senses in any way you like, the sign of the answer will show you the sense of the actual variables. The important thing is for you to show them on your networks and use them in your analyses.

The inductor admittance forcing is shown in Fig. 6-4b. Note that a voltage source is here applied to the inductor to model the voltage forcing of admittance analysis.

The Bode plots in Fig. 6-4c come from the frequency analysis of the inductor. For example, impedance forcing requires that the input be current

$$i = \text{i}_m \sin(\omega t). \qquad (6\text{-}3)$$

This input creates the model equation

$$e = \text{L}\, di/dt. \qquad (4\text{-}13)$$

(Note that the input is on the right-hand side of this equation, according to the rules for model form. This form will shortly be very important in evaluating the element's phase response.)

The inductor's output response is thus

$$e = \text{L}\, \text{d}[\text{i}_m \sin(\omega t)]/dt, \qquad (6\text{-}8)$$

$$= L\omega i_m \cos(\omega t). \tag{6-9}$$

This last cosine function must be changed into a sine function in order to compare the output with the original sine input. This is easily done using the laws of trigonometry,

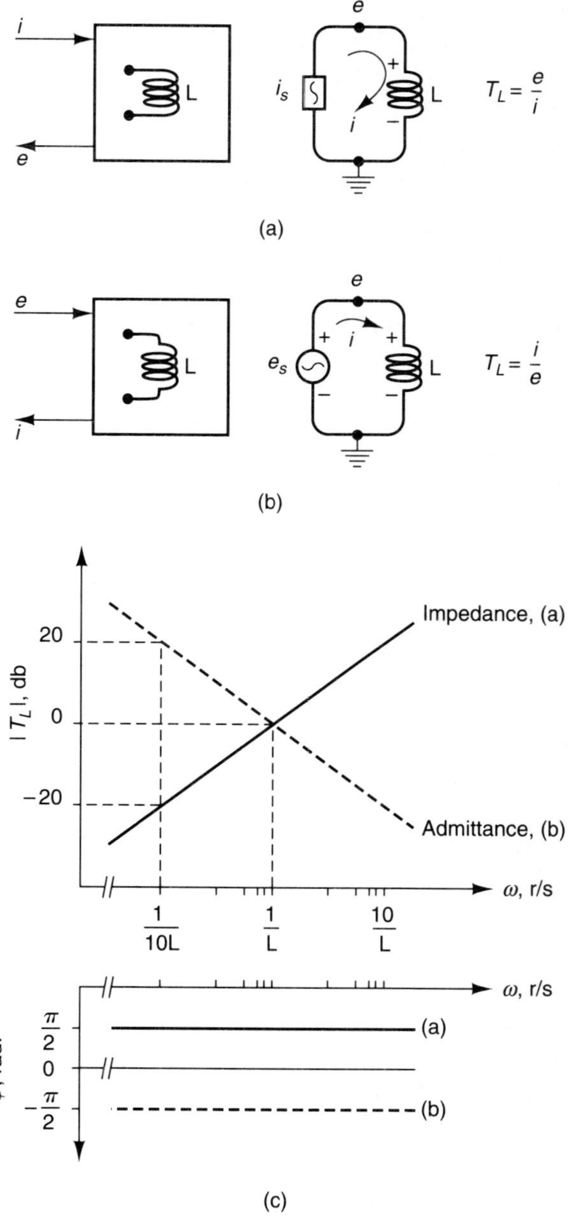

Figure 6-4. The Ideal Inductor. (a) Impedance forcing, (b) admittance forcing, (c) Bode diagram

$$e = L\omega i_m \sin(\omega t + \pi/2). \tag{6-10}$$

The plus phase angle, $\pi/2$, is chosen here because the model (Eq. 4-13) has the derivative on the *input* side of the equation. This comes from a physical situation where the element acts to differentiate the input. All such derivatives make the output *lead* the input in time, so the phase angle must be positive.

Generally speaking, the phase relationship of output-to-input in ideal models is always determined by the location of the derivatives in the equation, and is expressed using the Phase Sign Rule.

The *Phase Sign Rule* rests on the observation that derivatives on the input-side of a model equation give positive phase angle, while derivatives on the output-side of the model give negative phase angle. If there are derivatives on both sides, then the frequency of the input is critical to the phase angle determination, and a more complete analysis of the model is required (this is discussed further in Sec. 6-5 of the text). The rule is that you should always choose the phase angle closest to 0° when computing trig. functions, according to the sign conventions given above.

Thus, when the impedance transfer ratio for the inductor is formed you find

$$T_{IL} = e/i = L\omega\,[\sin(\omega t + \pi/2)]/\sin(\omega t), \tag{6-11}$$

which is more compactly written as an amplitude and a phase angle,

$$T_{IL} = L\omega \angle \pi/2. \tag{6-12}$$

Admittance forcing of the inductor requires that the input be the voltage

$$e = e_m \sin(\omega t). \tag{6-6}$$

The inductor model thus becomes

$$L\, di/dt = e, \tag{6-13}$$

with the derivative now moved over to the output-side of the equation. The output current is thus

$$i = (1/L)\int e_m \sin(\omega t)\, dt, \tag{6-14}$$

$$= -(1/L\omega)\, e_m \cos(\omega t), \tag{6-15}$$

$$= (1/L\omega)\, e_m \sin(\omega t - \pi/2). \tag{6-16}$$

Note that the negative phase angle is chosen here because the derivative is on the output-side of the model equation.

The admittance transfer ratio is thus

$$T_{AL} = i/e = (1/L\omega)\ \angle -\pi/2. \tag{6-17}$$

These results are plotted in Fig. 6-4c. Note that the *amplitudes* of the transfer ratios, $|T|$, are plotted on a decibel scale (db) according to the relationships

$$|T_{IL}| = 20\ \log_{10}(L\omega), \tag{6-18a}$$

and

$$|T_{AL}| = 20\ \log_{10}(1/L\omega). \tag{6-18b}$$

The phase angle between the output and the input is plotted immediately below the amplitude plot so that you may see the correlation between the amplitude, frequency, and phase.

Figure 6-4c shows that the impedance of the inductor increases as the frequency increases. Recall from Chap. 4 that the inductor is a device which embodies the reluctance of the current to change. But higher frequency means a higher rate of current change, which should carry with it a higher impedance for the inductor, as shown in the figure.

Figure 6-3c also shows the phase angle, ϕ, between the output and the input. Notice that both phase angles remain constant for any frequency.

The two cases of forcing for the *ideal capacitor* are shown in Fig. 6-5. The admittance forcing case is shown in Fig. 6-5b, with a voltage input. In this case, the governing equation is

$$i = C\ de/dt. \tag{4-16}$$

Note that the derivative is associated with the input voltage

$$e = e_m \sin(\omega t). \tag{6-6}$$

So, the output is

$$i = C\omega\, e_m \cos(\omega t). \tag{6-19}$$

And the admittance transfer ratio for the ideal capacitor is

$$T_{AC} = i/e = C\omega\ \angle \pi/2. \tag{6-20}$$

Similarly, the impedance transfer ratio for the ideal capacitor is

$$T_{IC} = e/i = (1/C\omega)\ \angle -\pi/2. \tag{6-21}$$

The Bode Diagrams for the two capacitor transfer ratios are plotted in Fig. 6-5c. Here the impedance drops off as the frequency increases, but the phase angle is constant as in the inductor.

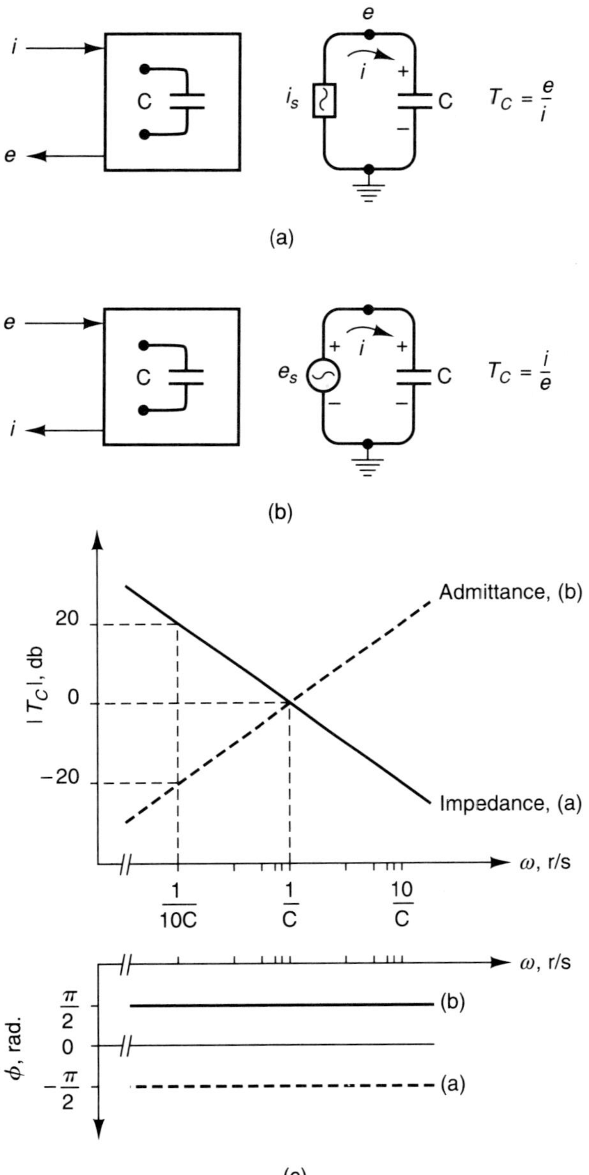

Figure 6-5. The Ideal Capacitor. (a) Impedance forcing, (b) admittance forcing, (c) Bode diagram

Once you have found the transfer ratios for an element, it is simple to compute the output for any sine input using

$$y = T(\omega)\, u. \tag{6-22}$$

For more complicated SISO networks, you have to do more physical modeling, but the frequency model appears just as in Eq. 6-22.

SISO Network Modeling. Transfer ratio network modeling is done using the same two steps as for the elements. First, the time-domain physical model is found, then the impedance and admittance ratios are found. The study of these networks is begun by considering an example.

Example 6-1

A designer is interested in the impedance of the real capacitor shown in Fig. 6-6a. This designer wants to study how the current leaks through the dielectric material. This leakage effect is modeled using the idealized resistance, R, in parallel with an ideal capacitance, C, as shown in Fig. 6-6b. Find the impedance of the idealized network.

Solution

First you must apply the impedance forcing to the ideal network as shown in Fig. 6-6c and d—note the current source which gives the current forcing in which the designer is interested. Also note the positive sign conventions which are chosen on the elements, and at node *b*.

The model may be obtained by first writing the Kirchhoff Current Law (KCL) for node *b*,

$$i = i_R + i_C. \tag{6-23}$$

Substituting element relations gives

$$i = e_R/\mathrm{R} + \mathrm{C}\ de_C/dt. \tag{6-24}$$

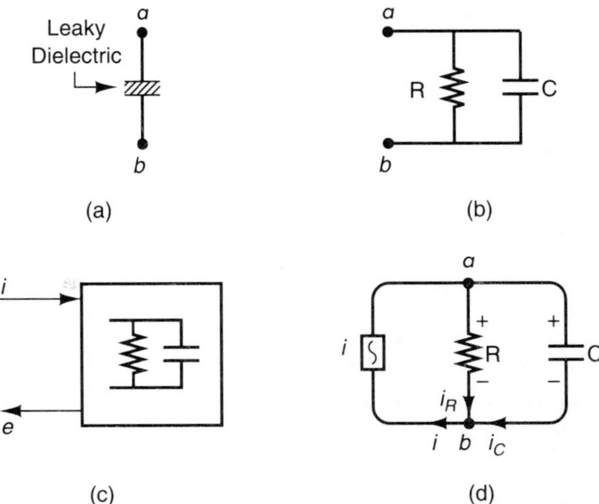

Figure 6-6. Leaky-capacitor Impedance. (a) Concept, (b) idealized model, (c) impedance forcing, (d) network model

Writing several KVL relations shows the parallel arrangement of the elements,

$$e_R = e_C = e. \tag{6-25}$$

Thus, the time domain model for the network is

$$\mathrm{RC}\, de/dt + e = \mathrm{R}i. \tag{6-26}$$

Notice that the only differential is on the output side of the equation, you can thus expect that the phase angle must be negative according to the Phase Sign Rule.

For the sinusoidal input

$$i = \mathrm{i}_m \sin(\omega t), \tag{6-3}$$

the general solution to the network model (Eq. 6-26) is

$$e = e_h + e_p, \tag{6-27}$$

$$= \mathrm{C}_1 \exp(-t/\mathrm{RC}) + \mathrm{C}_2 \sin(\omega t) + \mathrm{C}_3 \cos(\omega t). \tag{6-28}$$

The exponential term in Eq. 6-28 comes from the homogeneous solution to the differential equation. This is a transient term which you ignore in frequency-domain solutions.

For linear SISO differential equations you know that $\omega_{\mathrm{in}} = \omega_{\mathrm{out}}$, so you are able to assume a particular (forced) solution of the form shown in Eq. 6-28. Notice that you must include both the sine and cosine terms even though there is only a sine forcing function. This is true no matter what the order of the component model.

When you substitute the assumed solution you find that

$$\mathrm{RC}\omega\, [\mathrm{C}_2 \cos(\omega t) - \mathrm{C}_3 \sin(\omega t)] + \mathrm{C}_2 \sin(\omega t) + \mathrm{C}_3 \cos(\omega t) = \mathrm{Ri}_m \sin(\omega t). \tag{6-29a}$$

Regrouping yields

$$(\mathrm{RC}\omega \mathrm{C}_2 + \mathrm{C}_3) \cos(\omega t) + (\mathrm{C}_2 - \mathrm{RC}\omega \mathrm{C}_3) \sin(\omega t) = \mathrm{Ri}_m \sin(\omega t). \tag{6-29b}$$

Comparing like coefficients gives

$$\mathrm{RCC}_2\omega + \mathrm{C}_3 = 0, \tag{6-30}$$

$$\mathrm{C}_2 - \mathrm{RCC}_3\omega = \mathrm{Ri}_m. \tag{6-31}$$

Solving these two equations simultaneously gives the two frequency-domain coefficients for this forcing,

$$\mathrm{C}_2 = \mathrm{Ri}_m/[\mathrm{R}^2\mathrm{C}^2\omega^2 + 1], \tag{6-32}$$

$$\mathrm{C}_3 = -\mathrm{R}^2\mathrm{C}\omega\, \mathrm{i}_m/[\mathrm{R}^2\mathrm{C}^2\omega^2 + 1]. \tag{6-33}$$

The sine-forced solution is thus

$$e = \{Ri_m/[R^2C^2\omega^2 + 1]\}\,[\sin(\omega t) - RC\omega\cos(\omega t)]. \tag{6-34}$$

Again, the output expression must be converted into a phase-shifted sine since you need to compare the output to the input sine wave. That is, since

$$A\sin(\omega t + \phi) = B_1\cos\omega t + B_2\sin\omega t,$$

then

$$B_1 = A\sin\phi,$$

and

$$B_2 = A\cos\phi.$$

Applying the laws of algebra gives

$$A = (B_1^2 + B_2^2)^{1/2},$$

$$\phi = \tan^{-1}(B_1/B_2).$$

These results may be applied to the present problem to give

$$e = \{i_m R/[R^2C^2\omega^2 + 1]^{1/2}\}\sin(\omega t + \phi), \tag{6-35}$$

with

$$\tan\phi = -RC\omega, \tag{6-36}$$

or

$$\phi = -\tan^{-1}(RC\omega). \tag{6-37}$$

The impedance transfer ratio is thus

$$\boldsymbol{T = e/i = R/[R^2C^2\omega^2 + 1]^{1/2}\ \angle -\tan^{-1}(RC\omega).} \tag{6-38}$$

Comments

This problem was solved in two steps, as are all such similar problems. First, the physical time-domain network model is found (Eq. 6-28). The sine-forced solution to this model then gives the frequency-domain transfer ratio (Eq. 6-38).

An important difference between the models in this chapter and those of Chap. 4 is that the frequency-domain, transfer-ratio model is an expression of the forced solution *alone*. That is, the homogeneous, transient part of the general solution is ignored in order to focus on the particular, forced response in the sinusoidal steady state.

During the solution to the example, it was observed that $\omega_{out} = \omega_{in}$ for an ideal SISO network. This comes from the fact that the differential-equation model is linear with constant coefficients. This property thus exists for all ideal network models, whether they are SISO or MIMO. It is very useful, along with superposition, to build up the responses of the network models.

As a final observation, notice that this simple example required an exceptional amount of trigonometric manipulations. You will soon study a method which is far more efficient at converting the time-domain model into a frequency-domain model, thus avoiding these manipulations.

MIMO Network Modeling. In MIMO networks, you often need to compute the output at a port due to sine inputs at all the other ports. Transfer ratios may be used to do this, if *all* the inputs are pure sines. These inputs may oscillate at the same, or different frequencies. An output from an ideal MIMO component is thus composed of superposed responses due to each of the inputs, as if they act alone.

That is, since every output, y_n, is generally a function of all m inputs to the component,

$$y_n = f(u_1, u_2, \ldots, u_m), \tag{6-39}$$

then it may be expanded using a first-order Taylor series,

$$y_n = (\partial y_n/\partial u_1)\, u_1 + (\partial y_n/\partial u_2)\, u_2 + \cdots + (\partial y_n/\partial u_m)\, u_m, \tag{6-40}$$

where u_1 oscillates at ω_1, u_2 oscillates at ω_2, etc.

By way of review, recall that the partial derivative $(\partial y/\partial u)$ is computed in the same manner as an ordinary derivative, except that all other independent variables in the equation (the other inputs in this case) are treated as constants while it is computed. Thus, if $y = 3u_1 + 5u_2$, then $\partial y/\partial u_1 = 3$, and $\partial y/\partial u_2 = 5$.

Now, since the input-output variables are assumed to be *small-amplitude* sinusoids, $\partial y/\partial u = y/u = T =$ constant. The output y_n may thus be computed for any set of inputs using Eq. 6-40, if all the transfer ratios are known.

The way to find the MIMO transfer ratios is to first find the physical network model, then to find the ratios by solving the model using sine-input forcing to mimic the partial derivatives of Eq. 6-40. An example of this is shown in Fig. 6-7.

The multiport diagram of Fig. 6-7a shows the network and causality of interest. Note in Fig. 6-7b that two forcing sources have been added at the ends of the network to respresent the two inputs. These two inputs are assumed to be at the same phase angle, namely 0°, in order to compute the *relative* phase shift between the port variables for the transfer ratios. In the most general case, the inputs are also assumed to be at different frequencies, ω_a and ω_b.

The solution for a particular transfer ratio of interest makes use of the principle of superposition. That is, a sinusoidal input is assumed for the input of interest, *while all the other inputs are set equal to zero.* So, in the example of Fig. 6-7, if you are interested in $T_2 = \partial i_a/\partial i_b$, then you would assume that the inputs are

$$e_a = 0 \text{ (e.g., a \textit{short} between } a \text{ and } c), \tag{6-41a}$$

and

$$i_b = i_{bm} \sin \omega_b t, \tag{6-41b}$$

as you solve the network for i_a. Then $T_2 = i_a/i_b$.

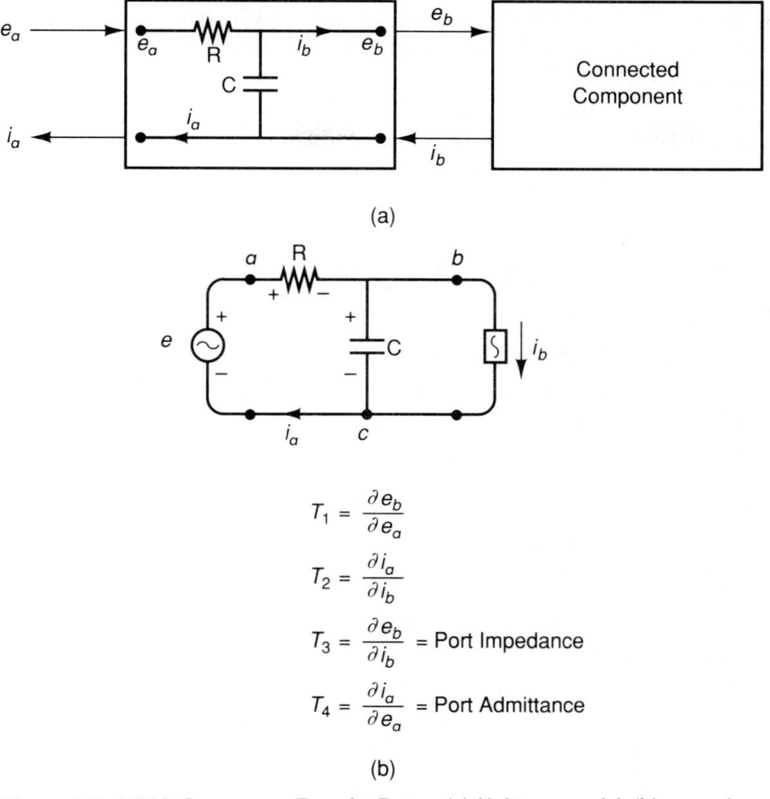

Figure 6-7. MIMO Component Transfer Ratios. (a) Multiport model, (b) network model and transfer ratios

If you are interested in $T_1 = \partial e_b/\partial e_a$ for the example, then you would assume that the inputs are

$$e_a = e_{am} \sin \omega_a t, \tag{6-42a}$$

and

$$i_b = 0 \text{ (e.g., an } open \text{ between } b \text{ and } c), \tag{6-42b}$$

as you find e_b, and $T_1 = e_b/e_a$.

The repeated solution of the network model with the proper inputs thus yields all the transfer ratios. The set of transfer ratios may then be combined to yield the pure-sine output model in the form of Eq. 6-40.

All possible transfer ratios for the two inputs in the example problem are shown in Fig. 6-7b. Since there are two outputs and two inputs in the network, there must be four input transfer ratios, as shown in the figure. The last two transfer ratios are recognized as the impedance of the *b* port, and the admittance of the *a* port. You could also compute the admittance of the *b* port and the impedance of the *a* port—these ratios are not shown in the figure.

Note that the impedance of the *b* port, T_3, must match the impedance of the connected component at that port in order for the design to function properly. This observation is often used in system design when these two dynamic quantities are matched by designers. This technique is called *impedance matching*, and is one reason why designers are so interested in impedances.

HOMEWORK

6-1. For the component shown in Fig. 6-8a:

a) Find the impedance of the network.

b) Sketch the analogous structural-translational component, and the analogous fluid component, assuming impedance forcing.

6-2. For the component shown in the figure Fig. 6-8b, what is e_{ab} if $e_s = 10 \cos 5t$ volts?

6-3. For the component shown in Fig. 6-8c, what is the impedance?

6-4. For the component shown in Fig. 6-8d:

a) Find the transfer ratio $T = e_{cd}/e_{ab}$.

b) Plot the magnitude and phase of T on a Bode diagram for the case when R = 10Ω, L = 1h, C = 0.01f, and $1 \leq \omega \leq 100$ r/s.

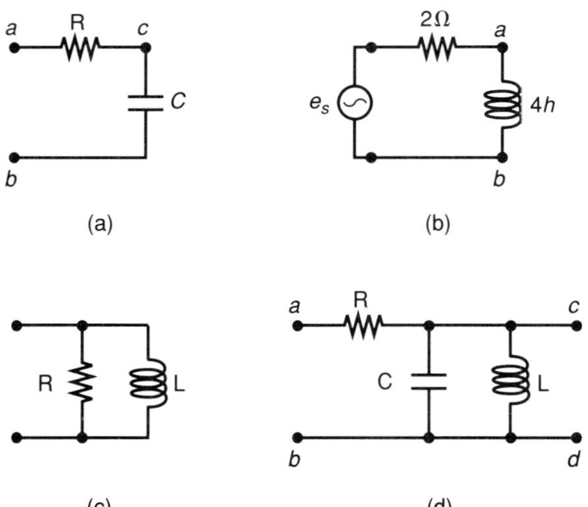

Figure 6-8. Networks for the Homework

6-2 PHYSICAL ANALYSIS OF COMMON AC-ELECTRONIC COMPONENTS

There are many electrical hardware components which have been specifically designed to operate in an alternating-current (AC) environment. In this section, representative components of this type are discussed which are either *amplifiers* or *transducers*. However, only high-power transducers are discussed here since low-power transducers (e.g., sensors) are usually assumed to behave in an ideal fashion in preliminary physical modeling; that is, they are assumed to have negligible AC dynamics. Thus, the same model as that found in Sec. 4-3 may be used here for the sensors.

AC-electronic components function by using the manipulation of AC power, which is where we begin the analysis.

AC Power. In general, the port of an AC-electronic component has

$$e = \mathrm{e}_m \sin(\omega t + \alpha), \tag{6-43}$$

and

$$i = \mathrm{i}_m \sin(\omega t + \beta). \tag{6-44}$$

Note that the phase of each of the two port variables is at some arbitrary angle, depending on the selection of the central system time, t. Also note that the causality of the port variables is not important for the analysis of power. Further, the following comments apply in an analogous way to the power at any frequency-domain port, not just the electrical ones in the discussion.

The power at the port is thus

$$\mathcal{P} = e\, i, \tag{6-45}$$

$$= \mathrm{i}_m\mathrm{e}_m \sin(\omega t + \alpha)\sin(\omega t + \beta), \tag{6-46}$$

$$= \mathrm{i}_m\mathrm{e}_m[(\sin\omega t \cos\alpha + \sin\alpha \cos\omega t)(\sin\omega t \cos\beta + \sin\beta \cos\omega t)], \tag{6-47}$$

$$= \mathrm{i}_m\mathrm{e}_m[\sin^2\omega t \cos\alpha \cos\beta + \cos^2\omega t \sin\alpha \sin\beta$$
$$+ (\sin\omega t \cos\omega t)(\cos\alpha \sin\beta + \sin\alpha \cos\beta)]. \tag{6-48}$$

But the laws of trigonometry require

$$\sin^2\omega t = (1/2)(1 - \cos 2\omega t), \tag{6-49}$$

and

$$\cos^2\omega t = (1/2)(1 + \cos 2\omega t). \tag{6-50}$$

Substituting these and rearranging gives

$$\mathcal{P} = \mathrm{i}_m\mathrm{e}_m[(1/2)(\cos\alpha\cos\beta + \sin\alpha\sin\beta) + (1/2)\cos 2\omega t\,(\sin\alpha\sin\beta$$
$$- \cos\alpha \cos\beta) + (\sin\omega t \cos\omega t)(\cos\alpha\sin\beta + \sin\alpha \cos\beta)]. \tag{6-51}$$

Moreover, the laws of trigonometry also require

$$\sin\omega t \cos\omega t = (1/2)\sin 2\omega t, \tag{6-52}$$

$$\cos\alpha\sin\beta + \sin\alpha\cos\beta = \sin(\alpha + \beta), \tag{6-53}$$

$$\sin\alpha \sin\beta - \cos\alpha \cos\beta = -\cos(\alpha + \beta), \tag{6-54}$$

and

$$\cos\alpha \cos\beta + \sin\alpha \sin\beta = \cos(\alpha - \beta). \tag{6-55}$$

When these are substituted into Eq. 6-51,

$$\mathcal{P} = \mathrm{i}_m \mathrm{e}_m[(1/2) \cos(\alpha - \beta) - (1/2) \cos 2\omega t \cos(\alpha + \beta) + (1/2) \sin 2\omega t \sin(\alpha + \beta)]. \tag{6-56}$$

The last two terms can be further combined using the relationship in Eq. 6-54 to give the final form

$$\mathcal{P} = \mathrm{i}_m \mathrm{e}_m[(1/2) \cos(\alpha - \beta) - (1/2) \cos(2\omega t + \alpha + \beta)]. \tag{6-57}$$

Note that the frequency-domain power thus has two terms which are added together in the form

$$\mathcal{P} = \mathcal{P}_{\text{avg}} + \mathcal{P}_{\text{dyn}}, \tag{6-58}$$

where

$$\mathcal{P}_{\text{avg}} = (\mathrm{i}_m \mathrm{e}_m/2) \cos(\alpha - \beta). \tag{6-59}$$

This may be written in terms of the *relative* phase angle between the current and voltage, $\phi = (\alpha - \beta)$,

$$\mathcal{P}_{\text{avg}} = (\mathrm{i}_m \mathrm{e}_m/2) \cos\phi. \tag{6-60}$$

Note that this *average power*, $\mathcal{P}_{\text{avg}}$, is thus a *constant* value associated with the sinusoidal steady state. Also note that $\mathcal{P}_{\text{avg}}$ is proportional to the amplitudes of the port variables *and* the phase angle between them.

The other term in the frequency-domain power, the *dynamic power*, $\mathcal{P}_{\text{dyn}}$, is a time-varying term which oscillates about zero at twice the frequency of the port variables,

$$\mathcal{P}_{\text{dyn}} = -(\mathrm{i}_m \mathrm{e}_m/2) \cos(2\omega t + \alpha + \beta). \tag{6-61}$$

The units of AC power are v-a (for volt-amps). These units are used to distinguish AC power from DC power (which has the units of Watts).

Note that the average power has a maximum magnitude whenever the phase is such that

$$\phi = \pm n\pi, \; n = 0, 1, 2, \ldots . \tag{6-62}$$

Similarly, the average power has zero magnitude whenever

$$\phi = \pm n\pi/2, \; n = 1, 3, 5, \ldots . \tag{6-63}$$

Recall back to the inductor and capacitor analyses of Sec. 6-1 where it was shown that both of these devices produce a phase of $\pm\pi/2$ under current or voltage forcing. Thus, the *ideal* capacitor and inductor give phase shift without power consumption. Another way to say this in light of Eqn. 6-63 is that these devices store energy without power consumption.

The ideal resistor, on the other hand, has zero phase shift and thus maximum power consumption. However, resistors are necessary to provide damping and stability to the component. The cost of these desirable features is power consumption.

These principles of AC power are next used to help to understand the more complicated devices in AC components, beginning with AC amplifiers.

AC Amplifiers. Electronic AC amplifiers come in two forms. The *passive amplifiers* have no externally connected power source. But the *active AC amplifiers* are connected to an external power source which they use to amplify inputs, to isolate network parts, and to protect portions of the component network, just as in the DC amplifiers of Sec. 4-3.

Passive AC electronic amplifiers are also called *transformers*. They work by inducing a magnetic field to modify the electronic power, as shown in Fig. 6-9.

The transformer is made by winding two uninsulated coils, of turns n_a and n_b, about a common magnetic-permeable core. The core acts as a conductor for the magnetic field which is created in one winding and is transmitted to the other. So, each winding acts like an inductor.

If the windings are in opposite directions around the core, then they will have the high-side of the voltage as shown in Fig. 6-9a by the large dots over the coils. If the windings are in the same directions around the core, then the high-voltage side would be at the bottom end of the coil on the right, as would the large dot on that coil.

All transformers have a designated *primary* coil for *input*, which is indicated by the subscript p. They also have a secondary coil for output, indicated by the subscript s. The designation of primary or secondary depends on the design causality, and the way in which the voltage is used in the connected circuit.

The *pure* transformer does transformation of voltage without exhibiting the effects of resistance, capacitance, or inductance (e.g., current reluctance). This notion is shown at the right of Fig. 6-9a. Note also that the *ideal* transformer is both *pure* and *linear*.

The only two possible cases of causality are shown in Fig. 6-9b. At the left, a is the primary coil because its voltage is directed into the device. At the right, b is the primary coil.

The ideal transformer works to convert the input (primary) voltage *amplitude*, e_{pm}, into the output (secondary) voltage *amplitude*, e_{sm}, according to

$$e_{sm} = N\, e_{pm}. \tag{6-64}$$

Whether this steps-up or steps-down the sinusoidal voltage amplitude depends on the number of turns in the primary and secondary windings.

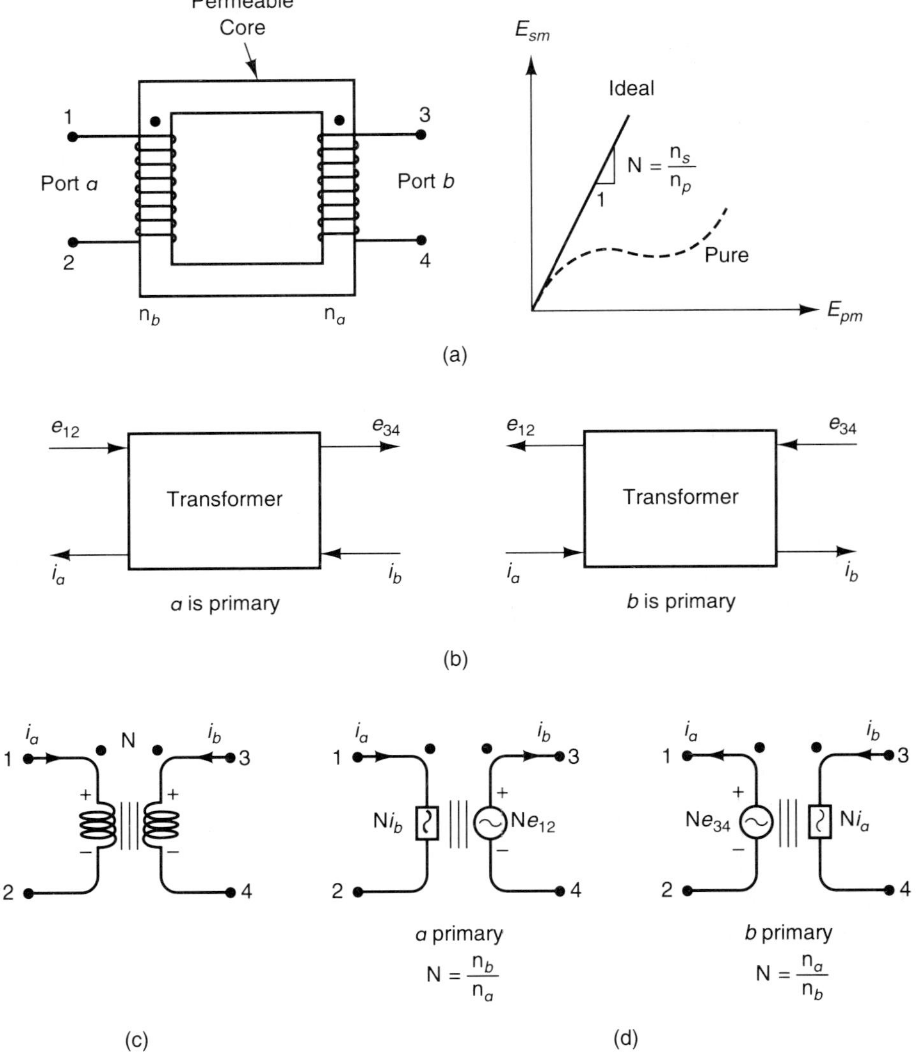

Figure 6-9. The Ideal Transformer. (a) Physical realization, (b) multiports, (c) schematic, (d) idealized network models

In addition to the voltage relationship, the passive transformer must have

$$\text{average power in} = \text{average power out}, \tag{6-65}$$

or

$$e_{pm} i_{pm} \cos \phi_p = e_{sm} i_{sm} \cos \phi_s. \tag{6-66}$$

However, since there is no phase shift taking place in the *ideal* transformer, $\phi_p = \phi_s$, and

$$e_{pm} i_{pm} = e_{sm} i_{sm} . \tag{6-67}$$

And, using Eq. 6-64, the transformer current relationship becomes

$$i_{pm}/i_{sm} = N. \tag{6-68}$$

Thus, if voltage *amplitude* is stepped-up through the transformer, then current *amplitude* must be stepped-down and vice versa.

The common circuit schematic for the transformer is shown in Fig. 6-9c. The large dots at the ends of the coils show the design selection of the voltage "high" sides. So, in the circuit analysis, the positive senses of currents and voltage drops must be assigned consistent with the dots as shown. The confusing part of this symbol is that the coils can easily be confused with the inductor symbol when this is not intended. Consequently, the ideal circuit model is slightly different, depending on the causality cases, as shown in Fig. 6-9d. Note that in both cases, the secondary port acts as a voltage source in the secondary circuit.

Active AC amplifiers come in many forms, but this text only considers the same ones that were studied in Sec. 4-3, these being the *voltage-isolation* and *current* amplifiers. These devices have dynamic responses, some of which were studied by Bode when he developed his famous Bode diagrams. In any case, you may neglect their dynamics in this introductory study for the sake of simplicity. The two active AC amplifiers of interest thus appear as in Fig. 6-10. The model of their performance is the same as in Sec. 4-3 (e.g., $e_{34} = K_1 e_{12}$, and $i = K_2 e_{12}$), which must be satisfied at every instant of time.

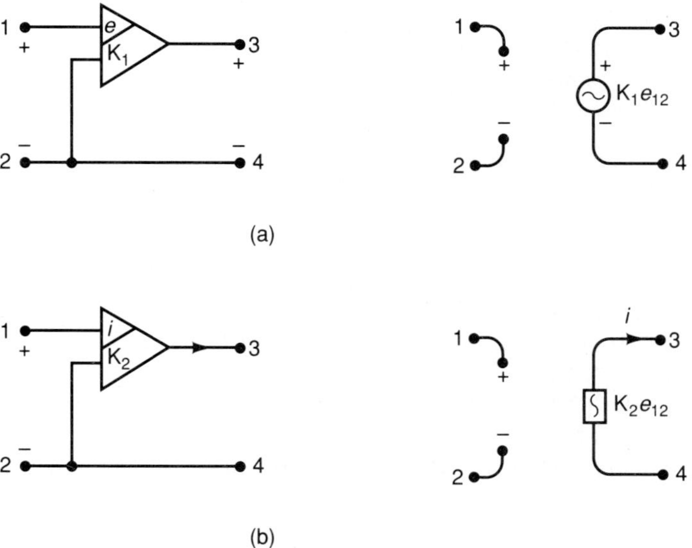

Figure 6-10. AC Power Amplifiers. (a) Voltage-isolation, (b) current

AC Transducers. A wide variety of AC electrical transducers exists. However, in this section only a few representative devices of this type are considered. These convert AC electrical power into either "DC" structural-rotational power, into "AC" fluid-acoustic power, or into "DC" thermal power. We begin with the last type, commonly called the electric heater.

The heat production of the ideal *electric heater* is easily computed using the model of a heater as a single-port resistor, much like that shown in Fig. 4-5a, connected to a single heat-flow port. The physical relationship between these ports may be developed beginning with the First Law of Thermodynamics for the heater,

$$\text{work in} = \text{heat available.} \tag{6-69a}$$

But,

$$\text{work in} = \int (\text{power in})\, dt, \tag{6-69b}$$

where the power input is the average power being supplied by the electronic circuit. So, combining these relations,

$$\text{heat available} = \int (\mathrm{e}_{12m}\mathrm{i}_m \cos\phi)/2\, dt, \tag{6-69c}$$

or

$$h = (\mathrm{e}_{12m}\mathrm{i}_m \cos\phi)/2, \tag{6-70}$$

where e_{12m} is the *amplitude* of the voltage drop across the heater, i_m is the *amplitude* of the current flow through the heater, and ϕ is the phase angle between this voltage and current. Further, since the heater is modeled as a simple *ideal* resistor, $\mathrm{e}_{12m} = \mathrm{i}_m \mathrm{R}$, and $\phi = 0$. Thus,

$$h = [\mathrm{R}(\mathrm{i}_m)^2]/2. \tag{6-71}$$

Note that the heat output computed by Eq. 6-71 does not take into account the *efficiency* of the process of heat transfer to the medium being heated (e.g., air or water in home heating applications). A 100% efficient process (e.g., an ideal process) is assumed by neglecting losses in Eq. 6-69. Further, to be more precise in the heater model, you must also consider the convective fluid mechanics of the heated medium. The total heat transfer problem thus becomes much more theoretically complicated, and must be dealt with through the input/output methods of Chap. 7.

AC electric motors transduce AC electric power into "DC" structural-rotational power. Virtually all AC electric motors work on the principle of induction through magnetic fields. This is similar to the magnetic power transfer that was studied in the transformer previously. It is the idea that a loop of conductor moving in a magnetic field has a current induced in it. Or, if the conductor loop already has a current in it, then forces are exerted on the loop as it moves through the field. If you are careful about the orientation of the loop in the field, and the amount of current in the loop, then you can construct a motor based on this idea. In Chap. 4 this idea was the basis for the DC motor, using DC electric signals. Here you have the same idea, this time with AC electric signals.

The AC motor is often constructed as shown in Fig. 6-11a, in the form of a *two-phase induction motor*. This type of motor is very common for precision applications and is very rugged and reliable. It is called a "two-phase" motor because two sinusoidal signals at different phase angles are needed to drive the motor. Three-phase induction motors also exist, with operation very similar to that of the two-phase motor.

The two input voltage signals to the two-phase motor must be 90° apart in time and space. One phase, e_f, the so-called "fixed phase," is held at a fixed amplitude. The other phase, called the "control phase," e_c, is *amplitude modulated* (AM) in the electronic circuitry to produce the variable motor torque desired.

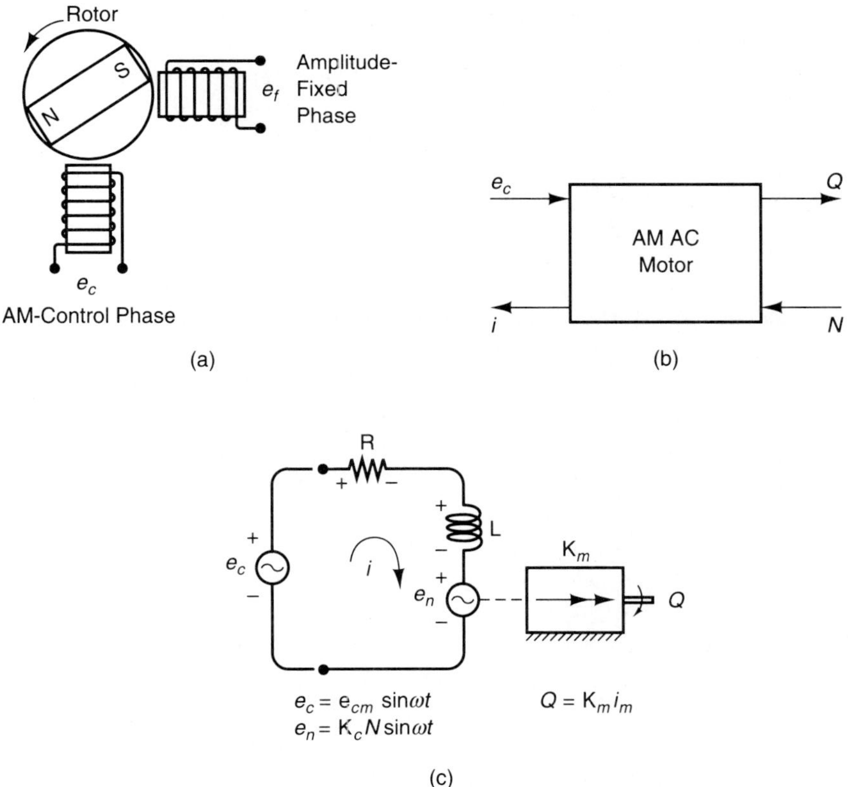

Figure 6-11. The Two-phase Induction Motor. (a) Construction, (b) casuality, (c) idealized model

The motor torque interacts with the load to produce the speed as shown in the multiport diagram of Fig. 6-11b. Note that the multiport does not show the amplitude-fixed phase since this is not allowed to vary. The amplitude-fixed power is thus treated like the power which is supplied to an active amplifier—it is ignored in the circuit analysis.

In other variable-speed induction motors, the *frequency* of the control signal is varied to control the motor torque. These motors are more efficient, but are more suitable for heavy-duty applications. They are used in such designs as the electric drive for ships and cars. This type of motor is not considered here.

The network model for the two-phase AC motor is shown in Fig. 6-11c. Note that a lossy model of the motor electronics is necessary for AC analysis since the control *voltage* is the input and you need to compute the control current in order to find the output torque of the motor.

The motor analysis begins with a KVL equation for the control circuit,

$$e_c - e_R - e_L - e_n = 0. \tag{6-72}$$

After rearranging and substituting element relations, this becomes

$$\mathrm{L}\, di/dt + \mathrm{R}i = e_c - e_n. \tag{6-73}$$

However, the two terms on the right-hand side can be further expanded to

$$e_c = \mathrm{e}_{cm} \sin(\omega t + \alpha), \tag{6-74}$$

and

$$e_n = \mathrm{K}_c N \sin(\omega t + \alpha). \tag{6-75}$$

This last equation shows an assumption that the back-emf due to the load, e_n, occurs at the same phase angle, α, as the input control voltage, e_c. Further, it is assumed that the amplitude of the back-emf is proportional to the motor speed, N, with a constant of linear proportionality K_c. Moreover, since α is arbitrary, you may take $\alpha = 0$ for simplicity in the following analyses.

When these relations are inserted into Eq. 6-73, you find that

$$\mathrm{L}di/dt + \mathrm{R}i = (\mathrm{e}_{cm} - \mathrm{K}_c N) \sin \omega t, \tag{6-76}$$

which gives

$$i = \mathrm{i}_m \sin(\omega t + \phi). \tag{6-77}$$

Also,

$$q = \mathrm{K}_m \mathrm{i}_m \tag{6-78}$$

is used to model the torque output as in the DC motor. These last three equations comprise the time-domain model for the two-phase induction motor.

Synchronous AC generators are transducers which convert DC structural power into AC electrical power. They function in a similar manner to the AC induction motor, except that they have a speed input from the structural side and a phase-synchronized voltage output on the AC electrical side. However, due to their complexity, they are not analyzed in this introductory text.

Audio speakers are transducers for changing AC-electrical signals into AC-fluid signals (acoustics). The decibel scale, *dbs,* which is on the Bode diagram was specifically developed for use in acoustic analyses, resting solidly on the idea of superposed, small-amplitude, harmonic pressure signals. Such signals are analogous to those in the electrical networks which you are now studying. Consequently, you would rightly expect that the production or reproduction of sound signals is virtually always accomplished in AC electrical networks. Of course, audio speakers are a critical part of those networks.

A common form of the audio speaker is shown in Fig. 6-12a. This device takes an input current to a coil around a structurally fixed permanent magnet. The

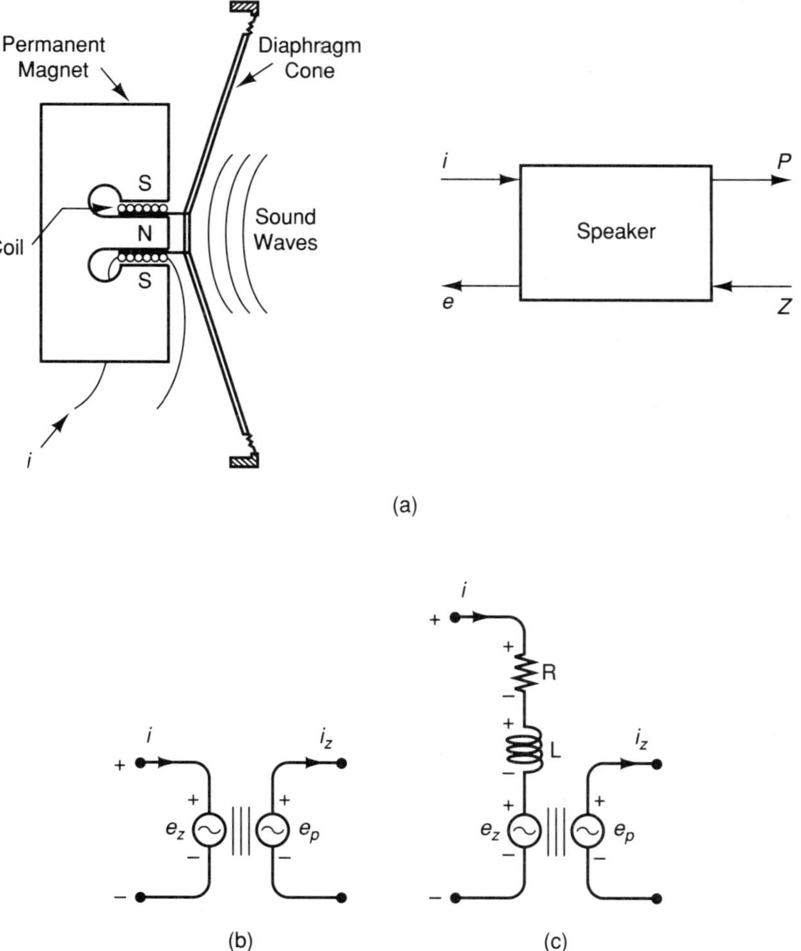

Figure 6-12. The Acoustic Speaker, (a) Construction and causality, (b) idealized analog, (c) lossy analog

coil is attached to a speaker cone in such a way that the coil-and-cone are free to move with respect to the magnet. Thus, as the current varies in a periodic fashion, a periodic magnetic force is exerted on the coil-and-cone. This force produces motion of the cone, which produces pressure waves that are heard as sound. Thus, the causality of the speaker is as shown in the multiport diagram in the figure.

The performance of the ideal speaker is like that of the ideal DC motor (Sec. 4-3), except that the output is in fluid variables rather than structural variables. That is, the output amplitudes are

$$p_m = K_s i_m, \tag{6-79}$$

and

$$e_{zm} = CK_s z_m. \tag{6-80}$$

Again, these equations are relations between the *amplitudes* of AC signals. And once more, it would be easy to show from these last two equations that (average power in = average power out) for this passive device.

The electrical analog of the ideal speaker is shown in Fig. 6-12b. The analog model equations are

$$e_{pm} = K_s i_m, \tag{6-81}$$

and

$$e_{zm} = CK_s i_{zm}. \tag{6-82}$$

The lossy model of the speaker is shown in Fig. 6-12c. Note that the structural side of the model ignores the vibration damping which must occur from the structure moving in air. If this is deemed as significant, then you would add a damper (e.g., analog resistance) to account for the effect. This amounts to modeling the fluid admittance of the atmospheric load.

AC microphones work in a very similar manner to the audio speaker to transduce input sound pressure waves into AC electrical current signals. The same relationships as Eqs. 6-81 and 6-82 apply.

HOMEWORK

6-5. Given the port impedance

$$T_I = R/(R^2C^2\omega^2 + 1)^{1/2}\angle -\tan^{-1}(RC\omega),$$

compute the power consumed at this port when

$$i = 10\sin(5t + 10°)\text{ amps},$$

$$R = 100\Omega,$$

$$C = 0.01\text{ f}.$$

6-6. A designer has created the electric component shown in Fig. 6-13. For this network:

a) Sketch a multiport diagram, and

b) Find i, i_b, and e_{34}.

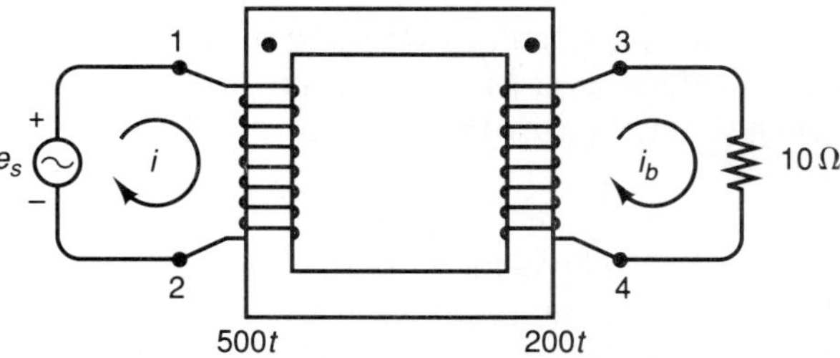

Figure 6-13. Transformer Network for Prob. 6-6

6-7. What is the magnitude of the input electrical admittance to the two-phase AC motor when the load on the motor is a rotational friction, B, such that, $N/\mathrm{B} = q$? (Hint: Use $q = \mathrm{K}_m \mathrm{i}_m$.)

6-8. A designer wants to create the speaker system shown in Fig. 6-14. Why is current forcing of the lossy speaker a better way to design this system?

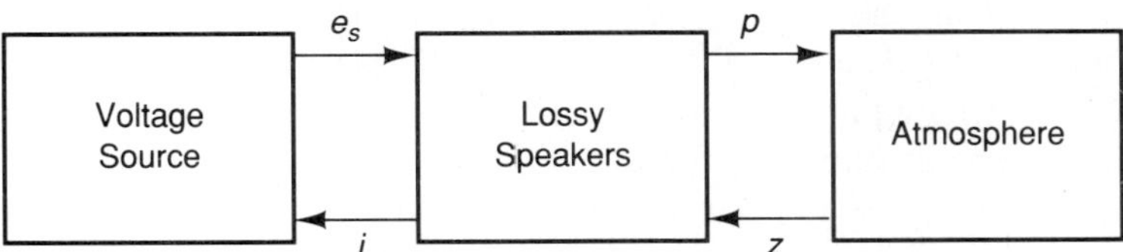

Figure 6-14. Multiport for Prob. 6-8

6-3 PHYSICAL ANALYSIS OF STRUCTURAL AND FLUID COMPONENTS

Many structural and fluid components are designed to respond to frequency-domain input forcing. Such inputs create *forced* vibrations in the components, and these must be distinguished from the time-domain *free* vibrations which were studied in the last chapter.

Both types of structural vibrations can be destructive, and designers create *dampers* and *absorbers* to minimize them. An example of the latter type, to be studied in this section, is the engine mounts on an automobile which are designed to isolate and block the engine vibrations from propagating into the passenger compartment.

Forced vibrations in fluid systems are studied in this text only in the form of small-amplitude audible waves (acoustics). Such small waves allow you to build and study an ideal model of the acoustic vibration. These vibrations in acoustic systems may be desirable or undesirable. In an auditorium, desirable acoustic vibrations need to be enhanced while undesirable vibrations are reduced. This is done through acoustic system design which optimizes such factors as geometrics, electronic speaker size and placement, and surface reflective treatments. However, this text only considers the propogation of the waves without regard for interactions with the boundary. This simplified introductory analysis thus forms the basis for more advanced follow-on study.

Forced Structural Vibrations. Frequency-domain idealizations of structural components are based upon the same fundamental physical elements as those studied in Chap. 4. These are the M, b, and k elements in translational components, and the J, B, and K elements in rotational components. The time-domain methods used to identify and compute these elements in Chap. 4 are the same for these effects in the frequency domain, except that significant coulomb friction or other discontinuous nonlinearites are not allowed in this linear study.

In addition to these elemental effects, there are structural effects which act as vibration source elements. A common case of these is the operation of motors and engines, which is next studied by way of example, to illustrate the full transfer-ratio modeling effort for structural components.

Engines and motors create vibrations due to imperfect balancing of moving parts and imperfect balancing of motive forces. These imbalances act as periodic force sources on structural parts, as shown in Fig. 6-15. A designer is often interested in the motions arising from the force imbalances in such a system.

At the left of Fig. 6-15a, an motor of mass M_1 is shown rigidly attached to a support of mass M_2. The support, in turn, is connected by springs of net stiffness k_2 to the rigid foundation. This construction creates a *single-isolator* form of the absorber.

When the motor is operating at the speed N (rpm, or ω, rad/s), the motor mass acts as if it were subjected to a periodic force input,

$$F_1 = F_{1m} \sin \omega t. \tag{6-83}$$

Note that this is a very simple applied force. A more likely case is that the force above is simply the input at a fundamental frequency, with higher harmonics also present. In such a case, you would simply superpose the output responses to these higher harmonics to determine the total response. In any case, the physical analysis of the fundamental input alone is fairly insightful for this introductory study.

The electrical analog of the motor with a single isolator is shown at the right of Fig. 6-15a. Note that the two masses rigidly connected together thus act as a net single mass, with a single velocity ($e_1 = e_2$, in analog terms). The problem thus is that you know the masses and the spring stiffness, and you wish to find the vibrational motion of the motor.

The analysis of the analog dynamics starts with KCL for the top node in Fig. 6-15a

$$i_{F1} - i_{M1} - i_{M2} - i_{k2} = 0. \tag{6-84}$$

The element relations are inserted to this to give

$$i_{F1} - \mathrm{M}_1\, de_1/dt - \mathrm{M}_2\, de_2/dt - \mathrm{k}_2 \int e_2 dt = 0. \tag{6-85}$$

Rearranging this gives

$$\mathrm{M}_1\, d^2e_1/dt^2 + \mathrm{M}_2\, d^2e_2/dt^2 + \mathrm{k}_2 e_2 = di_{F1}/dt. \tag{6-86}$$

From KVL you obtain the voltage relationships,

$$e_1 = e_2 = e, \tag{6-87}$$

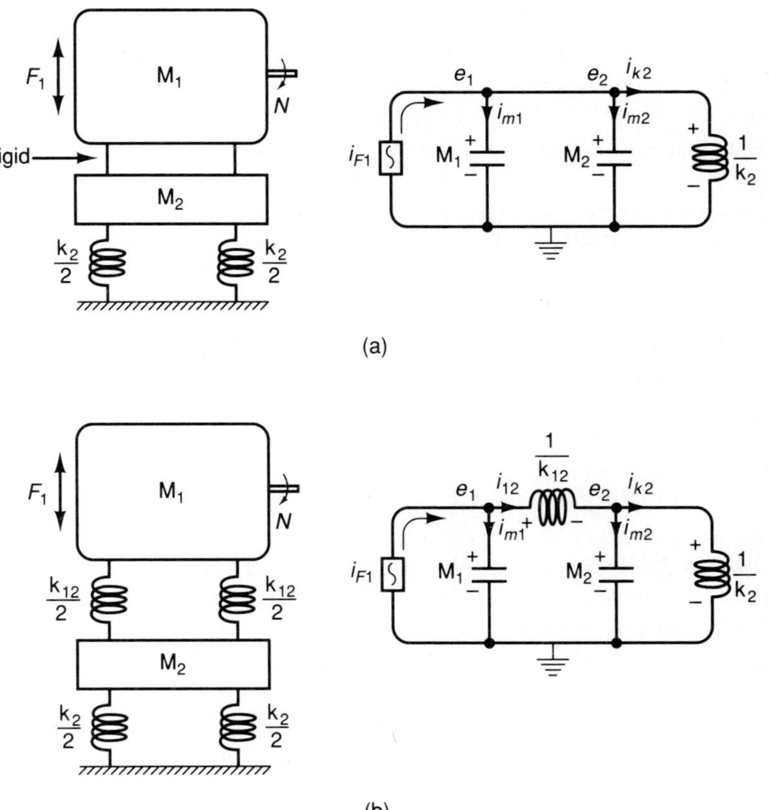

Figure 6-15. Vibration Isolation. (a) Single isolator, (b) double isolator

and the model becomes

$$(\mathrm{M}_1 + \mathrm{M}_2)\, d^2e/dt^2 + \mathrm{k}_2 e = di_{F1}/dt. \tag{6-88}$$

Reversing the analogy gives the time-domain model

$$\mathbf{(M_1 + M_2)}\, \boldsymbol{d^2v/dt^2} + \mathbf{k_2}\boldsymbol{v} = \boldsymbol{dF_1/dt}. \tag{6-89}$$

When the force input described by Eq. 6-83 is input to the above model (Eq. 6-89), and you solve for the frequency-domain transfer ratio, you find

$$T_1 = v/F_1 = 1/[(\mathrm{M}_1 + \mathrm{M}_2)\omega^2 - \mathrm{k}_2] \angle -\pi/2. \tag{6-90}$$

Further, if you assign

$$\omega_{n1} = [\mathrm{k}_2/(\mathrm{M}_1 + \mathrm{M}_2)]^{1/2}, \tag{6-91}$$

which is the natural frequency of the component, then

$$\boldsymbol{T_1 = 1/[(\mathrm{M}_1 + \mathrm{M}_2)(\omega^2 - \omega_{n1}^2)] \angle -\pi/2.} \tag{6-92}$$

Note that the presence of the springs has created a phase shift of $-\pi/2$ rad. between the force and velocity of the motor. Thus, the average power of the motor vibration is zero (Eq. 6-60). In detail design, you would find that there is really some small damping in the component which leads to a phase slightly different than $-\pi/2$. But for preliminary design, you can estimate the performance acceptably with the above methods.

Also note that the vibration transfer ratio becomes undesirably infinite whenever $\omega = \omega_{n1}$. This "resonance" condition must be avoided in the steady operation of the vibration absorber. So, given the two masses, the design problem thus becomes one of choosing the spring stiffness so that the operating range of the motor speed, ω, is either all-above or all-below the critical resonance speed, ω_{n1}.

Notice that when ω is very small, the transfer ratio is nearly constant in magnitude. However, when ω is very large the transfer ratio decreases as ω increases. Thus, the designers know to design the absorber so that $\omega > \omega_{n1}$. Moreover, since T_1 decreases with increasing ω, the designers use the idle speed, ω_{idle}, as a worst-case to choose the design value of the spring stiffness.

Example 6-2

Given a motor and mounting plate with the following properties

$$\mathrm{M}_1 = 1 \text{ sl},$$

$$\mathrm{M}_2 = 0.1 \text{ sl},$$

$$F = 5 \sin \omega t \text{ lb}_f,$$

$$30 \le \omega \le 300 \text{ rad/s}.$$

Design a single vibration absorber so that its transfer ratio is

$$T_1 \le 0.01 \text{ ft/lb}_f\text{-s}.$$

Solution

Using Eq. 6-90, and the low-end (idle) motor speed,

$$0.01 = 1/[(1.1)(900) - k_2].$$

That is,

$$k_2 = 890 \text{ lb}_f/\text{ft.} = 74.2 \text{ lb}_f/\text{in.}$$

Comments

Note that in vibration-absorber problems, as the speed goes above idle, the transfer ratio goes down. Consequently, choosing the design point at the idle speed is a good example of worst-case design.

Also note that, at some time, the motor speed must go through the resonant frequency as it goes to or from the operating range. However, since this is a transient condition, it only results in a very slight shaking of the component as the speed passes through the natural frequency.

Whenever the vibration absorber has another set of springs between the vibration source and the second mass, as in Fig. 6-15b, a *double-isolator* form of the absorber is created. Automobile engine mounts are of this type of absorber. Here, the two masses are known, the spring-rate to ground is known from the tires, and the force input is assumed to be known. The output speeds of the two masses are desired. (Note that these two speeds are different in general, since they are now separated by the connecting spring.)

The new analog for the system with the added springs is shown at the right of Fig. 6-15b. Again, you begin with KCL at nodes 1 and 2,

$$i_{F1} - i_{M1} - i_{12} = 0, \tag{6-93}$$

and

$$i_{12} - i_{M2} - i_{k2} = 0. \tag{6-94}$$

When the element relations are substituted,

$$i_{F1} - M_1 de_1/dt - k_{12} \int e_{12}\, dt = 0, \tag{6-95}$$

and

$$k_{12} \int e_{12}\, dt - M_2 de_2/dt - k_2 \int e_2\, dt = 0. \tag{6-96}$$

So, when the integrals are cleared and the analogy reversed, you obtain the time-domain model

$$\mathbf{M}_1 d^2 v_1/dt^2 + \mathbf{k}_{12} v_1 = dF_1/dt + \mathbf{k}_{12} v_2, \tag{6-97}$$

and

$$\mathbf{M}_2 d^2 v_2/dt^2 + (\mathbf{k}_{12} + \mathbf{k}_2) v_2 = \mathbf{k}_{12} v_1. \tag{6-98}$$

Note that this model has *one* input (F_1), and *two* outputs (v_1 and v_2). The transfer ratios of interest are thus, $T_1 = v_1/F_1$, and $T_2 = v_2/F_1$.

The time-domain model also shows that now there are *two* natural frequencies to be concerned about. These are

$$\omega_{n1} = (\mathrm{k}_{12}/\mathrm{M}_1)^{1/2}, \tag{6-99}$$

and

$$\omega_{n2} = \left[(\mathrm{k}_{12} + \mathrm{k}_2)/\mathrm{M}_2\right]^{1/2}. \tag{6-100}$$

The first natural frequency is like that from the single-isolator model equation, it comes from Eq. 6-97 and is associated with the motor and its connected springs. The second natural frequency comes from Eq. 6-98, and is associated with the second mass and its connected springs. The vibration absorber is now designed to operate in a motor-speed range higher than *both* of these natural frequencies. That is, you find operating speed

$$\omega > \omega_{n1}, \tag{6-101}$$

and

$$\omega > \omega_{n2}. \tag{6-102}$$

And once more you find a situation where the motor must pass through natural frequencies as it goes to or from the operating speed range. In fact, the corresponding shaking of a car body and its engine on start up and shut down can be observed in most automobiles.

Forced Fluid Vibrations (Acoustics). Acoustics is the study of the propagation of pressure waves. In virtually all cases, these waves must pass through various forms of media before they are heard in the ear as "sound". These sound waves usually propagate through both solids and fluids (liquids and gases). However, in this text, the propagation of sound through solid matter is not discussed in order to concentrate on the physical phenomena of sound propogation through fluids.

Sound propagation is modeled using the idealized elements of fluid resistance, inertance, and capacitance (R_f, I_f, and C_f). Resistance to sound is computed from the two ways that sound is transmitted through matter: by conduction (with attenuation for distance), and by convection. Sound reflection also occurs as a type of propagation by conduction, and this also has a resistance. Note that these physical phenomena are very analogous to the ways that light and heat are propagated through media. Further, frequency-domain inertance arises when the acoustic propagation component stores energy by virtue of the sinusoidal fluid flowrate, and capacitance arises when the component stores energy by virtue of the sinusoidal pressure. However, all of these elements are very difficult to model in general and are left for advanced study.

Instead, for the purposes of this introductory study, the model of sound propagation is taken to be that of a simple delay from input to output, depending on the speed of sound in the fluid of interest.

So, a multiport diagram for an acoustic propagation component is shown in Fig. 6-16. This diagram is for a single emitter and a single receiver. Note that the emitter produces an output pressure wave in the frequency domain, and has an input sinusoidal flowrate from the fluid. Emitter examples are electronic speakers (as in Sec. 6-2), and the structural vibrations of motors and engines. Note that the propagation component is connected to the emitter in such a way that the admittance transfer ratio of the fluid is of interest at the port (output-Z_a/input-P_a is analog i/e, which is admittance). This admittance is generally dependent on the properties of the fluid, and is very difficult to compute due to the difficulties in resistance, capacitance, and inertance discussed above. However, in an idealized model of the physics at this port, the sinusoidal flowrate is ignored, as indicated by the dashed line in the figure. Similarly, at the receiving end of the propagation component, the "flowrate" of the receiver parts is usually assumed to be negligible.

The speed of a sound pressure wave in air (or in any ideal gas for that matter) is

$$V = (\mathrm{k}P_a/\rho)^{1/2}, \tag{6-103}$$

$$= (\mathrm{k}R\Theta)^{1/2}, \tag{6-104}$$

where k = 1.4 for ideal gases,
P_a = absolute ambient pressure ($\mathrm{lb}_f/\mathrm{ft}^2$),
ρ = mean density ($\mathrm{sl/ft}^3$),
R = Gas constant for the ideal gas (1545 ft^2/s^2–°R),
and, Θ = the absolute temperature (°R).

Using proper values for these numbers, you could compute the speed of sound in air at 68 °F to be 1,068 f/s.

Note that a sound wave is superposed on top of the ambient pressure level. That is,

$$P = P_a + p = P_a + \mathrm{p}_m \sin \omega t. \tag{6-105}$$

The part of the wave that you hear is the sinusoid, p.

If the sound source (say, an airplane wing) is moving through a compressible gas faster than the local speed of sound, then the small-amplitude sound waves can-

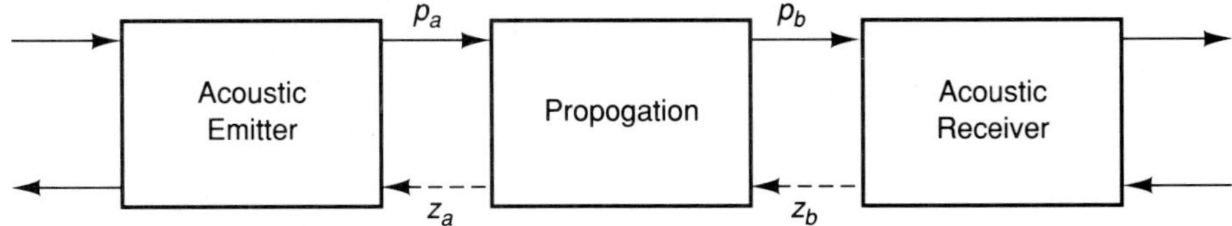

Figure 6-16. Acoustic Component Multiport Diagram

not move fast enough to propagate ahead of the wing. This creates a situation where the small waves reinforce each other to create a large-amplitude *shock wave*. Such a shock wave is heard by a stationary observer as a sonic boom. In fact, whenever the relative motion with respect to the gas is greater than 0.7 of the local speed of sound, then you should consider the effects of compressibility on the gas propagation component analysis. However, since such gross compressibility is a highly nonlinear, nonideal phenomenon, it will not be studied here.

The speed of sound in an incompressible fluid, like the liquids, is much higher than that for a gas. However, the speed in liquids is usually expressed through complicated empirical formulae, with many variables. For the present introductory purposes, a standard sound velocity of 5,100 ft/s in liquids is used.

As mentioned earlier, the decibel scale (db) is used to express the pressure level (PL) of a sound wave. This scale gives a measure of the useful strength of the wave. It makes use of the formula

$$PL = 20 \log_{10} (P_e/P_a), \tag{6-106}$$

where, P_e = the effective pressure = $P_a + p_{\text{rms}}$.

Further, the RMS value of a pressure wave is defined as

$$p_{\text{rms}} = \left[\frac{1}{\mathrm{T}}\int_t^{t+\mathrm{T}} p^2\, dt\right]^{1/2}, \tag{6-107}$$

where T is the period of the wave. The integration is computed over one period of the wave, thus giving the time-averaged (mean) value of the magnitude-squared. Hence the name, root-mean-squared, or RMS.

For example, if you apply the RMS formula to a pure-sinusoid in pressure,

$$p = \mathrm{p}_m \sin \omega t, \tag{6-108}$$

then you would find

$$p_{\text{rms}} = \left[(1/\mathrm{T}) \int \mathrm{p}_m{}^2 \sin^2 \omega t\, dt\right]^{1/2}, \tag{6-109}$$

$$= |\mathrm{p}_m| \left[(1/\mathrm{T}) \int \sin^2 \omega t\, dt\right]^{1/2}, \tag{6-110}$$

$$= |\mathrm{p}_m| \left[(1/\mathrm{T}) \int (1/2 - 1/2 \cos 2\omega t)\, dt\right]^{1/2}, \tag{6-111}$$

$$= |\mathrm{p}_m| \left[(1/\mathrm{T})(\mathrm{T}/2)\right]^{1/2}, \tag{6-112}$$

$$= |\mathrm{p}_m|/\sqrt{2}. \tag{6-113}$$

If you recall back to the average power of two electrical sinusoids at a port as defined earlier, you can see a similar idea again,

$$\mathcal{P}_{avg} = (e_m i_m/2) \cos \phi = (e_m/\sqrt{2})(i_m/\sqrt{2}) \cos \phi. \quad \text{(6-60 repeated)}$$

So, the RMS value is a measure of the useful strength of a sinusoid. It is expressed in Eq. 6-113 for a pressure wave, and in Eq. 6-60 for an electronic wave.

The first difficulty you face in computing RMS pressure values is when the periodic wave is not a pure sine. This may force you to use a numerical approach to the evaluation of Eq. 6-107. This approach is not difficult, it is merely tedious, and the numerical methods of Chap. 5 can be of good use once again.

The selection of the RMS integration time step, Δt, is based upon 1/10th of the period of the *highest* significant harmonic to come from a Fourier analysis of the periodic wave. As discussed in Chap. 3, at some point the amplitudes of the harmonics are small enough to be neglected. This *cutoff frequency* is the frequency that should determine the integration time step.

Once the RMS value of the sound (pressure) wave is computed, a pressure-level computation (Eq. 6-106) provides a way to estimate the strength of the wave for comparison purposes. Some typical sound pressure levels are listed in Table 6-1.

Table 6-1. Sound Pressure Levels (decibels)

	120	Threshold of feeling	60	Average factory
		Thunder, nearby artillery		Noisy home
Deafening	110	Nearby riveter	50	Average office
		Elevated train		Average discussion
	100	Boiler factory	40	Quiet radio
		Loud street noise		Quiet home
	90	Noisy factory	30	Auditorium
		Truck, unmuffled		Quiet talking
	80	Police whistle	20	Wind in leaves
		Noisy office		Whispering
Loud	70	Average street noise	10	Soundproof room
		Average radio		Threshold of hearing

HOMEWORK

6-9. Given a double-isolator vibration absorber in a car such that

$$k_2 = 1{,}000 \text{ lb}_f/\text{in} = 12{,}000 \text{ lb}_f/\text{ft},$$

$$k_{12} = 38{,}440 \text{ lb}_f/\text{ft},$$

$$W_1 = 500 \text{ lb}_f, \; M_1 = 15.5 \text{ slugs},$$

$$W_2 = 2{,}000 \text{ lb}_f, \; M_2 = 62.1 \text{ slugs}.$$

And such that the car will operate in the range

$$500 \le \omega \le 5{,}000 \text{ rpm } (52.4 \le \omega \le 209.0 \text{ rad/s}).$$

Is the operating range for this car above both natural frequencies of the vibration absorber?

6-10. Find the RMS value for $p = 2\cos 10t$ psia.

6-11. What could physically happen if an automobile were forced to steadily operate at the resonant frequency of the motor mounts? How does this show up mathematically?

6-12. What is the speed of sound in pure oxygen (an ideal gas) at 0° F?

6-4 MORE CAPABLE TRANSFER MODELS

So far, transfer ratios have been used as frequency-domain models of *linear* components. These models have worked well to predict the input-to-output response for pure-sine inputs. They gave the dependent transfer variables of amplitude and phase as they related to the independent variable of frequency. However, the mathematics of these ratio formulations was difficult and error-prone.

In this section, a more powerful, more insightful frequency-domain modeling method for *linear* components is introduced. It is called the *Laplace Transform Method*. When the mathematical transform in this method is applied to a SISO time-domain model, it yields a new model called the transfer *function* for the SISO component. When applied to a MIMO model, it yields the transfer *matrix* for the component.

The usefulness of the Laplace Transform rests mainly in two areas:

1. It is a way to transform linear, time-domain, *differential-equation* models into *algebraic-equation* models for manipulation, display, and solution; and
2. It is a way to store both the transient and frequency-domain models in a single concise form.

These advantages make it the method of choice over transfer ratios for modeling in the frequency domain.

The Laplace Transform This transformation converts a time-domain expression, $f(t)$, into its equivalent s-domain representation, $F(s)$, and is indicated by the symbol $\mathcal{L}$. Generally speaking, the power of the transform is that the Laplace variable, s, is used to keep track of the time derivatives which are present in $f(t)$. This is done through the defining transform

$$F(s) = \mathcal{L}[f(t)] = \int_{-\infty}^{\infty} e^{-st}[f(t)]\, dt. \tag{6-114}$$

That is, whatever function, $f(t)$, is to be transformed is inserted into the [ ] under the integral sign in the above relationship. And, since $\mathcal{L}$ is an *operation* on $f(t)$, it is thus called an "operator."

The function, $f(t)$, must have several properties in order for the integral to have meaning. These properties are:

1. The function must be piecewise continuous over the interval of integration; and
2. The function must converge to a finite value as $t \rightarrow \infty$.

These requirements are easily satisfied by signals that can be physically generated (except for some kinds of noise), and for variables in a stable model (e.g., those models having non-positive real eigenvalue parts).

The operator definition above makes the following properties obvious

$$\mathcal{L}[f_1(t) + f_2(t)] = \mathcal{L}[f_1(t)] + \mathcal{L}[f_2(t)], \tag{6-115}$$

and

$$\mathcal{L}[Af(t)] = A\mathcal{L}[f(t)]. \tag{6-116}$$

Therefore, the Laplace Transform is a linear operator.

When the Laplace transform of a time-domain input is computed, the result is symbollically indicated as

$$\mathcal{L}[u(t)] = U(s). \tag{6-117}$$

Applying the transformation to an output gives

$$\mathcal{L}[y(t)] = Y(s). \tag{6-118}$$

The *transfer function* of a SISO component is *defined* as the ratio of these two transformations,

$$T(s) = Y(s)/U(s). \tag{6-119}$$

Note the similarity of this definition to that of the transfer ratio of previous sections. However, the transfer function for a component is computed by applying the transform (Eq. 6-114) directly to the time-domain, differential-equation model of the component. There is no need to solve for the output *before* the transfer function is computed, as was the case with the transfer ratio. Thus you need, in general, to be able to compute the transforms of inputs, of outputs, and of derivatives of outputs since these are all parts of a typical differential-equation model.

In the next part of this section the transformation of some common inputs, $u(t)$, to find $U(s)$ is discussed. A little later in the section the transformation of derivatives is presented. Finally, the formulation of the s-domain transfer function, $T(s)$, for SISO components is presented, followed by a similar form for MIMO components.

Notice that once you have found $T(s)$, you may compute the general response of the SISO component by taking the *inverse Laplace Transform,*

$$y(t) = \mathcal{L}^{-1}\,[T(s)\;U(s)]. \tag{6-120}$$

This method is discussed further in Chap. 8. Here, the method is first presented as a general method for use in finding a general time-domain model, and then specialized to the frequency domain. This is done in much the same way as the undamped response is found as a particular case of the general response of a second-order component. Thus, you must first understand the Laplace transform as a time-domain method, to be used on time-domain equations, having both periodic and aperiodic inputs.

Transformation of Inputs. The aperiodic inputs of interest are the idealized unit ramp, unit step, and unit impulse, and modeled delay within a component. The periodic inputs are the superposed sines and cosines. So, each of these is next studied in order to build a menu of common input transformations to draw upon during modeling.

Recall that the *unit ramp* $u_r(t)$, has the form

$$\begin{aligned} u_r(t) &= 0,\ t < 0, \\ &= t,\ t \geq 0. \end{aligned} \tag{6-121}$$

The transformation of this input is done in the following steps, starting from Eq. 6-114,

$$U_r(s) = \mathcal{L}[u_r(t)] = \int_{-\infty}^{\infty} te^{-st}dt, \tag{6-122}$$

$$= \int_{-\infty}^{0} 0\,dt + \int_{0}^{\infty} te^{-st}dt, \tag{6-123}$$

$$= 0 + \left[t\frac{e^{-st}}{-s}\right]_0^{\infty} - \int_0^{\infty} \frac{e^{-st}}{-s}\,dt, \tag{6-124}$$

$$= \frac{1}{s}\int_0^{\infty} e^{-st}dt, \tag{6-125}$$

So,

$$U_r(s) = 1/s^2. \tag{6-126}$$

The *unit impulse* transformation is very easily analyzed once you realize that

$$u_i(t)\,e^{-st} = u_i(t). \tag{6-127}$$

This is true since the impulse is zero everywhere except at $t = 0$, where the exponential term is unity. Thus, the transformation gives

$$U_i(s) = \int_{-\infty}^{\infty} u_i(t)e^{-st}\,dt = \int_{-\infty}^{\infty} u_i(t)\,dt, \tag{6-128}$$

$$= \text{Area of the impulse}. \tag{6-129}$$

Thus, $U_i(s) = 1.$ (6-130)

The *unit step* is similarly computed to yield

$$U_s(s) = 1/s. \tag{6-131}$$

(The details of this transformation are reserved for a homework problem.)

The *sine wave* transformation is computed from

$$f(t) = 0,\ t < 0, \tag{6-132a}$$

and

$$= \mathrm{A}\ \sin\omega t,\ t \geq 0. \tag{6-132b}$$

This gives

$$F(s) = \mathcal{L}[f(t)] = \int_0^\infty \mathrm{A}\ \sin\omega t\ e^{-st}\ dt. \tag{6-133}$$

But, since $\sin\omega t = [e^{j\omega t} - e^{-j\omega t}]/2j$, the last equation becomes

$$F(s) = \frac{\mathrm{A}}{2j}\int_0^\infty (e^{j\omega t} - e^{-j\omega t})e^{-st}dt, \tag{6-134}$$

and

$$F(s) = \mathrm{A}/[2j(s - j\omega)] - \mathrm{A}/[2j(s + j\omega)], \tag{6-135}$$

or

$$F(s) = \mathrm{A}\omega/(s^2 + \omega^2). \tag{6-136}$$

A *cosine wave* input may be similarly computed from

$$f(t) = 0,\ t < 0, \tag{6-137a}$$

and

$$= \mathrm{A}\ \cos\omega t,\ t \geq 0. \tag{6-137b}$$

This gives

$$F(s) = \mathrm{A}s/(s^2 + \omega^2). \tag{6-138}$$

The modeling of *delay* in the time-domain inputs was discussed briefly in Chap. 5, and is described graphically for the frequency domain in Fig. 6-17a. This figure shows the finite delay, τ_d, between the time when the input, u, steps and the time when y responds. Modelers often ignore the time delay in time-domain modeling, but this effect is usually of more concern in frequency-domain components.

A common approach to the modeling of delay is shown in Fig. 6-17b. Here, the model is broken into two parts: the first part accomplishes the delay of the input, and the second part uses a model without delay, such as those in Chap. 4 and earlier in this chapter. Overall, the effect of the model is the same as the observed physical

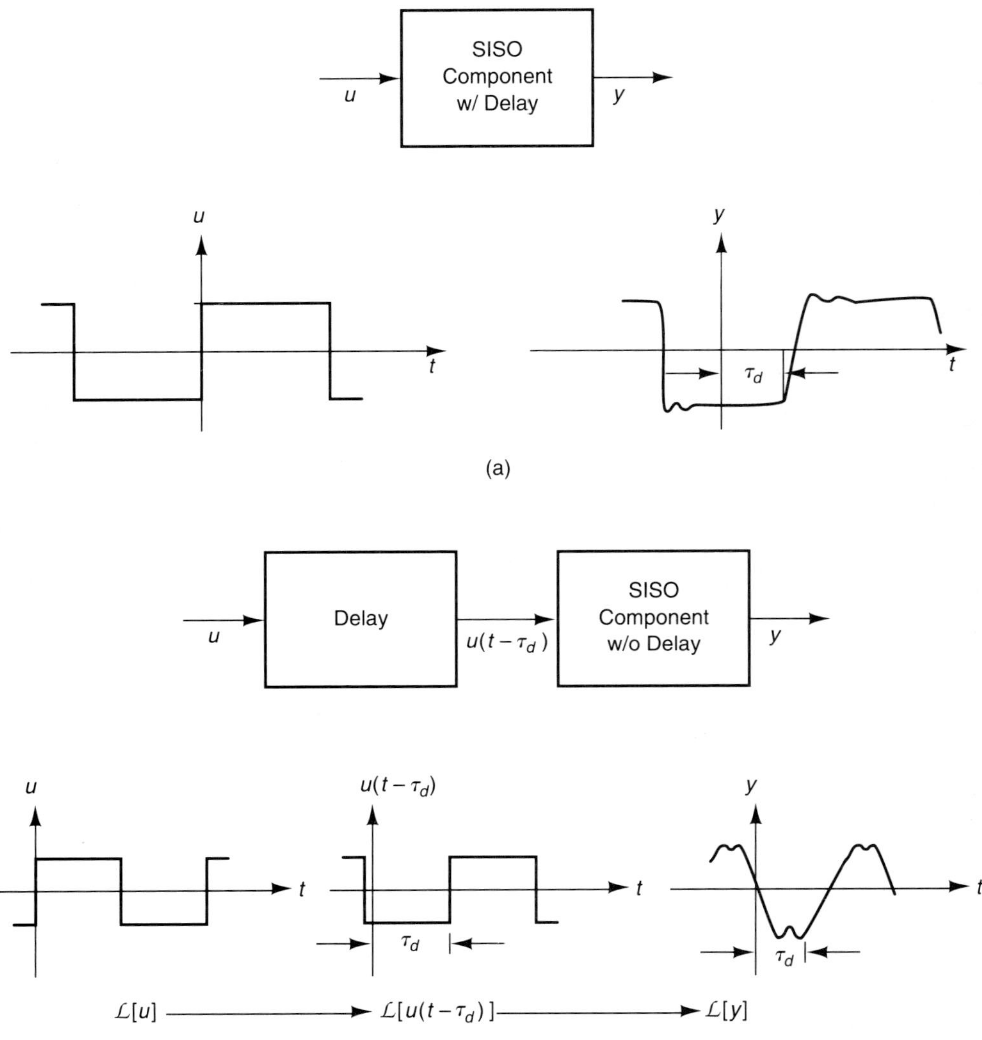

Figure 6-17. Frequency-domain Delay. (a) Physical response, (b) modeling concept

response. So, the first part of the Laplace modeling is to compute the transformation of the delayed input, $u(t - \tau_d)$.

The transformation of the delayed input signal comes from a broad understanding of the transform

$$U(s) = \int_{-\infty}^{\infty} u(\alpha)e^{-s\alpha}\, d\alpha, \tag{6-139}$$

which is re-interpreted in terms of a change of variable

$$\alpha = t - \tau_d, \tag{6-140}$$

and

$$d\alpha = dt. \tag{6-141}$$

These changes give a new representation of the transformation

$$U(s) = \int_{-\infty}^{\infty} u(t - \tau_d)e^{-s(t-\tau_d)}\, dt, \tag{6-142}$$

$$= e^{s\tau_d} \int_{-\infty}^{\infty} u(t - \tau_d)e^{-st}\, dt. \tag{6-143}$$

Rearranging this last equation gives the result of interest,

$$\mathcal{L}[u(t - \tau_d)] = \int_{-\infty}^{\infty} u(t - \tau_d)\mathrm{e}^{-st}\, dt = e^{-s\tau_d}U(s) \tag{6-144}$$

That is, the Laplace transform of a signal delayed by τ is equal to $e^{-s\tau}$ times the Laplace Transform of the undelayed signal.

Example 6-3

Given a model for a SISO fluid component

$$\mathrm{RC}\, dp/dt + p = \mathrm{R}z.$$

What is the Laplace Transform of the right-hand side of this equation (the input side) if there is an expected 7 sec. delay in the component between an input flowrate change and the start of the pressure response?

Solution

The transform of the right-hand side without delay is

$$\mathcal{L}[\mathrm{R}z] = \mathrm{R}\, Z(s).$$

Thus, the right side transform with delay becomes

$$\mathcal{L}[\mathrm{R}z(t - 7)] = e^{-7s}\mathrm{R}\, Z(s).$$

Comments

The fluid resistance, R, is a constant parameter which simply passes through the transformation. The 7 sec. delay modeling is thus accomplished by multiplying by the exponential term e^{-7s}. Also, you must leave the flowrate, $Z(s)$, in this general form until you know the specific function, $z(t)$, to be transformed.

Transformation of Derivatives. To find the Laplace Transform of the first derivative, you begin with a particular function, $f(t)$, which has the property

$$f(t) = 0,\ t \leq 0. \tag{6-145}$$

This allows you to ignore the integration interval $(-\infty, 0]$, and integrate the rest of the transform by parts to find

$$F(s) = \int_{-\infty}^{\infty} f(t)\mathrm{e}^{-st}\, dt = -\left[f(t)\frac{\mathrm{e}^{-st}}{s}\right]_0^{\infty} + \int_0^{\infty} \frac{df}{dt}\frac{\mathrm{e}^{-st}}{s}\, dt. \qquad (6\text{-}146)$$

Evaluation of the last equation gives

$$F(s) = f(0)/s + (1/s)\mathcal{L}[df/dt]. \qquad (6\text{-}147)$$

But, since the initial condition is zero, the transformation of the derivative is

$$\mathcal{L}[df/dt] = sF(s). \qquad (6\text{-}148)$$

Note that this last equation supports the previous work with linearized component models in Chap. 4 and the requirement that the linearized models have *zero initial conditions* in the time domain. And, when you pose a time-domain model in terms of lower-case variables, you are thus really describing excursions away from a steady-state linearization point. Such a linearized model near the steady operating point A is

$$Y = \mathrm{Y}_A + y, \qquad (6\text{-}149)$$

and

$$U = \mathrm{U}_A + u. \qquad (6\text{-}150)$$

This also gives

$$dY/dt = dy/dt, \qquad (6\text{-}151)$$

and

$$dU/dt = du/dt. \qquad (6\text{-}152)$$

The Laplace models are thus based on the notion that at time $t = 0$ the component and system are *at rest* at point A. This means

$$y(0) = 0,$$

$$Y(0) = \mathrm{Y}_A,$$

$$u(0) = 0,$$

and

$$U(0) = \mathrm{U}_A.$$

It also means that all derivatives of y are equal to zero at A because the system is "at rest" there. So, when you start with a time-domain model expressed in lower-case variables, you can always rely on zero initial conditions on all the states, and the simple transform of Eq. 6-148 may be used.

When you must transform higher-order derivatives into their Laplace function, you simply reapply the logic which led to Eq. 6-148 to find

$$\mathcal{L}[d^nf/dt^n] = s^nF(s). \tag{6-153}$$

Again, the requirement that the initial conditions must all be zero leads to the form of Eq. 6-153, for the same reasons as those for the first derivative.

As you use these results in practice, you do not recompute the Laplace integrals for each part of the time-domain model starting from Eq. 6-114. Instead, you resort to tabulations of common transforms such as those given in Table 6-2.

Table 6-2. Common Laplace Transforms

	$f(t)$	$F(s)$
Unit impulse	$u_i(t)$	1
Unit step	$u_s(t)$	$1/s$
Unit ramp	$u_r(t)$	$1/s^2$.
Delayed input	$u(t - a)$	$e^{-as}\,U(s)$
Sine	$\sin \omega t$	$\omega/(s^2 + \omega^2)$
Cosine	$\cos \omega t$	$s/(s^2 + \omega^2)$
Derivative	d^nf/dt^n	$s^nF(s)$

The Transfer Function. You now have the tools to transform all linear, constant-coefficient, time-domain models into the s-domain for further analysis. When you do this for SISO component models you create a transfer function, $T(s)$, for those components. This function is *defined* as the ratio of the transformed output over the transformed input, with zero initial conditions. That is, as discussed earlier in this section,

$$T(s) = Y(s)/U(s). \tag{6-119}$$

It is computed by applying the Laplace Transform to the time-domain model, and algebraically solving for the ratio.

Consider, for example, the transformation of

$$\tau_c\, dy/dt + y = 5u(t). \tag{6-154}$$

This is

$$(\tau_c s + 1)Y(s) = 5U(s). \tag{6-155}$$

And

$$T(s) = Y(s)/U(s) = 5/(\tau_c s + 1). \tag{6-156}$$

Note that, given a few tables of Laplace Transforms, the computation of the transfer function is a good deal simpler than the computation of the transfer ratios previously. This is the first benefit of the Laplace transform. In the next section, you will study how easy it is to approximate the plot of the transfer function on a Bode diagram in order to study a component design, which is the second benefit of the Laplace transform. Further, it is precisely this latter benefit which makes design-by-transfer-function such a powerful and popular tool.

Sometimes, the component time-domain model is expressed as a linear, SISO, state-space model

$$d\mathbf{x}/dt = \mathbf{A}\,x + \mathbf{B}\,\mathrm{u}, \tag{6-157}$$

and

$$y = \mathbf{C}\,x. \tag{6-158}$$

The transformation of this form is computed by applying the Lapalce transform to each equation in the equation set, the result of which may be written in vector form

$$s\mathbf{X}(s) = \mathbf{A}\,X(s) + \mathbf{B}\,U(s), \tag{6-159}$$

and

$$Y(s) = \mathbf{C}\,X(s). \tag{6-160}$$

Equation 6-160 is then solved for $X(s)$ to give

$$\mathbf{X}(s) = (s\mathbf{I} - \mathbf{A})^{-1}\mathbf{B}\,U(s). \tag{6-161}$$

When this solution is substituted into Eq. 6-160, the transfer function is computed to be

$$T(s) = Y(s)/U(s) = \mathbf{C}\,(s\mathbf{I} - \mathbf{A})^{-1}\mathbf{B}. \tag{6-162}$$

The Transfer Matrix. This form of the transformation arises whenever the Laplace transform is applied to a time-domain MIMO model. For example, consider the model first discussed in the introduction to Chap. 4,

$$\mathrm{RC}\,di_a/dt + i_a = \mathrm{C}\,de_a/dt - i_b, \tag{4-1}$$

$$\mathrm{R}i_a + e_b = e_a. \tag{4-2}$$

This is a two-output model (i_a and e_b) with two inputs (e_a and i_b).

When these two output equations are transformed you find

$$(\mathrm{RC}s + 1)\,I_a(s) = \mathrm{C}s\,E_a(s) - I_b(s), \tag{6-163}$$

$$\mathrm{R}I_a(s) + E_b(s) = E_a(s). \tag{6-164}$$

Further, these may be written in a matrix equation as follows

$$\begin{bmatrix} (\mathrm{RC}s + 1) & 0 \\ \mathrm{R} & 1 \end{bmatrix} \begin{bmatrix} I_a(s) \\ E_b(s) \end{bmatrix} = \begin{bmatrix} \mathrm{C}s & -1 \\ 1 & 0 \end{bmatrix} \begin{bmatrix} E_a(s) \\ I_b(s) \end{bmatrix}. \tag{6-165}$$

The transfer matrix comes from a matrix equation of the form

$$Y(s) = T(s)\,U(s). \tag{6-166}$$

Notice that you will have this form if you do a little rearranging of Eq. 6-165. That is, you need to multiply both sides of Eq. 6-165 by the inverse of the coefficient matrix as follows

$$\begin{bmatrix} I_a(s) \\ E_b(s) \end{bmatrix} = \begin{bmatrix} (\mathrm{RC}s + 1) & 0 \\ R & 1 \end{bmatrix}^{-1} \begin{bmatrix} \mathrm{C}s & -1 \\ 1 & 0 \end{bmatrix} \begin{bmatrix} E_a(s) \\ I_b(s) \end{bmatrix}. \tag{6-167}$$

When the inversion is computed and multiplied you find

$$\begin{bmatrix} I_a(s) \\ E_b(s) \end{bmatrix} = \begin{bmatrix} \dfrac{\mathrm{C}s}{(\mathrm{RC}s + 1)} & \dfrac{-1}{(\mathrm{RC}s + 1)} \\ \dfrac{1}{\mathrm{RC}s + 1} & \dfrac{\mathrm{R}}{(\mathrm{RC}s + 1)} \end{bmatrix} \begin{bmatrix} E_a(s) \\ I_b(s) \end{bmatrix}. \tag{6-168}$$

The elements of the transfer matrix are easily recognized from Eq. (6-168),

$$T\,(1,\,1) = \partial Y_1/\partial U_1 = \mathrm{C}s/(\mathrm{RC}s + 1), \tag{6-169}$$

$$T\,(1,\,2) = \partial Y_1/\partial U_2 = -1/(\mathrm{RC}s + 1), \tag{6-170}$$

$$T\,(2,\,1) = \partial Y_2/\partial U_1 = 1/(\mathrm{RC}s + 1), \tag{6-171}$$

$$T\,(2,\,2) = \partial Y_2/\partial U_2 = R/(\mathrm{RC}s + 1). \tag{6-172}$$

If the time-domain MIMO model is expressed in state-space form,

$$dx/dt = \mathbf{A}\,\boldsymbol{x} + \mathbf{B}\,u, \tag{6-173}$$

$$\boldsymbol{y} = \mathbf{C}\,\boldsymbol{x}. \tag{6-174}$$

Then the transfer matrix comes from the exact same manipulations which you used to find Eq. 6-162 and the SISO transfer function earlier. That is,

$$\boldsymbol{Y}(s) = \mathbf{C}(s\mathbf{I} - \mathbf{A})^{-1}\mathbf{B}\,\boldsymbol{U}(s), \tag{6-175}$$

so

$$\boldsymbol{T}(s) = \mathbf{C}(s\mathbf{I} - \mathbf{A})^{-1}\mathbf{B}. \tag{6-176}$$

HOMEWORK

6-13. Show that $\mathcal{L}[10u(t)] = 10/s$.

6-14. Find the transfer function (z is input) for

$$10\,dp/dt + p = 6z.$$

6-15. Find the transfer function (f is input) for

$$\mathrm{M}\ d^2v/dt^2 + \mathrm{b}\ dv/dt + \mathrm{k}v = \mathrm{k}\ df/dt.$$

6-16. Given the SISO model

$$dx_1/dt = 2x_2,$$

$$dx_2/dt = -5x_1 - 2x_2 + 3u,$$

and

$$y = x_1 + x_2.$$

Find the transfer function.

6-17. Find the transfer matrix for the following MIMO component (inputs are on the right)

$$10\ dp_1/dt + (p_1 - p_2) = 3\ dz_1/dt,$$

$$4\ dp_2/dt + p_2 = 7z_2.$$

6-5 EVALUATING THE TRANSFER FUNCTION MODEL

As seen in the last section, a transfer function is computed from a time-domain, differential-equation model. In this section, the transfer function is used to characterize a frequency-domain component on a Bode diagram, and thus achieve a quick graphical understanding of the component response. Such a quick estimate of response is often very helpful and appropriate during preliminary design.

However, this method depends on an interpretation of the general s-domain transfer function, $T(s)$, into a form which shows only the frequency-domain behavior of the component. This interpretation begins with a study of the Laplace variable, s.

The Laplace Variable, s. In the frequency domain, you are interested in periodic inputs only. The simplest form of these is the sine function

$$u(t) = \mathrm{A}\ \sin\omega t. \tag{6-177}$$

The Laplace transform of this input is

$$U(s) = \mathrm{A}\omega/(s^2 + \omega^2), \tag{6-136}$$

as studied in the last section. Notice that this last equation is the same transform as that for the differential equation

$$d^2u/dt^2 + \omega^2 u = \mathrm{A}\omega\ u_i(t), \tag{6-178}$$

with zero initial conditions. So, the Laplace transform sees Eqs. 6-177 and 6-178 as *exactly* the same, having the same transfer function. Now, to find the homogeneous, *characteristic* solution to Eq. 6-178, you let

$$u = Ce^{st}, \tag{6-179}$$

from which you find the characteristic values of s to be

$$s = \pm j\omega. \tag{6-180}$$

Further, the superposed characteristic solutions in the form of Eqs. 6-179 and 6-180 are equivalent to the form of Eq. 6-177, which is easily demonstrated using Euler's equations (Eqs. 5-51 and 5-52).

Also, note that $s = \pm j\omega$ is characteristic of the cosine function ($As/[s^2 + \omega^2]$, as in Table 6-2), for exactly the same reasons. In fact, Eq. 6-180 is characteristic of *all* periodic functions since these functions may be transformed into superposed sines and cosines, as Fourier has shown.

This means that the frequency-domain model of the component may be extracted from the transfer function simply by setting $s = \pm j\omega$ in the following forms

$$T(j\omega) = Y(j\omega)/U(j\omega), \tag{6-181a}$$

or

$$T(-j\omega) = Y(-j\omega)/U(-j\omega). \tag{6-181b}$$

Notice that each of the variables in Eqs. 6-181 must be a complex number (e.g., set $s = j\omega$ in Eq. 6-136 above). So, when $s = j\omega$, T is of the form

$$T(j\omega) = M(\cos\phi + j\sin\phi), \tag{6-182}$$

where M is the magnitude of the transfer function, and ϕ is the angle that the transfer-function vector makes with the real axis, measured counter-clockwise as shown in Fig. 6-18. Further, it can be shown that the angle ϕ is also the phase angle of the transfer function, which is stored here in the complex-variable form of $T(j\omega)$.

When $s = -j\omega$, $T(-j\omega)$ yields exactly the same magnitude as $T(j\omega)$, but with $-\phi$ as the phase angle. So, system modelers only evaluate $s = j\omega$ (e.g.,

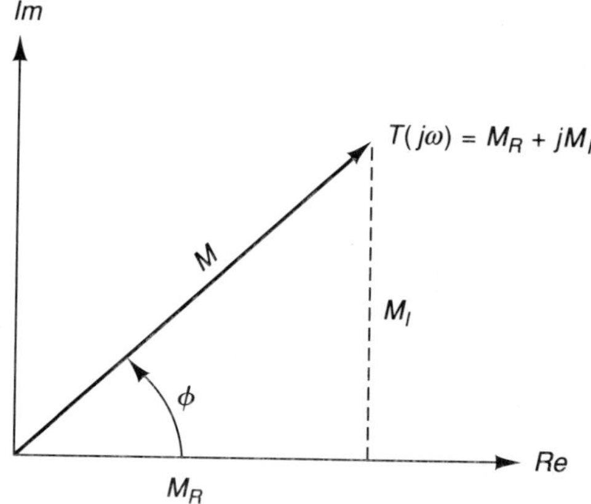

Figure 6-18. $T(j\omega)$ on the Complex Plane

Eq. 6-181a) when they determine the component's frequency-domain, M-and-ϕ response. These values are then applied in exactly the same way as the transfer ratio of the previous sections in order to compute the component's forced periodic response.

The SISO Forced Response. The periodic response of a SISO component is computed from a superposed set of sine inputs transferred to a superposed set of sine outputs. Any one of these inputs may be at an arbitrary phase angle, α,

$$u = \mathrm{U}\sin(\omega t + \alpha), \tag{6-183}$$

and thus create a phase-shifted steady sinusoidal response of

$$y_f = M\mathrm{U}\sin(\omega t + \alpha + \phi). \tag{6-184}$$

In this manner, a forced response is created for each of the harmonics, n, that are input

$$y_{fn}(t) = M_n \mathrm{U}_n \sin(n\omega_0 t + \alpha_n + \phi_n). \tag{6-185}$$

And the net forced response comes from the superposition of all of the harmonic responses,

$$y_f(t) = \sum y_{fn}(t) = \sum M_n \mathrm{U}_n \sin(n\omega_0 t + \alpha_n + \phi_n). \tag{6-186}$$

The key to using all of this effectively is to understand how to extract the magnitude, M, and phase, ϕ, from the transfer function, $T(j\omega)$.

Note that, in general, when you substitute $s = j\omega$ in $T(s)$, you create a complex ratio of the form

$$T(j\omega) = (B_R + jB_I)/(A_R + jA_I). \tag{6-187}$$

This ratio is clarified when you multiply numerator and denominator by the complement of the denominator. That is,

$$T(j\omega) = [(B_R + jB_I)(A_R - jA_I)]/[(A_R + jA_I)(A_R - jA_I)], \tag{6-188}$$

$$= [(A_R B_R + A_I B_I) + j(A_R B_I - A_I B_R)]/(A_R^2 + A_I^2) \tag{6-189}$$

Now Eq. 6-189 shows the specific terms that led to the plot of Fig. 6-18. These are

$$M_R = (A_R B_R + A_I B_I)/[A_R^2 + A_I^2], \tag{6-190}$$

$$M_I = (A_R B_I - A_I B_R)/[A_R^2 + A_I^2]. \tag{6-191}$$

And they give

$$M = (M_R + M_I)^{1/2}, \tag{6-192}$$

and

$$\phi = \tan^{-1}(M_I/M_R). \tag{6-193}$$

Example 6-4

Suppose

$$T(s) = 1/s.$$

Find M and ϕ of $T(j\omega)$. Plot the results on a Bode diagram for the range $0.1 \leq \omega \leq 100$ rad/s.

Solution

Simple substitution gives

$$T(j\omega) = 1/j\omega = -(1/\omega)\,j.$$

Consequently,

$$M = 1/\omega,$$

and

$$\phi = -\pi/2 \text{ rad}.$$

These results are plotted in Fig. 6-19.

Comments

Note that the Bode diagram has uniform vertical scales and logarithmic horizontal

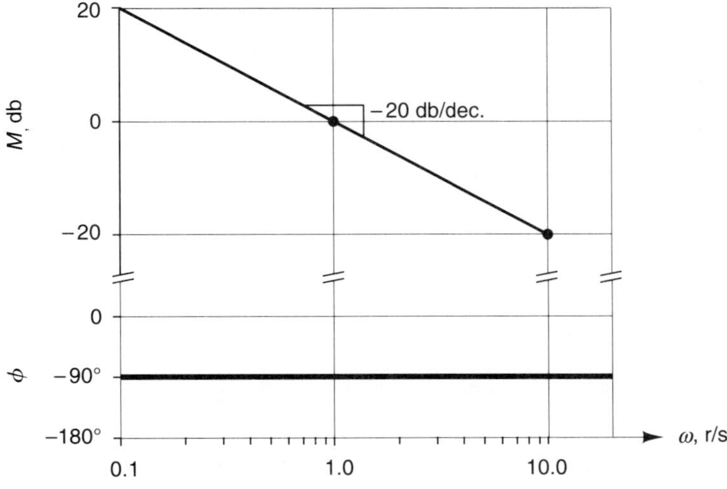

Figure 6-19. Bode Diagram of $T(j\omega) = 1/(j\omega)$

scales. Such a plot is called a semi-log plot. Also, recall that the vertical scale on the magnitude plot is a decibel value of the magnitude ($20 \log_{10} M$).

The slope of the magnitude plot has a value of -20 decibels/decade for this lone s in the denominator. This *rolloff rate* of magnitude is characteristic of an s-factor in the denominator. Furthermore, the phase angle of $-90°$ is also characteristic of such an s factor. Note that if the s were in the numerator ($T(j\omega) = j\omega$), then the slope would be $+20$ db/dec., and the phase angle would be $+90°$.

If $T(s) = \mathrm{K}(1/s)$, then the previous magnitude plot would be shifted up or down according to the constant K since

$$20 \log_{10}(\mathrm{K}/\omega) = 20 \log_{10} \mathrm{K} + 20 \log_{10}(1/\omega). \tag{6-194}$$

In other words, multiplication of numerator and denominator factors in $T(j\omega)$ results in their decibel values being added or subtracted in the magnitude plot, according to the rules of logarithms.

All factors in the transfer function, $T(j\omega)$, need to be added (superposed) to find the total forced response. Thus, it is important to factor the numerator and denominator of the transfer function in order to find the terms which are contributing to the total forced output response. In the next part of this section, the typical factor forms are studied for their magnitude and phase contributions.

Delay Factors. The transfer function for delay is

$$T(s) = e^{-\tau s}. \tag{6-195}$$

When $s = j\omega$, this becomes

$$T(j\omega) = e^{-j\omega\tau}, \tag{6-196}$$

$$= \cos\omega\tau - j \sin\omega\tau, \tag{6-197}$$

from Euler's formulas (Eqs. 5-51, 5-52). So, for the delay factor,

$$M = 1 \text{ for any } \omega, \tag{6-198}$$

and

$$\phi = -\omega\tau. \tag{6-199}$$

Thus, pure delay has no effect on the transfer magnitude, but it does affect the phase shift according to the formula shown.

First-order Factors. The general form for a first-order factor in the denominator is

$$T(s) = 1/(\tau s + 1). \tag{6-200}$$

In the frequency domain $s = j\omega$, and this model becomes

$$T(j\omega) = 1/(\tau j\omega + 1),$$
$$= (1 - \tau j\omega)/(1 + \omega^2\tau^2). \tag{6-201}$$

This allows you to compute

$$M = (1 + \omega^2\tau^2)^{-1/2}, \tag{6-202}$$

and

$$\phi = \tan^{-1}(-\omega\tau). \tag{6-203}$$

When this factor is plotted on the Bode diagram, the result is Fig. 6-20. Equation 6-202 shows that very small frequencies give $M \approx 1 =$ constant. Very large frequencies give $M \approx 1/\omega\tau$, which has the same rolloff of -20 db/dec. that was seen in the last example. The transition between these two regions occurs at frequencies near the *break frequency*, ω_b, (also called the "corner frequency") where

$$\omega_b = 1/\tau. \tag{6-204}$$

Note that the phase angle, ϕ, passes through $-45°$ at the break frequency, and is reasonably steady a decade above and a decade below the break frequency. These observations help in sketching the phase curves.

Figure 6-20 also shows that the exact plot of the db magnitude may be very well approximated by two straight lines. This sort of *asymptotic approximation* introduces a small error in the region of the break frequency, yet it captures the char-

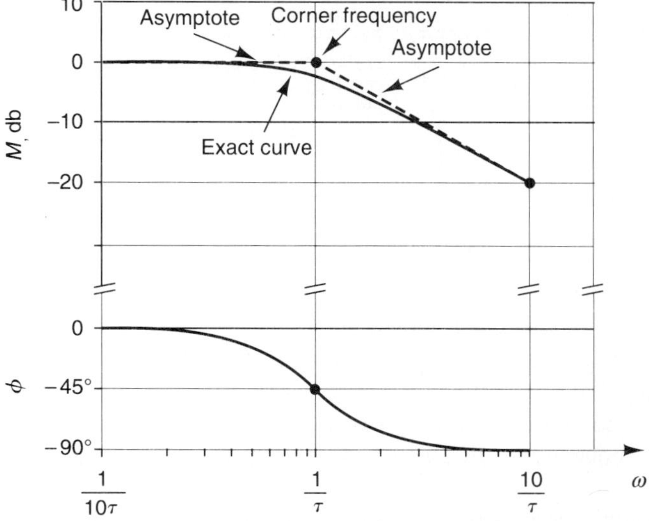

Figure 6-20. Bode Diagram of $T(j\omega) = 1/(1 + j\omega\tau)$. [Katsuhiko Ogata, MODERN CONTROL ENGINEERING, 2/e, 1990, pg. 436. Reprinted by permission of Prentice Hall, Englewood Cliffs, New Jersey.]

acter of the frequency response quite well. Thus, system designers and modelers often resort to these asymptotic approximations of $T(j\omega)$, especially during preliminary design.

If the first-order factor is in the numerator ($T(s) = \tau s + 1$), then the magnitude plot would break up rather than down at a rate of +20 db/dec., at the same break frequency, $1/\tau$. And the corresponding phase angle would be positive rather than negative, $\phi = \tan^{-1}(\omega\tau)$. Again, multiplying by a constant ($T(s) = K(\tau s + 1)$) merely shifts the magnitude curve up or down on the Bode diagram, without affecting its slope or the phase plot.

Second-order Factors. The general form for a second-order factor is

$$T(s) = \frac{1}{\left[\dfrac{s}{\omega_n}\right]^2 + \left[\dfrac{2\zeta}{\omega_n}\right]s + 1}. \tag{6-205}$$

Substituting $s = j\omega$ gives the frequency-domain form

$$T(j\omega) = \frac{1}{1 - \left[\dfrac{\omega}{\omega_n}\right]^2 + j\left[\dfrac{2\zeta\omega}{\omega_n}\right]}, \tag{6-206}$$

from which you can extract,

$$M = \frac{1}{\sqrt{\left\{1 - \left[\dfrac{\omega}{\omega_n}\right]^2\right\}^2 + \left[\dfrac{2\zeta\omega}{\omega_n}\right]^2}}, \tag{6-207}$$

$$\phi = \tan^{-1}\left[\frac{-2\zeta\left(\dfrac{\omega}{\omega_n}\right)}{1 - \left(\dfrac{\omega}{\omega_n}\right)^2}\right]. \tag{6-208}$$

The Bode diagram for this factor is shown in Fig. 6-21.

The second-order magnitude equation and the Bode diagram show that $M \approx 1$(0 db) for $\omega << \omega_n$. Further, the magnitude plot has a break frequency similar to that seen for the first-order factor. However, a peak is now observed in the region of the break frequency, with a height dependent on the all-important damping ratio, ζ.

The peak in the second-order frequency response is called the *resonant peak*. It occurs at the *resonant frequency*, ω_r. It comes about when $\zeta \to 0$ and $\omega \to \omega_n$ *simultaneously* (e.g., see Eq. 6-207). When this occurs, it creates a situation where $M \to 1/0$. The magnitude of the frequency response thus gets very large and may

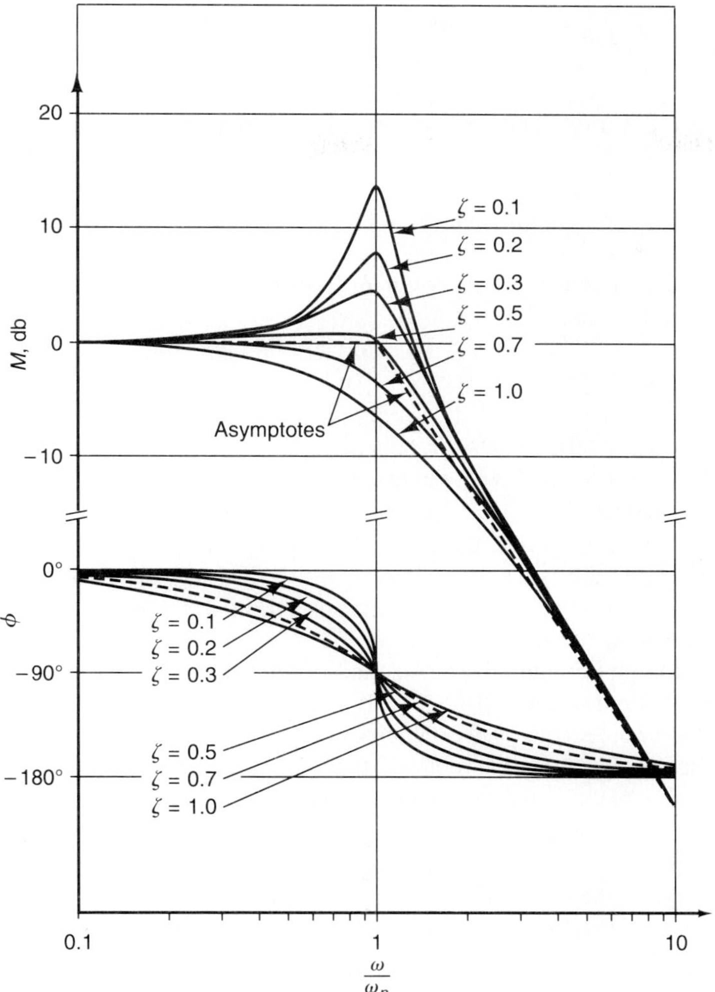

Figure 6-21. Bode Diagram of $T(j\omega) = 1/[1 - (\omega/\omega_n)^2 + j(2\zeta\omega/\omega_n)]$. [Katsuhiko Ogata, MODERN CONTROL ENGINEERING, 2/e, 1990, pg. 441. Reprinted by permission of Prentice Hall, Englewood Cliffs, New Jersey.]

stimulate a destructive response from the component (like an opera singer destroying a wine glass with her voice). This condition is called *resonance*.

Often, it is important to find the frequency at which resonance occurs. This occurs when the denominator of M is minimum. So, if you let the denominator of M be D_M, then you can find a minimum in D_M to locate the resonant frequency. Start with

$$D_M(\omega) = \left[1 - \left(\frac{\omega}{\omega_n}\right)^2\right]^2 + \left[2\zeta\left(\frac{\omega}{\omega_n}\right)\right]^2. \tag{6-209}$$

And rewrite this into the form

$$D_M(\omega) = \left[\frac{\omega^2 - \omega_n^2(1 - 2\zeta^2)}{\omega_n^2}\right]^2 + 4\zeta^2(1 - \zeta^2). \tag{6-210}$$

Now, since ζ is a fixed number for any given design, you find that D_M is a minimum when

$$\omega = \omega_r = \omega_n(1 - 2\zeta^2)^{1/2}. \tag{6-211}$$

This is the defining equation for the resonant frequency. Note that this equation only has a real meaning when

$$0 \le \zeta \le 0.707,$$

which comes from the term under the radical sign.

If $\zeta > 0.707$, resonance will *not* occur.

Moreover, with $0.707 < \zeta \le 1$, the amplitude plot looks like the first-order plot, with two exceptions. First, the rolloff rate at high frequencies is now -40 db/dec. since the magnitude is $M \approx 1/s^2$ as $\omega \to \infty$. Second, the phase angle, ϕ, is now transitioning from 0° to $-180°$ as ω increases through the break frequency, in a manner determined by the damping ratio.

Note that the phase angle for the second-order factor is $-90°$ at the break frequency, for all damping ratios, $0 \le \zeta \le 1$, and that the phase transition is mostly completed in the two decades of frequencies centered on the break frequency.

Recall that for $\zeta > 1$, the roots of the second-order characteristic equation (the $T(s)$ denominator) are real and unequal. In other words, for $\zeta > 1$ you should be able to reduce any second-order factor into two first-order factors.

In fact, as seen in the previous chapter, all numerator and denominator polynomials of $T(s)$ higher than second order can be reduced into products of first- and second-order factors. And they *should* be factored in order to use the preceding development to quickly sketch the frequency-response Bode diagram.

Combined Factors. When you factor the transfer function, in general, you find

$$T(s) = \frac{(s + z_1)(s + z_2) \cdots (s + z_m)}{(s + p_1)(s + p_2) \cdots (s + p_n)} = \frac{N(s)}{D(s)}. \tag{6-212}$$

That is, when you set $N(s) = 0$, you find the values of s which make the transfer function zero. These are called the *zeroes* of the transfer function and are symbolized by $z_1, z_2, \ldots, z_m$. When you set $D(s) = 0$, you find the values of s which make the transfer function infinite. These are called the *poles* of the transfer function and are symbolized by $p_1, p_2, \ldots, p_n$. The zeroes are associated with the derivatives of the input, while the poles are associated with the derivatives of the output. In both cases, the factors may be real or they may be complex. If complex, they always occur in complex conjugate pairs.

Note that $D(s) = 0$ is another statement of the characteristic equation. Once again, you see the importance of the characteristic equation, this time as it governs the location of the break frequencies. And once again, you are forced to find the roots of the characteristic equation. However, in order to understand what the denominator factors are telling you about the magnitude and phase, you need to resort to the mathematics of complex variables once more.

Following the substitution of $s = j\omega$ in the above factored transfer function, you find the numerator factors to be of the form

$$Z_m = j\omega + z_m, \tag{6-213}$$

and you find the denominator factors to be of the form

$$P_n = j\omega + p_n. \tag{6-214}$$

Note that any two numerator factors may also be written in their complex-variable form as a magnitude and a phase angle,

$$Z_1 = Z_{m1}(\cos\phi_1 + j\,\sin\phi_1), \tag{6-215}$$

and

$$Z_2 = Z_{m2}(\cos\phi_2 + j\,\sin\phi_2). \tag{6-216}$$

When these two are multiplied,

$$\begin{aligned} Z_1 Z_2 = Z_{m1} Z_{m2}[&(\cos\phi_1\,\cos\phi_2 - \sin\phi_1\,\sin\phi_2) \\ &+ j(\sin\phi_1\,\cos\phi_2 + \cos\phi_1\,\sin\phi_2)], & \text{(6-217)} \\ = Z_{m1} Z_{m2}&[\cos(\phi_1 + \phi_2) + j\,\sin(\phi_1 + \phi_2)] & \text{(6-218)} \end{aligned}$$

And you may correctly conclude that when factors are multiplied, the magnitudes are multiplied while the phase angles are added.

For the case of division,

$$\begin{aligned} Z_1/P_3 &= Z_{m1}/P_{m3}(\cos\phi_1 + j\,\sin\phi_1)/(\cos\phi_3 + j\,\sin\phi_3). & \text{(6-219)} \\ &= Z_{m1}/P_{m3}[\cos\phi_1\,\cos\phi_3 + \sin\phi_1\,\sin\phi_3) \\ &\quad + j(\sin\phi_1\,\cos\phi_3 - \cos\phi_1\,\sin\phi_3)], & \text{(6-220)} \\ &= Z_{m1}/P_{m3}\,[\cos(\phi_1 - \phi_2) + j\,\sin(\phi_1 - \phi_2)] & \text{(6-221)} \end{aligned}$$

Thus, division of the complex factors results in division of magnitudes and subtraction of phase angles.

So, when you compute the magnitude of the transfer function $T(j\omega)$, in decibels, you find

$$M = \sum 20 \log_{10} Z_i + \sum 20 \log_{10} (1/P_i),$$

$$= \sum 20 \log_{10} Z_i - \sum 20 \log_{10} P_i. \tag{6-222}$$

And a corresponding phase angle of

$$\phi = \sum \phi_{Zi} - \sum \phi_{Pi}. \tag{6-223}$$

Example 6-5

A designer has estimated a transfer function to be

$$T(s) = \frac{100e^{-0.02s}(10s + 1)}{(0.1s + 1)(s^2 + 0.4s + 1)}.$$

Sketch the Bode diagram for this model over the range $0.1 \leq \omega \leq 100$ rad/s.

Solution

The factors of $T(j\omega)$ for this transfer function are

$$Z_1 = 100,\ M_1 = 20 \log_{10}100 = 40 \text{ db} = \text{constant},$$

$$Z_2 = e^{-0.02j\omega},\ M_2 = 0 \text{ db},\ \phi_2 = -0.02\omega \text{ rad} = -1.146\omega \text{ deg},$$

$$Z_3 = (10j\omega + 1),\ \omega_b = 0.1 \text{ rad/s},$$

$$P_1 = 1/(0.1j\omega + 1),\ \omega_b = 10 \text{ rad/s},$$

$$P_2 = 1/[(1 - \omega^2) + 0.4j\omega],\ \omega_n = 1 \text{ rad/s},\ \zeta = 0.2.$$

These factors are plotted and added together in Fig. 6-22.

Comments

Each of the four factors which contribute to the magnitude response are shown in Fig. 6-22 (Z_2 only contributes to the phase response). Z_1 is a constant 40 db, and the other factors have break frequencies at 0.1, 1, and 10 r/s as shown. Note that Z_3 is a break upward at 20 db/dec. since it is in the numerator, while the others break downward in their role as denominator factors.

The net magnitude, M, is obtained by adding all of the contributions at each frequency. Such a calculation is shown by the letters a-b-c-d in the figure. At this frequency, P_1 and P_2 have not yet begun to contribute (point a), but Z_1 and Z_3 are adding together (points b and c) to give M at point d. Note that the break upward caused by Z_3 is more than cancelled out by the break downward caused by P_2 for frequencies above 1 r/s. The resonant peak is shown for the second-order factor more as a reminder than as a quantitative answer (remember, this is preliminary design). When the frequencies increase past 10 r/s, the rolloff rate goes to -40 db/dec. which can be verified from the transfer function, letting $s \to \infty$ and examining the exponent of s (2 in this case).

Similarly, the phase angles are added together at each frequency. The best way to do this is to first make the estimate without the delay contribution (the dashed line in the figure), then add it in later. When the break frequencies are separated by a decade or more, as they are here, the phase sketch is very easy since the factors may be considered separately. The numbers on the phase figure are the kinds of points of interest. The point 1 corresponds to the phase angle at the break frequency of Z_3 (45°). The point 2 is where the break up is almost done, but the P_2 pole has not

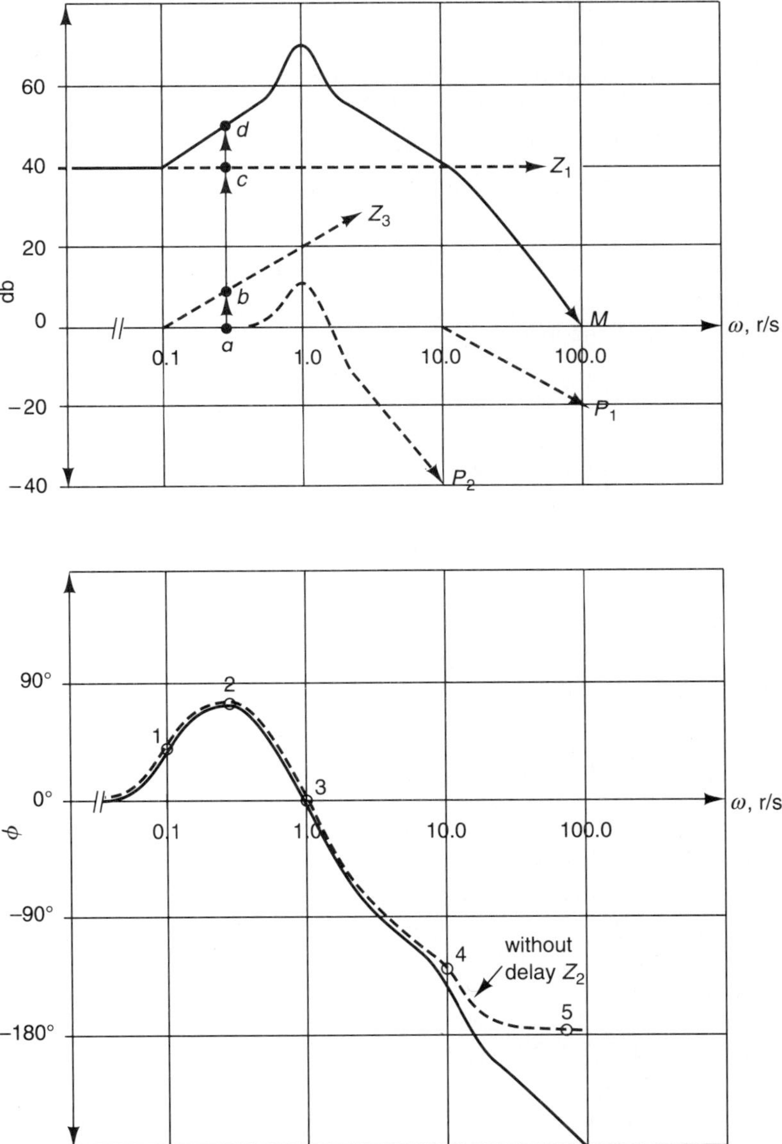

Figure 6-22. Bode Diagram of Ex. 6-5

significantly started yet. The point 3 is the break frequency of the second-order factor, which should be $-90°$ from the phase at which it started. At point 4 the last break down is adding $-45°$ from the end of the previous factor. Point 5 is $-45°$ from the last break frequency, and is settling to a constant value.

The delay is simply added to the undelayed curve, at the end of the analysis. For example, at 10 r/s, the delay factor adds in $-10(0.02) = -0.2\text{r} = -11.46°$.

The Bode diagrams for each of the factors thus shows the SISO transfer function response at any possible frequency. Moreover, since the input may also be a superposition of many frequencies, you are able to use the transfer function Bode diagrams to compute the response at any of these frequencies. Superposition gives you the forced response as discussed earlier in this section,

$$y_f(t) = \sum y_{fn}(t) = \sum M_n U_n \sin(n\omega_0 t + \alpha_n + \phi_n). \tag{6-186}$$

The MIMO Forced Response. In MIMO components you are dealing with a *matrix*, $\boldsymbol{T}(j\omega)$. Each entry in this matrix is a transfer function. This means that the transfer matrix entries are of the form

$$T_{ni}(j\omega) = Y_{fn}(j\omega)/U_i(j\omega), \tag{6-224}$$

where n and i are the indicies of the matrix locations, and all of the previous computations and plots for $T(j\omega)$ may be applied here as well.

HOMEWORK

6-18. Given $T(s) = 1/(5s + 1)$. Find $y_f(t)$ when $u(t) = 10 \sin 3t$.

6-19. Sketch the Bode diagram (magnitude and phase) for the SISO component

$$T(s) = \frac{1{,}000(10s + 1)(s + 1)}{(s^2 + 10s + 100)(100s + 1)}.$$

Sketch this function from one decade below the lowest break frequency to one decade above the highest break frequency.

6-20. List two improvements that a transfer-function model offers over a transfer-ratio model.

6-6 SUMMARY, BIBLIOGRAPHY, AND REVIEW PROBLEMS

This chapter was concerned with finding a suitable frequency-domain model for a component which is designed to operate in the frequency domain. It began with a look at the R, L, and C electronic elements and how they respond to a pure-sine input. This response was then used to introduce the transfer-ratio model, which is a sine-amplitude transfer from input to output, and a sine-phase shift from input to output. The model development was continued by studying simple networks and their transfer-ratio characteristics.

The general modeling plan-of-attack was to first develop a fundamental, time-domain model, then to find the forced response of this model in the time domain, and finally to express the frequency-domain transfer-ratio model.

These principles were then extended to more complicated components of the electrical, structural, and fluid types which are also designed to operate in the frequency domain. Again, a fundamental, time-domain model was first sought, then solved to yield the frequency-domain, transfer-ratio model.

The Laplace transform was next introduced as a helpful tool to easily transform the time-domain model into a more useful form. This transform was shown to convert the time-domain model into the s-domain, and it allowed you to treat the differential equations as if they were algebraic equations. Better yet, it allowed you to skip the step of solving the time-domain model before the transfer model was computed. This transform was thus shown to yield a transfer-function model for a SISO component, $T(s)$, and a matrix of transfer functions for a MIMO component, $\boldsymbol{T}(s)$.

The frequency-domain model of interest was found from the transfer function by simply letting $s = j\omega$ in the transfer function. Forced solutions in the frequency domain were then found by extracting the magnitude and phase from $T(j\omega)$ and applying these exactly as in the transfer-ratio model.

At the end of the chapter the Bode diagram was studied as a means to display the frequency-domain model. Bode approximation methods were studied which allow quick and easy sketching of the transfer-function, frequency-domain response.

The conclusion that you may reach from all of this is that $T(s)$ or $\boldsymbol{T}(s)$ are all that you need to know to be able to characterize a linear component since the transfer function carries both the time-domain response and the frequency-domain response in a single concise form. While the time-domain aspect of the Laplace transform was not studied in this chapter, it will be presented in Chap. 8, where the inverse Laplace transform is discussed in some detail.

In all of this, the linearity of the component is a critical assumption leading to the corresponding superposition of solutions. This is, perhaps, the greatest weakness in the transfer function model. When you begin modeling in earnest, you soon find that you can only bend this linearity assumption a little bit before solutions become untrustworthy. Again, you find yourself in conflict with the Fundamental Truth of Modeling, that all physical effects are nonlinear. When this happens, you must then return to modeling in the time domain to continue.

BIBLIOGRAPHY

Abrie, P. L. D., *The Design of Impedance-Matching Networks for Radio-Frequency and Microwave Amplifiers,* Dedham, Mass.: Artech House, Inc., 1985.

Cannon, R. H., Jr., *Dynamics of Physical Systems,* New York, N.Y.: McGraw-Hill Book Co., 1967.

Chalmers, B. and A. Williamson, *A.C. Machines,* New York, N.Y.: J. Wiley & Sons, 1991.

Hemond, C. J., jr., *Engineering Acoustics and Noise Control,* Englewood Cliffs, N.J.: Prentice-Hall Inc., 1983.

Hubert, C. I., *Electric Circuits, AC/DC an Integrated Approach,* New York, N.Y.: McGraw-Hill Book Co., 1982.

Lalanne, M., Berthier, P. and J.D. Hagopian, *Mechanical Vibrations for Engineers,* New York, N.Y.: J. Wiley & Sons, 1983.

Ogata, K., *Modern Control Engineering,* Englewood Cliffs, N.J.: Prentice-Hall Inc., 1990.

Pippard, A. B., *The Physics of Vibration,* New York, N.Y.: Cambridge Univ. Press, 1989.

Shearer, J. L., Murphy, A.T. and H.H. Richardson, *Introduction to System Dynamics,* Reading, Mass.: Addison-Wesley Publ. Co., 1967.

Smith, B. J., *Acoustics,* London, Eng.: Longman Group, Ltd., 1971.

REVIEW HOMEWORK

6-21. Given the network shown in Fig. 6-23,

a) What is the transfer ratio, i_R/i_s?

b) Plot the phase angle of the transfer ratio on a Bode diagram for $1/(10RC) \leq \omega \leq 10/(RC)$.

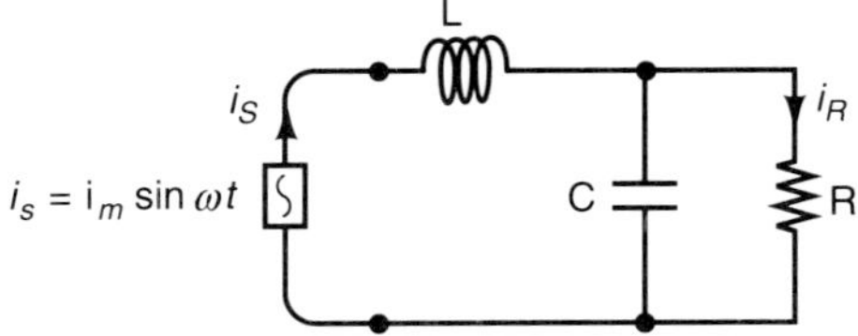

Figure 6-23. Network for Prob. 6-21

6-22. Plot the impedance transfer ratio and phase angle for the network shown in Fig. 6-24 on a Bode diagram over the range $0.01 \leq \omega \leq 1$ r/s.

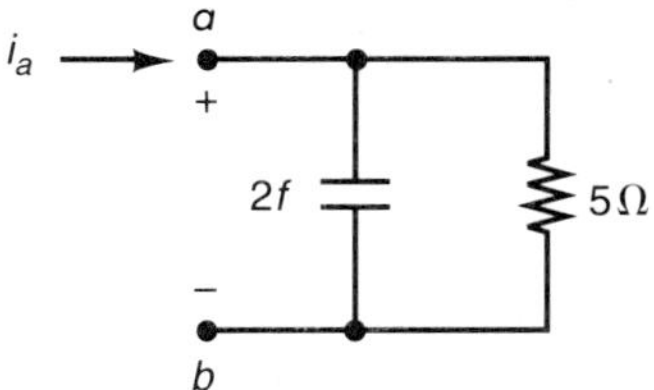

Figure 6-24. Network for Prob. 6-22

6-23. Given the port admittance transfer ratio $T_A = 0.1 \angle 10°$. What is the power consumed at this port when

$$e = 100 \sin(10t - 70°)?$$

6-24. Given the transformer network shown in Fig. 6-25,

a) What is i_a?

b) Compare your answer with that of Prob. 6-6.

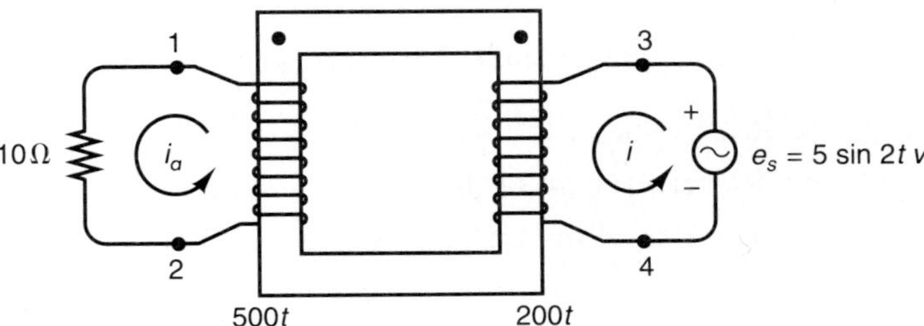

Figure 6-25. Network for Prob. 6-24

6-25. The following parameters have been measured for the motor of Fig. 6-11c

$$k_m = 1{,}000 \text{ ft-lb}_f\text{/amp},$$

$$k_c = 0.5 \text{ v-s/rad},$$

$$R = 10\ \Omega,$$

and

$$L = 1 \text{ h}.$$

How much torque will this motor produce at 10 rad/s if

$$e_c = 15 \sin 10t \text{ v}?$$

6-26. The multiport diagram in Fig. 6-26 shows a component design for an audio system. Sketch the analog circuit schematic.

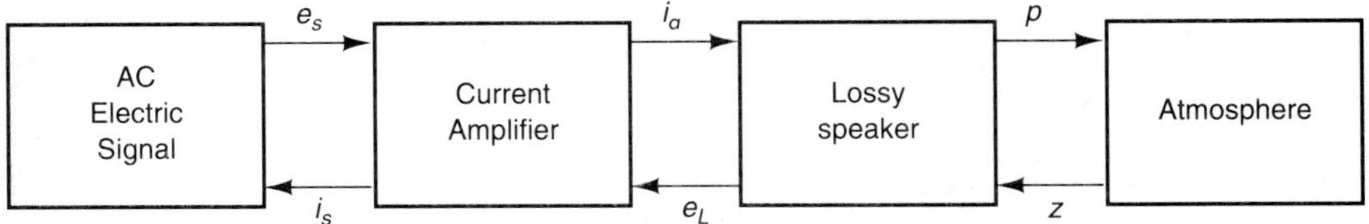

Figure 6-26. Multiport for Prob. 6-26

6-27. An electric motor is to be mounted on a vibration absorbing spring platform in a submarine after the manner shown in Fig. 6-15b. The following constants apply in this problem

$$k_2 = 1{,}000 \text{ lb}_f/\text{in},$$

$$W_1 = 200 \text{ lb}_f,$$

and

$$W_2 = 500 \text{ lb}_f.$$

Compute the net connecting-spring stiffness, k_{12}, so that highest natural frequency of the component is 1/2 of the planned operating speed of the motor (1,000 rev/min).

6-28. A sound wave is passing through the atmosphere, it is described by its level above atmospheric pressure

$$p = 3 \sin 2t + 4 \cos 2t \text{ psi.}$$

a) Compute the RMS value of this wave,

b) Compute the decibel strength of the wave.

6-29. Use the integral definition of the Laplace transform to find $\mathcal{L}[f(t)]$ if

$$f(t) = 0,\ t < 0,$$

and

$$= 3 \cos 2t,\ t \geq 0.$$

6-30. Find the transfer function for

$$C\, de/dt + (1/R)\, de/dt + (1/L)e = di/dt.$$

6-31. Find the transfer matrix for the following state-space model

$$\boldsymbol{A} = \begin{bmatrix} 0 & 1 \\ -1 & -2 \end{bmatrix},\ \boldsymbol{B} = \begin{bmatrix} 0 & 0 \\ 0 & 5 \end{bmatrix},\ \boldsymbol{C} = \begin{bmatrix} 1 & 2 \\ 3 & 6 \end{bmatrix}.$$

6-32. Given the following transfer function, find $y_f(t)$ when $u(t) = 3 \sin(3t + 0.1 \text{ rad.})$,

$$T(s) = \frac{12}{s^2 + 6s + 12}.$$

6-33. Sketch the Bode diagram of the transfer function

$$T(s) = \frac{100(10s + 1)e^{-0.001s}}{(s + 1)(s + 10)(s + 100)}.$$

Chapter 7

Component Physical Analysis: The Input-Output Approach

7-0 INTRODUCTION

You will often encounter situations where a model is needed for a component piece of hardware, yet little is known about the hardware construction. This may happen in those cases where proprietary considerations require concealment of the hardware details. And it may happen in cases where the hardware is so complicated that an element-by-element model is simply impractical or unsuitable for preliminary design. These considerations usually require you to treat the hardware as a kind of unknown "black box" for which you must find a suitable model, as suggested by Fig. 7-1.

In this chapter, it is assumed that you are given the hardware component with concealed internal details, and you need to build a model for it that is part of a larger system context.

The greatest difficulty in such black-box, input-output modeling is the unknown *internal connectivity* of the component. In fact, it is precisely this connectiv-

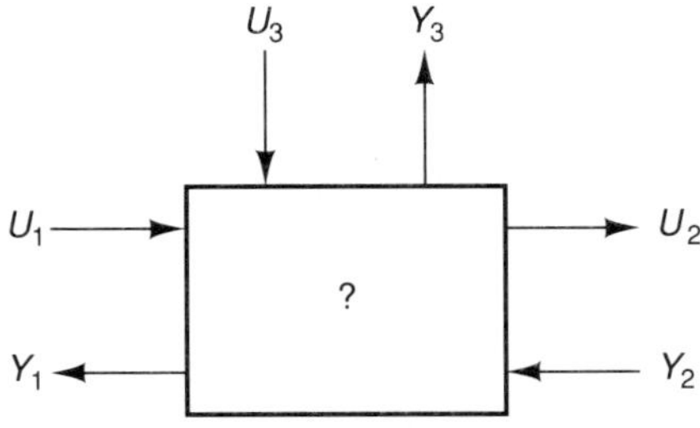

Figure 7-1. A Component with Unknown Internal Connectivity

ity which gives great power to the fundamental approaches studied in Chaps. 4 and 6. Without this connectivity you must build upon what you know to be true. That is, even if *all* the details of the hardware are not known, some concepts about its dynamics usually are known. Thus, it is always a good idea to attempt to build a fundamental model using what is known first, before input-output modeling is attempted, in order to gain some insight for the black-box approach. Input-output modeling should always be your final try for a model, your court of last resort.

Again, as in the previous physical analysis studies, the objective here is to capture the response of the component without overmodeling the dynamics. This idea is critical now because it is very easy to overmodel a component using the input-output method. You desire the simplest model possible, without undue complexity, to show the designer the important effects alone. Remember the First Law of Modeling: The simpler, the better.

The input-output model is started after the multiport diagram in order to place the component inputs and outputs within the context of the system. Once this is done, the linear, time-domain, state-space, MIMO model is of the form

$$dx/dt = \mathbf{A}\, x + \mathbf{B}\, u,$$
$$y = \mathbf{C}\, x + \mathbf{D}\, u.$$

This model has the *steady-state* outputs

$$y = [\mathbf{D} - \mathbf{C}\mathbf{A}^{-1}\mathbf{B}]u.$$

In other words, each output signal is potentially a function of all the inputs to the component in the steady state (or in the transient state, for that matter). So, starting from this observation, this chapter presents various input-output experiments to generate data for modeling purposes. This is followed by a discussion of the analyses which help you to identify the all-important characteristics of model *form* and model *fit*, which are known individually and collectively as *model identification*.

Model form has to do with the linearity of the component, its order, and a proper model equation with a set of unknown coefficients. This information comes from studying how the component experimentally responds to ideal inputs, and it may be expressed in state-space form, in transfer-function form, or in transfer-matrix form. Of course, this sort of qualitative modeling must rely heavily on your understanding from previous chapters.

Given the model form, the Least-Squared-Error method is used to precisely fit the unknown coefficients in the model to the data.

The last two sections of this chapter deal with two types of components which are so complicated that they are almost exclusively modeled with input-output methods. The first of these is the convective component which provides a means for fluid convection of thermal dynamics. The second type is the digital component. You should find these components fairly easy to model using the input-output modeling method. We begin with a discussion of the experimental approach.

7-1 EXPERIMENT DESIGN AND THE MODELING APPROACH

Obviously, to fit input-output models, you must first create input-output data. This data comes from input-output experiments which are carefully designed to reveal the nature of the component under study.

In this section, two approaches to input-output experimentation are presented, along with the error analyses which guide experimental selections and modeling evaluations, and the general procedure which is used in input-output modeling.

Experimental Approach. The two experimentally based approaches that are used to model hardware components are the *characteristic* and the *in-system* experiments. It is very important to be clear about the difference between these two experiments because they lead to very different modeling methods.

The *characteristic* experiment removes the component from its system setting. This is a form of bench testing where the component is isolated and its external ports are carefully controlled to ensure the independence of all inputs and outputs. The objective is to stimulate the inputs one at a time and measure the component's characteristic output response(s). The use of singularity inputs is common in this type of experiment, as is the use of pure-sine inputs.

The *in-system* experiment measures the performance of the component while it remains connected within its working system. Consequently, the measured component data in these tests usually has all the inputs and all the outputs varying at the same time.

The difference between the two types of tests may be illustrated graphically. Consider, for example, the two components shown in Fig. 7-2a which are connected together in an electrical system. Note that $e_1 = e_2$ and $i_1 = i_2$ from the port connection. Also note that the internal sign conventions are consistent between the port variables, and the internal details are shown in order to illustrate the effect of coupling.

The fundamental physics of the components is shown in Fig. 7-2b and c. If the internal connectivity of the components is known, then the plots shown could be found using the methods of Chap. 4 or 6 to find the *characteristic* input-output response. Or, these curves could also be measured independently using characteristic experimentation.

In Fig. 7-2d, the coupled, *in-system,* measured performance of the two components is shown. Note that the measurement of the variables in the connected port (e_p and i_p) gives data which is common to *both* characteristics. Furthermore, the physical connection of the two components amounts to mathematically combining the characteristic equations, and leaves you in a situation where the in-system measured data shown in Fig. 7-2d suggests that $e_p = Ci_p$ is the same characteristic model for *both* components at that port, yet this is an error.

So, characteristic experiments are more expensive, but they can give very detailed data and excellent models. In-system experiments are usually cheaper since the system prototype tests are run in any case, or a component may have been tested

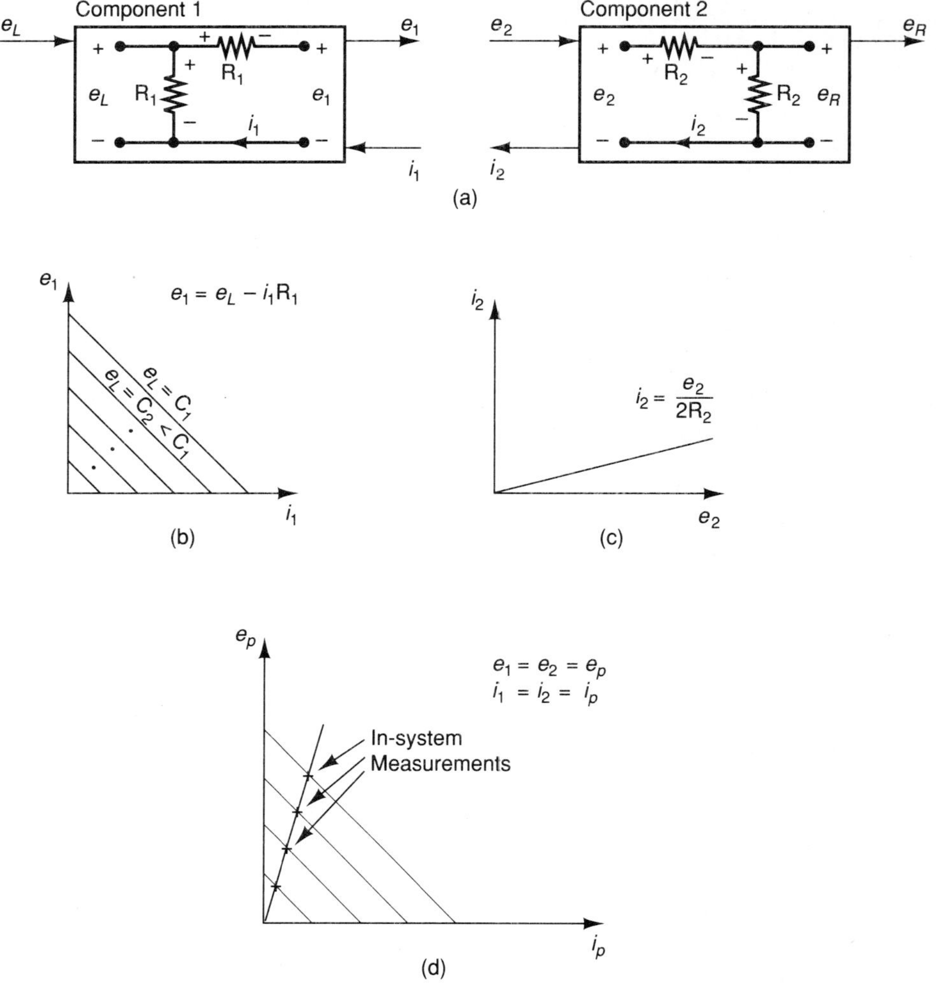

Figure 7-2. A Connected Port. (a) Schematic, (b) characteristic of 1, (c) characteristic of 2, (d) in-system data

in an earlier system. However, the in-system data requires a substantial mathematical processing using advanced statistical methods to accomplish model form and fit evaluations. Consequently, in-system experiments and data will not be discussed in this text.

In any case, you may sometimes be given some steady and transient data for a component, along with a design calling for the component to be included in a larger system, and a request for a system model. In view of the foregoing discussion, the first question to be asked in such a case is: Did the data come from a characteristic experiment or was it measured in an in-system experiment? If the former, then the methods of this text may be used. If the latter, then you must use more advanced analysis methods.

Your ability to translate the measured characteristic data into a useful model must first be guided by your understanding of the error that the data carries within it.

Experimental Error. The three types of errors which are of concern here are the measurement error, the relative error, and the engineering error.

Measurement error, e_M, is associated with the amount of error incurred in the measurement of variables. It is determined from the smallest division of the meter, ΔY, from which a measurement, Y_M, is made. That is,

$$e_M = \pm 1/2\ (\Delta Y/Y_M)\ 100\%. \tag{7-1}$$

The error formula above is a mathematical way of saying that you can only read a meter as close as its nearest, smallest division. Further, since you may make an error in estimating this, the error may be as large as $\pm 1/2$ of that smallest division, as shown in Eq. 7-1.

Of course, the measurement error assumes that the meter has been properly calibrated, so that an accurate reading is a measurement of an exact value. This is rarely true. In order to quantify this *relative gage error*, e_{GR}, you use the following relation

$$e_{GR} = [(Y_M - Y_{\text{EXACT}})/Y_{\text{EXACT}}]\ 100\%. \tag{7-2}$$

Note the difficulty in determining the exact value of the output in order to compute this relative error. Well-calibrated meters have the relative gage error tabulated for the user based upon very carefully controlled characteristic experiments.

The *relative prediction error*, e_{PR}, is often used to describe model results,

$$e_{PR} = [(Y_{\text{MODEL}} - Y_M)/Y_M]\ 100\%. \tag{7-3}$$

In this error, you compare the model output to the measured output for the *same input*.

The final type of error, the *engineering error*, e_E, helps guide you in what is acceptable in your models. That is, each of the above errors is generally present to some degree when you compare model predictions to actual measurements. Thus,

$$e_E = e_M + e_{GR} + e_{PR}. \tag{7-4}$$

Acceptable engineering error in preliminary modeling is less than about $\pm 10\%$. Typically, you give up a couple of percentage points of error in the output data for comparison. Thus, your model needs to be about $\pm 8\%$ compared to the data. This gives you a convenient measure of acceptable nonlinearity for linearization purposes, and the acceptable range of model performance.

Characteristic Modeling Procedure. The modeling procedure always moves from the set-up, to steady-data modeling, to transient-data modeling. You do the steady-state model before the transient model to eliminate the added confusion of the rate effects on steady-model form.

The *set-up* consists of choosing all the input generators and all the output measuring equipment to meet accuracy and data-processing needs. Ideal or near-ideal input generators should always be used if available. (Remember: Ideal sources show no droop upon loading.)

The selection of an input generator is guided by the requirement that the input source must have a known characteristic and be of sufficient variability so that the unknown component characteristic at that port can be found. This is made a great deal easier if a variable *ideal* source is used to generate the input since it forces the input *independent* of the component output. However, variable ideal sources are very uncommon for structural, fluid, and thermal ports. Thus, you must often settle for knowing the variable source characteristic and being cautious about the experimental procedures in mechanical modeling.

Steady output data is usually taken with meters, gages, or computers. Transient data is taken with analog strip-chart recorders or with a digital sampling recorder (e.g., a digital computer). Seven distinct steps can be identified in the modeling procedure following the set-up.

Step 1: Generate the steady-output data. The steady data is generated in the time domain or in the frequency domain, depending on the needs of the component. Care must be taken that the steady input-output data pairs are *rich* enough (e.g., adequate in terms of number and span) to give a good set of data points for form and fitting evaluations. (Specific guidelines on steady-data richness in the time domain and the frequency domain are presented throughout this chapter.)

Step 2: Choose the steady-model form. The first concern in this is to evaluate the component linearity from the steady data, be it in the steady state or in the sinusoidal steady state. These evaluations are critical in order to lay a basis for the mathematical superposition of terms in a built-up, linear, mathematical model. (Specific methods for the evaluation of component linearity using steady characteristic data are discussed at several places later in the text.)

If the steady data shows a linear trend or if a linearization is acceptable, a linear algebraic model may be used. In the time domain, such a linear steady model is of the form

$$y_1 = b_m u_m + b_{m-1} u_{m-1} + \cdots .$$

If the steady time-domain data shows a nonlinear trend that must be reproduced for accuracy reasons, then a nonlinear form must be used,

$$y_1 = f(u_m, u_{m-1}, \ldots).$$

In the frequency domain, the component *must* be linear so that you may work with the transfer function. Nonlinear frequency-domain components are not considered in this text.

Step 3: Fit the steady model to the steady data. This means that you use a numerical procedure (to be discussed in Sec. 7-4) to find the best set of coefficients for the given data. In the time domain, you are here fitting coefficients only to the right-hand side of the output equation, and there are no derivatives in the equation. In the frequency domain, you are fitting a transfer function to the steady frequency-domain data.

This step completes the frequency-domain model, but more work must be done to find the derivative terms in the time-domain model.

Step 4: Choose the transient inputs. Again, the *inputs* must now be rich enough to stimulate the component dynamics. Physically generated singularity functions are used to do this in characteristic time-domain testing. However, sometimes the experimenter may run into some difficulty when trying to generate these inputs for structural, fluid, and thermal ports. If that is the case, then a frequency-domain approach may work, using sine inputs, aimed at finding the transfer function and working backwards to the differential equation.

Step 5: Generate the transient output data. Now, the *output* data measurements must again be rich enough in terms of number and span for evaluations of model form and fitting.

Step 6: Choose the model form for the dynamics. This is done by adding time-derivative terms to either side of the steady time-domain equation found in step two. For example, supposing that the time domain component of step two shows an exponential response to a step input in a characteristic experiment, this may be modeled in form by adding the first-order derivative term,

$$\tau\, dy_1/dt + y_1 = b_m u_m + b_{m-1} u_{m-1} + \cdots .$$

Note that this equation achieves the same steady state as that in step two, only the rate term is added here to model the transient. The shape of the exponential is embodied in the time constant, τ, which is found in the last modeling step.

Step 7: Fit the time-domain dynamic equation(s) to the transient data. Sometimes, the derivative coefficients are simply estimated for the preliminary design model based on a visual inspection of the data. However, in this text, a more rigorous numerical procedure is recommended.

HOMEWORK

7-1. Given steady input-output data for an existing component which is to be part of a system model.

a) What should you first ask about the experiment(s) which were used to generate the data?

b) Why should you be concerned about output data richness?

7-2. For the meter in Fig. 7-3, what is the measurement error for this reading?

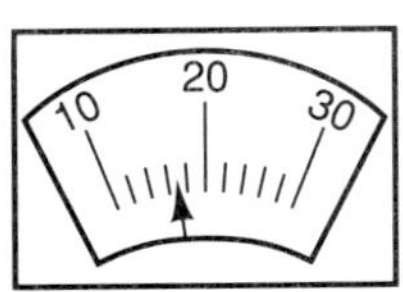

Figure 7-3. Gauge for Prob. 7-2

7-3. Why do system modelers conduct fits to steady data *before* they conduct transient experiments?

7-4. A system has 3 time-domain inputs. Their *ranges* are as follows: $U_1^* = (-2, 2)$, $U_2^* = (0, 10)$, and $U_3^* = (0, 5)$. In order to conduct a thorough steady test of the system, it is decided that these inputs should be varied in 10 equally spaced steady settings. How many distinct steady input combinations must be tested?

7-2 IDEAL STEADY-CHARACTERISTIC MODELING

Steady input-output modeling is usually begun during steady characteristic experimentation using ideal input sources. That is, the steady data from a component is usually evaluated as the steady experiment proceeds, with suitable adjustments to ensure an adequate, rich data set on completion. However, for the purposes of this text, it is assumed that all the data is generated *prior* to evaluation and it needs to be considered for model *form* selection before the model is fit to the data.

Steady Data Evaluations. These evaluations must accomplish three tasks: data richness evaluations, the plotting of the data, and data linearity evaluations.

Richness evaluations are concerned with determining if there are enough data points which are suitably distributed over the input-output range of interest to reach a valid conclusion about model form. The main concern here is one of coverage; that is, that a significant physical effect is not overlooked. Some quantitative guidelines for these evaluations are presented in the next section of the text, where the fitting of the steady model to the data is discussed.

Plotting of the input-output data is used to graphically show trends in the data. Thus, plots reveal patterns which may be difficult to recognize in the numbers alone. And, in fact, this evaluation of the data draws upon the latent pattern-recognition capability which is quite strong in human beings to build a basis for the selection of a model form.

Linearity evaluations are of particular interest during preliminary analyses. These evaluations are based on plots of input-output pairs in a rich data set, and they fall into three classes:

1. *Linear and Near Linear*. These types are characterized by a plot which is nearly linear to the eye. However, since nothing is exactly linear, the response is linear within a few percentage-error points—it is approximately linear.
2. *Nonlinear and Linearizable*. This type of response appears in some way to be nonlinear, with at least part of the response being linearizable. A "linearizable" component is one which has outputs which are single-valued functions of the input(s), and the outputs are smooth and differentiable in the linearizable region(s). Into this class falls the higher-order algebraic nonlinearities, called the *soft nonlinearities*. Many of this type of response were discussed in the study of idealized fundamental elements in Chaps. 4 and 6.

   Also included in the linearizable type of response are the neglectable discontinuities which so often accompany digital components with very small input-output resolution, and the piecewise-linearizable components with mechanical coulomb friction between sliding surfaces.
3. *Nonlinear and Nonlinearizable*. Some components, such as switches or relays, give outputs which are multiple-valued functions of the input(s); these are not linearizable. Also excluded are those regions of an input-output plot which contain nondifferentiable discontinuities. These are the so-called *hard nonlinearities*. They are not modeled in this text. They are quite evident from inspection of the steady input-output data plots.

Thus, you use inputs to provoke outputs so that you can classify the component response for further analysis and design. Note that the classification of the component is the same as the classification of its response. A linear response means a linear component and a linear model. A near-linear response means a nonlinear component, but a linear (linearizable) model. A nonlinearizable response means a nonlinear component and a nonlinear model.

You will study the building and performance of linear models in a linearizable region in this text, to meet preliminary design evaluation needs. These linear models have the most powerful analysis tools and they lead to the most general design conclusions. It is essential to thoroughly understand this sort of linear input-output modeling before going on to nonlinear systems.

Time-Domain Modeling. Whenever a component is tested in-system, it has a time-domain steady state if the system has a steady state. However, when some components are isolated for characteristic time-domain testing, they may not have a time-domain steady state. For this type of component, either the sinusoidal steady state (e.g., the frequency domain) should be used for testing and modeling or the component should be tested and modeled in-system.

For those components which do have a stand-alone time-domain steady state, you should begin by finding the steady operating point about which the component (and the model) is designed to function. This steady operating point is designated in vector form as $(\mathbf{U}_o, \mathbf{Y}_o)$, such that

$$\boldsymbol{U} = \mathbf{U}_o + \boldsymbol{u}, \tag{7-5}$$

and

$$\boldsymbol{Y} = \mathbf{Y}_o + \boldsymbol{y}. \tag{7-6}$$

Note that a data meter usually measures $\boldsymbol{U}$ and $\boldsymbol{Y}$, but your linear model must be expressed in terms of $\boldsymbol{u}$ and $\boldsymbol{y}$ in order to be consistent with the models of previous chapters, and to aid in the mathematical assembly of all the component models into a system model.

Next, use the fact that all the time derivatives of $\boldsymbol{y}$ are zero in the steady state. Thus, a nonlinear algebraic equation exists for *each* output (there are L of them, in general) in terms of *all* the inputs to the component. That is, for

$$\boldsymbol{U} = (U_1, U_2, \ldots, U_K)^T, \tag{7-7}$$

there exists an output

$$Y_i = f_i(\boldsymbol{U}). \tag{7-8}$$

In order to linearize this output equation, you expand it using a first-order Taylor series,

$$dY_i \approx (\partial f_i/\partial U_1)\, dU_1 + (\partial f_i/\partial U_2)\, dU_2 + \cdots + (\partial f_i/\partial U_K)\, dU_K. \tag{7-9}$$

Moreover, since

$$dY_i \approx \Delta Y_i = y_i, \tag{7-10}$$

and

$$dU_i \approx \Delta U_i = u_i, \tag{7-11}$$

you may write the linear, steady-state model as

$$y_i \approx (\partial f_i/\partial U_1)\, u_1 + (\partial f_i/\partial U_2)\, u_2 + \cdots + (\partial f_i/\partial U_K)\, u_K. \tag{7-12}$$

Note that each of the partial derivatives in Eq. 7-12 is really the sensitivity of the output to an input *while all the other inputs are held constant* at the steady-state operating point. This meaning is carried along with the nature of the partial derivative, and it must guide you in your experimental work and data evaluations.

Example 7-1

A time-domain electrical component with three ports has the steady operating point

$$\mathrm{U}_{o1} = 3 \text{ v},\ \mathrm{U}_{o2} = 2 \text{ v},\ \mathrm{U}_{o3} = 6 \text{ v},$$

$$\mathrm{Y}_{o1} = 1 \text{ a},\ \mathrm{Y}_{o2} = 3 \text{ a},\ \mathrm{Y}_{o3} = 10 \text{ a}.$$

The steady data of Fig. 7-4 was taken by holding U_{o1} and U_{o2} constant and varying $U_3 = U_{o3} + u_3$. The plots are of the outputs after they reached steady state.

Find the sensitivities which this data expresses.

Solution

First, note the data richness which gives confidence that the input-output physical phenomena are thoroughly covered by input-output measurement pairs.

A straight line has been sketched in for each of the plots near zero. These lines do not pass exactly through all of the data points, yet they capture the linear trend of the data near zero input. Such lines are called an "eyeball best-fit" to the data.

The sensitivities of the lines are

from Fig. 7-4a, $\partial f_3/\partial U_3 \approx \Delta y_3/\Delta u_3 = 1$,
from Fig. 7-4b, $\partial f_2/\partial U_3 \approx \Delta y_2/\Delta u_3 = 2$,
and from Fig. 7-4c, $\partial f_1/\partial U_3 \approx \Delta y_1/\Delta u_3 = -1$.

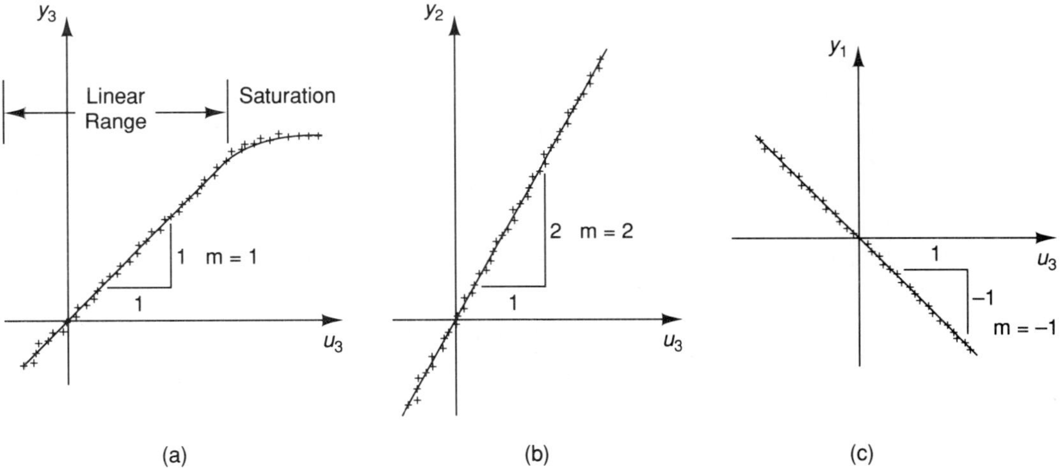

Figure 7-4. Sample Steady-state Plots for u_3 Variations. (a) y_3, (b) y_2, (c) y_1

Comments

It is critical that the component be allowed to come to steady state before each of the steady data points is taken. The disappearance of the time derivatives is critical to this method.

The eyeball-fits of the data are often surprisingly good. In fact, you will often use them preferentially in preliminary design since greater accuracy is seldom warranted at that time. Should more accuracy be desired, you may use a method to find the straight line which minimizes the net error associated with missing the data points. This approach is called the Least-Squared-Error fit to the data, it is discussed in the next section of this chapter.

Note the region of saturation in Fig. 7-4a. This is very common in all components, it bounds the range of linearity as shown in the figure, and is determined by acceptable engineering accuracy.

Frequency-Domain Modeling. The first consideration in the frequency domain is that the sinusoidal steady state linearization point must be identified for the component and the model. For clarity, the component is now assumed to be a SISO component, but the results are later generalized into the MIMO form. The linearization point thus is a linearization frequency, ω_o, and a single input linearization amplitude, U_o.

As studied in the previous chapter, the frequency-domain transfer function model that you now seek rests heavily on the property of superposition. Thus, you must establish the linearity of the component using data in the vicinity of U_o and ω_o before you try to build the transfer-function model.

As in time-domain modeling, the first evaluations are the richness of the data and the plots which are used to select model form. Four data plots are used to evaluate the frequency-domain linearity:

1. The plot of the Y-versus-time *waveform* for a sine input at U_o and ω_o. You know that a linear component produces a sine output for a sine input. So, if the output waveform for a sine input is full of lumps, notches, and other irregularities, then you know it is a nonlinear component. However, care must be taken in this because the component may be *close enough* to linear that a linear model will work well. The remaining plots help you to answer this concern.

2. The plot of output-versus-input *amplitude* at the linearization frequency. This sort of plot is shown in Fig. 7-5a. In this test, the frequency of the input is held constant at the linearization frequency, ω_o, while the input amplitude, u_m, is varied about U_o. Again, the component exhibits saturation when the input amplitude becomes excessive.

3. The plot of output-versus-input *frequency* near the linearization point. This plot is shown in Fig. 7-5b. Here the input amplitude is held constant at U_o as the input frequency is varied around the linearization frequency. In a linear component, the input and output frequencies are the same, which shows up as a slope of 1:1 on this plot.

4. The Bode diagram of the *Fourier analysis* of the output response to a pure-sine input. This plot is created by letting $U = U_o \sin\omega_o t$ and measuring the output waveform. The output waveform is then analyzed by the methods of Fourier in order to calculate the strength of the harmonics. The strength of these harmonics guides your evaluation.

For example, if you observe the periodic output of Fig. 7-5c in response to an input sine wave, then you may wonder if you can ignore these harmonics and linearize the component by simply neglecting the nonlinearity. The test for this is that the largest amplitude of any higher harmonic, y_{mn}, must be small compared to the amplitude of the fundamental frequency, y_{mo}. Numerically,

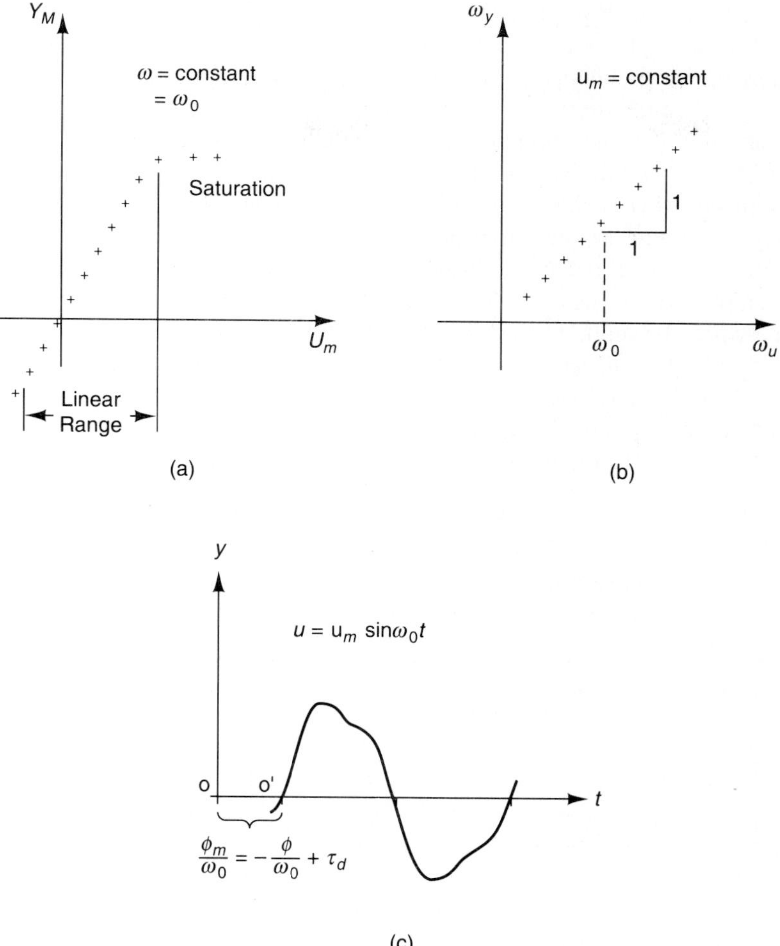

Figure 7-5. Sample Steady Frequency-domain Plots. (a) Amplitude, (b) frequency, (c) shape, phase and delay

$$y_{mn} \leq 0.05\, y_{mo}. \tag{7-13}$$

If the test of Eq. 7-13 is true, then you can usually ignore all higher harmonics in favor of the fundamental frequency alone, and a linearization will work quite well.

Given these data and component evaluations, the main part of frequency-domain modeling is in choosing the *form* of the transfer-function model according to the magnitude-transfer and phase data on a Bode diagram. Note that this data is taken by letting the input be $U = U_o \sin\omega t$, varying the input frequency in the vicinity of ω_o, and recording the amplitude and phase of the output, y. The Bode diagram is then drawn after some slight calculations to compute the magnitude of the transfer function in decibels.

The transfer-function form is next evaluated by working on the transfer-magnitude plot, starting at the low-frequency end of the plot, and looking for the transfer-function factors which were discussed in the last chapter. These are located by fitting straight-line asymptotes to the plot in increments of ±20 db/dec., or ±40 db/dec., and allowing for the damping in second-order factors as you go.

Once the magnitude factors are chosen, the no-delay form of the transfer function is set. Thus, a corresponding no-delay phase may be sketched on the measured-phase plots in order to further check the linearity of the component. If the component is linear, then good agreement should exist at low frequencies between the no-delay phase and the measured phase.

The component delay is indicated by the phase shift at high frequencies. This is shown graphically in the time domain in Fig. 7-5c. Here the $t = 0$ axis (point O) is the time when the input cycle starts, and the time when the output cycle starts is indicated by O'. The difficulty is that the time $O - O'$ is partly due to the delay of the component and partly due to the phase shift of the component. In any case, these constants are easily separated using the experimental data on a Bode phase-shift diagram, and looking at the high-end of the frequency range.

Example 7-2

Given the measured response shown in Fig. 7-6, and knowledge that the component is linear, determine the transfer function of the component.

Solution

The first thing to note about the measured responses in Fig. 7-6 is that there are no

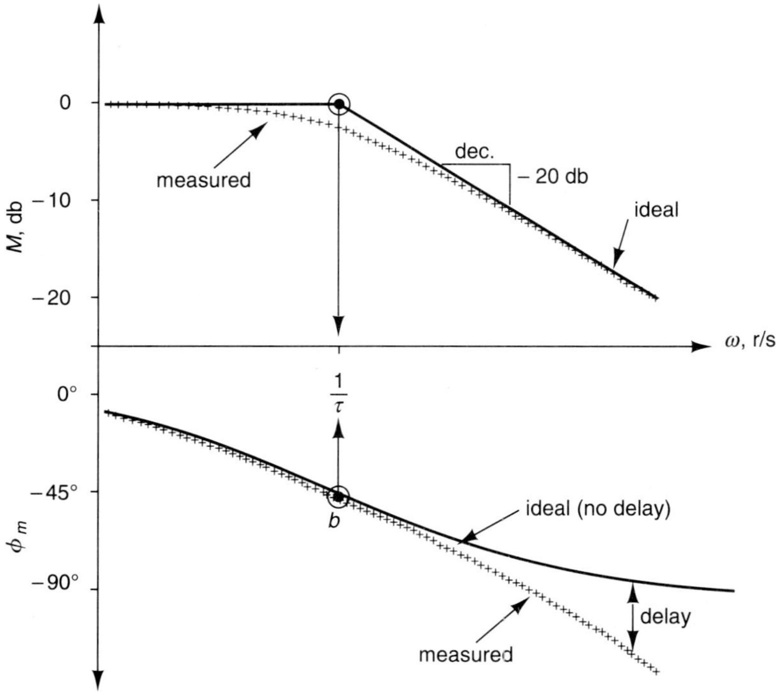

Figure 7-6. Example Modeling in the Frequency Domain

obvious break frequencies in the measured data. This is true in general, and the frequency data usually presents a smooth and rounded contour. Consequently, you must first locate the break frequency by fitting a straight-line asymptote to the amplitude data at a slope of -20 db/dec. This locates an approximation to the break frequency, $1/\tau$, for the transfer-function model and leads to the form

$$T(s) = 1/(\tau s + 1). \tag{7-14a}$$

The next thing that you must do is to evaluate the measured-phase data. This is done by first sketching in the ideal no-delay phase line (e.g., the ideal phase without delay), as shown in the figure. This shows excellent agreement everywhere with the measured data except at high frequencies, where a delay begins to exert itself. Thus, you know that the model form must be

$$T(s) = e^{-Cs}/(\tau s + 1), \tag{7-14b}$$

where the exponential factor is added to account for the delay.

Comments

Note that the ideal no-delay phase angle may often be estimated directly from the magnitude plot (as long as the break frequencies are well separated), as shown in the figure. Thus, the points at a break frequency (e.g., *a-b* in the figure) must be in agreement with each other, and the divergence of the measured data from the ideal no-delay data shows the effect of the delay.

In order to assemble a transfer matrix for a MIMO component, the above methods need to be applied to each input-output pair, as each input is independently and sinusoidally varied. Following this, the matrix is formed. That is, for the MIMO component,

$$\mathbf{y} = \mathbf{T}(j\omega)\, u, \tag{7-15}$$

which is, for the L^{TH} output,

$$y_L = T_{L1}(j\omega)u_1 + T_{L2}(j\omega)u_2 + \cdots + T_{LK}(j\omega)u_K. \tag{7-16}$$

You thus identify the transfer functions in each of the output equations, one input at a time. You do this by first choosing the steady linearization point where *all* the **u** and **y** are zero. You then choose *one* u and vary it sinusoidally (u_m fixed and ω varying for each data point). You measure *all* the outputs (M and ϕ), plot the Bode diagrams and choose a form for each of the transfer functions. In this way, you generate data to identify all the transfer functions in each *column* of $\mathbf{T}(j\omega)$ which correspond to each input to the component acting alone.

HOMEWORK

7-5. The following steady-state characteristic data has been measured for a two-input, two-output component:

$u_1 = 0$		
u_2	y_1	y_2
0	0	0
1	1	1
2	4	2
3	9	3
4	16	4

$u_2 = 0$		
u_1	y_1	y_2
0	0	0
1	2	2
2	4	8
3	6	18
4	8	32

a) Conduct an "eyeball-fit" of a straight line to the data such that the line stays within $\pm 10\%$ of the data near zero.

b) What ranges of the inputs are permissible to stay within the error of (a)?

c) What is your linear steady-state model?

7-6. A certain frequency-domain SISO component has the measured characteristic output response shown in Fig. 7-7. The amplitude response at constant frequency is shown at the left and the frequency response at constant amplitude is shown on the right.

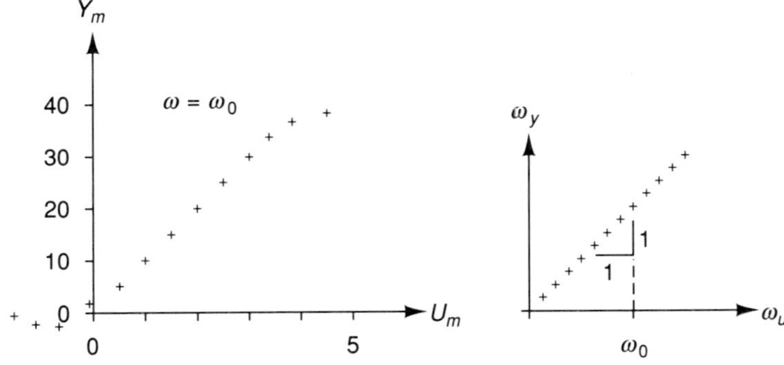

Figure 7-7. Frequency Data for Prob. 7-6

a) Over what approximate range in input amplitude will superposition hold?

b) What input linearization amplitude would you recommend for a Bode diagram?

c) What sort of agreement are you looking for in the Bode diagram as a check on linearity?

7-7. Why is it incorrect to simply try to measure ideal phase shift from the input and output time histories? (e.g., what confusing factor besides phase angle may also be in the data, especially at high frequencies?)

7-8. Which of the steady time-domain characteristic responses shown in Fig. 7-8 are linearizable near the origin. If they are not, explain why not.

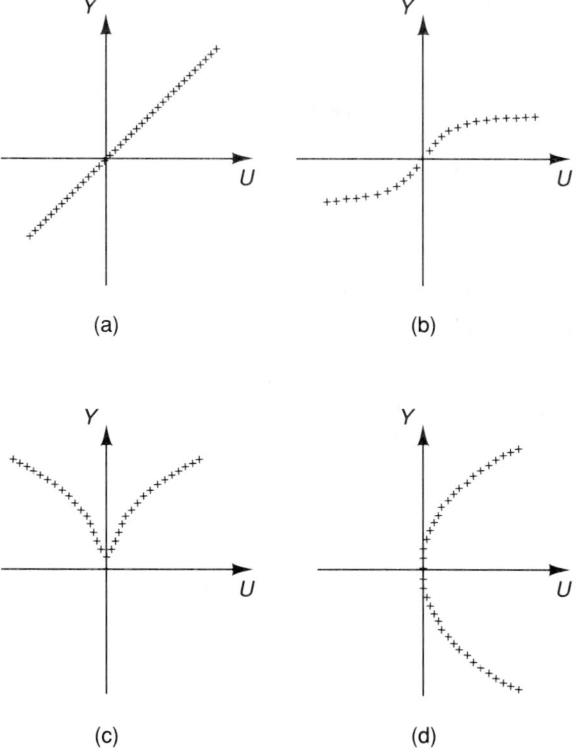

Figure 7-8. Time-domain Data for Prob. 7-8

7-3 THE LEAST-SQUARED-ERROR (LSE) FITTING METHOD

Many methods exist to fit equation models to data. None of these methods works all the time, but the LSE method is, perhaps, the most successful and widely used.

The LSE method is used after you have determined the *form* of your model from the qualitative examination of the steady experimental data. This qualitative model form must be expressed in terms of a number of unknown, constant coefficients. The LSE method will then size the coefficients in the model using the measured output data such that the model output has minimum error compared to the data, provided that a few requirements are first met.

In this section the general theory of Least-Squared-Error data fitting is presented along with its application to the fitting of steady data for both time-domain and frequency-domain component models.

Fitting Requirements. The LSE method rests on two requirements: first, the model equation to be fit must be linear in its *coefficients*; and second, the data must be *rich* enough to provide an accurate fit.

All algebraic functions of the form

$$Y = \Sigma \mathrm{B}_m U^m, \tag{7-17}$$

are acceptable for fitting since these expressions for Y are linear in their coefficients. For example, the quadratic equation

$$Y = \mathrm{B}_0 + \mathrm{B}_1 U + \mathrm{B}_2 U^2, \tag{7-18}$$

is linear in its coefficients and can be fit to a set of data with the LSE method.

Note that a change of variable can sometimes convert an improper function for fitting into a proper one. That is, suppose

$$Y = \mathrm{B}_0 U^{B_1}, \tag{7-19}$$

which is not linear in the unknown constants B_0 and B_1. However, if you take the log-base-10 of both sides you find

$$\log_{10} Y = \log_{10} \mathrm{B}_0 + \mathrm{B}_1 \log_{10} U. \tag{7-20}$$

Further, if you redefine

$$\log_{10} Y = y,$$

$$\log_{10} \mathrm{B}_o = \mathrm{b}_0,$$

$$\log_{10} U = u,$$

and

$$\mathrm{B}_1 = \mathrm{b}_1.$$

Then Eq. 7-20 is converted into the form

$$y = \mathrm{b}_0 + \mathrm{b}_1 u, \tag{7-21}$$

which is proper for LSE fitting. (Note, however, that the data needs to be processed according to the redefinitions before the fit takes place.)

Also, before the fit can take place, it is important to have a *rich* set of input-output data. This means that there be enough data points to ensure that the form is correct before the coefficients are calculated. For example, for a linear algebraic equation in one independent variable, only two data points are necessary to compute the two coefficients in the equation (e.g., Eq. 7-21). Yet, only two data points are insufficient to be able to characterize any operating range as linear. Ten evenly spaced data points over the same range would be more desirable, and 100 evenly spaced points would be richer still. Your objective is to find the "best" straight line through the 100 data points. In this text, the best line is the one with the least error through the rich data set.

The LSE Approach. The approach which follows is for a function of a single independent variable; that is, it is for a SISO model. This simple case is used here to present the theory; it is generalized later for the MIMO case.

Suppose that 20 steady-state data measurements are taken as shown in Fig. 7-9. A glance at the data suggests that a linear fit would work well since that is the general trend of the data. But, since there are 20 data points, a 19TH order polynomial could be fit *exactly* through the 20 points. However, that would violate the "simpler is better" rule of modeling and would confuse the basic trend of the response to someone who only has access to the model—a designer, for example. Thus, modelers start with low-order polynomials in their modeling, and they only add higher-order terms if they are required to do so to capture the overall data trend. Consequently, you should always question the need for high-order polynomial fits to data, *especially* during preliminary design.

As shown in Fig. 7-9, there is generally some error, e_i, between a predicted output, y_i, and a measured output, y_{mi}, at the same input, u_i, such that

$$e_i = y_i - y_{mi}. \tag{7-22}$$

When the linear model is inserted into this, you find

$$e_i = (\mathrm{b}_0 + \mathrm{b}_1 u_i) - y_{mi}. \tag{7-23}$$

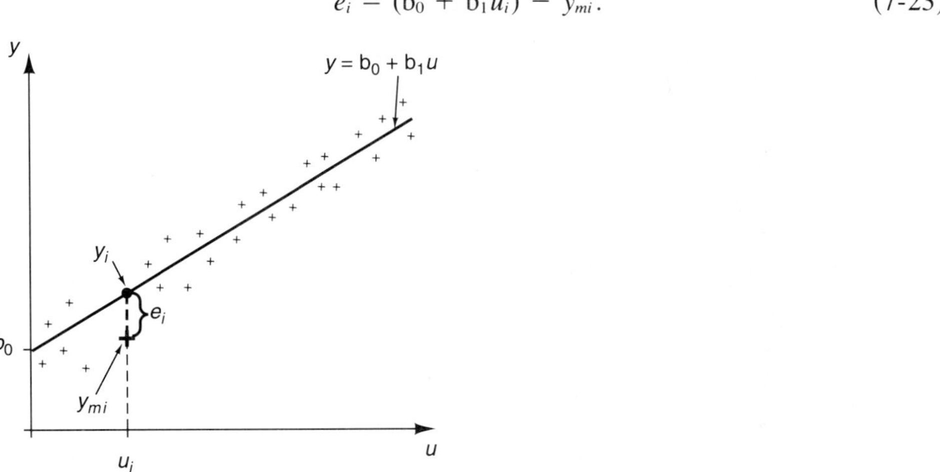

Figure 7-9. Steady-state Linear LSE Fitting

Note that if you simply add up the errors at this point, some are positive, and some are negative. Adding now thus leads to cancellation of errors, and some information about the fit is lost. On the other hand, if you *square* the errors before you add them, then no error information is lost to cancellation.

The squared error is thus

$$(e_i)^2 = \mathrm{b}_0^2 + \mathrm{b}_1^2 u_i^2 + y_{mi}^2 + 2\mathrm{b}_0\mathrm{b}_1 u_i - 2\mathrm{b}_0 y_{mi} - 2\mathrm{b}_1 u_i y_{mi}. \tag{7-24}$$

The net error of concern now is called the mean-squared-error, E. It is defined by

$$E = (1/N)\ \Sigma(e_i)^2, \tag{7-25}$$

where N is the total number of data points and the summation goes from $i = 1$ to N. E is thus the average, or mean, of the sum of the errors squared.

When the equation for the individual errors is inserted into the mean error,

$$E = b_0^2 + (b_1^2/N)\ \Sigma u_i^2 + (1/N)\ \Sigma y_{mi}{}^2 + (2b_0 b_1/N)\ \Sigma u_i - (2b_0/N)\ \Sigma y_{mi} - (2b_1/N)\ \Sigma u_i y_{mi}. \tag{7-26}$$

Equation 7-26 is the equation which must be minimized to give the least net error in predicted output values. Notice that N and all the summations may be computed from the given data points. So, the only unknowns in Eq. 7-26 are the two coefficients, b_0 and b_1. Thus, since the error is squared, $\partial E/\partial b_0 = 0$ and $\partial E/\partial b_1 = 0$ force the *minimization* of E. This gives

$$\partial E/\partial b_0 = 0 = 2b_0 + (2b_1/N)\ \Sigma u_i - (2/N)\ \Sigma y_{mi}, \tag{7-27}$$

or

$$b_0 N + b_1\ \Sigma u_i = \Sigma y_{mi}. \tag{7-28}$$

Similarly,

$$\partial E/\partial b_1 = 0 = (2b_1/N)\ \Sigma u_i^2 + (2b_0/N)\ \Sigma u_i - (2/N)\ \Sigma u_i y_{mi}, \tag{7-29}$$

or

$$b_0\ \Sigma u_i + b_1\ \Sigma u_i^2 = \Sigma u_i y_{mi}. \tag{7-30}$$

Equations 7-28 and 7-30 thus make two linear equations in the two unknowns, b_0 and b_1. Further, letting $\mathbf{b} = [b_0\ b_1]^T$, then these two equations for the linear model fit may be written in the vector form

$$\mathbf{A\ b} = \mathbf{C}, \tag{7-31a}$$

with

$$\mathbf{A} = \begin{bmatrix} N & \sum_{i=1}^{N} u_i \\ \sum_{i=1}^{N} u_i & \sum_{i=1}^{N} u_i^2 \end{bmatrix}, \qquad \mathbf{C} = \begin{bmatrix} \sum_{i=1}^{N} y_{mi} \\ \sum_{i=1}^{N} u_i y_{mi} \end{bmatrix}.$$

Notice that the general LSE method thus creates a 2×2 **A** matrix for a linear fit. It will create a 3×3 **A** matrix for a quadratic fit, a 4×4 for a trinomial, and so on. The solution for the coefficients is

$$\mathbf{b} = \mathbf{A}^{-1}\,\mathbf{C}. \tag{7-31b}$$

Thus, you must be able to invert the **A** matrix in order to find the **b** vector.

Recall that the **A** matrix is invertible if its determinant is nonzero (when **A** is also called "nonsingular"). However, this is really a remark about the richness of the given data. In other words, if the data is not rich enough, then the determinant of **A** is zero (**A** is "singular"), and no solution exists for the coefficients (the model does not exist). However, if you take N much larger than the order of the polynomial to be fit, and measure y at N evenly spaced u values, then the data will be rich enough for fitting and the model will exist.

Validation of the Fit. It is important to always validate a proposed model with experimental data. In this validation, there are many mathematical tools which can be used to quantify the goodness of fit (e.g., error distribution, net error, etc.). However, in this text a plot of the model with the data is used as the single tool to validate the fit. Such a plot contains a great deal of quantitative and qualitative information about the goodness of fit which may only be concealed by mathematical goodness-of-fit expressions. Fig. 7-9 is such a plot. Again, this is done to draw on your latent ability to recognize the pattern of a well-fitted polynomial.

Example 7-3

Given the five measured data points

$$(u, y_m) = (0, 0), (1, 1), (2, 4), (3, 9), (4, 16),$$

which perfectly fit the quadratic $y_m = x^2$. Find the straight line through these points which has the least-squared error.

Solution

First compute the summation values for the five data points ($N = 5$)

$$\Sigma u_i = 10, \qquad \Sigma u_i^2 = 30,$$

$$\Sigma y_{mi} = 30, \qquad \Sigma u_i y_{mi} = 100.$$

Next form the matrix solution,

$$\begin{bmatrix} 5 & 10 \\ 10 & 30 \end{bmatrix} \begin{bmatrix} \mathrm{b}_0 \\ \mathrm{b}_1 \end{bmatrix} = \begin{bmatrix} 30 \\ 100 \end{bmatrix},$$

which you can use to find the b values. Solving gives,

$$\begin{bmatrix} \mathrm{b}_0 \\ \mathrm{b}_1 \end{bmatrix} = \begin{bmatrix} 5 & 10 \\ 10 & 30 \end{bmatrix}^{-1} \begin{bmatrix} 30 \\ 100 \end{bmatrix}.$$

Expanding the inverse in the last equation gives

$$\begin{bmatrix} \mathrm{b}_0 \\ \mathrm{b}_1 \end{bmatrix} = \frac{1}{50} \begin{bmatrix} 30 & -10 \\ -10 & 5 \end{bmatrix} \begin{bmatrix} 30 \\ 100 \end{bmatrix}.$$

Multiplying the last equation out gives the coefficients for the answer,

$$y = -2 + 4u.$$

The validation plot is given in Fig. 7-10.

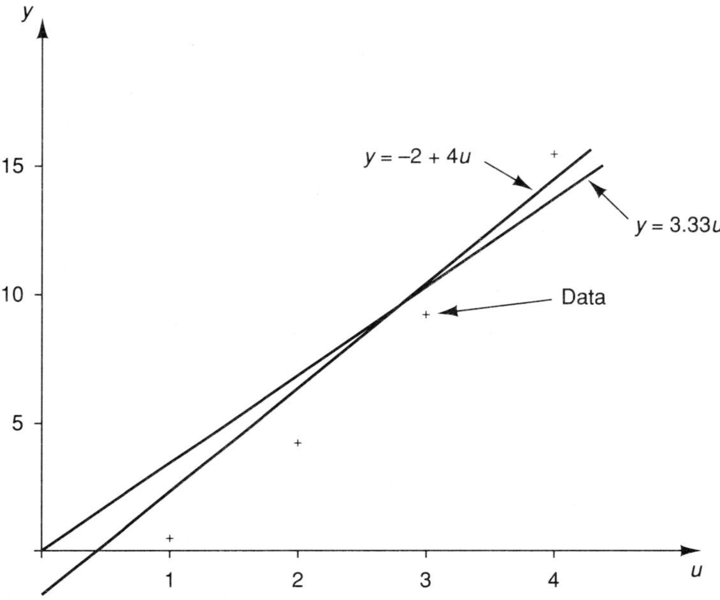

Figure 7-10. Example LSE Linear Fits

Comments

Note that the LSE fit above does go through the data field in an averaging manner which seems correct to the eye.

However, also note that the fit found above does *not* go through the origin. This is a serious problem since the linearizations discussed in Chaps. 4 and 6 are specifically chosen to do this. That is, it is important that $y = 0$ when $u = 0$ in a time-domain component model in order to account for the system steady state. Thus, you need to check the model form to ensure a correct time-domain *component* model. The other straight line in Fig. 7-10 is for one of these special component models—they are discussed next.

Steady Time-domain Component Fitting. The simplest form for a steady, SISO, time-domain component model is

$$y = b_1 u. \tag{7-32}$$

Note that this form forces the model to give $y = 0$ when $u = 0$. The theory for this fit is as follows, beginning with the error at the data points,

$$e_i = b_1 u_i - y_{mi}. \tag{7-33}$$

The error-squared thus becomes

$$e_i^2 = b_1^2 u_i^2 + y_{mi}^2 - 2b_1 u_i y_{mi}. \tag{7-34}$$

The mean-squared-error follows directly,

$$E = (b_1^2/N)\,\Sigma u_i^2 + (1/N)\,\Sigma y_{mi}^2 - (2b_1/N)\,\Sigma u_i y_{mi}. \tag{7-35}$$

Now, since E is a function of only one unknown, b_1, the minimization of E is accomplished by $dE/db_1 = 0$. Or,

$$dE/db_1 = 0 = (2b_1/N)\,\Sigma u_i{}^2 - (2/N)\,\Sigma u_i y_{mi}. \tag{7-36}$$

Solving this gives

$$b_1 = \Sigma u_i y_{mi}/\Sigma u_i. \tag{7-37}$$

Using the data of the last example in Eq. 7-37 gives the line shown in Fig. 7-10.

Occasionally, the curvature of the steady-state output data is so severe over the range of inputs that a nonlinear model form must be fit for accuracy reasons. These nonlinear polynomials may be quadratics, or cubics, or other high-order functions. Again, the steady-component polynomial model must pass through the u-y origin to accurately model the system steady state, and it must have certain attributes to give a good component model.

There are two possible forms of the quadratic model that may be used to force a passage through the origin. The first form is

$$y = b_1 u + b_2 u^2. \tag{7-38}$$

This form has the LSE matrix equation,

$$\begin{bmatrix} \sum_{i=1}^{N} u_i^2 & \sum_{i=1}^{N} u_i^3 \\ \sum_{i=1}^{N} u_i^3 & \sum_{i=1}^{N} u_i^4 \end{bmatrix} \begin{bmatrix} b_1 \\ b_2 \end{bmatrix} = \begin{bmatrix} \sum_{i=1}^{N} u_i y_{mi} \\ \sum_{i=1}^{N} u_i^2 y_{mi} \end{bmatrix}. \tag{7-39}$$

The other possible form for the component quadratic fit is

$$y = b_1 u^2. \tag{7-40}$$

This has the fitting equation

$$b_1 = \Sigma u_i^2 y_{mi} / \Sigma u_i^4. \tag{7-41}$$

The main difference between the two forms of the quadratic fit is that Eq. 7-38 may allow the predicted output to be negative for positive inputs when this is not intended. A validation plot of the predicted response is critical to detect this problem.

Often an asymmetric curvature fit is desired, say to fit saturation data. A common model for this type of curvature is

$$y = b_1 u + b_3 u^3. \tag{7-42}$$

Notice that this polynomial has no u^2 term in it, thus giving the necessary asymmetry, and thus doing a good job of modeling saturation in steady-state data (with b_1 positive and b_3 negative).

The LSE matrix equation that goes with Eq. 7-42 is

$$\begin{bmatrix} \sum_{i=1}^{N} u_i^2 & \sum_{i=1}^{N} u_i^4 \\ \sum_{i=1}^{N} u_i^4 & \sum_{i=1}^{N} u_i^6 \end{bmatrix} \begin{bmatrix} b_1 \\ b_3 \end{bmatrix} = \begin{bmatrix} \sum_{i=1}^{N} u_i y_{mi} \\ \sum_{i=1}^{N} u_i^3 y_{mi} \end{bmatrix}. \tag{7-43}$$

MIMO fitting is very similar to SISO fitting, with the addition that you must be sure that *all* inputs have been properly and sufficiently varied during the taking of the steady-state data. If this is not true, then the LSE matrix inversion may be singular (zero determinant), and/or the fit will not represent the true component performance. Validation plots again are critical.

For the given linear form

$$y = b_1 u_1 + b_2 u_2. \tag{7-44}$$

The error at each data point now becomes

$$e_i = b_1 u_{1i} + b_2 u_{2i} - y_{mi}. \tag{7-45}$$

And this leads to an LSE matrix equation

$$\begin{bmatrix} \sum_{i=1}^{N} u_{1i}^2 & \sum_{i=1}^{N} u_{1i} u_{2i} \\ \sum_{i=1}^{N} u_{1i} u_{2i} & \sum_{i=1}^{N} u_{2i}^2 \end{bmatrix} \begin{bmatrix} b_1 \\ b_2 \end{bmatrix} = \begin{bmatrix} \sum_{i=1}^{N} u_{1i} y_{mi} \\ \sum_{i=1}^{N} u_{2i} y_{mi} \end{bmatrix}. \tag{7-46}$$

Of course, you could also fit a higher-order polynomial for a MIMO model, but this is really too complicated to be very instructive. The basic method is the same as that for the SISO component.

Steady Frequency-domain Component Fitting. In the discussion which follows, it is assumed that the frequency-domain component was first checked for linearity. This linearity allows you to fit a transfer function to the data once you are assured of the richness of the frequency-domain data. The development which follows assumes that the form of the transfer function has already been selected.

The richness determination of the frequency data comes from the observation that you need sufficient points to fit asymptotes and second-order resonant peaks. In this, it is assumed that both magnitude and phase data were used to find the transfer function model form. Before LSE fitting of the model, however, the frequency data should be checked for richness: it should be well-spaced at high frequencies and more closely grouped at lower frequencies. This is done because the fit is accomplished on a log scale of frequency. A maximum-spacing rule of thumb for the data is

$$\Delta\omega \leq \omega/10, \tag{7-47}$$

where $\Delta\omega$ is the spacing between data points in the vicinity of the frequency ω, and with more closely spaced data points taken in the vicinity of any resonant peaks.

The LSE method is used to fit transfer function models to frequency-domain data, working entirely with the magnitude, M, data. This is possible since the component phase data was already used to verify the model form, as in the last section of the text. For example, in the transfer function

$$T(s) = \frac{\mathrm{B}\left(\dfrac{s}{\omega_a} + 1\right)}{s\left(\dfrac{s}{\omega_b} + 1\right)\left(\dfrac{s^2}{\omega_n^2} + 2\zeta\dfrac{s}{\omega_n} + 1\right)}, \tag{7-48}$$

the magnitude evaluation of $T(j\omega)$ gives

$$\begin{aligned} M = {} & 20 \log_{10} (\mathrm{B}/\omega) \\ & + 20 \log_{10} [(\omega/\omega_a)^2 + 1]^{1/2} \\ & + 20 \log_{10} [(\omega/\omega_b)^2 + 1]^{-1/2} \\ & + 20 \log_{10} [(1 - \{\omega/\omega_n\}^2)^2 + (2\zeta\omega/\omega_n)^2]^{-1/2}. \end{aligned} \tag{7-49}$$

The LSE fitting of the transfer function thus proceeds from the left of the Bode-magnitude diagram to the right; that is, from low frequencies to high. Notice that the amplitude constant, B, *cannot* be determined from steady *time-domain* data since the steady-state does not occur when this component is isolated due to the s factor in the denominator. In any case, the magnitude data which goes with this transfer function and the fitting of Eq. 7-49 may look something like that shown in Fig. 7-11.

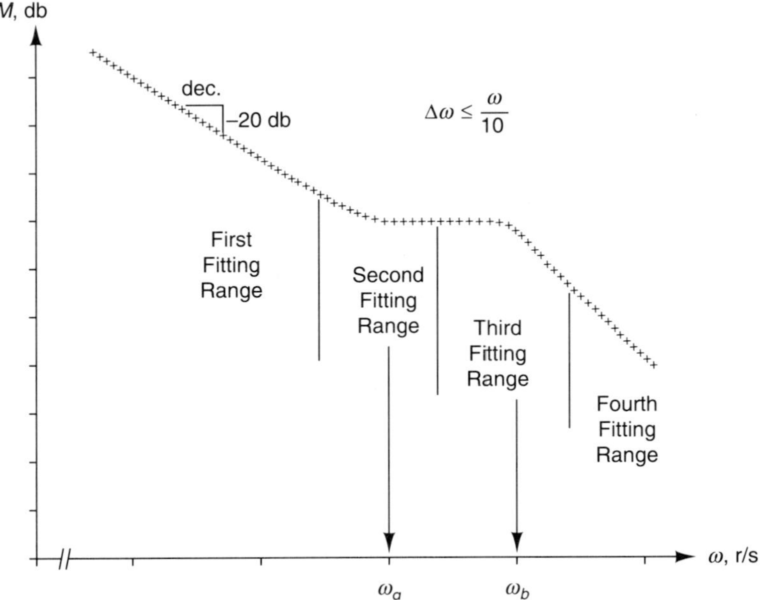

Figure 7-11. LSE Frequency-domain Fitting

The first fitting range contains data in the lowest frequencies. In this range for the example

$$M_i = 20 \log_{10} (\mathrm{B}/\omega_i), \tag{7-50}$$

$$= 20 \log_{10} \mathrm{B} - 20 \log_{10} \omega_i. \tag{7-51}$$

Notice that the factors from the higher frequencies do not exert any significant influence over this factor. This observation leads to the general plan of attack of working from the lowest frequency to the highest, and quantifying the model through a series of fitting ranges.

The error between the low-frequency factor and the data is

$$e_i = M_i - M_{mi},$$

or

$$= (20 \log_{10} \mathrm{B} - 20 \log_{10} \omega_i) - M_{mi}. \tag{7-52}$$

Recall that the magnitude, M_{mi}, is the decibel value of the ratio of the amplitudes of the measured input and output sine waves at the frequency ω_i. Thus,

$$M_{mi} = 20 \log_{10}(y_{mi}/u_{mi}), \tag{7-53}$$

and

$$e_i = 20 \log_{10} \mathrm{B} - 20 \log_{10}(\omega_i y_{mi}/u_{mi}). \tag{7-54}$$

If you now make the substitutions

$$f_i = 20 \log_{10}(\omega_i y_{mi}/u_{mi}), \tag{7-55}$$

and

$$\mathrm{C} = 20 \log_{10} \mathrm{B}, \tag{7-56}$$

then the error is

$$e_i = \mathrm{C} - f_i, \tag{7-57}$$

which is an appropriate form for LSE fitting to find the constant, C. This gives

$$E = \Sigma e_i^2/N = \mathrm{C}^2 - (2\mathrm{C}/N)\, \Sigma f_i + (1/N)\, \Sigma f_i^2. \tag{7-58}$$

Minimizing this expression gives

$$dE/d\mathrm{C} = 0 = 2\mathrm{C} - (2/N)\, \Sigma f_i, \tag{7-59}$$

and

$$\mathrm{C} = \Sigma f_i/N. \tag{7-60}$$

The constant, B, is found by reversing the substitution,

$$\mathrm{B} = 10^{\mathrm{C}/20}, \tag{7-61}$$

which completes the LSE fit for the first section of data.

The second fit should take place using the data near the lowest break frequency, ω_a, as you continue to quantify the transfer function. Near this break frequency, a Taylor's series expansion is used for the log function, which is expanded near $x = 1$ (that is, $x = \omega/\omega_a = 1$). So,

$$20 \log_{10} [1 + x^2]^{1/2} = 10 \log_{10} [1 + x^2] = f(x).$$

Thus, $f(x)$ is linearized to

$$[f]_{x=1} + [df/dx]_{x=1}(x - 1) = 10 \log_{10} 2 + 10(x - 1).$$

And

$$20 \log_{10}[1 + x^2]^{1/2} \approx -6.99 + 10x \quad \text{near } x = 1. \tag{7-62}$$

Notice that *all* first-order factors must occur in the form of Eq. 7-62, with signs reversed for a denominator factor.

As the magnitude fit of the frequency data proceeds, all earlier fits are carried along, including their known coefficients. This means that, for the second fitting in the example,

$$M(\omega) \approx 20 \log_{10} (\mathrm{B}/\omega) - 6.99 + 10(\omega/\omega_a), \tag{7-63}$$

which can be fit directly to the measured data, M_{mi}, *near the break frequency* since this function is linear in the single unknown coefficient $(1/\omega_a)$. Once ω_a is found, then $T(s) = \mathrm{B}(s/\omega_a + 1)/s$ until the next break frequency when the next factor is fit, and so on.

The LSE transfer-function fitting process is thus typically very laborious and usually unwarranted in the early stages of design. Instead, the required data richness is used to "eyeball" fit the graphical asymptotes which were presented in the last section of the text. Such quick estimates of the break frequencies are very appropriate to the early stages of design.

HOMEWORK

7-9. Given the data points for a SISO component
$(u, y_m) = (0, 0), (1/2, 1/8), (1, 1), (1.5, 3.375), (2, 8)$.

a) Fit a linear LSE model to the data that passes through $(u, y) = (0, 0)$.

b) Plot your result with the data, and provide validation comments.

7-10. Use the data from the previous problem to,

a) Fit a quadratic LSE model to the component in the form

$$y = \mathrm{b}_1 u + \mathrm{b}_2 u^2.$$

b) Plot your results with the data and comment on the acceptability of the prediction in the range $0 \leq u \leq 0.5$.

7-11. Start from the definition of the error, e_i, for Eq. 7-42 and derive Eq. 7-43.

7-12. Convert the following nonlinear equation

$$Y = \mathrm{C}_1 U_1 U_2^{\mathrm{C}_2},$$

into a form which is suitable for LSE fitting.

7-13. Given the frequency data shown in Fig. 7-12 for a linear component. *Estimate* the transfer function which goes with the data.

7-4 IDEAL TRANSIENT-CHARACTERISTIC MODELING

Transient experiments and modeling always follow steady experiments and modeling, and are always used to develop a steady time-domain model into a dynamic time-domain model. In this section the ideal-source characteristic modeling procedure is continued to find the rate terms for the dynamic model. The procedure thus assumes that the transient input selection and transient data generation have been completed, and rate-form selection is necessary. This selection prepares for the final step of fitting the dynamic model to the transient data, which is discussed in the next section of the text.

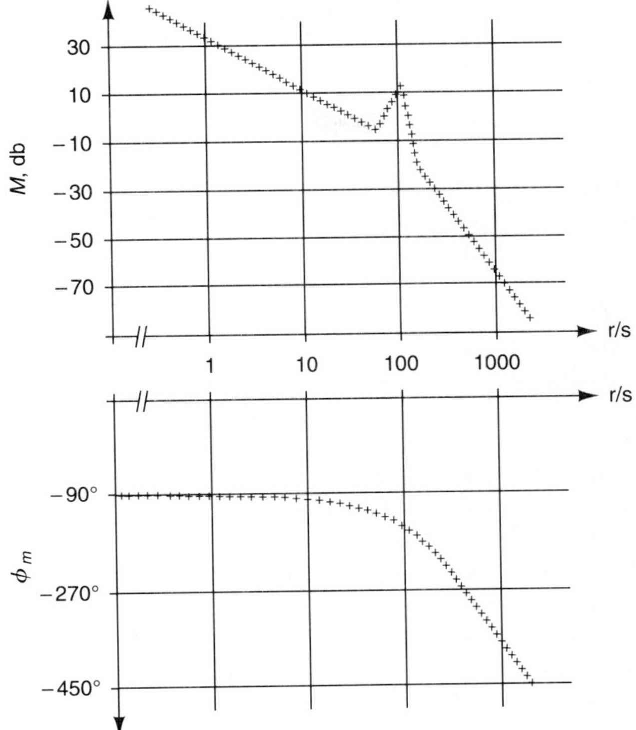

Figure 7-12. Bode Diagram for Prob. 7-13

Transient Modeling Procedure. So far you have found a characteristic, steady-state, MIMO model of the form

$$y_L = \mathrm{b}_{L1}u_1 + \mathrm{b}_{L2}u_2 + \cdots + \mathrm{b}_{LK}u_K, \tag{7-64}$$

where the b's are the steady-state sensitivities of the output y_L to each input u_i while the other inputs are held to zero. Collectively, these output equations may be written in vector form as

$$\boldsymbol{y} = \mathbf{b}\ \boldsymbol{u}. \tag{7-65}$$

Note that the model form is now assumed to be *linear*. This restriction is necessary since the modeling procedure shortly makes use of the principle of superposition. In fact, the failure of superposition greatly complicates the modeling procedure and further study in this area is left for other texts.

The next step in the modeling procedure is to determine the time-derivative terms which also belong in the linear model. Some of these derivatives are input derivatives, some are output derivatives. Generally speaking, the output-side derivatives are found first, followed by the input-side rate terms. This is done on an input-output pair basis, as suggested by the terms in Eq. 7-64.

The general model form for each input-output pair is

$$a_n d^n y/dt^n + a_{n-1} d^{n-1} y/dt^{n-1} + \cdots + y$$
$$= b_0 u + b_1\, du/dt + \cdots + b_m\, d^m u/dt^m. \qquad (7\text{-}66)$$

Note the unity coefficient on the undifferentiated output, and the allowance for a highest-differentiation of the input of a different order from that of the output. Again, subscripting of these coefficients is according to the order of the associated differentiation.

Now, certain experiments are useful to identify certain coefficients in the model of Eq. 7-66, and others are useful for the remaining terms. To sort out the data from these experiments, it is necessary to recall the fundamentals of model responses from Chap. 5.

For example, consider the step-singularity input and the MIMO response as shown in Fig. 7-13. Here an input is given to channel 1 while the other inputs are held at $u = 0$, and the three responding outputs are recorded as shown.

Note that all the data in Fig. 7-13 is taken as a function of the time, t. This is necessary since you are going to build a model based on the time as the independent variable. Such data recordings are usually done with a strip-chart recorder (giving continuous data) or with a digital-computer sampling scheme (giving discrete data). In any case, you need tabular data taken at *uniform* time steps for an LSE model fitting. This suggests a computer-sampling approach since the quantification of a strip-chart recording by hand can be a very tedious task if a large number of data points are desired, which is usually the case.

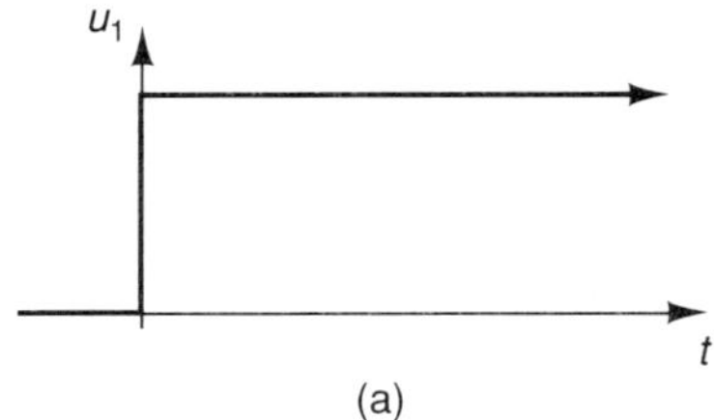

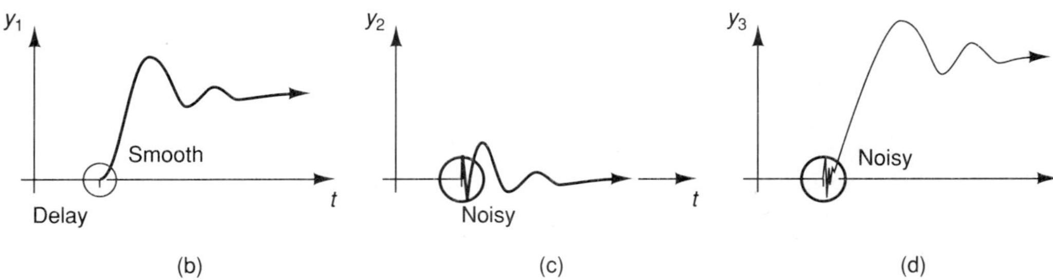

Figure 7-13. Step Responses in a MIMO Component. (a) Input, (b), (c), and (d) are output responses

The first quality to observe in the responses in Fig. 7-13 is that *delay* is very obvious in all of the output transients. The delay time constant, τ_d, is thus well-identified and easy to model.

The next quality to observe is the *noisiness* at the start of the step-responses. This noisiness is characteristic of a component which acts to differentiate this input. Thus, if you see a noisy response at the start of a step or impulse response, then you know that input derivatives are likely to be present. Notice that this information does not tell you the order of the input derivatives, only that they are present.

For example, in Fig. 7-13b, the channel 1 output is smooth at the beginning of its response, thus suggesting that there are no input derivatives. The other two channels both show noise and the probable presence of input derivatives. Moreover, channel 2 (Fig. 7-13c) shows an impulse response to the step input. This can only occur if the step input is differentiated in the output channel. Thus you know that channel 2 has at least a first-order derivative of input.

The next quality to observe is that an initial guess at the *order* of the output model is suggested by its transient curve. For the curves in Fig. 7-13, all the outputs look like they may follow underdamped second-order models. They may fit a higher-order model, but they are not of lesser order. Curves *b* and *d* look like typical step responses, while *c* looks like an impulse response. These are all good first choices for the models of these input-output pairs. Remember, you want the simplest model possible.

The next problem you face is putting all the input-output responses together into a single output model. When you do this, you must resort to the linearity of the component and use a frequency-domain method to supplement the transient modeling.

Combined Transient and Frequency Modeling. The linearity of the component means that the input-output response of each channel in the component may be modeled using a transfer-function model. It also means that an output model may be found by superposing all the transfer-function input-output models.

By way of example, suppose that a two-port component is subjected to two tests as shown in Fig. 7-14 a and b. That is, a unit step is first input at u_1, while $u_2 = 0$. The resulting first-order response is measured at y_1. In the second test, a unit step is input at u_2, while $u_1 = 0$. The result is a second-order response at y_1.

Note that the apparent order of the two responses in Fig. 7-14 appear to be different, and you may become confused about the order of the component in channel 1. However, this poses no problem since these responses are independent of one another in characteristic testing, you may thus resort to the demonstrated linearity of the component to superpose the two responses using a frequency-domain approach,

$$y_1 = T_{11}(s)\, u_1 + T_{12}(s)\, u_2 . \tag{7-67}$$

This last equation can be further identified when you insert the forms of the two transfer functions which are suggested by the data of Fig. 7-14,

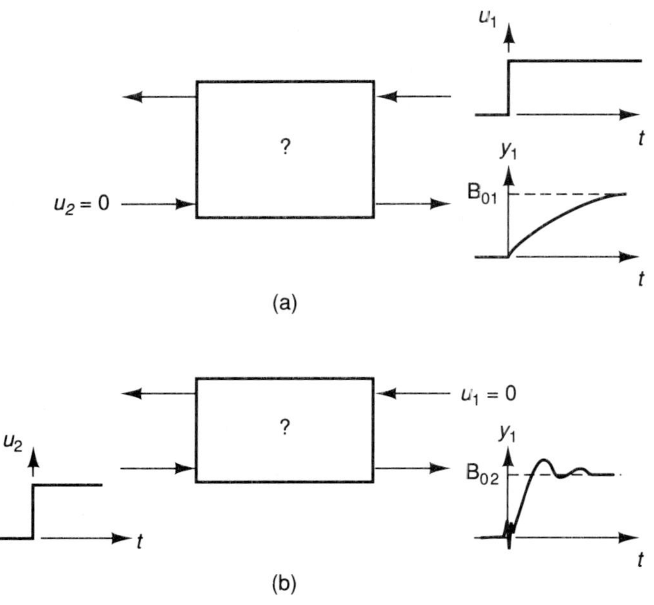

Figure 7-14. Two Transient Experiments for y_1 in a MIMO Component. (a) u_1 step, $u_2 = 0$, (b) $u_1 = 0$, u_2 step

$$y_1 = \frac{B_{01}}{(\tau s + 1)} u_1 + \frac{(B_{12}s + B_{02})}{\left(\dfrac{s^2}{\omega_n^2} + \dfrac{2\zeta}{\omega_n} s + 1\right)} u_2. \tag{7-68}$$

Note that a numerator factor has been added in the second transfer function to account for the noisiness that is observed in the response, and the constants B_{01} and B_{02} may be read directly from the strip-chart recordings ($s = 0$ is the steady state corresponding to zero derivatives for all variables). So, B_{12}, ζ, ω_n, and τ need to be quantified from the transient data. Further, if you clear the fraction in Eq. 7-68 and compute the inverse Laplace transform, you see that the component acts like a third-order component, with derivatives of each of the inputs.

Now, if the forms of the denominators in all the channels is the same, and the constants all appear roughly equal (e.g., ± 10%), then you may greatly simplify the model. That is, suppose

$$y_1 = \frac{B_{01}}{(\tau s + 1)} u_1 + \frac{B_{02}}{(\tau s + 1)} u_2, \tag{7-69}$$

then you may clear the common fraction and invert the Laplace form to find

$$\tau dy_1/dt + y_1 = B_{01}u_1 + B_{02}u_2. \tag{7-70}$$

Otherwise, you must clear all the Laplace fractions to find the order of the output equation.

Thus, the frequency domain may be used to build the output side of the model, but the input side remains unclear at this point. To remedy this, you may resort to either frequency-domain test data, or you may use singularity-input data, or you may use both.

Occasionally, frequency-domain data is available to plot Bode diagrams for the time-domain component. These diagrams can be used to identify both sides of the time-domain model through transfer functions, as discussed in the previous sections of the text. However, some time-domain components are too frail to stand up to sinusoidal inputs and are not good candidates for this testing (e.g., some types of robotic mechanisms). Still, there are quite a few time-domain components which do accept frequency testing quite well, especially electrical components.

For those components which do not accept frequency testing, you must use singularity-input data to find the input-derivative coefficients.

Singularity-input Modeling. As discussed previously, step- and impulse-input data are very useful in identifying the *output-side* coefficients, but they are simply not *rich* enough to allow you to observe much about the input derivatives from the response. Fortunately, there are other singularity inputs which are very well-suited to identifying input rate terms.

The various cases of singularity input richness are shown in Table 7-1. Note that a step or impulse has not got enough richness to yield a du/dt value, other than at the start of the response when it shows up as noise. Thus, if the component acts to differentiate the input, then these two forms will *not* be able to generate data which can be used to fit the input derivative coefficients. However, as you move down the list of singularity functions, you see that higher and higher input derivative coefficients can be identified.

Table 7-1. Dynamic Input Richness

Input	*To identify*	*Can't identify*
Steady	Linearity, b_0	Derivative coefficients
Step or impulse	Output coefficients, τ_c, ζ, ω_n, etc.	du/dt or higher input Coefficients
Ramp	du/dt coefficient, b_1	d^2u/dt^2 or higher
Parabolic	d^2u/dt^2 coefficient	d^3u/dt^3 or higher

The approach most often used is to identify dynamic models by successive use of singularity inputs in the order shown in Table 7-1. This creates a situation where the inputs are ideally suited to identify the constants shown. The functions thus act to surgically extract each order of input derivative and its model coefficient.

So, for the linear time-domain component, the output model order is first evaluated through the use of step and/or impulse inputs for each channel of output. This data also indicates the presence of input-side derivative terms. The input-side rate coefficients are then found through the use of higher singularity inputs, as shown in Table 7-1.

HOMEWORK

7-14. A step input results in the responses shown in Fig. 7-15 for five different SISO components. Evaluate the form of an initial model for these components according to: what order model you would first use, whether the component acts to differentiate the input, and what constants need to be evaluated.

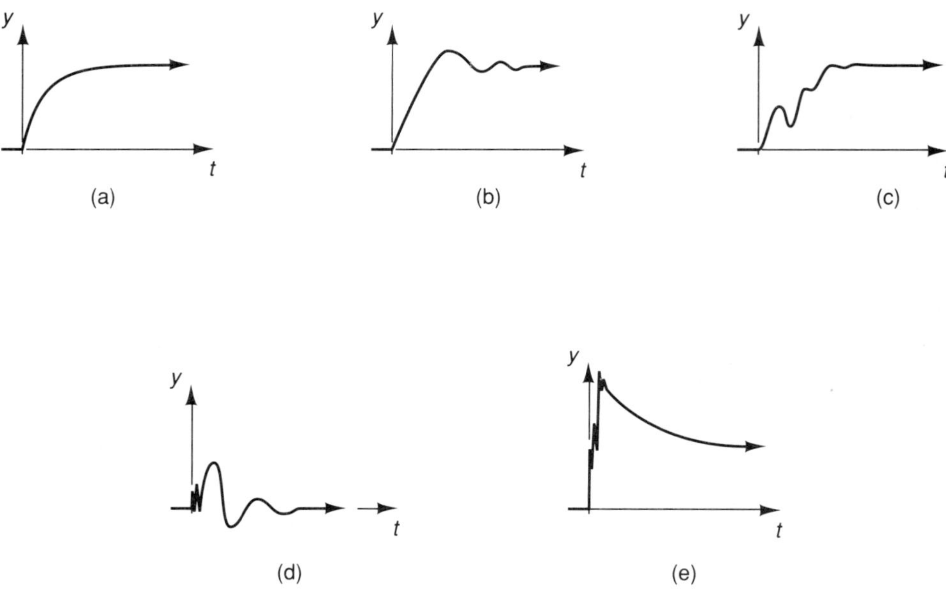

Figure 7-15. SISO Step Responses for Prob. 7-14

7-15. Given a linear component, and that its output response to a step input shows initial noise (indicating the presence of input derivatives), what other input functions can you use to identify the input derivative coefficients?

7-11. Given the Bode diagram shown in Fig. 7-16 for a SISO component [$M = 20 \log_{10}(y_m/u_m)$]. What time-domain model form is suggested by the diagram, if $y = 5u$ was found during steady modeling?

7-5 LSE FITS TO TRANSIENT DATA

The approach used here is very similar to that which was used for steady-data LSE fits. It begins by examining the data measurements to ensure sufficient data richness,

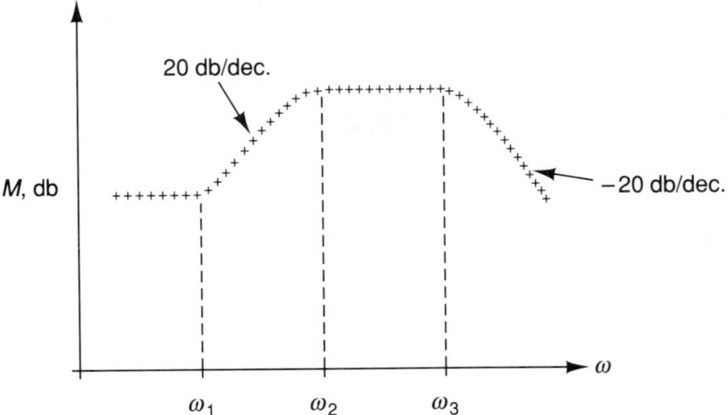

Figure 7-16. Magnitude Data for Prob. 7-16

and it assumes that the dynamic model form to be fit is given. The identification of the dynamic model is then done in two steps: first the calculation of the rate coefficients, then the plotting of the model with the data to validate the fit.

Transient Data Evaluations. The first issue in transient fitting is the same as that for steady fitting: is the richness of the input-output data points adequate for fitting? Now, however, there may be two different needs to be met. There may be an interest in the model coefficients, or there may be an interest in the numerical model output prediction. If the need is for an estimate of the model coefficients only, then the LSE method may be applied to only a few well-chosen data points. If the need is for an accurate output prediction, then many more points are needed to accurately fit the coefficients.

Since we are focused here on preliminary design, the need is often for an initial estimate of the model coefficients in order to characterize the design. Thus, the following *Transient Fitting Rule* is used as the sole guideline:

> At least 10 evenly spaced output data points are needed to fit the characteristic curvature of a transient.

This rule is applied as shown in Fig. 7-17 for data from linear components a digital sampling component.

In characteristic experiments showing a first-order response, the time constant, τ_c, establishes the curvature. Thus, the fitting rule gives

$$\Delta t \leq \tau_c/10. \tag{7-71}$$

In characteristic experiments showing an underdamped second-order response, the damped period, T, of the response is of interest. This gives

$$\Delta t \leq T/10. \tag{7-72}$$

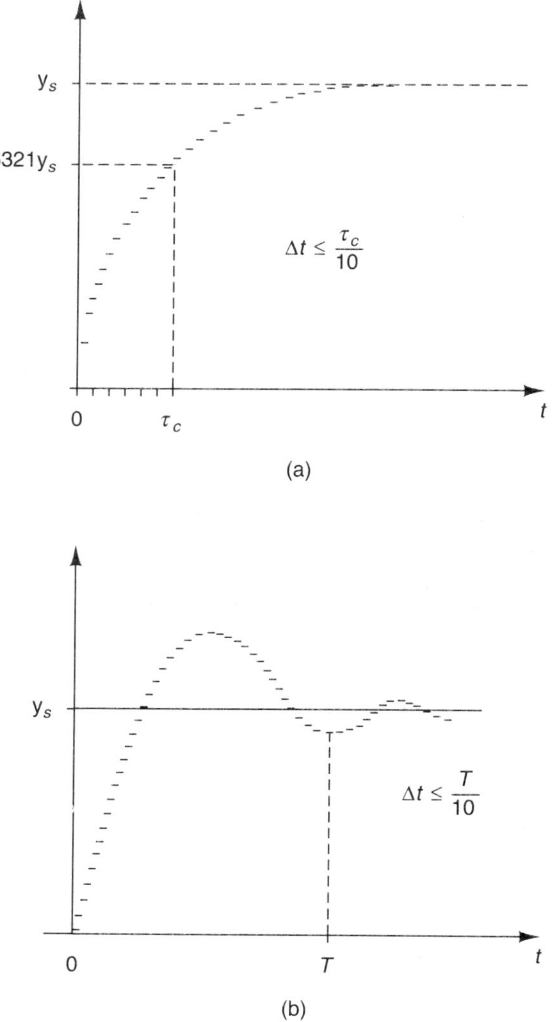

Figure 7-17. Typical Transient Data Histories. (a) First-order response, (b) second-order response

You will find that the fitting rule gives you good estimates of the linear-model *coefficients* for preliminary design which are well within engineering accuracy ($\pm$ 10%). However, more data points are necessary if high accuracy is desired in *output* predictions. For example, if you desire to predict y to within $\pm 2\%$ for a second-order linear component (as opposed to the model coefficients), then you may need as many as 300 data points per curvature. (This issue is discussed again in Sec. 7-7 of the text where digital sampling is considered in more detail.)

Thus, an extremely poor fit of a linear model to a limited data set using LSE coefficients is probably due to an incorrect model *form* rather than too few data points, provided that the fitting rule is used.

Transient Fitting. The LSE transient fitting method rests squarely upon the matching of the given data to a finite-difference approximation of the model. This method is most commonly used for both the first- and second-order models, although higher-order models are also commonly fit. In the following, the theory for the first- and second-order models is developed to demonstrate the LSE fitting approach.

The *first-order* form of the linear model is

$$\tau_c\, dy/dt + y = b_0 u. \tag{7-73}$$

And the finite-difference approximation to this model is

$$\tau_c(y_{i+1} - y_{i-1})/2\Delta t + y_i = b_0 u_i. \tag{7-74}$$

Note that this method uses a *central-difference* formula for the derivative approximation. Also note that b_0 is assumed to be known from the steady-data experiments which were conducted earlier. Further, Δt is also assumed known, as are the values of the output and input at the sample intervals indicated by Δt and the i subscripts. Thus, the only unknown in the model is the time constant, τ_c.

The error, e_i, at each measured output, y_{mi}, is

$$e_i = y_i - y_{mi}, \tag{7-75}$$

$$= b_0 u_i - \tau_c(y_{i+1} - y_{i-1})/2\Delta t - y_{mi}, \tag{7-76}$$

which is linear in the single unknown, τ_c, so the LSE method may be applied.

Thus, the mean-squared-error is

$$\begin{aligned} E &= (1/N)\Sigma(e_i)^2, \\ &= (b_0^2/N)\,\Sigma u_i^2 + (\tau_c^2/4\Delta t^2 N)\,\Sigma(y_{i+1} - y_{i-1})^2 + (1/N)\,\Sigma y_{mi}^2 \\ &\quad - (b_0\tau_c/\Delta t N)\,\Sigma u_i(y_{i+1} - y_{i-1}) - (2b_0/N)\,\Sigma u_i y_{mi} \\ &\quad + (\tau_c/\Delta t N)\,\Sigma y_{mi}(y_{i+1} - y_{i-1}). \end{aligned} \tag{7-77}$$

Since τ_c is the only unknown in this equation, $dE/d\tau_c = 0$ gives the minimum net error. That is,

$$\begin{aligned} dE/d\tau_C = 0 = (2\tau_c/4\Delta t^2 N)\,\Sigma(y_{i+1} - y_{i-1})^2 - (b_0/\Delta t N)\,\Sigma u_i(y_{i+1} - y_{i-1}) \\ + (1/\Delta t N)\,\Sigma y_{mi}(y_{i+1} - y_{i-1}). \end{aligned} \tag{7-78}$$

This last equation may be solved for the fitted time constant

$$\tau_c = 2\Delta t\,[b_0\Sigma u_i(y_{i+1} - y_{i-1}) - \Sigma y_{mi}(y_{i+1} - y_{i-1})]/\Sigma(y_{i+1} - y_{i-1})^2. \tag{7-79}$$

The *second-order* transient LSE fit comes from approximating the second-order model

$$(1/\omega_n^2)\, d^2y/dt^2 + (2\zeta/\omega_n)\, dy/dt + y = b_0 u, \tag{7-80}$$

with the central finite-difference formula

$$\left(\frac{1}{\omega_n^2}\right)\left[\frac{\dfrac{\{y_{i+2} - y_{i+1}\}}{\Delta t} - \dfrac{\{y_{i-1} - y_{i-2}\}}{\Delta t}}{3\Delta t}\right] + \left(\frac{2\zeta}{\omega_n}\right)\left[\frac{y_{i+1} - y_{i-1}}{2\Delta t}\right] + y_i = b_0 u_i. \tag{7-81}$$

Note that the second-order derivative is here approximated in terms of the difference of two slopes, one at $i + (3/2)$ and one at $i - (3/2)$. Consequently, the separation between these slopes is $3\Delta t$, as the formula above shows.

When Eq. 7-81 is solved for y_i, a more useful form is obtained,

$$y_i = b_0 u_i - C_1(y_{i+2} - y_{i+1} - y_{i-1} + y_{i-2}) - C_2(y_{i+1} - y_{i-1}), \tag{7-82}$$

where, $C_1 = 1/(3\Delta t^2 \omega_n^2)$, (7-83)

and $C_2 = \zeta/(\omega_n \Delta t)$. (7-84)

When you apply the LSE method to Eq. 7-82 above, you find that

$$\begin{bmatrix} \sum_{i=1}^{N} g_i^2 & \sum_{i=1}^{N} g_i h_i \\ \sum_{i=1}^{N} g_i h_i & \sum_{i=1}^{N} h_i^2 \end{bmatrix} \begin{bmatrix} C_1 \\ C_2 \end{bmatrix} = \begin{bmatrix} -\sum_{i=1}^{N} f_i g_i \\ -\sum_{i=1}^{N} f_i h_i \end{bmatrix}, \tag{7-85}$$

where the entries in the matrices are determined from the factors in the starting equation,

$$f_i = b_0 u_i - y_i, \tag{7-86}$$

$$g_i = -y_{i+2} + y_{i+1} + y_{i-1} - y_{i-2}, \tag{7-87}$$

$$h_i = -y_{i+1} + y_{i-1}. \tag{7-88}$$

So, given C_1 and C_2 from the LSE fit, you can compute ω_n and ζ, in that order, from Eqs. 7-83 and 7-84.

The general procedure for an LSE transient fit is thus as follows:

1. Choose the uniform spacing of the data points, Δt, to meet the fitting rule or other accuracy requirements. Set $t = 0$ at the start of the *output* response, and eliminate any delay factors from the model inputs.
2. Quantize the u_i, y_{mi} data at the Δt sample times. If the data is given as one or more continuous strip-chart recordings, this may mean that a good deal of manual data processing awaits you. Data from a digital computer is normally taken in fractions of the Δt increments, and is stored in a file on a data base or floppy disk.

3. Compute the summation terms for the LSE equations.
4. Solve the LSE equations for the unknown coefficients.
5. Validate the fit by plotting the data and the model predictions.
6. Add in any delay effects to the model inputs to synchronize the model output predictions with input stimulation.

Example 7-4

Given the model form

$$a_2\, d^2y/dt^2 + a_1\, dy/dt + y = b_0 u + b_1\, du/dt, \tag{7-89}$$

and the knowledge that $y = 6u$ from a steady analysis. Describe how you would use the following quantized data sets and the LSE method to estimate the four coefficients in this model:

1. Evenly spaced y measurements at $\Delta t = 0.001$ s. in response to a unit-*step* in u. The apparent period of the response is 1 sec.
2. Evenly spaced y measurements at $\Delta t = 0.01$ s. in response to a unit-*ramp* in u. The apparent period of the response is again observed to be 1 sec.

Solution

The solution to this problem is conducted in three steps, beginning with the steady data. That is, the first step is to recognize that $b_0 = 6$ by comparing the given result of the steady analysis to the given model form with all derivatives equal to zero, as is true in the steady state.

In the second step, the steady result is combined with the step-input data to fit the model with $du/dt = 0$, as is true for the step input. That is, for data set 1 in this example,

$$a_2\, d^2y/dt^2 + a_1\, dy/dt + y = 6u. \tag{7-90}$$

This model is next fit to the given step data in order to determine the two output-side coefficients using the LSE fitting method. Before this fit can be accomplished, however, the fit step size must be chosen. And, since the apparent period of the response is 1 sec., you may use any step size smaller than 0.1 sec., which is clearly met by the given data. The six-step procedure for an LSE fit as described above may then be accomplished.

In the last step to the solution, data set 2 is used to fit the full model form with all coefficients known except for b_1. The finite-difference equation to be used for the LSE fit to this model may be found using the same finite-difference approach as for the other model forms discussed previously in the text. This gives

$$\begin{aligned} b_1 = [\Delta t/\Sigma(u_i - u_{i-1})^2]\,[&\Sigma y_{mi}(u_i - u_{i-1}) \\ &+ C_1\,\Sigma(y_{i+2} - y_{i+1} - y_{i-1} + y_{i-2})(u_i - u_{i-1}) \\ &+ C_2\,\Sigma(y_{i+1} - y_{i-1})(u_i - u_{i-1}) \\ &- b_0\,\Sigma(u_i)(u_i - u_{i-1}], \end{aligned} \tag{7-91}$$

where C_1 and C_2 are the same as defined in Eqs. 7-83 and 7-84. Again, the six-step LSE fitting procedure is then followed to find the best estimate of b_1 using data set 2.

Comments

Note the similarity of the time-domain model build-up process which is described here to that for the frequency-domain model using the LSE method. In the frequency method, you work from low to high input frequencies, adding frequency richness to identify the factors (coefficients) in the transfer function. Here you add input-singularity richness to determine the rate coefficients in the time domain.

HOMEWORK

7-17. Start with the second-order central-difference approximation (Eq. 7-82) and use the definitions of Eqs. 7-86 through 7-88 to derive Eq. 7-85.

7-18. Given the following characteristic step response data, $y_m = 0$ for $t < 0$ and $y = 2u$ for the steady model, with $u = 5u_s(t)$ for the transient,

t	0.0	0.1	0.2	0.3	0.4	0.5	0.6	0.7	0.8	0.9	1.0	1.1	1.2
y	0.0	1.4	3.0	5.1	7.1	8.9	11.2	11.8	11.6	11.	9.8	9.1	8.9

a) Use the LSE fitting method to estimate ζ and ω_n for a second-order model of the response without input derivatives.

b) Plot the response with the data and comment on the validation of your model.

7-19. What is the least number of evenly spaced data points in the interval $0 \leq t \leq 5$ sec. that should be used to fit the transient data shown in Fig. 7-18 for the estimates of the model constants in a second-order model?

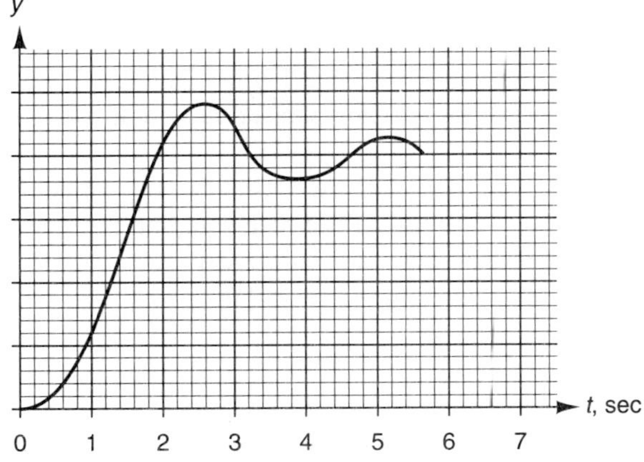

Figure 7-18. Output History for Prob. 7-19

7-6 NON-IDEAL CHARACTERISTIC MODELING: CONVECTIVE COMPONENTS

Some components, including the convective components, do not lend themselves well to ideal characteristic testing and modeling. The primary reason for this is that there are few, if any, ideal input generators for these components; hence, the "non-ideal" name for this sort of modeling. A complicating factor is that their fundamental, internal physics is so complicated that developing a simple preliminary model from internal details (e.g., as done in Chaps. 4 and 6) is seldom possible.

The usual systems approach to these components in preliminary modeling is to work towards finding an input-output model *form* which has a reasoned input-output connectivity based either on physical fundamentals, or on the multiport diagram. An example of this is the requirement that the net mass flowrate into a fluid component's ports must equal the change of mass within the component, otherwise known as the Law of Mass Conservation. Other physical laws or the connectivity of the multiport diagram may be similarly used to reason out the input-output form of the component model. And, given this form, it is possible to estimate the model coefficients using bench-test experiments and data.

The alternative to this approach is to use advanced statistical methods to determine the model form and fit simultaneously from in-system data. However, this alternative is not a preliminary design approach, and is left for more advanced texts so that the fundamental input-output modeling method may be discussed here for convective components, and in the next section for digital components.

Recall from earlier chapters that the convective components occur whenever fluid mechanics and thermodynamics simultaneously dominate the physics of a system component. This occurs most often in compressible-fluid flows or in fluid flows with heat transfer to the environment or to another component. Common examples of these flows are found in automotive engines and their cooling systems, household cooling and heating, refrigeration, and engine power systems. Common convective components in these systems include passive resistance-inertance-and-capacitance (RIC) components, compressors and pumps, and heat exchangers. The preliminary modeling of each of these components is discussed in this section.

The section begins with an examination of the fluid-thermal model, which is the basis of all convective components. The performance of the most common convective components is next discussed, and the section ends with a brief study of several common convective systems.

The Fluid-thermal Model. In this introductory study, it is assumed that no phase change takes place within a convective component. That is, a liquid is never heated enough to change it into a vapor, and a vapor is never cooled enough to change it into a liquid. Thus, you study only *single-phase* components in this section. This assumption greatly simplifies the study of thermodynamics for these components, yet it allows you to create many useful and common preliminary models.

In addition, it is assumed that all gases are ideal, obeying the Ideal Gas Law, $P/\rho = R\Theta$; and all liquids are incompressible, so that their density, ρ, is constant.

The set of variables which are needed to describe the power at a convective port must now account for both fluid and thermal power. As discussed in Chap. 3, the fluid power flowing through a port is the product of the average fluid pressure at the port, P (lb_f/ft^2, n/m^2), and the net volume flowrate through the port, Z (ft^3/s, m^3/s). The thermal power flowing through a port (e.g., heat flowrate, H) is the product of the net mass flowrate through the port, W (sl/s, kg/s), the average temperature of the fluid across the plane of the port, Θ (°R, °K), and the average constant-pressure heat capacity of the fluid at the port, C_p (Btu/sl-°R, Cal/kg-°K), according to

$$H = W\, C_p\, \Theta. \tag{7-92}$$

Typical values for the constant-pressure heat capacity are given in Table 7-2; note that these values are temperature dependent, but they may be considered constant over small ranges of temperature (e.g., for change less than 200° F).

Table 7-2. Specific Heat Capacity and Gas Constants[1] (Between 32°F and 212°F)

Liquids (C_p, Btu/sl–°F)		*Gases (C_p; R, ft-lb_f/sl–°R)*	
Alcohol	18.7	Air	7.76; 1716
Ethylene Glycol	19.38	Ammonia	16.84; 2930
Petroleum	16.1	Butane (N)	12.72; 853
Sea Water	30.3	Carbon Dioxide	6.6; 1130
Water	32.2	Carbon Monoxide	7.82; 1777
		Ethane	12.43; 1658
		Methane	19.09; 3104
		Natural Gas	18.03; 2547
		Oxygen	6.99; 1555
		Propane	12.65; 1127

[1]Note: Values for steam and common refrigerants are not given since those fluids are usually encountered in two-phase components.

Thus, combining the above with the fluid-flow model

$$W = \rho\, Z, \tag{7-93}$$

(with the pressure, P, forcing the flowrate), the set of port variables which system modelers use for convective ports are W, Θ, and P, since these variables may be used in combination to compute the power flowing through any convective port.

Further, since the fluid *convects* the heat along, W and Θ are chosen with the same causality on the multiport diagram, and P is chosen with an opposing causality to the flowrate to allow the connected component to interact.

The Convection Modeling Approach. The convection modeling approach is begun with a multiport analysis to identify all components and their input-output causality in the convective system. The four possible types of components which arise from this analysis are the fluid RIC, the pump, the compressor, and the heat exchanger. The modeling steps for these components are then basically the same as those discussed earlier for the ideal characteristic approach, with a few exceptions.

The *first*, and most critical, step in convective modeling is that a steady-component operating point must be chosen to serve as a reference point for the component linearizations which will be developed. This point should be specified by the designer. The methods which follow are used to create a model of the component's dynamics in the time domain near this operating point. In other words, the convective components are usually modeled using offset linearizations in the same manner as those discussed in Chap. 4.

The *second* step is to generate the steady-output data near the operating point. As before, this data should be rich enough to satisfy form and fitting evaluations.

The *third* step is to choose the steady-model form. Now, because the input sources are nonideal, the form selection is not guided by the data alone, but is more dependent on the multiport diagram for connectivity rationalizations.

The steady coefficients are fit in the *fourth* step (e.g., with all rate terms equal to zero). Sometimes, these coefficients may be estimated from the designer's experience, or the LSE method may be used on steady bench-test data for the component. Note that this does not require the rigorous ideal-source input forcing of the previous sections since the model form is known *before* the testing begins.

In the *fifth* and last step, the dynamic effects must be evaluated. This is done using two ideas: first, all the fluid dynamics may be lumped into fundamental building-block RIC components, to be used as desired; and second, all the steady models found for the other components are assumed to be dynamic models that have no rate terms. In addition, delay constants are added to the inputs as necessary to simulate the observed or estimated delayed responses.

Thus, the only rate coefficients in the convective model are those in the fundamental RIC model. These are seldom found from LSE fits since proper dynamic data for such fits is *extremely* hard to generate. (Can you imagine the difficulty of trying to generate a step in pressure, or in heat flowrate?) When they arise, the rate coefficients are usually estimated based on previous experience (e.g., from the designer, or using the methods of Chap. 4).

RIC Components. All convective systems demonstrate fluid resistance, inertance, and capacitance somewhere in the flow. When these are known to dominate or are of suspected importance, the RIC component of Fig. 7-19 is used to represent their idealized effects in the system model.

First, note that the idealized RIC model in Fig. 7-19 assumes that no heat is transferred from the fluid to the outside of this component. If such heat flow does occur, say due to poor insulation accompanied by a large interior-exterior temperature difference, then the heat-transfer effects should be lumped into a connected heat-exchanger model. In other words, the RIC model is for fluid dynamics with internal thermodynamics only.

So, the thermal transfer model for the RIC is based upon the assumption of zero heat flow to the exterior. For liquid flows, this simply means that

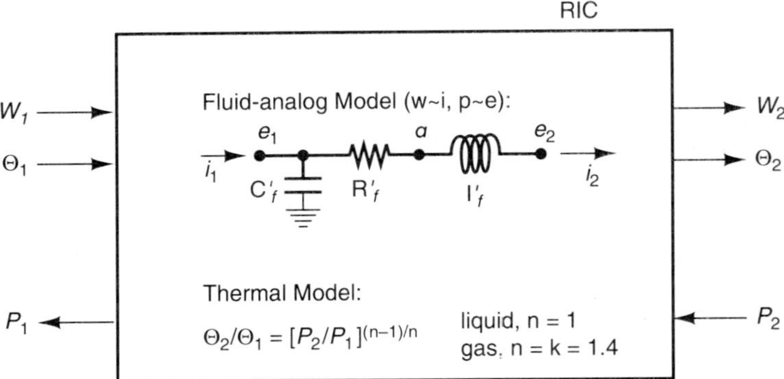

Figure 7-19. The Resistance-inertance-capacitance (RIC) Model

$$\Theta_2 = \Theta_1 \tag{7-94}$$

for both steady and transient conditions.

However, a *steady gas* flow through a restriction without heat transfer to the exterior is generally seen to change temperature according to the thermodynamic law

$$\Theta_2/\Theta_1 = (P_2/P_1)^{(k-1)/k}, \tag{7-95}$$

where the variables are absolute measurements, and where $k = 1.4$ for all ideal gases.

Thus, if you take the logarithm of both sides of Eq. 7-95 and differentiate both sides, you obtain a useful linear model of the changes in gas output temperature, θ_2, near the operating point Θ_{o1}, Θ_{o2}, P_{o1}, and P_{o2}. This gives

$$\theta_2 + [(\Theta_{o2}/P_{o1})(k - 1)/k]\, p_1 = (\Theta_{o2}/\Theta_{o1})\, \theta_1 + [(\Theta_{o2}/P_{o2})(k - 1)/k]\, p_2. \tag{7-96}$$

The fluid-flow equations which model the RIC component may be easily derived from the fluid-network analogy

$$C_f'\, dp_1/dt + w_2 = w_1, \tag{7-97}$$

and

$$I_f'\, dw_2/dt + R_f' w_2 - p_1 = -p_2. \tag{7-98}$$

Here, Eq. 7-97 is recognized as a statement of the Law of Mass Conservation near the operating point, it is expressed in terms of the unknown fluid capacitance of the RIC idealization. Note that this capacitance may be related to that which is discussed in Chap. 4 according to the simple relation

$$C_f' = \rho\, C_f, \tag{7-99}$$

and the methods of Chap. 4 can be used to estimate the capacitive constant here as well.

Similarly, Eq. 7-98 is a statement of Newton's Second Law near the operating point for this component idealization. It, too, is expressed in terms of coefficients which multiply the inertia term (I_f') and the frictional loss term (R_f'). These terms may also be estimated using the methods of Chap. 4,

$$I_f' = I_f/\rho, \tag{7-100}$$

and

$$R_f' = R_f/\rho. \tag{7-101}$$

Where the *operating-point density* in the component must be estimated and used in Eqs. 7-99 through 7-101.

The primary limitation of the RIC model occurs if a gas flow through a restriction (such as a valve) is subjected to a *sustained* absolute-pressure ratio (P_1/P_2) which is outside of the range

$$0.654 \leq P_1/P_2 \leq 1.528. \tag{7-102}$$

When this happens, the fastest speeds of the molecules in the flow are near the speed of sound, which is determined solely by the *upstream* pressure. Any relative decrease in downstream pressure is ineffective in further changing the flowrate since the molecules cannot go any faster; hence the condition is called "choked" flow. This is a distinct effect due to gas compressibility—it changes the causality and connectivity of the model and is very problematic for preliminary analysis.

The usual approach to this is to use the RIC model during preliminary analysis, then to check the prediction of pressure ratios. If choked flow is indicated, then the model is changed to accommodate this effect and the predictions are recomputed. However, most systems are designed to avoid the limitations of internal gas compressibility so the RIC approach is usually successful.

Thus, the RIC dynamic model is the three linear equations 7-96, 7-97, and 7-98. It is expressed in terms of the three unknown variables w_2, θ_2, and p_1 which represent *changes* in the component variables near an operating point. These equations must be modified as necessary to account for delay in the input-output response, by delaying one or more inputs. Thus, the model is readied for system assembly.

Example 7-5

Two common situations for the use of the RIC model are shown in Fig. 7-20. Fig. 7-20a shows the geometry for a well-insulated pipe flow of liquid of density, ρ, in a pipe of diameter, d, of a very long length, ℓ. In Fig. 7-20b, a gas is flowing through a short well-insulated pipe, and through a flat-plat orifice of area, A_o. Find a RIC model for each of these flows.

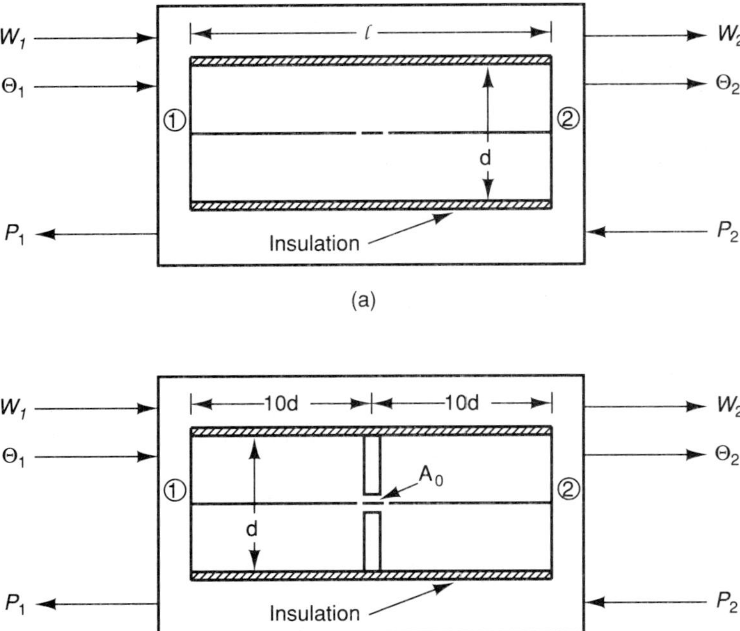

Figure 7-20. Examples of RIC Application. (a) Pipe flow, (b) flow restriction

Solution

(a) Since the pipe is very long, and this is a liquid flow, the flow experiences mainly inertance and resistance effects. Thus, you may set $C_f = 0$ in the preliminary RIC model above and find

$$\theta_2 = \theta_1, \tag{7-103a}$$

$$w_2 = w_1, \tag{7-103b}$$

and

$$p_1 = I_f' \, dw_2/dt + R_f' w_2 + p_2. \tag{7-103c}$$

Again, since this is a liquid flow, the delay is usually insignificant, and is ignored. The constants are calculated using the methods of Sec. 4-8. And, if the length of the pipe is short, then the resistance may also be ignored, just as in Sec. 4-8.

(b) Now, since this is a gas flow, the inertance of the flow may be ignored. The pipe length is so chosen that the entire resistive affect of the orifice plate is captured. Further, since the pipe is so short, there is little volume to create a capacitive affect. Thus, this component acts like a resistance with a classic fluid-thermal affect. That is, the model is

$$w_2 = w_1, \tag{7-104a}$$

$$p_1 = R_f' w_2 + p_2, \tag{7-104b}$$

and

$$\theta_2 = (\Theta_{o2}/\Theta_{o1})\theta_1 + [(\Theta_{o2}/P_{o2})(k - 1)/k]\, p_2 - [(\Theta_{o2}/P_{o1})(k - 1)/k]\, p_1. \tag{7-104c}$$

The constants in these equations must be measured or estimated near the operating point of interest, which may involve some bench-test experimentation. In this, it is possible to estimate a theoretical fluid resistance for this flow by linearizing the orifice-flow equation in Sec. 4-8. Also, the short length of this component makes it possible to neglect the delay which is so often seen in longer pipes and larger volumes with gas flows.

Comments

The equations in the solution models above are written in a form which makes it clear that there are enough knowns to solve for the outputs. This is not necessary in general, however, as the system assembly method of Chap. 8 makes these rearrangements as needed.

The RIC model is thus shown as a convenient tool to lump and idealize the passive fluid-dynamic effects in various parts of a convective system. It is used effectively to model long or short pipes, distributed RIC effects, valves or other restrictions, accumulators or capacitances; in short, everything discussed in Sec. 4-8. It is also used effectively in combination with the other convective component models to create a useful system model. These other components include those needed to force the convective fluid through the system. So, pumps are used to move liquids, while fans and compressors are used to move gases. Each of these components thus also requires a RIC-compatible convective model.

Pumps. The idealized models for positive-displacement pumps and centrifugal pumps are discussed in Sec. 4-9. However, those idealized models are of little use here, in view of the combined thermal and fluid physics. Instead, a more functional approach to these devices is taken.

The multiport diagram for *all* types of convection pumps is as shown in Fig. 7-21; note the separation of the physical "pump" into a lumped "inertia" component, and a lumped "fluid-thermal" component. The flowrate is always taken as the output now since the pump is either a flow-producing device, or it is a pressure-producing device with a strong dependency on the connected system. In the latter case, the pump may also be viewed as a flowrate-producing device with a strong dependency on the system. Thus, the same output causality is chosen for both cases for convenience.

The changes in the steady-output flowrate of the pump in Fig. 7-21 near an operating point are assumed to fit a linear model,

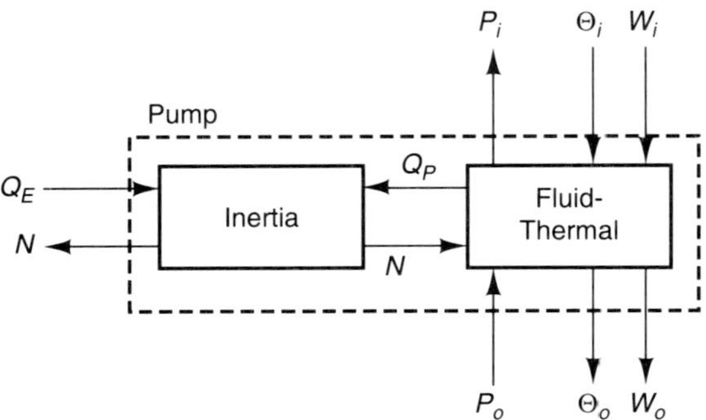

Figure 7-21. The Pump Multiport Diagram

$$w_o = C_1 \, p_o + C_2 \, n, \tag{7-105}$$

where C_1 and C_2 are constants from the steady operating data, and n is the change in speed from the operating-point speed. Note that this output form neglects the connectivity to the w_i and θ_i inputs. This is done to clearly establish the *sources* of the flowrate to be the speed, n, and the system back-pressure, p_o, which is observed in application.

However, you must be careful in choosing steady data to fit this model form since it is not a fundamental law, it merely comes from reasoning using the multiport diagram connectivity. For example, in one experiment the speed should be fixed, and a fluid resistance in the test system should be varied in order to find C_1 (e.g., $n = 0$ for this experiment). A similar experiment, with the outlet pressure fixed, must be conducted to find C_2.

Another output of this pump is the temperature, θ_o, which comes from a constant-temperature model of the liquid flow through the pump,

$$\theta_o = \theta_i. \tag{7-106}$$

A third output of this pump is the inlet pressure, p_i. If there is a connected inlet reservoir, then this pressure is usually neglected. If some other component is connected at the inlet,

$$p_i = C_3 \, w_i + C_4 \, n. \tag{7-107}$$

The form of this model is chosen for the same reason as that for Eq. 7-105; namely, that the inlet back-pressure should mainly be a function of the inlet flowrate and the pump speed, according to observation.

The final output of this fluid-thermal component is the torque, q_p. Note that this must be the *output* of the fluid-thermal pump since the inertia demands that this torque be its input according to the Second Law of Newton. When the causality of the multiport diagram is used to give the model form, it is

$$q_p = C_5\, n + C_6\, p_o + C_7\, w_i, \tag{7-108}$$

which models the changes in pump torque near a component operating point. Now, the only variable showing a negligible connectivity is the inlet temperature, θ_i. Again, the same caution about the experimental data for fitting these coefficients as that for Eq. 7-105 is recommended since this is not a fundamental form.

Note that the steady fluid-thermal model above is also the dynamic model since this *liquid-flow* responds almost instantaneously to changes in any input variable, and delays are usually not of concern in a pump. However, should you desire to add fluid-dynamic effects to this model, then you may connect a RIC component at the inlet and/or outlet to do so.

The inertia part of the pump model is analyzed using the Second Law of Newton for rotation,

$$\mathrm{J}\; dn/dt = q_E - q_p. \tag{7-109}$$

Thus, the input engine torque, q_E, creates a positive speed, while the pump torque, q_p, acts to resist the engine. The inertia, J, is the structural inertia of the pump. If you desire to model fluid inertia as well, you should use a RIC component, not this inertia.

Fans and Compressors. In the discussion which follows, only the gas compressor is evaluated. However, fans operate in the same manner, and the discussion is equally valid if "fan" is substituted in the text for "compressor."

The function of gas compressors is shown in the multiport diagram of Fig. 7-22. Note that the compressor is usually assumed to interact with an ideal gas reservoir, which is the atmosphere in the case of an air compressor. This means that the flowrate demand on the inlet side of the compressor, W_1, does not affect the constant-temperature and constant-pressure supplied by the reservoir at the inlet. Hence,

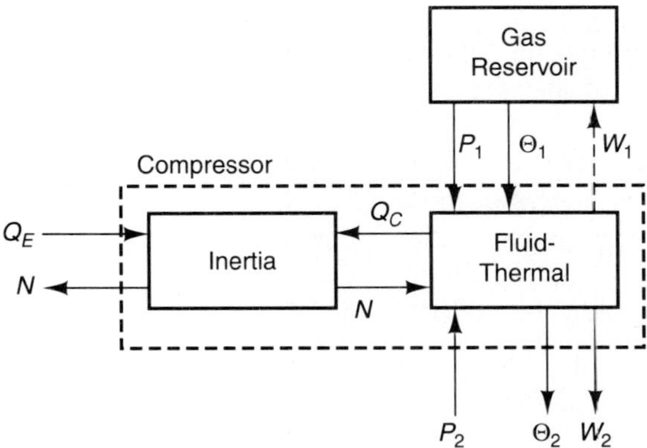

Figure 7-22. The Compressor Multiport Diagram

this ineffectiveness is indicated by the dashed line representing the inlet flowrate demand, and it means that the linearized inlet variables, p_1 and θ_1, are *both zero at all times*. Also note that this is the *only* time when the temperature and flowrate have opposite causalities at a convective port. For all other inlet connections, the same type of forcing as that for the pump should be used.

Figure 7-22 also shows the lumping of the structural inertia separately from the fluid-thermal effects, just as done for the pump earlier. And again, the compressor model is posed in terms of variables which measure the change from a component operating point. Thus, for the inertia,

$$\mathrm{J}\, dn/dt = q_E - q_c. \tag{7-110}$$

And the compressor inertia, J, may be estimated or measured from the structural properties of the device.

The change in mass flowrate out of the compressor, w_2, is of the same steady form as that for the pump,

$$w_2 = \mathrm{C}_1\, n + \mathrm{C}_2\, p_2, \tag{7-111}$$

for the same reasons. Further, the change in steady temperature output from the pump is computed according to a thermodynamic model of gas compression without heat loss to the environment, Eq. 7-95. When this model is linearized about a compressor operating point, and the zero inlet-change variables are inserted, you find

$$\theta_2 = [(\Theta_{o2}/\mathrm{P}_{o2})(k - 1)/k]\, p_2. \tag{7-112}$$

The change in steady-torque output of the fluid-thermal compressor is assumed to fit the linear form

$$q_c = \mathrm{C}_3\, n + \mathrm{C}_4\, p_2. \tag{7-113}$$

Note that q_c is not taken to be a function of either inlet change variable since they are both zero, thanks to the ideal-reservoir supply.

Again, the steady characteristic experimental data which is used to estimate the unknown coefficients in the compressor model (e.g., those in Eqs. 7-111 and 7-113) must be taken from steady experiments which are properly designed to isolate the effects of the variables.

Fluid-dynamic effects may be added to the above model by using the RIC component discussed earlier. However, the fluid-dynamic lags and the delays in a compressor are seldom large enough to be of concern.

Thus the dynamic compressor model (Eqs. 7-111, 7-112, and 7-113) is ready for assembly into the system model.

Heat Exchangers. These components are the main ingredient in many common convective systems. Their purpose is to heat up or cool down a fluid as shown in Fig. 7-23.

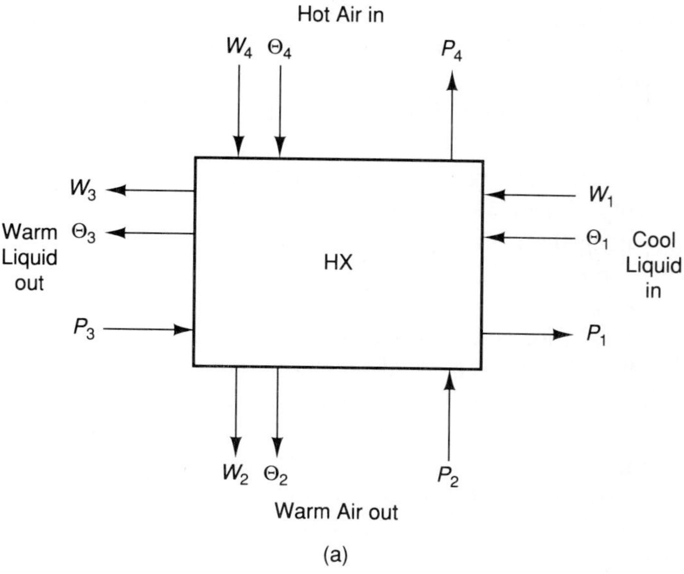

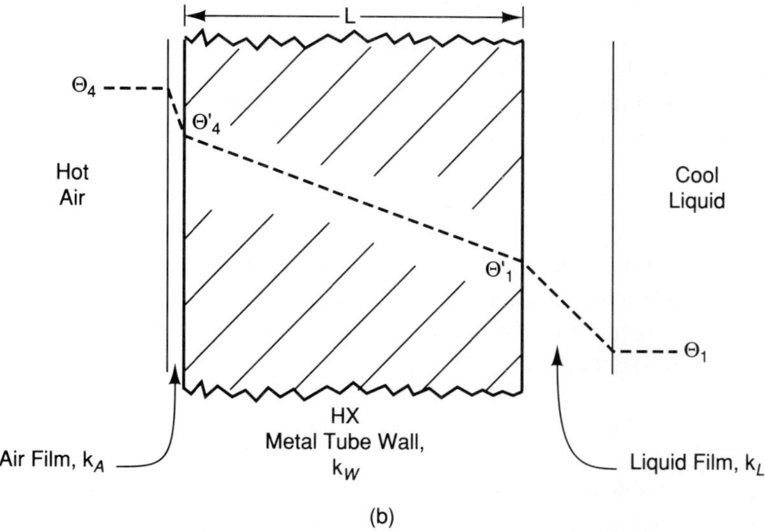

Figure 7-23. The Heat Exchanger. (a) Multiport diagram, (b) detail of the heat transfer

The multiport diagram of Fig. 7-23a shows that the construction of a heat exchanger is for the purpose of exchanging heat, which it does by directing one fluid over a tube array which contains the other fluid. It has a hot side, where a hot fluid enters and leaves as a warm fluid. And it has a cool side, where a cool fluid enters and leaves as a warm fluid. In the present introductory study, a preliminary model is built of the response of the heat exchanger for small dynamic changes in the flow variables near an operating point for a *single-phase* heat exchanger. That is, the heat transfer is assumed to be of such a nature that neither fluid changes state by boiling or condensing.

The geometry of the heat path from the hot air through a tube wall to the cool liquid is shown in Fig. 7-23b. On top of this geometric sketch is a plot of the temperature at points along the path, shown as a dotted line in the figure. Most of the temperature change in the fluids takes place in a thin film near the tube wall, while the change in the wall is spread uniformly through its thickness. The difficulty here is that the fundamental thin-film heat-transfer coefficients, k_A and k_L, are strongly dependent on the fluid flowrate, the fluid pressure, and the fluid temperature. They are thus far from constant and the fundamental approach thus introduces more confusion than it resolves, even for a single tube.

Worse yet, the heat exchanger presents a geometrical design problem which can be varied infinitely using different tube arrangements, multiple passes of fluids, different directions of passes, different materials, and so on, in order to increase the efficiency. So, each heat exchanger may be different. In short, the fundamental approach to modeling the heat exchanger has little chance of success.

However, there are only two input-output responses needed to fashion an effective preliminary heat exchanger model.

The first necessary heat-exchange response is the overall *thermal efficiency*, η. This parameter is the ratio of the amount of heat gained on the cool side, over the amount of heat lost on the hot side. The rest of the heat is lost to inefficiency; hence, the parameter name. In symbolic terms,

$$\eta = H_{\mathrm{COOL}}/H_{\mathrm{HOT}}. \tag{7-114}$$

These heat flowrates in the last equation may be further evaluated using the observation that the heat exchange usually takes place at, or near, constant pressure on each side of the exchanger. Thus, using Eq. 6-92 for Fig. 7-23, and assuming a constant-pressure heat-transfer model,

$$\eta = [W_3C_{P3}\Theta_3 - W_1C_{P1}\Theta_1)]/[W_4C_{P4}\Theta_4 - W_2C_{P2}\Theta_2]. \tag{7-115}$$

Note that the terms have been put into Eq. 7-115a to give a positive efficiency based on assumed relative temperature sizes (e.g., $\Theta_3 \geq \Theta_1$ is assumed).

Also note, in general, $\eta < 1$ since there are always thermal losses. Yet, in preliminary modeling, you may safely assume that

$$\eta = 1. \tag{7-115b}$$

The heat-exchange model also includes the idea that fluid mass does not accumulate in the exchanger, so that

$$W_3 = W_1 = W_C, \tag{7-115c}$$

and

$$W_2 = W_4 = W_H. \tag{7-115d}$$

Also, near any given operating point,

$$C_p = \text{constant}. \tag{7-115e}$$

Further,

$$C_{p1} = C_{p3} = C_{pC}, \tag{7-115f}$$

and

$$C_{p4} = C_{p2} = C_{pH}. \tag{7-115g}$$

The second overall heat-exchange response to be modelled is that the transfer of heat from the hot fluid through the exchanger walls to the cool fluid must fit the conduction heat-flow model discussed in Chap. 4. Thus,

$$W_H C_{pH}(\Theta_4 - \Theta_2) = (\mathrm{k}A/\ell)\,(\Theta_H - \Theta_C). \tag{7-116}$$

The *overall* conduction constants in parenthesis on the right side of Eq. 7-116 are determined by the particular heat exchanger design, and are computed or estimated by the system designer. As a system modeller, you may assume that they are known. The driving conduction temperature difference, $(\Theta_H - \Theta_C)$, may be computed during preliminary design using simple averages of the port temperatures,

$$(\Theta_H - \Theta_C) = [(\Theta_4 + \Theta_2)/2 - (\Theta_1 + \Theta_3)/2]. \tag{7-117}$$

The idealized heat-exchanger model for dynamic *changes* near an operating point is thus found first by linearizing, then combining Eqs. 7-115, 7-116, and 7-117:

$$w_2 = w_4, = w_H, \tag{7-118a}$$

$$w_3 = w_1, = w_C \tag{7-118b}$$

$$p_4 = p_2, \tag{7-118c}$$

$$p_1 = p_3, \tag{7-118d}$$

$$\theta_2 = C_1 w_H - C_2 w_C + C_3\theta_4 + C_4\theta_1, \tag{7-118e}$$

and

$$\theta_3 = C_5 w_H - C_6 w_C + C_7\theta_4 + C_8\theta_1, \tag{7-118f}$$

Note that the constant coefficients, $C_1, \ldots, C_8$, in the last two equations are found from manipulating the linear equation set, and are thus evaluated using the system temperatures and flowrates at the operating point.

If you desire to add fluid dynamics to the preliminary single-phase heat-exchanger model in Eq. 118, then you should use the RIC component to do so. Delays are usually not of concern in the heat exchanger, but if you wish to add them, then they appear by delaying the inputs, as done earlier.

Common Convective Systems. Two common convective systems and their multiport diagrams are shown in Fig. 7-24.

Figure 7-24a shows a typical schematic for a water-cooling convective system for an internal-combustion engine. Figure 7-24b is the corresponding multiport diagram. Here, the atmosphere at 1 provides a constant-temperature cool-air source for the system, and is pushed across a heat exchanger (called the "radiator") by a fan which is not shown in the multiport.

The cooling liquid is then pushed by a centrifugal pump at 4-6 which is driven by torque, Q, drawn from the engine. The cooling liquid passes through the hot engine block, where it removes heat. Note that the temperature difference places a heat demand on the block, so the causality arrows are directed as shown.

The engine block is modeled here as a zero-resistance thermal conductor for heat which is connected between the constant-temperature combustion source at 7, and the fluid-flow heat removal.

The plumbing through the engine block and back to the heat exchanger is shown in the multiport diagram, but the fluid dynamics of this is usually neglected. Those modelers who wish to model these effects may use a RIC component to do so.

The radiator heat exchanger then provides the means for the rejection of the heat from the engine to the atmosphere, thus completing the cycle.

The schematic of Fig. 7-24c shows a common arrangement for a home- or building-heating system; and Fig. 7-24d shows the accompanying multiport diagram. The fan inlet in this system is connected to an atmospheric reservoir (not shown), which supplies the inlet conditions. The fan produces a certain flowrate in response to the electrical power that is input and the back-pressures reflected from all downstream components.

The heat exchanger in this system is really a kind of transducer, converting electrical power into heat flowrate. The effectiveness of this conversion is represented by the efficiency, ε, which models the net electrical power transferred to the flow by the heated elements. After realizing that $W_3 = W_2 = W$, this model may be shown to be

$$\Theta_3 = (1/C_{P3})\,[(\varepsilon I^2 R/W) + C_{P2}\Theta_2]. \tag{7-119}$$

The duct provides a means of directing the heated flow to the area of need, and insulation is provided to limit heat loss. The register is usually located in a wall, and is adjusted to fine tune the amount of heated air entering a room. Both the duct and the register may be well-modelled using the RIC component.

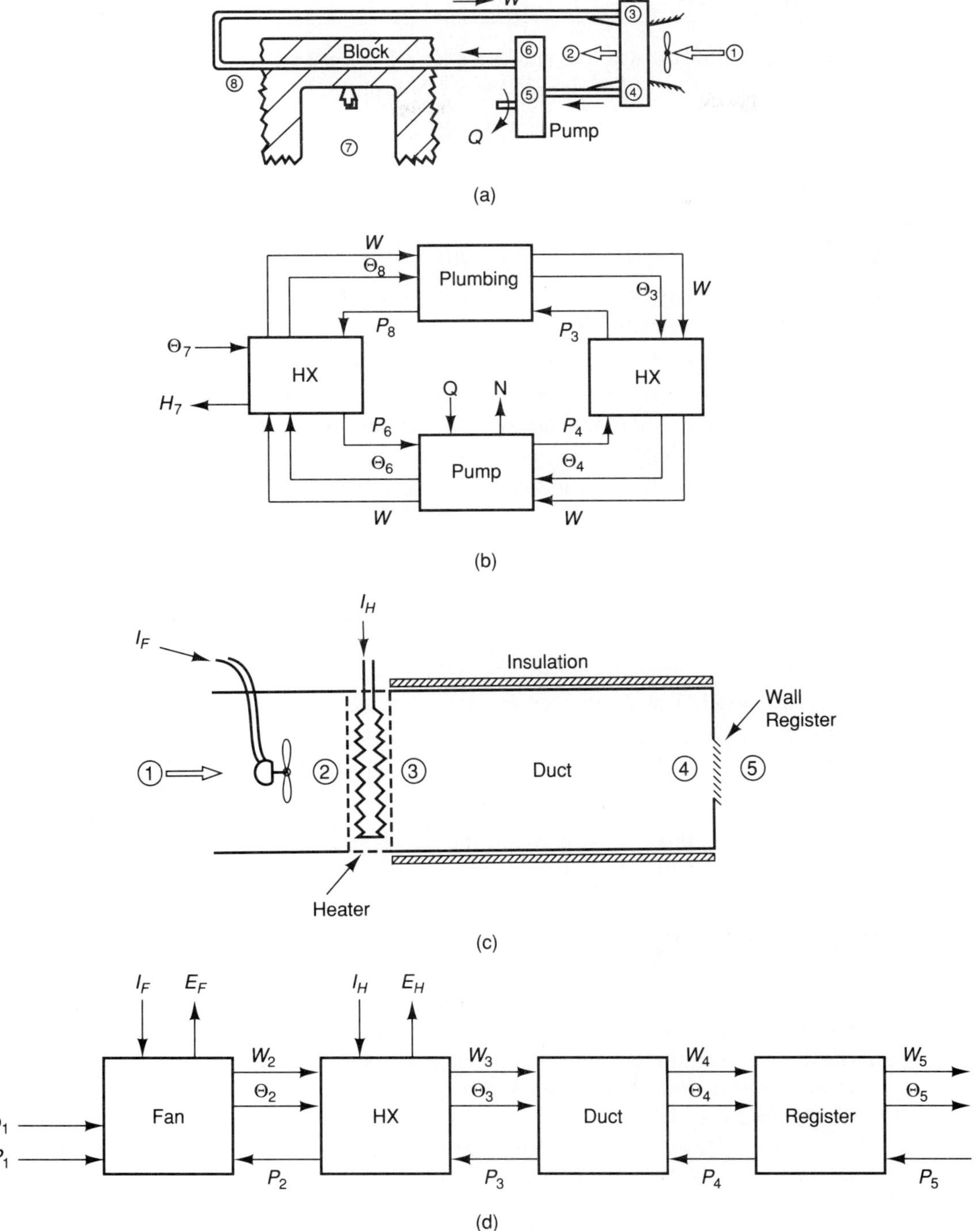

Figure 7-24. Common Convective Systems. (a) Liquid-cooled engine, (b) multiport for *a*, (c) space-heating, (d) multiport for *d*

HOMEWORK

7-20. The following steady-compressor data is measured near the operating point $(Q_c, P_o, N) = (20 \text{ ft-lb}_f, 25 \text{ psia}, 1000 \text{ rpm})$.

$$\text{At } N = 1{,}000 \text{ rpm:}$$

$$(Q_c, P_o) = (22, 20), (20, 25), (18, 30);$$

$$\text{at } P_o = 25 \text{ psia:}$$

$$(Q_c, N) = (18, 900), (20, 1{,}000), (22, 1{,}100).$$

a) Explain why this data is sufficient to begin to build a dynamic model of the compressor torque.

b) Use an eyeball best-fit to the data to find the model of the torque.

7-21. Given the electric-resistance heater of Fig. 7-24c. Starting from the definition of the heat-transfer efficiency for this heater, derive the model in the text.

7-22. Given the liquid-cooling system of Fig. 7-24a, the multiport of Fig. 7-24b, and a fluid system of the following nature:

$$W = 1 \text{ sl/min},$$

$$\Theta_8 = 160 \text{ °F},$$

$$\Theta_6 = 80 \text{ °F},$$

$$\text{fluid} = \text{Ethylene Glycol}.$$

How much heat is removed from the engine?

7-23. Given a well-insulated convective pipe with a large ℓ/d and a flow of air.

a) What are the equations for a RIC model of this flow?

b) If the operating point is at $P = 30$ psia, $\Theta = 600$ °R, and $R_f = 35 \text{ s-lb}_f/\text{ft}^5$ for this pipe. What is R_f' for the RIC flow model?

7-7 DIGITAL COMPONENTS

From a system point-of-view, it may well be that the most difficult part of modeling a digital component is to understand the three steps that this component uses to perform its role and interface with the analog components in the system. These steps are: analog-to-digital conversion (ADC), programmed computation, and digital-to-analog conversion (DAC). These steps must all be completed in a timely way so that the digital component does not introduce unacceptable delay into the system. However, the ability of the component to sample a large amount of data at uniform time steps, and to process that data automatically (say, for LSE fits), makes the digital component a very common part of modern systems. In this section, each of the three digital steps is considered in order to develop a useful model of the digital component.

While this study does go a little beyond simple input-output modeling, it does not go so far as to model the *elements* which comprise the heart of all digital components. These elements include the microprocessor(s), the memory elements, the clock, the data buses, and the registers, among others. It is assumed that these all work in a transparent way to provide smooth digital functioning so that you may concentrate on the overall digital component modeling problem.

Digital Operations. The operation of the digital component that is studied here is shown in Fig. 7-25.

Note that the multiport diagram in Fig. 7-25a shows a component governed by a clock which synchronizes event timing in the computer every Δt seconds. The dashed arrows in this figure show the presence of electrical currents in true ports, yet the essence of good digital design is that these are not of much concern. Thus, the digital computer is generally modeled as a *voltage*-processing device.

The input, u, at the left of the multiport is a continuous analog signal coming from the connected analog component, usually a voltage from a sensor. This is converted by the ADC into a discrete sampled signal, u_i, when signaled by the computer. Often, the computer program (called an "algorithm") contains a "READ" statement which actually triggers the start of ADC.

The computer uses the sampled input to compute and store the corresponding discrete output, y_i. This digital number is sent immediately to the DAC by the com-

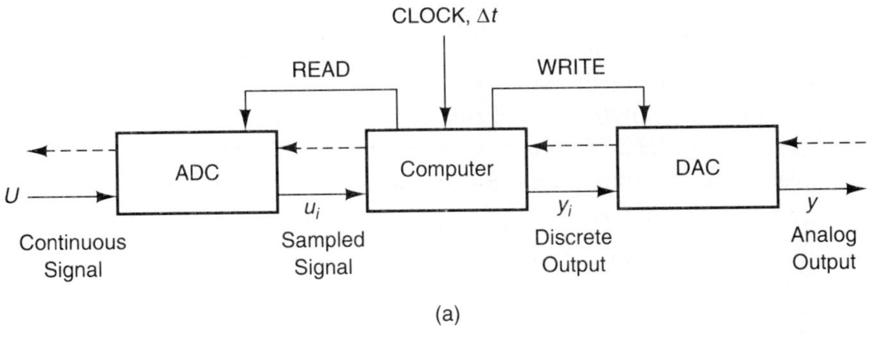

(a)

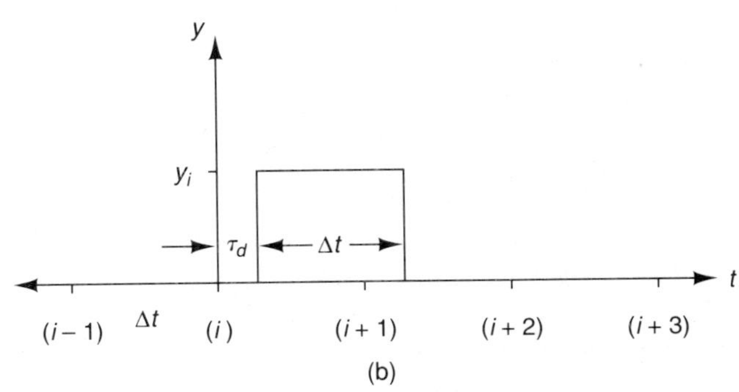

(b)

Figure 7-25. Digital Computer Operations. (a) Multiport diagram, (b) output history

puter (say, using a "WRITE" statement) so that it may be converted into a continuous analog output for the connected analog component. Thus, the DAC usually sends a voltage to an actuator of some sort.

The timing of this sequence of events is shown in Fig. 7-25b. Here the cycle starts at time step "i" when the computer algorithm is started. The computer begins by telling the ADC to read, it then computes the digital value of y for the time i, it tells the DAC to write, and the DAC converts the digital output into an analog output. Collectively, the time for all of this is called the *computational delay*, τ_D. In a well-designed system, the delay is very small compared to the sample interval, Δt.

The most common type of DAC is called a *zero-order hold* because it creates a steadily held, constant output. Higher-order holds include the first-order hold (which outputs a constant slope), the second-order hold (which has a constant curvature), etc.

So, each of the three digital steps in the computational delay (ADC, computation, DAC) must be modeled as you build a model of the digital component.

Analog-to-Digital Conversion (ADC). Analog-to-digital converters are also called "samplers," and digital components which use them are thus called "sampled-data" components. The operation of these samplers centers around the two issues of *resolution* and *sampling rate*.

An example of the resolution of a three-bit sampler is shown graphically in Fig. 7-26. Here the sampler is set to convert an analog input signal, U, into its binary equivalent for a computer interface. The input signal is expected to be in the closed interval between U_{min} and U_{max}, as shown on the horizontal axis. The output must be one of the eight binary numbers shown on the vertical axis since these are all the possible combinations of the three bits in the converter. Thus, note that the vertical axis is really a binary counting of the fractions allowed by the three-bit sampler. That is, since there are only three bits, you only have $2^3 = 8$ unique com-

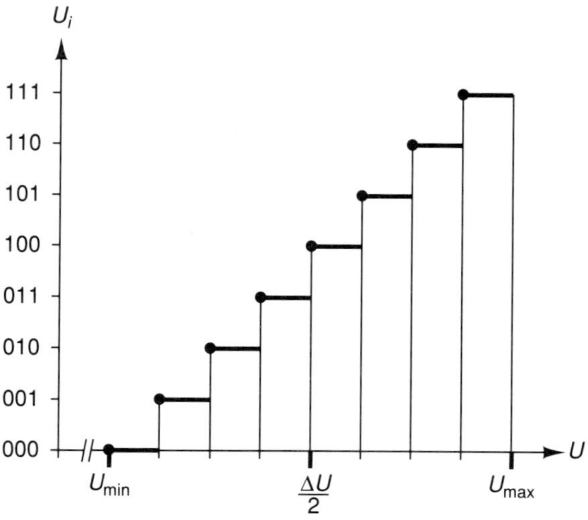

Figure 7-26. Three-bit Sampler Discretization

binations of the bits. These combinations are shown in ascending order in the figure, each of which accounts for 1/8th of the total input spread, which is

$$\Delta U = (U_{\max} - U_{\min}). \tag{7-120}$$

The usual designer's interpretation of *exact* binary conversion takes place only at the extreme left ends of each input range shown in Fig. 7-26. That is, when the analog signal is in the range $U_{\min} \leq U < U_{\min} + \Delta U/8$, then the conversion value is $u_i = 000$; and $u_i = 000$ is exact when $U = U_{\min}$. In this, the maximum measurement error, e_M, is equal to the size of the smallest division. Or,

$$e_M = \pm(1/2)^k \, \Delta U, \tag{7-121}$$

where k is the number of bits in the sampler. So, for a three-bit sampler, the maximum measurement error is $\pm 12.5\%$; for an eight-bit sampler, it is $\pm 0.391\%$.

In addition to a fine resolution in the sampler, it must sample at a rate, ω_s (samples/sec.), fast enough to capture any transients in the input and to preserve input dynamic richness. This input richness occurs in two forms: that for time-domain transients and that for superposed frequencies.

For time-domain transients, it is assumed that the inputs are characterized by the time constant, τ (sec.), or the natural frequency, ω_n (rad/sec.). Thus the sampling guidelines become

$$\omega_s \geq 10 \text{ samples}/\tau \text{ (sec)}, \tag{7-122}$$

or

$$\omega_s \geq 10 \text{ samples/(cycle period)}, \tag{7-123a}$$

which is

$$\omega_s \geq 10\omega_n/2\pi \text{ (samples)(rad/sec)/(rad/cycle)}. \tag{7-123b}$$

If the sampled analog signal is a periodic signal, then it has the harmonic-frequency content

$$u = \Sigma \, U_n \sin(n\omega_o t + \phi_n). \tag{7-124}$$

That is, the periodic input has various harmonics from $n = 0$ to ∞ in a Fourier sense, and the amplitudes of the harmonics may be plotted verses ω (or $n\omega_o$) as shown in Fig. 7-27.

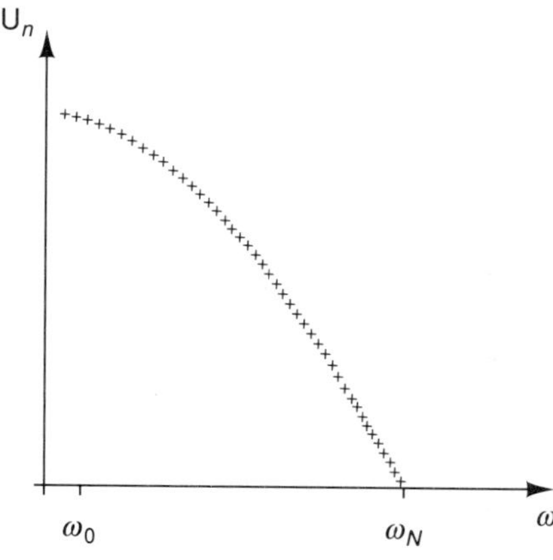

Figure 7-27. Harmonic Amplitudes versus Frequency

Note that the input harmonic amplitudes, U_n, must decrease to zero in the limit $n \to \infty$ since the Fourier series is convergent. However, the frequency at which the amplitudes are *acceptably* zero is $\omega_N = N\omega_o$. The input harmonic that corresponds to this highest-significant frequency is

$$u_N = U_N \sin(\omega_N t + \phi_N). \tag{7-125}$$

Since there are two unknowns in this equation (U_N and ϕ_N), at least two data points are required per cycle in order to fit the above equation (say, in an LSE manner), and give a rich enough representation of the input. That is, you need a sample rate, ω_s (samples/sec), compared to the highest significant frequency, ω_N (rad/sec), which meets

$$\frac{\omega_s \text{ (samples/sec)}}{\omega_N \text{ (rad/sec)}} 2\pi \text{ (rad/cycle)} > 2 \text{ (samples/cycle)}. \tag{7-126}$$

Or,

$$\omega_s > (1/\pi)\, \omega_N. \tag{7-127}$$

A sampling frequency that is high enough for the maximum significant frequency is high enough to capture all the lesser frequencies as well. Moreover, this analysis can be proven in more rigorous mathematics, which is presented elsewhere under the title of *Shannon's Sampling Theorem.*

The converse of the sampling relationship is also true, a slow sampling frequency (large Δt) can act as a filter and determine the highest identifiable frequency.

Thus, the sampler must be accurate and fast to be useful, and you must look to resolution and sampling rate in order to model how the sampling is performed.

The next step in the digital process is the computer algorithm—it too must be fast to keep pace with the analog components.

Algorithm Timing. The computing part of a digital component is where the program logic is executed. This logic includes arithmetic operations, logic operations, and information transportation and storage, all presented within a string of machine instructions called the "algorithm."

Example 7-6

A designer wants to implement a digital filter of the form

$$\tau\, dy/dt + y = u,$$

such that the sample rate is $\Delta t = 0.01$ sec., and the time constant for the filter is $\tau = 0.1$ sec. What is the algorithm, in FORTRAN, that will accomplish this task?

Solution

The filter is implemented through a finite-difference form of the analog equation above,

$$\tau\,(y_i - y_{i-1})/\Delta t + y_{i-1} = u_{i-1},$$

or,

$$y_i = y_{i-1} + \Delta t(u_{i-1} - y_{i-1})/\tau.$$

For the given constants, this becomes

$$y_i = y_{i-1} + 0.1(u_{i-1} - y_{i-1}).$$

The program works by running a variable named "CLOCK" and testing it against the computer time, stored under the reserved variable named "TIME," to test if the time step has elapsed. If the step has elapsed, then the CLOCK is reset, and the finite-difference equation for the filter is computed.

The program is as follows:

```
C       INITIALIZE THE PROGRAM FOR THE FIRST PASS:
01   YIM1 = 0.0
02   UIM1 = 0.0
03   CLOCK = TIME
C       COMPUTE THE TIME ELAPSED:
10   TEST = TIME - CLOCK
C       CHECK THE SYSTEM INTERRUPTS:
11   READ ( , ) INT
12   IF (INT .EQ. 1) PAUSE
C       TEST THE TIME ELAPSED AGAINST THE TIME STEP:
13   IF (TEST .GE. 0.01) GO TO 30
C       RECYCLE IF THE TEST FAILS:
20   GO TO 10
C       RESET THE CLOCK IF THE TEST PASSES:
30   CLOCK = TIME
C       THE ADC SAMPLER READS:
32   READ ( , ) UI
C       THE OUTPUT VALUE IS COMPUTED:
33   YI = 0.1*(UIM1 - YIM1) + YIM1
C       THE DAC WRITES:
34   WRITE ( , ) YI
C       THE VALUES ARE STORED FOR THE NEXT CYCLE:
35   YIM1 = YI
36   UIM1 = UI
C       RECYCLE THE CLOCK:
37   GO TO 10
40   STOP
     END
```

Comments

The program is set up as an infinite loop through the use of the unconditional GO TO statements. The operator, or the computer, may halt the program through the interrupts, INT. While only one interrupt is shown in this example program, many may actually be required to safely control the component. However, each logic test takes time—too many will make the component too slow.

The computer stores the value of the current time under the reserved variable TIME. Sometimes this is done in a floating point number, as in the example. In any case, the programmer (and modeler) must be careful to use consistent number representations for CLOCK, TIME, and TEST.

There are several ways to determine the execution speed of an algorithm. One way is to resort to the manufacturer's literature for the microprocessor(s) to be used. This literature gives the execution times for each of the building-block operations (add, subtract, multiply, IF, GO TO, storage assignment, etc.). You then apply these times to the algorithm to find the net execution time. This may be the only approach possible for proposed new designs. However, if the component has an existing microprocessor, then a simpler approach is possible.

The simpler method is to program the chip to execute the algorithm such that all the IF and GO TO statements always go through the most time-consuming path, for every CLOCK step. Then the chip is programmed to run, say, 1,000 or 10,000 cycles of the algorithm so that you may measure the net execution time. Division then gives a good estimate of how long the chip will take to run a single pass-through of the worst-case computations of the algorithm.

For the above example, the simple method may be done in three steps. First, change the TEST statement so that it reads

```
13      IF(TEST .GE. 0.0) GO TO 30.
```

Then, put the entire loop, from statement 10 to 37, into a DO loop with the counter running from 1 to 1,000. Finally PRINT out TIME before and after the execution. Division of the difference of TIME by 1,000 then gives an estimate of the single, worst-case pass-through time.

Modeling of a computer is usually done with another computer. Consequently, the same algorithm is programmed into the model as that in the component to be modeled. The issue of timing then has to do with the time step on which the outputs actually occur.

Digital-to-analog Conversion (DAC). The final step of the digital component is to convert its digital output into an analog output which is suitable for the connected analog components. By far, the most common method of DAC is through a zero-order hold.

The model for the zero-order hold is a step at time $(t + \tau_d)$, with a canceling step at $(t + \tau_d + \Delta t)$. In the time domain this is

$$y_i(t) = y_{mi} [u_s(t - \tau_d) - u_s(t - \tau_d - \Delta t)], \tag{7-128}$$

where y_{mi} is the constant magnitude of the step in the interval.

The Digital Component Model. You are now in a position to find the model for a digital component. This model is based upon the evaluation of the resolution and timing of the component. One of three cases may arise: first, the component may be so fast, and of such fine input-output resolution, that it *appears* to the connected components as if it is an analog component; second, it may have fast ADC and very fine resolution, but it also has a large computational delay; and third, it has a coarse input *or* output resolution.

In the first case, the computational delay is very small compared to the sample rate, and the sample rate is very small compared to the time constants of the connected components. Further, the digital resolution at input and output are small enough to give negligible errors and discontinuities in the output(s). This leads to a situation where the digital component may be modeled with an analog model of whatever is programmed. Such components are called *analog-mimic* components. In Ex. 7-6, an implementation of this kind had as its time-domain model the filter equation

$$\tau\, dy + y = u, \tag{7-129}$$

while its frequency-domain model is

$$T(s) = 1/(\tau s + 1). \tag{7-130}$$

In the second case, the input-output resolution is fine, but the computational delay is significant. This time, you model the component as an analog component with delay. Again, this is an analog-mimic component. For an implementation of this sort for Ex. 6-7, the models are

$$\tau\, dy + y = u(t - \tau_d), \tag{7-131}$$

and

$$T(s) = \frac{e^{-\tau_d s}}{(\tau s + 1)}. \tag{7-132}$$

In the third case, the input or output resolution is too coarse to ignore. You must model these digital components, and their system, from step-to-step in the time domain. They are usually modeled by simulation on another digital computer. Building their model starts with modeling the ADC sampler as a device which converts an analog signal into a digital number at certain system sample times. The delay of the computer algorithm must next be estimated. The output of the computer is then modeled using the exact same algorithm which the computer uses to compute y from u at the sample times. Finally, the DAC is modeled as a zero-order hold at a delayed system time, as indicated by Eq. 7-128.

HOMEWORK

7-24. What is the maximum measurement error in a 12-bit ADC?

7-25. Program the computer code given in Ex. 7-6 and estimate the worst-case execution time by making it cycle 10,000 times. Will this program execute quickly enough on your computer to be a small fraction of the given Δt?

7-26. What minimum sample rate (samples/sec.) is necessary to capture all of the frequencies in

$$u = 10 \sin(3t) + 6 \sin(12t + 10°) + 4 \sin(48t + 20°)?$$

7-27. What minimum sample rate (samples/sec.) is necessary to capture the dynamics of the following:

a) $0.01\, dy/dt + y = 5u$,

b) $0.01\, d^2y/dt^2 + 0.1\, dy/dt + y = 0.01\, u$?

7-8 SUMMARY, BIBLIOGRAPHY, AND REVIEW PROBLEMS

Some components are so complicated that the best approach to their model is through input-output modeling. These complications are sometimes seen in the form of a fundamentally complicated process, such as that seen in fluid-thermal or compressible flows. Sometimes, the complications are seen in the form of a very large number of elements which are at work within the component, such as in digital components.

In either case, the nature of preliminary design requires you to capture the component performance in the simplest model possible, while preserving the characteristic dynamics. The modeling approach that engineers use to do this entails two tasks: first, finding the *form* of the model, then conducting a *fit* of that form to experimental data.

These tasks are conducted in one of two ways: using either a characteristic modeling approach, or using an in-system modeling approach. In the former, the component is isolated so that carefully controlled, input-output, bench-test experiments may be conducted to complete the form and fit evaluation tasks. In the latter, the component is tested while it is connected into its operating system.

This chapter discussed the characteristic modeling approach as it appears during preliminary design. This approach was studied in two forms, separated by whether the experimental inputs are generated by near-ideal or by nonideal sources.

Given ideal sources, the steady input-output linearity of the component was first checked experimentally, and the steady model form was determined. This was done for both time-domain and frequency-domain components. The steady-model coefficients were then fit to the experimental data using the Least-Squared-Error (LSE) method.

Next, the ideal singularity inputs were used to give data to find the rate effects for the time-domain model. These rate effects were then modeled using terms which were added as necessary to the steady-model form to account for the observed transient characteristics. Finally, the unknown rate coefficients in the model were fit using the LSE method.

At the end of the chapter, the nonideal-source characteristic modeling method was applied to the convective and digital components. This modeling procedure was similar to that for the component with ideal-source inputs, except that the input-output causality was reasoned from the multiport diagram and knowledge of the component behavior, rather than from experimental data.

So, having completed study in the previous chapters, you thus find yourself at a point where you have an array of modeling tools which can handle the vast majority of engineering components. The next modeling task is the assembly of these component models into a system model, and the prediction of the system performance. This is discussed in the next chapter.

BIBLIOGRAPHY

Astrom, K.J. and B Wittenmark, *Computer Controlled Systems*, Englewood Cliffs, N.J.: Prentice-Hall, Inc., 1984. There are excellent discussions of digital sampling and computerized LSE fits in this text, as well as further general reading on the design of computer-controlled systems.

Baumeister, T. and L.S. Marks (eds.), *Standard Handbook for Mechanical Engineers*, New York, N.Y.: McGraw-Hill Book Co., 1967.

Boyce, M.P., *Gas Turbine Engineering Handbook*, Houston, Texas: Gulf Publ. Co., 1982.

Doeblin, E.O., *System Modeling and Response*, New York, N.Y.: J. Wiley & Sons, 1980. This book has a nice discussion of experimental set-up and data use.

Fraas, A.P. and M.N. Ogisik, *Heat Exchanger Design*, New York, N.Y.: J. Wiley & Sons, Inc., 1965.

McNaughton, K.J., *The Chemical Engineering Guide to Heat Transfer, Volume I—Plant Principles*, New York, N.Y.: McGraw-Hill Book Co., 1986. This book describes many practical aspects of heat exchangers.

Natke, H.G. (ed.), *Application of System Identification in Engineering*, New York, N.Y.: Springer Verlag, 1988.

Norton, H.N., *Sensor and Analyzer Handbook*, Englewood Cliffs, N.J.: Prentice-Hall, Inc., 1982.

Schouken, J. and R. Pintelon, *Identification of Linear Systems*, New York, N.Y.: Pergamon Press, 1991. This is a thorough text, but a little heavy in the mathematics. It is advanced reading.

Singh, J., *Heat Transfer Fluids and Systems for Process and Energy Applications*, New York, N.Y.: Marcel Dekker Inc., 1985.

Tse, F.S. and I.E. Morse, *Measurement and Instrumentation in Engineering*, New York, N.Y.: Marcel Dekker Inc., 1989.

Walker, G., *Industrial Heat Exchangers*, New York, N.Y.: Hemisphere Publ. Co., 1990.

REVIEW HOMEWORK

7-28. A MIMO component has the following actual performance

$$y_{1A} = 3u_1^2 + 7u_2.$$

A modeler has fit a linear model to data from this component of the form

$$y_{1L} = 6u_1 + 7u_2.$$

a) How much relative error is there in y_{1L} when

$$u_1 = 3, \text{ and } u_2 = 5?$$

b) Is this error within engineering error?

7-29. A modeler has completed the validation of the linearity of a SISO component using an ideal input source. Judging from the strip-chart recordings of characteristic transient data for this component shown in Fig. 7-28, comment on the model order, delay, damping ratio, natural frequency, and whether this component may act to differentiate the input.

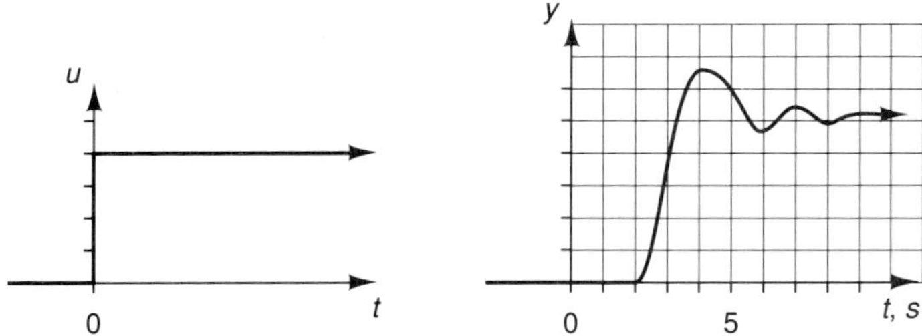

Figure 7-28. Time Histories for Prob. 7-29

7-30. Following the determination of SISO component linearity, a modeler has generated the steady ideal-characteristic frequency data shown in Fig. 7-29. Given this data, estimate the model order, delay, any time constants, and whether this component acts to differentiate the input.

7-31. Given the following steady ideal-characteristic input-output data for a SISO component: $(u, y_m) = (0, 0), (1, 1), (2, 4), (3, 9)$. Find the linear steady LSE model for this component which passes through $(u, y) = (0, 0)$.

7-32. Given the following steady ideal-characteristic input-output data for a SISO component: $(u, y_m) = (0, 0), (1, 1), (2, 3), (3, 5)$. Find the quadratic steady LSE model for this component which passes through $(u, y) = (0, 0)$.

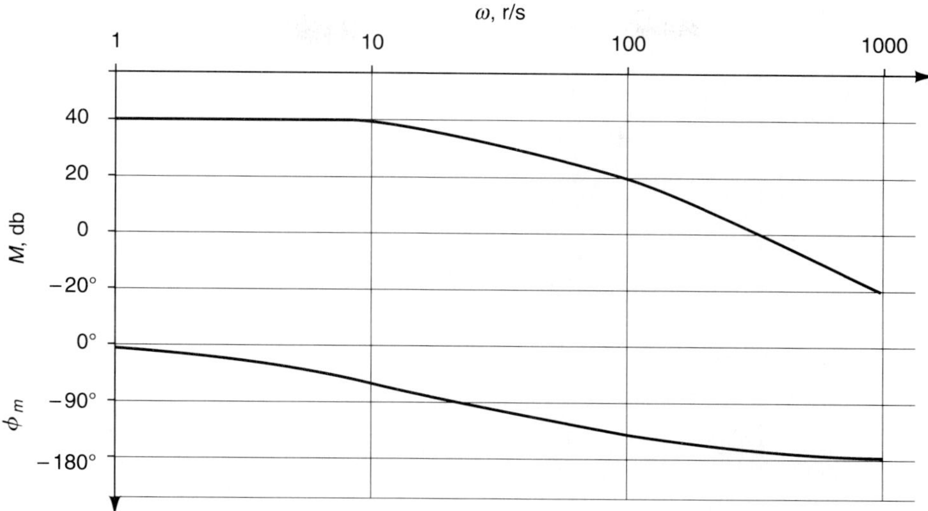

Figure 7-29. Bode Diagram for Prob. 7-30

7-33. A modeler has checked the linearity of a time-domain SISO component and found that

$$y = 5u.$$

Some measured ideal-characteristic transient data for the component in response to an input step $u = 3u_s(t)$ is

t(sec.)	0.0	0.1	0.2	0.3	0.4	0.5	0.6	0.7	0.8	0.9	1.0	1.1
y_m	0.0	1.0	1.9	2.8	3.7	4.5	5.0	5.4	5.6	5.8	5.9	6.0

Fit a first-order dynamic LSE model to this data.

7-34. A modeler has verified the linearity of a SISO frequency-domain component. The Bode diagram in Fig. 7-30 shows the data. Use the visual method to estimate the transfer function for this data.

7-35. Often, when an inlet design is used with a compressor in a jet engine system, it is more convenient to functionally combine the inlet with the compressor, as shown in Fig. 7-31. Use this multiport diagram to find the form of a steady, linear model for the output speed, n, of the combined component.

7-36. A fit of a first-order dynamic model to the transient data for the compressor of the previous problem has large error in the predicted speed transient. To remedy this, the modeler wants to use a second-order dynamic model ($\zeta = 0.05$, $\omega_n = 1$). What form of the model is required to combine these constants with those from the previous problem?

7-37. A periodic analog signal is expected to have the frequency content shown in Fig. 7-32. This analog signal is to be fed into a digital component. What minimum sample rate for an ADC is needed to capture this signal for digital processing?

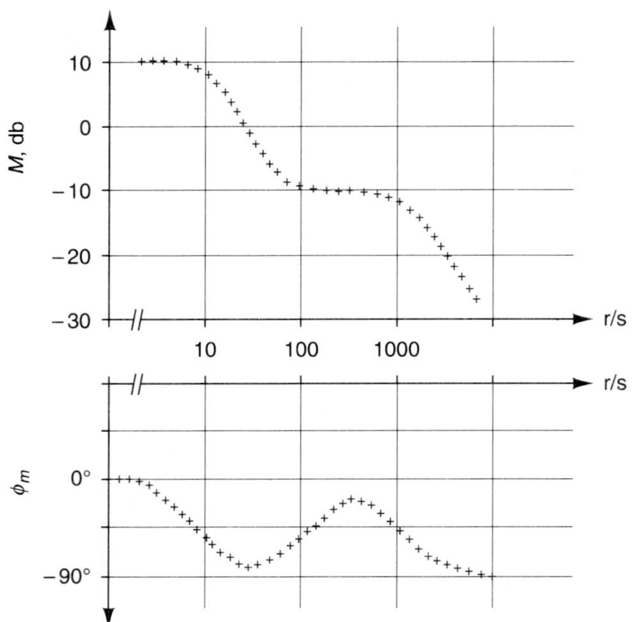

Figure 7-30. Bode Diagram for Prob. 7-34

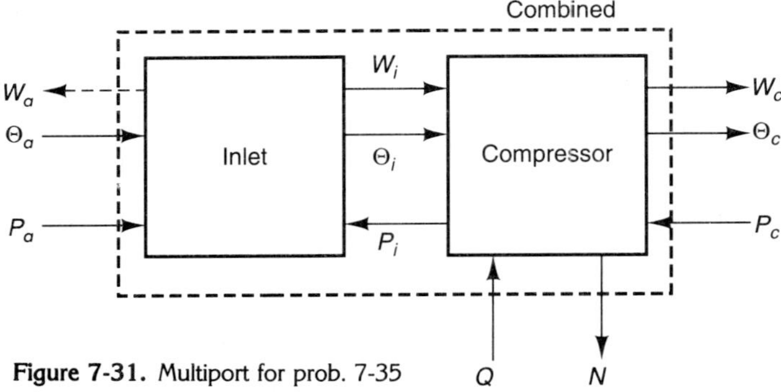

Figure 7-31. Multiport for prob. 7-35

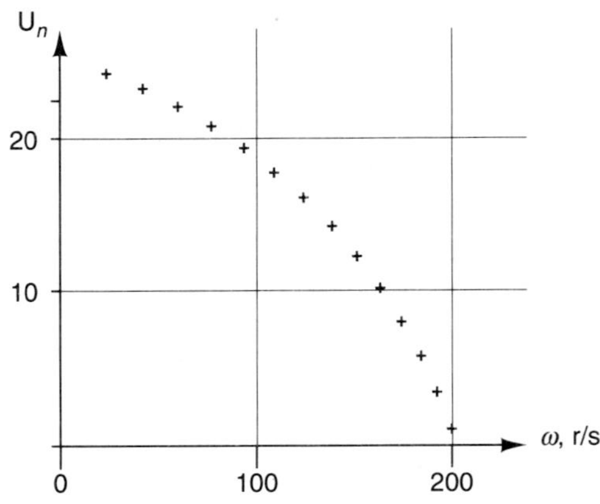

Figure 7-32. Frequency Content for Prob. 7-37

7-38. A certain steady input signal to a digital component is expected to have $U_{min} = 10$, and $U_{max} = 18$. A 3-bit ADC will be used to translate the input into a digital number. What digital value will be given by the ADC when $U = 15.5$?

7-39. A modeler wants to use a RIC component to model a convective liquid flow through a flat-plat orifice:

a) Find the model equations.

b) Use the theoretical orifice-flow model of Sec. 4-8 to find R_f' for the RIC model.

7-40. The following steady-pump data is measured near the operating point $(W_o, P_o, N) = (9\text{sl/s}, 100\text{psi}, 500\text{rpm})$:

$$\text{At } N = 500 \text{ rpm},$$

$$(W_o, P_o) = (10, 90), (9, 100), (8,110).$$

$$\text{At } P_o = 100 \text{ psi},$$

$$(W_o, N) = (8, 400), (9, 500), (10, 600).$$

a) Explain why this data is sufficient to start a dynamic model of the pump outlet flowrate,

b) Use an eyeball best-fit to the data to find a model of the flowrate output?

Chapter 8

Mathematical Analysis: System Model Assembly and the Predicted Response

8-0 INTRODUCTION

At this point in the system model-building process you have accomplished two tasks: You have constructed a cause-and-effect multiport diagram of the system in order to identify functional components within the system and you've built models of every component. Your task now is to use the multiport diagram to assemble the system components and accurately predict the system response.

Note that every engineering system may be composed of four different types of components, all discussed previously:

1. The time-domain analog component. These are discussed at length in Chaps. 4 and 7. They are modeled with a set of linear or nonlinear differential equations.

2. The frequency-domain analog component. These are discussed in Chap. 6. In this introductory study only the linear variety of this component is considered, which comes from a fundamental modeling effort. They are modeled by a transfer function (SISO components), or by a transfer matrix (MIMO components).

3. The analog-mimic digital component. These digital components work fast and with a fine resolution, so that they may be considered to be analog components. They are briefly considered at the end of Chap. 7. They are modeled with the time-domain algebraic or differential equations which they are programmed to execute.

4. The large-step digital component. These components work with a very coarse input or output resolution. They must be modeled as step-producers. They are discussed briefly at the end of Chap. 7. They are modeled with the time-domain set of logic and arithmetic operations which they are programmed to execute in producing their step outputs.

Note that some of the component models may be expressed in the frequency domain through transfer ratios or transfer matrices, while other component models may be expressed in the time domain through differential equations. But model assembly requires consistent form and you must now express all the components in one domain or the other.

The usual approach to MIMO system assembly requires that all component models be expressed in the time domain. Thus, all frequency-domain models must first be converted into the time domain by clearing all the transfer-function ratios and computing the inverse Laplace transform of all the s-domain model equations.

Other texts assemble system models in the frequency domain, using a method of component block-diagram algebra, but this approach is only recommended for further reading here. This method is most commonly used for linear SISO components, and it faces severe difficulties in nonlinear or MIMO systems.

An example of a four-component system to be assembled is shown in Fig. 8-1. Note that each internal-component output matches with another component input. These are, in fact, the basic relationships which you will use later to assemble the model. However, the types of components in the system also guides your assembly approach.

In this text, we will follow the procedure for system model assembly and simulation shown in the decision tree of Fig. 8-2. Here, you enter the top of the flowchart with a set of component models and a system multiport diagram. The components are then converted into the time domain as discussed above, and assembled using the port definitions in the multiport diagram. The rest of the mathematical analysis then follows the decision tree shown in the figure. The organization of this chapter also follows this analysis tree, as indicated by the chapter section numbers above the boxes in the figure.

The first decision in the analysis is whether the set of "analog" equation models is linear or not. If any equation in the set is nonlinear, then the whole set is nonlinear. Nonlinear modeling is not a focus of this text. However, the nonlinear branch

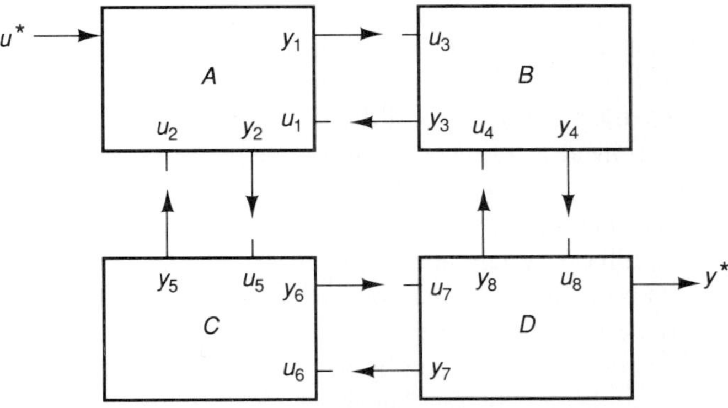

Figure 8-1. Component Connections in a Simple System

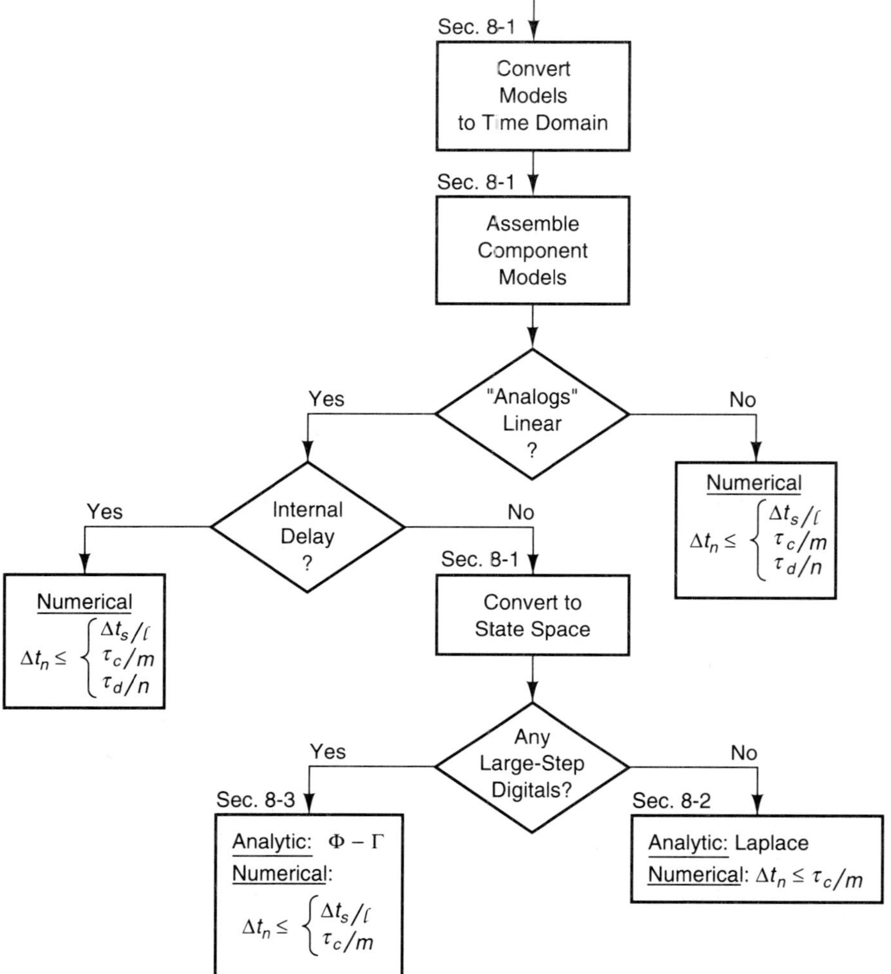

Figure 8-2. Mathematical Analysis Flow Chart

usually leads to a numerical simulation of the model, as seen in Chap. 5. This simulation must use a time step, Δt_n, which is an integer fraction of the smallest time constant(s), τ_c, the smallest delay, τ_d, and the smallest ADC step-size, Δt_s.

For systems with linear analog components, the evaluation of internal-component delay is the next critical task. That is, significant internal-component delays usually force you to use a numerical approach, with an integration step size chosen in a similar way to the nonlinear system above. However, if the internal delays in a linear system may be ignored, then an extension of the state-space transformations of Chap. 5 is used on the analog system parts.

The final analysis decision is guided by the presence of large-step digital components. If such are connected to the linear analog system, then the **Φ-Γ** (phi-gamma) method works well to predict the analytic system response. If a numerical simulation of such a system is desired, then you may use a numerical method, with an integration step size chosen to meet ADC sample-rate and time constant restrictions.

For a linear system without large-step digital components, you may resort to the Laplace method to find an analytic response in either the frequency domain or the time domain. Or you may numerically simulate the system response in the time domain, choosing the integration time step based upon the smallest system time constant.

Notice across the bottom of Fig. 8-2 that you may *always* use a numerical method on the digital computer to simulate the response of *any* system model.

Furthermore, when you predict the system response, you find that both the system transient and the system steady state are shaped by the constants in the model. If you trace the origin of the model constants, you find that they come from the fundamental physical parameters of the system. These are the masses, resistances, capacitances, spring rates, etc. They are the fundamental design parameters in which the designer is interested. They are the things which the designer will change to shape the system response to meet the design objectives. Consequently, the sensitivity of the system response to the model parameters is usually of great interest to the designer.

So, in this chapter you first study how to assemble the system model. You then develop tools to simulate the system response using the model. And you finally study how to compute the system model parameter sensitivities to improve the system design.

8-1 ASSEMBLING THE SYSTEM MODEL

The system model is assembled using the system multiport diagram. This diagram is the "top down" functional basis for the component models which are to be assembled—it is the best possible basis for a "bottom up" reassembly of the system. In particular, the port definitions are the keys to mathematically assembling the system model.

The system assembly takes place in two steps: first, the component models are all expressed in time domain form; and second, the port input-output pairings are used to mathematically connect the equation set. These two steps will work for linear or nonlinear systems.

Component Model Form. Converting all component models into the time domain means that frequency-domain components require special evaluations.

The frequency-domain components considered here are modeled using a transfer function. These transfer functions are a ratio of polynomials in the Laplace variable, s, which are found by transforming a time-domain differential equation. Further, as seen in Chap. 6,

$$s^n\, Y(s) = d^n y/dt^n. \tag{8-1}$$

Thus, to convert the transfer function back into the time domain you must take the inverse of the Laplace notation. This is easily done by clearing the transfer function ratio and reversing the notation as suggested by Eq. 8-1.

Example 8-1

Convert the transfer function

$$T(s) = Y(s)/U(s) = 5(s + 1)/(s + 3)(s + 2),$$

back into the time domain.

Solution
First, clear the ratio

$$(s + 3)(s + 2)Y(s) = 5(s + 1)U(s).$$

Then reverse the notation

$$d^2y/dt^2 + 5\ dy/dt + 6y = 5\ du/dt + 5u.$$

Connecting the Components. The components are "connected" into a system-model equation set in three steps. First, the mathematical equalities which represent the port connections are defined. Second, the port equalities are substituted into the *right-hand sides* of the component output equations. And third, the equation set is simplified to eliminate all algebraic equations for the internal outputs, y. (Note that *external* variables are asterisked—as in u^* and y^*.)

Example 8-2

Consider the simple two-component system shown in Fig. 8-3. The two components are to be connected at ports 2 and 3 in the manner shown. Find the system model.

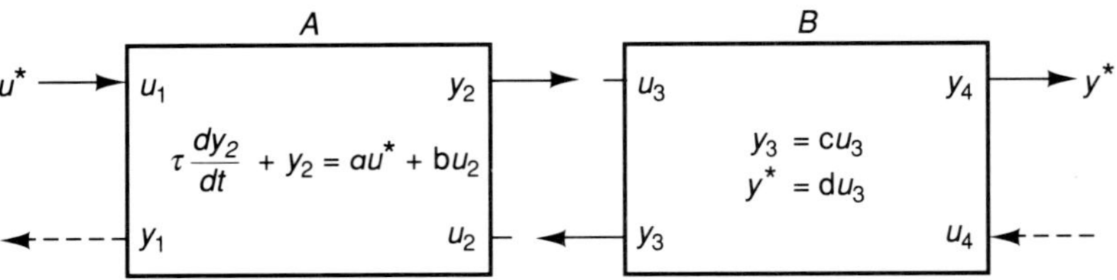

Figure 8-3. Example Two-component System

Solution
Both of the component models are already in the time domain, so the three-step connecting procedure may be performed.

The first step is to write down the port equalities for all port connections. These are

$$u_2 = y_3,$$

and

$$u_3 = y_2.$$

The second step is to substitute these relations into the *right sides* of all component equations. This gives

$$\tau\, dy_2/dt + y_2 = au^* + \mathrm{b}y_3,$$

$$y_3 = \mathrm{c}y_2,$$

and

$$y^* = \mathrm{d}y_2.$$

The third step is to eliminate all algebraic equations for the internal outputs. This is done by substituting the second relation into the *right side* of the first,

$$\tau\, dy_2/dt + y_2 = au^* + \mathrm{bc}y_2.$$

Simplifying this equation leaves you with the *system* model,

$$\tau\, dy_2/dt + (1 - \mathrm{bc})y_2 = au^*,$$

$$y^* = \mathrm{d}y_2.$$

Comments

Note first that there are no *internal* inputs in the system model. This is always the case.

Further, there are exactly as many differential equations in the system model as there are in all of the component models.

Also, if there are any large-step digital components in the system, then the elimination of internal inputs by substitution is as far as you can go for those components. They must be modeled with a numerical procedure on a digital computer. The state-transformations which follow are for the remaining analog parts of the system.

The above procedure will work quite well for either linear or nonlinear equation sets. However, further understanding of the linear, all-analog MIMO system may be achieved by transforming this system model into state space for design and simulation studies.

State-space Transformation. Recall that two methods were presented in Chap. 5 for system transformations. Method 1 is used for those linear or nonlinear systems which have no differentials of input. Method 2 is used for those linear systems which have differentials of inputs. In this section, an extension of the Method 2 transformation is presented to account for the MIMO nature of the system.

The state-space transformation of Chap. 5 is for a single output differential equation, with a single input. So, in general, there are n first-order state equations in the transformation in terms of n states for the nth-order differential equation. The basic form of this equation is

$$d^n y/dt^n + C_{n-1}\, d^{n-1}y/dt^{n-1} + \cdots + C_1\, dy/dt + C_0 y$$
$$= b_n\, d^n u/dt^n + b_{n-1}\, d^{n-1}u/dt^{n-1} + \cdots + b_1\, du/dt + b_0 u. \qquad (8\text{-}2)$$

The proper state selections for the state transformation are

$$x_1 = y - \beta_0 u, \qquad (8\text{-}3)$$
$$x_2 = dy/dt - \beta_0\, du/dt - \beta_1 u = dx_1/dt - \beta_1 u, \qquad (8\text{-}4)$$
$$x_3 = d^2y/dt^2 - \beta_0\, d^2u/dt^2 - \beta_1\, du/dt - \beta_2 u = dx_2/dt - \beta_2 u, \qquad (8\text{-}5)$$
$$\ldots = \ldots,$$
$$x_n = d^{n-1}x_{n-1}/dt^{n-1} - \beta_{n-1} u. \qquad (8\text{-}6)$$

And the β's are computed from the following

$$\beta_0 = b_n, \qquad (8\text{-}7)$$
$$\beta_1 = b_{n-1} - C_{n-1}\beta_0, \qquad (8\text{-}8)$$
$$\beta_2 = b_{n-2} - C_{n-1}\beta_1 - C_{n-2}\beta_0, \qquad (8\text{-}9)$$
$$\beta_3 = b_{n-3} - C_{n-1}\beta_2 - C_{n-2}\beta_1 - C_{n-3}\beta_0, \qquad (8\text{-}10)$$
$$\ldots = \ldots,$$
$$\beta_n = b_0 - C_{n-1}\beta_{n-1} - \cdots - C_1\beta_1 - C_0\beta_0. \qquad (8\text{-}11)$$

The nth state equation comes from the given differential equation (Eq. 8-2),

$$dx_n/dt = \beta_n u - C_{n-1}x_n - C_{n-2}x_{n-1} - \cdots - C_1 x_2 - C_0 x_1. \qquad (8\text{-}12)$$

Notice that this transformation works for an equation with or without input derivatives. It is thus the most general form of transformation for a *linear* differential equation. But the form expressed in Eq. 8-2 is for a SISO equation, and you are now interested in a MIMO transformation. The study of this transformation is begun with some observations about the model equation form.

The system assembly procedure discussed earlier in this section requires you to put all of the output differential equations into the form

$$d^n y/dt^n + C_{n-1}\, d^{n-1}y/dt^{n-1} + \cdots + C_1\, dy/dt + C_0 y$$
$$= b_n\, d^n u_1^*/dt^n + b_{n-1}\, d^{n-1}u_1^*/dt^{n-1} + \cdots + b_1\, du_1^*/dt + b_0 u_1^*$$
$$+ d_n\, d^n y_m/dt^n + \cdots + d_0 y_m. \qquad (8\text{-}13)$$

Note that the only derivative of input which may occur in this *system* equation is of the *external* input(s) since all the internal inputs were eliminated in the assembly. In addition, there may also be one or more internal outputs which have differentials on the right-hand side of the equation (only one is shown in Eq. 8-13 by way of example).

The n state selections that you now use are similar to those used in the SISO transformation, except that they now accommodate the MIMO system form. That is, the state selections are now

$$x_1 = y - \beta_0 u_1{}^* - \delta_0 y_m, \tag{8-14}$$

$$\beta_0 = \mathrm{b}_n,\ \delta_0 = \mathrm{d}_n, \tag{8-15}$$

$$x_2 = dx_1/dt - \beta_1 u_1{}^* - \delta_1 y_m, \tag{8-16}$$

$$\beta_1 = \mathrm{b}_{n-1} - \mathrm{C}_{n-1}\beta_0,\ \delta_1 = \mathrm{d}_{n-1} - \mathrm{C}_{n-1}\delta_0, \tag{8-17}$$

with the nth state selection

$$x_n = dx_{n-1}/dt - \beta_{n-1} u_1{}^* - \delta_{n-1} y_m, \tag{8-18}$$

$$\beta_{n-1} = \mathrm{b}_1 - \mathrm{C}_{n-1}\beta_{n-2} - \cdots - \mathrm{C}_2\beta_1 - \mathrm{C}_1\beta_0, \tag{8-19}$$

$$\delta_{n-1} = \mathrm{d}_1 - \mathrm{C}_{n-1}\delta_{n-2} - \cdots - \mathrm{C}_2\delta_1 - \mathrm{C}_1\delta_0. \tag{8-20}$$

The nth first-order, state-derivative equation always comes from the given differential equation (Eq. 8-13),

$$dx_n/dt = \beta_n u_1{}^* + \delta_n y_m - \mathrm{C}_{n-1}x_n - \mathrm{C}_{n-2}x_{n-1} - \cdots - \mathrm{C}_1 x_2 - \mathrm{C}_0 x_1. \tag{8-21}$$

Where

$$\beta_n = \mathrm{b}_0 - \mathrm{C}_{n-1}\beta_{n-1} - \cdots - \mathrm{C}_1\beta_1 - \mathrm{C}_0\beta_0, \tag{8-22}$$

and

$$\delta_n = \mathrm{d}_0 - \mathrm{C}_{n-1}\delta_{n-1} - \cdots - \mathrm{C}_1\delta_1 - \mathrm{C}_0\delta_0. \tag{8-23}$$

Note that you can extend this pattern of state substitution for as many derivatives on the right-hand side as you like. Further, if there are no right-side derivatives, then the β and δ terms are used to adjust accordingly.

So, each output differential-equation transformation will have at least one algebraic state definition like Eq. 8-14. When these are all grouped together, they may be rewritten in vector form as

$$\mathbf{y} = \mathbf{C}_o \boldsymbol{x} + \mathbf{D}_o \boldsymbol{u}^*. \tag{8-23}$$

Moreover, note that this set is only for the state transformation of a *single* output differential equation.

When all of the state-differential equations (those like Eqs. 8-16 through 8-21) are written together for the entire *system*,

$$d\boldsymbol{x}/dt = \mathbf{A}_o\boldsymbol{x} + \boldsymbol{\delta y} + \boldsymbol{\beta u}^*. \tag{8-24}$$

Substitution of all the output equations like Eq. 8-23 gives

$$d\boldsymbol{x}/dt = \mathbf{A}_o\boldsymbol{x} + \boldsymbol{\delta}[\mathbf{C}_o\boldsymbol{x} + \mathbf{D}_o\boldsymbol{u}^*] + \boldsymbol{\beta u}^*, \tag{8-25}$$

which may be rearranged into the desired state-space form of the system model

$$d\boldsymbol{x}/dt = \mathbf{A}\boldsymbol{x} + \mathbf{B}\boldsymbol{u}^*, \tag{8-26}$$

with

$$\mathbf{A} = \mathbf{A}_o + \boldsymbol{\delta}\mathbf{C}_o, \tag{8-27}$$

and

$$\mathbf{B} = \boldsymbol{\delta}\mathbf{D}_o + \boldsymbol{\beta}. \tag{8-28}$$

The state-space system model is completed when you write the system output equation in terms of the states and external inputs,

$$\boldsymbol{y}^* = \mathbf{C}\boldsymbol{x} + \mathbf{D}\boldsymbol{u}^*. \tag{8-29}$$

Thus, you have a general method for state-space transformation of a set of linear differential equations with derivatives on the right-hand side, but without significant internal delay. This method results in a state-space model which is fully specified by the **A, B, C,** and **D** matrices.

Example 8-3

Convert the following SISO system equations into a state-space model. Express your answer in matrix form.

$$d^2y/dt^2 + 3\ dy/dt + 4y = 2\ du^*/dt + 7u^* + 3\ dy_1/dt + 6y_1. \tag{8-30}$$

$$2\ dy_1/dt + y_1 = 3u^*.$$

$$y^* = 10y. \tag{8-31}$$

Solution

The first equation is for a second-order output ($n = 2$), with input derivatives. The second equation represents a first-order output, with no input derivatives. The last equation is an algebraic external output equation; save it until last.

The first state is selected for the first-order output equation,

$$x_1 = y_1, \tag{8-32}$$

since it has no input derivatives. The corresponding differential equation for this state is thus

$$dx_1/dt = -0.5x_1 + 1.5u^*. \tag{8-33}$$

The second-order output equation has *two* state selections, as follows

$$x_2 = y - \beta_0 u^* - \delta_0 y_1,$$

$$\beta_0 = b_2 = 0, \ \delta_0 = d_2 = 0,$$

so

$$x_2 = y. \tag{8-34}$$

Also,

$$x_3 = dx_2/dt - \beta_1 u^* - \delta_1 y_1,$$

$$\beta_1 = b_1 - C_1\beta_0 = 2 - (3)(0) = 2,$$

$$\delta_1 = d_1 - C_1\delta_0 = 3 - (3)(0) = 3,$$

and

$$dx_2/dt = x_3 + 2u^* + 3y_1. \tag{8-35}$$

The final state-derivative equation comes from substituting the states into the given second-order differential equation,

$$dx_3/dt = \beta_2 u^* + \delta_2 y_1 - C_1 x_3 - C_0 x_2,$$

$$\beta_2 = b_0 - C_1\beta_1 - C_0\beta_0 = 7 - (3)(2) - (4)(0) = 1,$$

$$\delta_2 = d_0 - C_1\delta_1 - C_0\delta_0 = 6 - (3)(3) - (4)(0) = -3,$$

and

$$dx_3/dt = u^* - 3y_1 - 3x_3 - 4x_2. \tag{8-36}$$

Collecting the algebraic equations (Eqs. 8-32 and 8-34) gives

$$\mathbf{y} = \begin{bmatrix} y_l \\ y \end{bmatrix} = \begin{bmatrix} 1 & 0 & 0 \\ 0 & 1 & 0 \end{bmatrix} \begin{bmatrix} x_1 \\ x_2 \\ x_3 \end{bmatrix} + \begin{bmatrix} 0 \\ 0 \end{bmatrix} [u^*]. \tag{8-37}$$

Note that the output vector for y must be written in terms of *all* the states x, and *all* the external inputs u*. This last equation allows you to identify the two matrices

$$\mathbf{C}_o = \begin{bmatrix} 1 & 0 & 0 \\ 0 & 1 & 0 \end{bmatrix}, \qquad \mathbf{D}_o = \begin{bmatrix} 0 \\ 0 \end{bmatrix}.$$

The differential state equations (Eqs. 8-33, 8-35, and 8-36) are next collected into matrix form,

$$\begin{bmatrix} \dfrac{d\mathbf{x}}{dt} \end{bmatrix} = \begin{bmatrix} \dfrac{dx_1}{dt} \\ \dfrac{dx_2}{dt} \\ \dfrac{dx_3}{dt} \end{bmatrix} = \begin{bmatrix} -0.5 & 0 & 0 \\ 0 & 0 & 1 \\ 0 & -4 & -3 \end{bmatrix} \begin{bmatrix} x_1 \\ x_2 \\ x_3 \end{bmatrix} + \begin{bmatrix} 0 & 0 \\ 3 & 0 \\ -3 & 0 \end{bmatrix} \begin{bmatrix} y_1 \\ y \end{bmatrix} + \begin{bmatrix} 1.5 \\ 2 \\ 1 \end{bmatrix} u^*. \tag{8-38}$$

This allows you to identify the remainder of the matrices

$$\mathbf{A}_o = \begin{bmatrix} -0.5 & 0 & 0 \\ 0 & 0 & 1 \\ 0 & -4 & -3 \end{bmatrix}, \boldsymbol{\delta} = \begin{bmatrix} 0 & 0 \\ 3 & 0 \\ -3 & 0 \end{bmatrix}, \boldsymbol{\beta} = \begin{bmatrix} 1.5 \\ 2 \\ 1 \end{bmatrix}.$$

You are now ready to solve for the **A** and **B** matrices,

$$\mathbf{A} = \mathbf{A}_o + \boldsymbol{\delta}\mathbf{C}_o = \begin{bmatrix} -0.5 & 0 & 0 \\ 3 & 0 & 1 \\ -3 & -4 & -3 \end{bmatrix},$$

and

$$\mathbf{B} = \boldsymbol{\delta}\mathbf{D}_o + \boldsymbol{\beta} = \begin{bmatrix} 1.5 \\ 2 \\ 1 \end{bmatrix}.$$

The final step in finding the state-space model is to express the output equation. For this example

$$y^* = \mathbf{C}\boldsymbol{x} + \mathbf{D}u^*,$$

$$= [0 \quad 1 \quad 0]\boldsymbol{x} + 0u^*,$$

which is a combination of Eqs. 8-31 and 8-34.

Comment

The hardest part of this procedure is in keeping track of all the state assignments, especially when the system has a large number of differential equations. Care must be taken that each state is used only once, and that none are skipped.

HOMEWORK

8-1. Given the following component models for the system shown in Fig. 8-4. Find the system state-space model.

$$\text{Component A: } y_1 = u^* - \mathrm{R}_1 u_1.$$

$$\text{Component B: } (\mathrm{L}/\mathrm{R}_2)\, dy^*/dt + y^* = (\mathrm{L}/\mathrm{R}_2)u_2.$$

$$\mathrm{L}\, dy_2/dt + \mathrm{R}_2 y_2 = u_2.$$

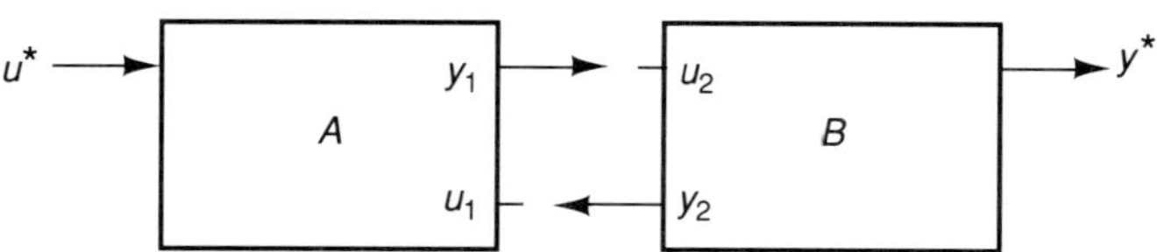

Figure 8-4. Multiport for Probs. 8-1 Through 8-6

8-2. Given the following component models for the system shown in Fig. 8-4. Find the system state-space model.
Component A: $\tau \, dy_1/dt + y_1 = au^* + bu_1$.
Component B: This is a large-step digital component with the following logic

$$\text{IF } (u_2 \text{ .LT. } 0.1) \; y_2 = 1.0, \; y^* = 0.0,$$

$$\text{IF } (u_2 \text{ .GE. } 0.1) \; y_2 = 0.0, \; y^* = 1.0.$$

8-3. Given the following component models for the system shown in Fig. 8-4. Find the system state-space model.

$$\text{Component A: } y_1 = 5u^* - 6u_1.$$

$$\text{Component B: } Y^*(s)/U_2(s) = 10(s + 1)/[(s + 2)(s + 3)],$$

$$Y_2(s)/U_2(s) = 5/(2s + 1)$$

8-4. Given the following component models for the system shown in Fig. 8-4. Find the system state-space model.

$$\text{Component A: } 3 \, dy_1/dt + 6y_1 = 2u_1 + 5u^*.$$

Component B: This is an analog-mimic component which is very fast acting, and with very fine resolution. The difference equations which the designer has used are

$$5(y_{2(i+1)} - y_{2i})/\Delta t + y_{2i} = 8u_{2i},$$

$$y_i^* = 3u_{2i}.$$

8-5. Given the following component models for the system shown in Fig. 8-4. Find the system state-space model.

$$\text{Component A: } y_1 = u^* + 2u_1,$$

$$\text{Component B: } y_2 = -3u_2,$$

$$y^* = 4u_2.$$

8-6. Given the following component models for the system shown in Fig. 8-4. Find the system state-space model.

Component A: This is a characteristic model for B:

$$\tau \, dy_1/dt + y_1 = 2u^*.$$

Component B: This model comes from in-system data,

$$y^* = 12u_2.$$

8-2 RESPONSE OF SYSTEMS WITHOUT LARGE-STEP COMPONENTS

Many common systems are designed without large-step components, as shown in Fig. 8-5—note that there are no clear ports in this figure. Instead, all the inputs are grouped together, as are all the outputs, in order to simplify the causality. The computation of $\mathbf{y}^*$ for these systems may be found as either an analytic response in terms of an algebraic function of time, or as a numerical output time history using integration on a digital computer.

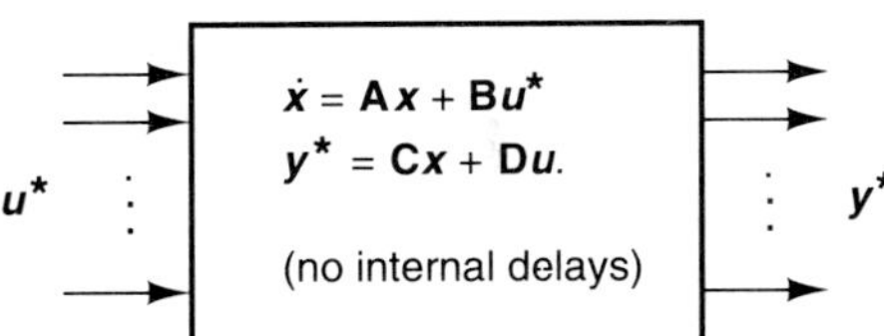

Figure 8-5. MIMO System without Finite-digital Components

Your first concern for computing an analytic output response now is to separate the frequency response from the time-domain response. To compute the analytic frequency response you may turn to the methods of Fourier, Laplace, and the Bode diagram. For the analytic time-domain response, you may also use the Laplace transform.

A numerical response is computed in the time domain using a finite-difference numerical integration of the state-space model. In this, all the frequency and time-domain inputs are integrated simultaneously to obtain the time history for $\mathbf{y}^*$.

Each of these approaches is discussed in this section, starting from the linear-system state-space model.

The Analytic Frequency Response. This approach is begun by first converting the time-domain state-space model into the frequency domain using the Laplace transform. That is, given

$$d\mathbf{x}/dt = \mathbf{A}\mathbf{x} + \mathbf{B}\mathbf{u}^*, \tag{8-39}$$

and

$$\mathbf{y}^* = \mathbf{C}\mathbf{x} + \mathbf{D}\mathbf{u}^*, \tag{8-40}$$

then the Laplace transform of this is

$$s\,\mathbf{X}(s) = \mathbf{A}\mathbf{X}(s) + \mathbf{B}\mathbf{U}^*(s), \tag{8-41}$$

and

$$\mathbf{Y}^*(s) = \mathbf{C}\mathbf{X}(s) + \mathbf{D}\mathbf{U}^*(s). \tag{8-42}$$

Solving Eq. 8-41 for $\mathbf{X}(s)$ yields

$$X(s) = (s\mathbf{I} - \mathbf{A})^{-1}\mathbf{B}U^*(s), \tag{8-43}$$

which may be substituted into Eq. 8-42 to give

$$Y^*(s) = [C(s\mathbf{I} - \mathbf{A})^{-1}\mathbf{B} + \mathbf{D}]\, U^*(s). \tag{8-44}$$

Thus, the *transfer matrix* is

$$T(s) = [\mathbf{C}(s\mathbf{I} - \mathbf{A})^{-1}\mathbf{B} + \mathbf{D}], \tag{8-45}$$

which reduces to the *transfer function* for SISO systems. In other words, $T(s)$ is a $(k \times m)$ matrix of transfer functions, where k is the number of outputs and m is the number of inputs.

Note that the transfer-matrix in Eq. 8-45, involves the matrix inversion $(s\mathbf{I} - \mathbf{A})^{-1}$. This is easy enough to compute by hand for a (1×1), a (2×2), or even for some (3×3) $\mathbf{A}$ matrices. However, for more complicated $\mathbf{A}$ matrices, appropriate software must be used (see the Bibliography).

Given the transfer matrix of the system, it is possible to compute the general frequency response of the system, as well as any specific frequency response of interest.

The *general frequency response* is $T(j\omega)$—this is also called the "characteristic" frequency response of the system. It is shown graphically on a *set* of Bode diagrams, one for each transfer-function entry in the transfer matrix.

Example 8-4

Estimate the general frequency response of the SISO system represented by

$$d\mathbf{x}/dt = \begin{bmatrix} 0 & 1 \\ -2 & -3 \end{bmatrix}\mathbf{x} + \begin{bmatrix} 0 \\ 1 \end{bmatrix}u^*,$$

$$y^* = [1 \quad 0]\mathbf{x}.$$

Solution

In order to find the transfer function, $T(s)$, for the system, you must first compute

$$(s\mathbf{I} - \mathbf{A})^{-1} = \begin{bmatrix} s & -1 \\ 2 & s+3 \end{bmatrix}^{-1} = \frac{1}{s^2 + 3s + 2}\begin{bmatrix} s+3 & 1 \\ -2 & s \end{bmatrix},$$

You may next substitute into Eq. 8-45 to find

$$T(s) = \frac{1}{s^2 + 3s + 2}[1 \; 0]\begin{bmatrix} s+3 & 1 \\ -2 & s \end{bmatrix}\begin{bmatrix} 0 \\ 1 \end{bmatrix}.$$

Multiplying this out gives the transfer function in the form

$$T(s) = 1/(s^2 + 3s + 2) = 1/[(s+1)(s+2)],$$
$$= 0.5/[(s+1)(s/2+1)].$$

The asymptotic plot of $T(j\omega)$ is shown in Fig. 8-6.

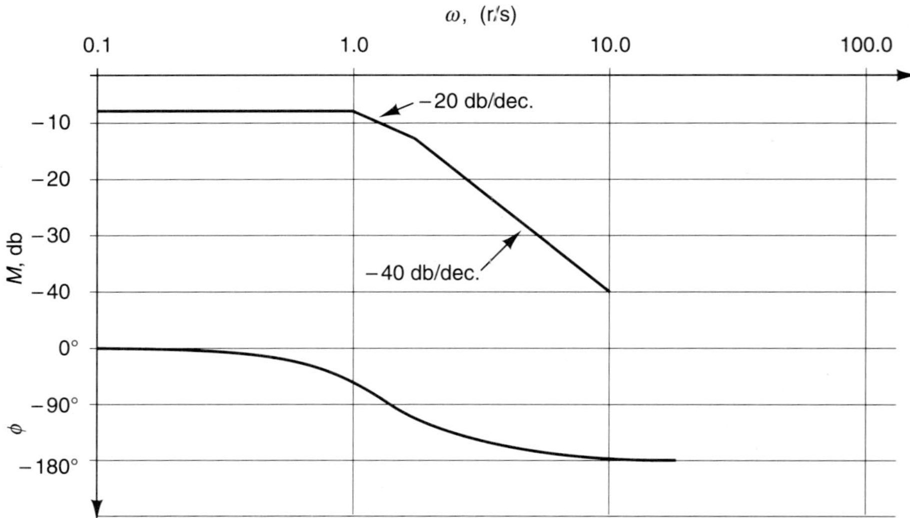

Figure 8-6. General Frequency Response of Ex. 8-4

Comments

Note that the factors in the transfer function are expressed in such a way that they have the form $(\tau s + 1)$. This is desirable so that the magnitude of the transfer function appears explicitly in the numerator at low frequencies, as s goes to zero.

The *specific frequency response* of a system is computed using the transfer matrix evaluated at the frequency of interest. That is, at a frequency ω_1 this evaluation is

$$\boldsymbol{Y}^*(j\omega_1) = \boldsymbol{T}(j\omega_1)\,\boldsymbol{U}^*(j\omega_1). \tag{8-46}$$

And, for a 2×2 transfer matrix this is

$$Y_1^*(j\omega_1) = T_{11}(j\omega_1)\,U_1^*(j\omega_1) + T_{12}(j\omega_1)\,U_2^*(j\omega_1), \tag{8-47}$$

and

$$Y_2^*(j\omega_1) = T_{21}(j\omega_1)\,U_1^*(j\omega_1) + T_{22}(j\omega_1)\,U_2^*(j\omega_1). \tag{8-48}$$

Example 8-5

Suppose a set of periodic inputs,

$$u_1^*(t) = 4\sin 3t + 2\sin 6t + \sin 9t,$$

and

$$u_2^*(t) = 15\cos 4t + 10\cos 8t + 5\cos 12t,$$

are input to a two-output system represented by $\boldsymbol{T}(j\omega)$. Show how the outputs may be computed from these inputs.

Solution

The frequencies that are in the two input signals are $\omega = 3, 4, 6, 8, 9$, and 12 rad/sec. At 3 rad/sec, the system model gives

$$Y_1^*(3j) = T_{11}(3j)\ U_1^*(3j) + T_{12}(3j)\ U_2^*(3j),$$

and

$$Y_2^*(3j) = T_{21}(3j)\ U_1^*(3j) + T_{22}(3j)\ U_2^*(3j).$$

But, since there is no 3 rad/sec component in $u_2^*(t)$, then

$$U_2^*(3j) = 0.$$

Thus,

$$Y_1^*(3j) = T_{11}(3j)\ U_1^*(3j),$$

and

$$Y_2^*(3j) = T_{21}(3j)\ U_1^*(3j),$$

where each of the two transfer functions have an M and a ϕ to apply to the input harmonic at the frequency of interest. So, in the time domain these expressions are

$$[\,y_1^*(t)]_3 = M_{11}(3)\ 4\ \sin[3t + \phi_{11}(3)],$$

and

$$[\,y_2^*(t)]_3 = M_{21}(3)\ 4\ \sin[3t + \phi_{21}(3)].$$

This analysis is continued on a frequency-by-frequency basis to build up the net analytic frequency response

$$y_1^*(t) = [y_1^*]_3 + [y_1^*]_4 + [y_1^*]_6 + [y_1^*]_8 + [y_1^*]_9 + [y_1^*]_{12},$$

$$y_2^*(t) = [y_2^*]_3 + [y_2^*]_4 + [y_2^*]_6 + [y_2^*]_8 + [y_2^*]_9 + [y_2^*]_{12}.$$

Comments

The numerical evaluation of M and ϕ at a frequency may come from a Bode diagram asymptotic approximation of the transfer matrix, or they may come from a numerical evaluation of $\boldsymbol{T}(j\omega_1)$. In preliminary design, the former approach is most common.

The Analytic Time-domain Response. In order to compute an analytic time-domain response for a system model, you may again turn to the Laplace transfer matrix

$$\boldsymbol{T}(s) = [\mathbf{C}(s\mathbf{I} - \mathbf{A})^{-1}\mathbf{B} + \mathbf{D}], \tag{8-45}$$

with

$$\boldsymbol{Y}^*(s) = \boldsymbol{T}(s)\ \boldsymbol{U}^*(s). \tag{8-49}$$

Now however, the *inverse* Laplace transform gives you the desired time-domain output,

$$\boldsymbol{y}^*(t) = \mathcal{L}^{-1}\,[\boldsymbol{Y}^*(s)] = \mathcal{L}^{-1}\,[\boldsymbol{T}(s)\ \boldsymbol{U}^*(s)]. \tag{8-50}$$

That is, the inverse Laplace transform is denoted by the symbol $\mathcal{L}^{-1}$. Thus, if $Y(s)$ is the Laplace transform of $y(t)$, denoted by $Y(s) = \mathcal{L}[\,y(t)]$, then $y(t)$ is the inverse Laplace transform of $Y(s)$, denoted by $\mathcal{L}^{-1}\,[Y(s)]$.

The conversion of the system external inputs, $\boldsymbol{u}^*(t)$, into their Laplace form, $\boldsymbol{U}^*(s)$, is discussed in Chap. 6, and the formation of $\boldsymbol{T}(s)$ from the system state-space model is given by Eq. 8-45. Thus, given a means to find the inverse Laplace transform, you could solve for the system outputs.

The mathematical *definition* of the inverse Laplace operation is

$$y^*(t) = \frac{1}{2\pi j}\int_{c-j\infty}^{c+j\infty} [Y^*(s)]\, e^{st}\, ds \; (t > 0). \tag{8-51}$$

Fortunately, however, there are easier ways to compute the inverse transform than the computation of the above integral. These include the use of transform tables for simple functions, and the use of partial fractions for more complicated functions.

The simplest way to compute the Laplace inverse is to merely look up a pre-computed integration in a tabulation, such as that shown in Table 8-1. Entire books have been devoted to the tabulation of such Laplace functions (see the Bibliography for examples).

More complicated functions than those in the tables can be reduced to superposed simple forms which are in the tables through a method called *partial-fraction expansion*. This method is based on the idea that a function of the form

$$Y^*(s) = \frac{N(s)}{D(s)} = \frac{B(s+z_1)(s+z_2)\cdots(s+z_m)}{(s+p_1)(s+p_2)\cdots(s+p_k)}, \tag{8-52}$$

may be expanded into the form

$$Y^*(s) = \sum_{i=1}^{k} \frac{a_i}{D_i(s)}, \tag{8-53}$$

where $D_i(s)$ are the factors of the denominator of $y^*(s)$, and a_i are real constants called the *residues* of D_i.

A critical restriction of the expansion method is that the order of $D(s)$ must be higher than the order of $N(s)$. This is easily met using synthetic division, as Ex. 8-6 shows.

Example 8-6

Express the following Laplace function in a way that is suitable for partial-fraction expansion

$$Y^*(s) = \frac{7s^3 + 5s^2 + 3s + 2}{s^2 + 3s + 1}.$$

Solution

You must first use synthetic division on this function since the numerator is of higher order than the denominator. This division gives

Table 8-1. Inverse Laplace Transforms

	$F(s)$	$f(t)$
1	1	unit impulse $\delta(t)$
2	$\dfrac{1}{s}$	unit step $1(t)$
3	$\dfrac{1}{s^2}$	t
4	$\dfrac{1}{s+a}$	e^{-at}
5	$\dfrac{1}{(s+a)^2}$	te^{-at}
6	$\dfrac{\omega}{s^2+\omega^2}$	$\sin \omega t$
7	$\dfrac{s}{s^2+\omega^2}$	$\cos \omega t$
8	$\dfrac{n!}{s^{n+1}}$	$t^n \quad (n = 1, 2, 3, \ldots)$
9	$\dfrac{n!}{(s+a)^{n+1}}$	$t^n e^{-at} \quad (n = 1, 2, 3, \ldots)$
10	$\dfrac{1}{(s+a)(s+b)}$	$\dfrac{1}{b-a}(e^{-at} - e^{-bt})$
11	$\dfrac{s}{(s+a)(s+b)}$	$\dfrac{1}{b-a}(be^{-bt} - ae^{-at})$
12	$\dfrac{1}{s(s+a)(s+b)}$	$\dfrac{1}{ab}\left[1 + \dfrac{1}{a-b}(be^{-at} - ae^{-bt})\right]$
13	$\dfrac{\omega}{(s+a)^2+\omega^2}$	$e^{-at} \sin \omega t$
14	$\dfrac{s+a}{(s+a)^2+\omega^2}$	$e^{-at} \cos \omega t$
15	$\dfrac{1}{s^2(s+a)}$	$\dfrac{1}{a^2}(at - 1 + e^{-at})$
16	$\dfrac{\omega_n^2}{s^2 + 2\zeta\omega_n s + \omega_n^2}$	$\dfrac{\omega_n}{\sqrt{1-\zeta^2}} e^{-\zeta\omega_n t} \sin (\omega_n \sqrt{1-\zeta^2}\, t)$
17	$\dfrac{s}{s^2 + 2\zeta\omega_n s + \omega_n^2}$	$\dfrac{-1}{\sqrt{1-\zeta^2}} e^{-\zeta\omega_n t} \sin (\omega_n \sqrt{1-\zeta^2}\, t - \phi)$ $\phi = \tan^{-1} \dfrac{\sqrt{1-\zeta^2}}{\zeta}$
18	$\dfrac{\omega_n^2}{s(s^2 + 2\zeta\omega_n s + \omega_n^2)}$	$1 - \dfrac{1}{\sqrt{1-\zeta^2}} e^{-\zeta\omega_n t} \sin (\omega_n \sqrt{1-\zeta^2}\, t + \phi)$ $\phi = \tan^{-1} \dfrac{\sqrt{1-\zeta^2}}{\zeta}$

$$
\begin{array}{r|l}
 & 7s \quad - 16 \\
s^2 + 3s + 1 & 7s^3 + 5s^2 + 3s + 2 \\
 & 7s^3 + 21s^2 + 7s \\
 & -16s^2 - 4s + 2 \\
 & -16s^2 - 48s - 16 \\
 & 44s + 18
\end{array}
$$

Thus you obtain a proper form for partial-fraction expansion

$$\mathbf{Y^*(s) = 7s - 16 + \frac{44s + 18}{s^2 + 3s + 1}}.$$

Comment

The ratio term in the last equation is the one which is to be expanded in fractions. Note that it has a higher order in the denominator and is thus in proper form.

Next, the partial-fraction expansion of a ratio term depends upon the nature of the factors of $D(s)$. For example, there are three types of factors that are possible: distinct real factors; complex-conjugate factors; and repeated real factors. Each of these requires a special form of expansion, and each results in a different time function for $y^*(t)$.

For *distinct real factors* in $D(s)$, the expansion is

$$Y^*(s) = \frac{a_1}{(s + p_1)} + \frac{a_2}{(s + p_2)} + \cdots + \frac{a_k}{(s + p_k)}, \tag{8-54}$$

where k is the order of the polynomial in $D(s)$.

The numerical value of a_k is found by multiplying $Y^*(s)$ by $(s + p_k)$ and letting $s = -p_k$ in the expression. This multiplication gives

$$[Y^*(s)(s + p_k)]_{s=-p_k} = [a_1(s + p_k)/(s + p_1) + a_2(s + p_k)/(s + p_2) + \cdots + a_k]_{s=-p_k}. \tag{8-55}$$

When the value of the root is substituted for s, all terms but the kth term must drop out of the right-hand side, giving

$$[Y^*(s)(s + p_k)]_{s=-p_k} = a_k. \tag{8-56}$$

This method is applied factor-by-factor to calculate all the residues of the expansion.

Finally, entry 4 of Table 8-1 is used to convert each fraction to its time function,

$$\mathcal{L}^{-1}[a_k/(s + p_k)] = a_k e^{-p_k t}, \tag{8-57}$$

and the results are superposed to give

$$y^*(t) = \mathcal{L}^{-1}[Y^*(s)] = a_1 e^{-p_1 t} + a_2 e^{-p_2 t} + \cdots + a_k e^{-p_k t}. \tag{8-58}$$

Example 8-7

Given the system model,

$$T(s) = \frac{10(s + 2)}{s^2 + 7s + 12},$$

and that the system is subjected to the step input $U^*(s) = 3/s$, find the analytic expression for the output $y^*(t)$.

Solution

Begin by finding $Y^*(s) = T(s)U^*(s)$,

$$Y^*(s) = \frac{10(s + 2)}{s^2 + 7s + 12}(3/s)$$

Next factor $D(s)$ to find the poles of $Y^*(s)$. These are easily shown to be s, $(s + 3)$, and $(s + 4)$.

These distinct real factors of $Y^*(s)$ can be expanded into the following partial-fraction form

$$Y^*(s) = \frac{a_1}{s} + \frac{a_2}{(s + 3)} + \frac{a_3}{(s + 4)},$$

where the a's are computed from the following expressions

$$a_1 = [s\,T(s)U^*(s)]_{s=0} = 30(2)/[(3)(4)] = 5,$$

$$a_2 = [(s + 3)\,T(s)U^*(s)]_{s=-3} = 30(-1)/[(-3)(1)] = 10,$$

$$a_3 = [(s + 4)\,T(s)U^*(s)]_{s=-4} = 30(-2)/[(-4)(-1)] = -15.$$

This gives the output

$$Y^*(s) = 5/s + 10/(s + 3) - 15/(s + 4).$$

Table 8-1 gives the time-domain analytic response

$$\mathbf{y^*(t) = 5u_s(t) + 10e^{-3t} - 15e^{-4t}.}$$

Comments

The most common mistake made here is to forget to multiply $T(s)$ by $U^*(s)$ before the expansion and the inverse. Such an error creates an incorrect time function for $y^*(t)$.

Note that you must be able to find the roots of $D(s)$ to apply this procedure. Of course, $D(s) = 0$ is satisfied by the poles and *eigenvalues* of $T(s)$. Thus, you cannot escape finding the eigenvalues of the system, and the eigenvalues determine the system response through the exponentials in the output time function.

The governing nature of the system eigenvalues is again demonstrated.

If there are *complex-conjugate factors* in $D(s)$, then the fraction expansion is of the form

$$Y^*(s) = \frac{a_1 s + a_2}{(s + p_1)(s + p_2)} + \frac{a_3}{(s + p_3)} + \cdots + \frac{a_k}{(s + p_k)}, \tag{8-59}$$

where $p_1 = p_R + jp_I$, and $p_2 = p_R - jp_I$.

The numerical values of the a_k are found in a manner just the same as previously. That is,

$$(a_1 s + a_2)_{s=-p_1} = [(s + p_1)(s + p_2)T(s)U^*(s)]_{s=-p_1}. \tag{8-60}$$

But, since p_1 is a complex number, this equation really generates two equations. One equation comes from equating the real parts, and one equation comes from equating the imaginary parts.

Example 8-8

Given the system model

$$T(s) = 4/(s^2 + 2s + 4).$$

What is the analytic system response to an input of $U^*(s) = 2/s$?

Solution

As in the last example, begin by computing

$$Y^*(s) = T(s)U^*(s) = 8/[s(s^2 + 2s + 4)].$$

The poles of this expression are at $s = 0, -1 \pm j\sqrt{2}$, so you must use the expansion form for a complex-conjugate pair,

$$Y^*(s) = (a_1 s + a_2)/(s^2 + 2s + 4) + a_3/s.$$

The numerator constants are computed as follows

$$(a_1 s + a_2)_{s=-p_1} = [(s^2 + 2s + 4)T(s)U^*(s)]_{s=-p_1}.$$

Or,

$$[a_1(1 - j\sqrt{2}) + a_2] = [8/(1 - j\sqrt{2})] = 8(1 + j\sqrt{2})/[(1 - j\sqrt{2})(1 + j\sqrt{2})]$$
$$= 8/3 + j8\sqrt{2}/3.$$

Equating the real parts gives

$$a_1 + a_2 = 8/3,$$

Equating the imaginary parts gives

$$-\sqrt{2}\, a_1 = 8\sqrt{2}/3.$$

Solving these two equations gives

$$a_1 = -8/3, \; a_2 = 16/3.$$

Also, $a_3 = [sT(s)U^*(s)]_{s=0} = 2.$

So, $Y^*(s) = 8(-s + 2)/[3(s^2 + 2s + 4)] + 2/s,$

$$= -(8/3)\, s/(s^2 + 2s + 4) + (4/3)\, 4/(s^2 + 2s + 4) + 2/s.$$

The last equation is written in this form to make it recognizable with those in Table 8-1. The first two terms have the following constants,

$$\omega_n = 2 \text{ rad/sec.},\ \zeta = 0.5,\ \phi = \tan^{-1}[(1 - \zeta^2)^{1/2}/\zeta] = 60°,$$

$$\zeta\omega_n = 1,\ 1/[(1 - \zeta^2)^{1/2}] = 1.155,\ \omega_n[(1 - \zeta^2)^{1/2}] = 1.732.$$

Thus, Table 8-1 gives the analytic time response

$$\mathbf{y^*(t) = 2u_s(t) + 3.08e^{-t}\sin(1.732t - 60°) + 3.08e^{-t}\sin(1.732t).}$$

If the factors of $D(s)$ involve a *repeated real pole*, which is repeated r times at p_1, then the partial-fraction expansion is of the form

$$Y^*(s) = \frac{a_1}{(s + p_1)} + \frac{a_2}{(s + p_1)^2} + \cdots + \frac{a_r}{(s + p_1)^r} + \frac{a_{r+1}}{(s + p_{r+1})} + \cdots + \frac{a_k}{(s + p_k)}. \qquad (8\text{-}61)$$

Notice that *all* of the powers of the repeated root must be present in the fraction expansion. The constants for the repeated roots are given by

$$a_r = [(s + p_1)^r T(s)U^*(s)]_{s=-p_1}, \qquad (8\text{-}62)$$

$$a_{r-1} = \{d/ds\, [(s + p_1)^r T(s)U^*(s)]\}_{s=-p_1}, \qquad (8\text{-}63)$$

$$a_{r-L} = 1/L!\, \{d^L/ds^L\, [(s + p_1)^r T(s)U^*(s)]\}_{s=-p_1}, \qquad (8\text{-}64)$$

$$\ldots = \ldots,$$

$$a_1 = 1/(r - 1)!\, \{d^{r-1}/ds^{r-1}\, [(s + p_1)^r T(s)U^*(s)]\}_{s=-p_1}. \qquad (8\text{-}65)$$

Here the derivative is used to extract the constants of interest for the repeated root. Note that you cannot simply multiply by $(s + p_1)^{r-1}$ to find a_{r-1} since this leaves an infinite expression in the rth term when you substitute the root,

$$[a_r/(s + p_1)]_{s=-p_1} = \infty.$$

The use of the derivative avoids this problem nicely.

The remaining fraction numerators are computed just as before. Thus, the time domain output for Eq. 8-61 is

$$y^*(t) = [a_1 + a_2 t + \ldots + a_r t^{r-1}/(r - 1)!]\, e^{-p_1 t} + a_{r+1}e^{-p_{r+1}t} + \cdots + a_k e^{-p_k t}. \qquad (8\text{-}66)$$

Example 8-9

Given the system model

$$T(s) = 10/(s^3 + 6s^2 + 12s + 8),$$

along with the system input $U^*(s) = 3/s$, find the analytic expression for $y^*(t)$.

Solution

The Laplace form for $Y^*(s)$ is found first,

$$Y^*(s) = 30/[s(s^3 + 6s^2 + 12s + 8)].$$

This form for $D(s)$ is factored into s, and $(s + 2)^3$. Thus, the $Y^*(s)$ may be expressed as

$$Y^*(s) = a_1/(s + 2) + a_2/(s + 2)^2 + a_3/(s + 2)^3 + a_4/s.$$

The numerators of the fractions are computed next,

$$a_3 = [(s + 2)^3\, T(s)U^*(s)]_{s=-2} = [30/s]_{s=-2} = -15,$$

$$a_2 = \{d/ds[30/s]\}_{s=-2} = [-30/s^2]_{s=-2} = -7.5,$$

$$a_1 = (1/2)\, \{d^2/ds^2[30/s]\}_{s=-2} = 1/2\ [60/s^3]_{s=-2} = -3.75,$$

$$a_4 = [s\ T(s)U^*(s)]_{s=0} = 30/2^3 = 3.75.$$

The fraction form of Y^* is converted into the time domain using Table 8-1. This gives

$$\mathbf{y^*(t) = [-15 - 7.5t - 1.875t^2]e^{-2t} + 3.75\ u_s(t).}$$

The Numerical Time-domain Response. Often, the designer only wants a time history of the output(s) for a given system. There are many ways to find this, but perhaps the easiest is to simply program the linear model equations,

$$dx/dt = \mathbf{A}x + \mathbf{B}u^*, \tag{8-26}$$

and

$$y^*(t) = \mathbf{C}x + \mathbf{D}u^*, \tag{8-29}$$

for numerical integration on the digital computer. This integration is very straightforward since all of the equations are either first-order linear differential equations or they are linear algebraic equations.

The usual approach is to use a five-step algorithm, based upon a forward-difference numerical integration. In a preliminary-design study a rectangular integration method is often used,

1. Initialize $\boldsymbol{x}_0$, and $\boldsymbol{u}_0^*$.

2. Compute the outputs at step i, $i = 0, N$
   $\boldsymbol{y}^*_i = \mathbf{C}\boldsymbol{x}_i + \mathbf{D}\boldsymbol{u}_i^*$, PRINT i, $\boldsymbol{u}^*_i$, $\boldsymbol{x}_i$, $\boldsymbol{y}^*_i$.

3. Integrate the states to the next step, $i + 1$,
   $x_{i+1} = (\Delta t\mathbf{A} + \mathbf{I})\,\mathbf{x}_i + \Delta t\mathbf{B}u_i$.
4. Increment the step,
   Store $x_i = x_{i+1}$, $i = i + 1$, READ u^*_i.
5. CONTINUE? GO TO 2.

The selection of the integration time step, Δt, is based upon the eigenvalues of the $\mathbf{A}$ matrix. The eigenvalues must be used in this determination because they govern the curvature of the output time response, and they must all have non-positive real parts for stability, as discussed in Chap. 5. When all of the eigenvalues are compared to each other, they have a smallest real part (e.g., the most negative), and they have a largest imaginary part. The integration step size must be chosen to capture the smallest of the *inverse* of these two limits, $\tau_{CS} = 1/|\lambda|$. The selection is then often made according to the rule of thumb

$$\tau_{CS}/10{,}000 \leq \Delta t \leq \tau_{CS}/500. \tag{8-67}$$

If Δt is chosen larger than this range, then poor accuracy in output curvature approximation may be seen using the rectangular method. If Δt is chosen smaller than this range, then error build-up due to excessive computation may occur.

Note that this rule-of-thumb for Δt is for the prediction of the system *output(s)*; it thus requires more time steps per curvature than the previous rules-of-thumb in Chap. 7 for a numerical method to estimate the model *constants*. That is, a 10% error in estimating output curvature for a model constant from system data is easily achieved using only a few data points since the errors are averaged out. However, a 10% error in *curvature* estimation during simulation usually leads to much more than a 10% error in output prediction. In other words, given exact constants, you must still use a small integration step size to ensure acceptable *curvature* estimation in an output history. The guideline of Eq. 8-67 helps you to achieve this.

HOMEWORK

8-7. Given the MIMO system model

$$dx_1/dt = -10x_1 + 2u_1^*,$$
$$dx_2/dt = x_3,\ dx_3/dt = -4x_1 - 5x_2 - 6x_3 + 7u_2^*,$$

and $y_1^* = 3x_2$, with $y_2^* = x_3$.

a) What is the transfer matrix?

b) Find the specific frequency response to

$$u_1^* = 8\cos 3t,$$

with

$$u_2^* = 4\sin 4t.$$

8-8. Two periodic signals,

$$u_1^* = 4 \sin 4t, \text{ and } u_2^* = 8 \cos 4t,$$

are input into the MIMO system,

$$dx_1/dt = x_2 + 2u_1^*, \qquad dx_2/dt = -4x_1 - 4x_2 + 6u_2^*,$$

$$y_1^* = x_1, \; y_2^* = x_2.$$

a) What is the general frequency response of this system, in transfer function form?

b) What is the specific frequency response to the given inputs?

8-9. Find the analytic time response of the system in Prob. 8-7 to the simultaneous singularity inputs

$$u_1^* = 10u_s(t), \text{ and } u_2^* = 2u_s(t).$$

8-10. Given the SISO system

$$dx_1/dt = x_2, \text{ and } dx_2/dt = -4x_1 - 4x_2 + 15u^*, \text{ such that } y^* = x_1.$$

Find the analytic time response to u* $= 5u_s(t)$.

8-11. Given the SISO system

$$dx_1/dt = x_2, \text{ and } dx_2/dt = -9x_1 - 3x_2 + u^*, \text{ such that } y^* = x_1.$$

Find the analytic time response of the system when $u^* = 4u_s(t)$.

8-12. Write a computer program which uses numerical integration to compute the response of the system given in Prob. 8-11. Use your program to compute the numerical response over the interval $0 \leq t \leq 5$ sec. Assume that all initial conditions are zero. Compare your answer to that of the previous problem.

8-3 RESPONSE OF SYSTEMS WITH LARGE-STEP DIGITAL COMPONENTS

There are many systems which are implemented with one or more large-step digital components. A typical version of this type of system is shown in Fig. 8-7.

Note that the usual input-output connectivity is *not* shown in Fig. 8-7a in the context of ports. This is done to show the overall causality of the system, as done earlier in Fig. 8-5. In any case, note that the computer inputs and outputs are usually specially designed to eliminate the port interactions at all connections. This is also assumed in the analysis which follows. The outputs are often isolated through the use of voltage-isolation amplifiers, while the inputs are isolated through proper sensor design.

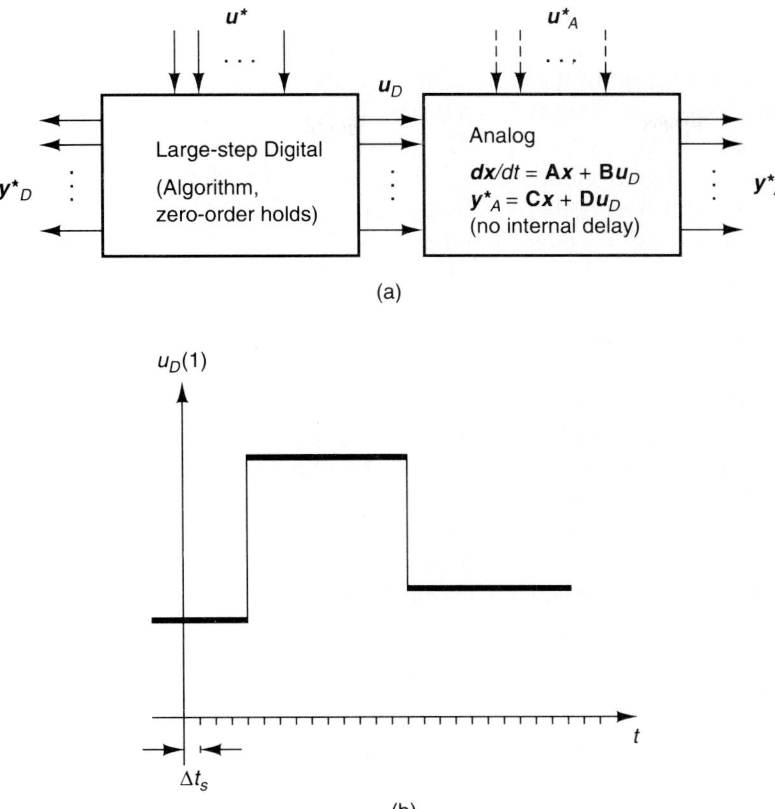

Figure 8-7. MIMO System with Finite-digital Components. (a) Multiport, (b) typical input in channel 1

The following analysis also assumes that the analog subsystem converts the analog inputs into analog outputs through linear dynamics, without significant internal-component delay. Further, the presence of the inputs which match the analog outputs is shown as a set of dashed lines in Fig. 8-7a. This is done to show that the outputs from the analog subsystem are not port-matched to the inputs from the digital component(s). Rather, the analog inputs, $\boldsymbol{u}_A{}^*$, are assumed negligible.

Also note in Fig. 8-7a that all the system external inputs, $\boldsymbol{u}^*$, are assumed to enter the system through the large-step digital component. In fact, these external inputs to the system are often entered by a human operator into a computer. The analog part of the system may be an engine, or some other type of machine.

The algorithm in the computer works to change the human inputs, $\boldsymbol{u}^*$, into digital-external outputs, $\boldsymbol{y}_D{}^*$, and analog inputs, $\boldsymbol{u}_D$. The digital external outputs are usually displays for the operator. The output of the computer to the analog subsystem is assumed to be through zero-order holds. This assumption is critical to the analysis, yet it may be relaxed somewhat when the analysis results are generalized later.

The step nature of the analog inputs is shown in Fig. 8-7b, for the analog input in channel 1, $\boldsymbol{u}_D(1)$. Here the inputs are shown to be typically constant-valued over many sample steps, with large changes in value. This is distinctly different from analog-mimic digital components, where the inputs often change from sample to sample, in a much smoother small-step fashion.

A classic use of a large-step digital component is for a minimum-time system. These systems are designed to move from point A to point B in as short a time as possible. The functional implementation of the system is to accelerate toward the target as fast as possible until a critical switching time, then to decelerate as fast as possible until the destination. The computer is used to keep track of the time and to accomplish the switching. Thus, the analog inputs have only two values. Further, the switch is commanded only once during an operation.

The usual approach to simulation of a minimum-time system is to use the state-space analog model in conjunction with the computer-switching algorithm to predict the system response. In this, it is assumed that the computer can execute the ADC, the algorithm, and the DAC within each sample step. The simulation of performance is thus done in five steps, following initialization:

1. Read the computer inputs, $\boldsymbol{u}^*$, the stored inputs, $\boldsymbol{u}_D$, and the stored states, $\boldsymbol{x}$, at the sample time.
2. Compute the analog outputs, $\boldsymbol{y}_A$, at the sample time.
3. Compute and store the states, $\boldsymbol{x}$, for the *next* sample time.
4. Compute and store the analog inputs, $\boldsymbol{u}_D$, and the digital outputs, $\boldsymbol{y}_D^*$, using the computer algorithm for the *next* sample step.
5. Increment the sample time, go to 1.

This section of the text discusses steps 2 and 3 in the simulation, under the assumption of a constant input vector, $\boldsymbol{u}_D(t)$, which is known for all time. This assumption leads to the development of the **Φ-Γ** (phi-gamma) method. This method can be used to predict either an analytic solution or a numerical solution—both are discussed below.

Note that you can *always* employ a numerical integration of the state-space equations, based upon some finite-difference numerical scheme (say, rectangular integration, as done elsewhere in this text). This requires the selection of a small integration time step, Δt_N, which is an integer fraction of the sample interval, Δt_s, and a small fraction of the smallest system time constant, τ_{CS}.

The Φ-Γ Method. The analysis begins with the state equations, which must be satisfied at all times, namely

$$d\boldsymbol{x}/dt = \mathbf{A}\boldsymbol{x} + \mathbf{B}\boldsymbol{u}_D, \tag{8-68}$$

and

$$\boldsymbol{y}_A = \mathbf{C}\boldsymbol{x} + \mathbf{D}\boldsymbol{u}_D. \tag{8-69}$$

The solution to Eq. 8-68 is composed of a homogeneous and a forced part,

$$\boldsymbol{x} = \boldsymbol{x}_h + \boldsymbol{x}_f. \tag{8-70}$$

The *homogeneous state solution* comes from the equation

$$d\boldsymbol{x}_h/dt = \mathbf{A}\boldsymbol{x}_h. \tag{8-71}$$

You may now assume that the solution has *non-zero initial conditions*, $\boldsymbol{x}(0)$, which will be justified shortly. The Laplace transform of Eq. 8-71 is thus

$$s\boldsymbol{X}_h(s) - \boldsymbol{x}(0) = \mathbf{A}\boldsymbol{X}_h(s), \tag{8-72}$$

or

$$\boldsymbol{X}_h(s) = (s\mathbf{I} - \mathbf{A})^{-1}\,\boldsymbol{x}(0). \tag{8-73}$$

The matrix inversion in the last equation may be analyzed using a series expansion about $s = 0$. That is,

$$(s\mathbf{I} - \mathbf{A})^{-1} = (1/s)\mathbf{I} + (1/s^2)\mathbf{A} + (1/s^3)\mathbf{A}^2 + \cdots. \tag{8-74}$$

So, when you compute the inverse Laplace transform in Eq. 8-73, you find that

$$\begin{aligned}\boldsymbol{x}_h(t) &= \mathcal{L}^{-1}[(s\mathbf{I} - \mathbf{A})^{-1}\,\boldsymbol{x}(0)] = \mathcal{L}^{-1}[(s\mathbf{I} - \mathbf{A})^{-1}]\,\boldsymbol{x}(0),\\ &= \mathcal{L}^{-1}[(1/s)\mathbf{I} + (1/s^2)\mathbf{A} + (1/s^3)\mathbf{A}^2 + \cdots]\,\boldsymbol{x}(0),\\ &= [\mathbf{I} + (t)\mathbf{A} + (t^2/2!)\mathbf{A}^2 + (t^3/3!)\mathbf{A}^3 + \cdots]\,\boldsymbol{x}(0).\end{aligned} \tag{8-75}$$

The term in the brackets in the last equation is *defined* as $e^{\mathbf{A}t}$. It is called the *state-transition matrix*, or the *matrix exponential*. It is given the symbol $\mathbf{\Phi}(t)$. Thus,

$$\boldsymbol{x}_h(t) = e^{\mathbf{A}t}\,\boldsymbol{x}(0) = \mathbf{\Phi}\boldsymbol{x}(0),$$

where

$$\mathbf{\Phi} = \mathcal{L}^{-1}[(s\mathbf{I} - \mathbf{A})^{-1}]. \tag{8-76}$$

The important properties of the matrix exponential include the following:

1. $$\begin{aligned} d/dt\,(e^{\mathbf{A}t}) &= [\mathbf{A} + (t)\mathbf{A}^2 + (t^2/2!)\mathbf{A}^3 + \cdots],\\ &= \mathbf{A}[\mathbf{I} + (t)\mathbf{A} + (t_2/2!)\mathbf{A}^2 + (t^3/3!)\mathbf{A}^3 + \cdots],\\ &= \mathbf{A}e^{\mathbf{A}t} = e^{\mathbf{A}t}\mathbf{A}.\end{aligned} \tag{8-77}$$
2. $$\mathbf{\Phi}(0) = e^{\mathbf{A}0} = \mathbf{I}. \tag{8-78}$$
3. $$\mathbf{\Phi}^{-1}(t) = (e^{\mathbf{A}t})^{-1} = (e^{-\mathbf{A}t}) = \mathbf{\Phi}(-t). \tag{8-79}$$

There are several ways that you may compute $\mathbf{\Phi}$, depending on the objectives of the analysis. You may resort to Eq. 8-76 when you are interested in an analytic solution, or you may approximate the series expansion in Eq. 8-75 when you are interested in a numerical solution. (More about this later.)

The *forced state solution* to the system comes from the equation

$$d\boldsymbol{x}_f/dt = \mathbf{A}x_f + \mathbf{B}\boldsymbol{u}_D. \tag{8-68}$$

The integrating factor for this equation is $e^{-\mathbf{A}t}$. Thus,

$$e^{-\mathbf{A}t}\, d\boldsymbol{x}_f/dt - e^{-\mathbf{A}t}\mathbf{A}\, \boldsymbol{x}_f = e^{-\mathbf{A}t}\, \mathbf{B}\boldsymbol{u}_D, \tag{8-80}$$

or

$$d/dt\, [e^{-\mathbf{A}t}\, \boldsymbol{x}_f] = e^{-\mathbf{A}t}\, \mathbf{B}\boldsymbol{u}_D, \tag{8-81}$$

and

$$\boldsymbol{x}_f(t) = e^{\mathbf{A}t} \int_0^t e^{-\mathbf{A}\tau}\, \mathbf{B}\boldsymbol{u}_D(\tau)\, d\tau. \tag{8-82}$$

This last formula is applied only over the interval when $\boldsymbol{u}_D$ is constant, thus accounting for the stepped nature of the input. This may involve one or several integration steps to capture the entire time history of the input. Furthermore, $t = 0$ for the integral is usually chosen at the start time of the constant inputs to the system, regardless of the value of the system time. The continuity in states between these forced steps is managed by the non-zero initial conditions in the earlier homogeneous solution. In fact, this is the reason that the initial conditions for the homogeneous solution were assumed to be non-zero.

Thus, since you anticipate a constant input, and since $\mathbf{B}$ is always constant, you can rewrite Eq. 8-83a,

$$\boldsymbol{x}_f(t) = \left[\int_0^t e^{\mathbf{A}(t-\tau)}\, d\tau \right] \mathbf{B}\, \boldsymbol{u}_D(0). \tag{8-83}$$

This last equation is more useful when you make the change of variable, $\sigma = t - \tau$. This gives

$$\boldsymbol{x}_f(t) = \left[\int_t^0 - e^{\mathbf{A}\sigma}\, d\sigma \right] \mathbf{B}\, \boldsymbol{u}_D(0) = \left[\int_0^t \Phi(\sigma)\, d\sigma \right] \mathbf{B}\, \boldsymbol{u}_D(0). \tag{8-84}$$

The term in brackets above is defined as the $\mathbf{\Gamma}$ matrix. That is,

$$\boldsymbol{x}_f(t) = \mathbf{\Gamma}\mathbf{B}\boldsymbol{u}_D(0). \tag{8-85}$$

The computation of the $\mathbf{\Gamma}$ matrix is closely related to the computation of the $\mathbf{\Phi}$ matrix. That is, for analytic solutions you may integrate the $\mathbf{\Phi}$ matrix directly according to Eq. 8-84, for numerical solutions you may compute $\mathbf{\Gamma}$ by integrating the series expansion of $\mathbf{\Phi}$.

Superposing the forced and homogeneous responses gives

$$x(t) = \Phi(t)\, x(0) + \Gamma(t)\, \mathbf{B} u_D(0), \tag{8-86}$$

$$y_A = \mathbf{C}x + \mathbf{D}u_D(0), \tag{8-87}$$

where $t = 0$ is the start of any constant $u_D(i)$.

Analytic Φ-Γ Solutions. These solutions are useful in those cases where u_D = constant for many sample steps, as in the minimum-time switching problem discussed earlier. In these cases, you use

$$\Phi = \mathcal{L}^{-1}[(s\mathbf{I} - \mathbf{A})^{-1}], \tag{8-76}$$

and the integral in brackets in Eq. 8-84 for Γ.

Example 8-10

An analog subsystem of a minimum-time, computer-controlled system is described by

$$\mathbf{A} = \begin{bmatrix} 0 & 1 \\ -4 & 0 \end{bmatrix}, \mathbf{B} = \begin{bmatrix} 0 \\ 1 \end{bmatrix}, \mathbf{C} = [1 \quad 0], \mathbf{D} = 0.$$

The zero-order hold input to the sub-system is to be

$$u_D(t) = 10,\ 0 \le t < t_1, \text{ and } u_D(t) = -30,\ t \ge t_1,$$

where t_1 is the switching time.

Find the analytical expression for the system output when $t \ge t_1$. All initial conditions are zero when $t = 0$.

Solution

The solution must be computed in two parts. The first part is up to the switching time, the second part is after the switching time. The solution at the switching time is needed to provide the initial conditions for the solution after the switching time.

The same Φ and Γ matrices are used in both solutions. Thus,

$$\Phi = \mathcal{L}^{-1}[s\mathbf{I} - \mathbf{A})^{-1}] = \mathcal{L}^{-1}\left\{\begin{bmatrix} s & 1 \\ -4 & s \end{bmatrix}^{-1}\right\} = \mathcal{L}^{-1}\begin{bmatrix} \dfrac{s}{s^2+4} & \dfrac{-1}{s^2+4} \\ \dfrac{4}{s^2+4} & \dfrac{s}{s^2+4} \end{bmatrix}$$

The inverse of the Laplace entries in this matrix gives,

$$\Phi(t) = \begin{bmatrix} \cos 2t & -\dfrac{1}{2}\sin 2t \\ 2\sin 2t & \cos 2t \end{bmatrix}$$

The Γ matrix is computed from

$$\mathbf{\Gamma} = \int_0^t \mathbf{\Phi}(\sigma)\, d\sigma .$$

Given the entries in the matrix above, this becomes

$$\mathbf{\Gamma}(t) = \begin{bmatrix} \frac{1}{2}\sin 2t & \frac{1}{4}(-1 + \cos 2t) \\ (1 - \cos 2t) & \frac{1}{2}\sin 2t \end{bmatrix}$$

The next step in the solution is to use the $\mathbf{\Phi}$ and $\mathbf{\Gamma}$ matricies to compute the states at the switching time using Eq. 8-86,

$$\begin{aligned} \boldsymbol{x}(t_1) &= \mathbf{\Phi}(t_1)\, \boldsymbol{x}(0) + \mathbf{\Gamma}(t_1)\, \mathbf{B}\boldsymbol{u}_D(0), \\ &= \mathbf{\Phi}(t_1)\, 0 + \mathbf{\Gamma}(t_1)\, \mathbf{B}\boldsymbol{u}_D(0), \\ &= \mathbf{\Gamma}(t_1)\, \mathbf{B}\boldsymbol{u}_D(0), \end{aligned}$$

This equation may be expanded in terms of its matrices,

$$\boldsymbol{x}(t_1) = \begin{bmatrix} \frac{1}{2}\sin 2t_1 & \frac{1}{4}(-1 + \cos 2t_1) \\ (1 - \cos 2t_1) & \frac{1}{2}\sin 2t_1 \end{bmatrix} \begin{bmatrix} 0 \\ 1 \end{bmatrix} [10].$$

Thus, the states at the switching time are

$$\boldsymbol{x}(t_1) = \begin{bmatrix} 2.5(-1 + \cos 2t_1) \\ 5 \sin 2t_1 \end{bmatrix}.$$

Equation 8-86 is next used again to find the states at the final time, now treating the conditions at $t = t_1$ as initial conditions,

$$\boldsymbol{x}(t) = \mathbf{\Phi}(t)\, \boldsymbol{x}(t_1) + \mathbf{\Gamma}(t)\, \mathbf{B}\boldsymbol{u}_D(t_1).$$

Substituting the matrices into this equation gives

$$\mathbf{x}(t_1) = \begin{bmatrix} \cos 2t & -\frac{1}{2}\sin 2t \\ 2\sin 2t & \cos 2t \end{bmatrix} \begin{bmatrix} 2.5(-1 + \cos 2t_1) \\ 5\sin 2t_1 \end{bmatrix} + \begin{bmatrix} \frac{1}{2}\sin 2t & \frac{1}{4}(-1 + \cos 2t) \\ (1 - \cos 2t) & \frac{1}{2}\sin 2t \end{bmatrix} \begin{bmatrix} 0 \\ 1 \end{bmatrix} [-30].$$

The output is then computed using Eq. 8-87.

Comments

Note that the "time" in $\mathbf{\Phi}$ and $\mathbf{\Gamma}$ is *always* measured from the last initial condition. This time corresponds to the last step change in any *one* of the inputs. This example was only for a SISO system, so care must be taken to watch for any input change in the MIMO cases.

A common source of error in this procedure is to fail to integrate the $\mathbf{\Gamma}$ matrix using the definite integral. Thus, the evaluation of the integral at $t = 0$ is overlooked. A helpful check on this is to realize that

$$\mathbf{\Phi}(0^+) = \mathbf{I}, \text{ and } \mathbf{\Gamma}(0^+) = \mathbf{0},$$

which are *always* true.

Numerical $\mathbf{\Phi}$-$\mathbf{\Gamma}$ Solutions. Numerical solutions employ the evaluation of $\mathbf{\Phi}$ and $\mathbf{\Gamma}$ over one sample step, τ_s. Further, the matrices are computed using the series approximations

$$\mathbf{\Phi}(\tau_s) = [\mathbf{I} + (\tau_s)\mathbf{A} + (\tau_s^2/2!)\mathbf{A}^2 + (\tau_s{}^3/3!)\mathbf{A}^3 + \cdots] \tag{8-88}$$

$$= \sum_{i=1}^{N} \mathbf{\Phi}_i = \sum_{i=1}^{N} \mathbf{A}\mathbf{\Gamma}_{i-1}. \tag{8-89}$$

Where N is the number of the last term retained before truncating the infinite series. Note that the terms in Eq. 8-88 are $\mathbf{\Phi}_1 = \mathbf{I}$, $\mathbf{\Phi}_2 = \tau_s\mathbf{A}$,· etc. as above. Also, $\mathbf{\Gamma}_0 = \mathbf{A}^{-1}$, $\mathbf{\Gamma}_1 = \tau_s\mathbf{I}$, $\mathbf{\Gamma}_2 = (\tau_s^2/2!)\mathbf{A}$, etc. Thus,

$$\mathbf{\Gamma}(\tau_S) = \int \mathbf{\Phi}(\sigma)\, d\sigma = [(\tau_s)\mathbf{I} + (\tau_s^2/2!)\mathbf{A} + (\tau_s^3/3!)\mathbf{A}^2 + \cdots], \tag{8-90}$$

$$= \sum_{i=1}^{N} \mathbf{\Gamma}_i = \sum_{i=1}^{N} \frac{\tau_s}{i}\mathbf{\Phi}_i. \tag{8-91}$$

The idea here is to use the computer to check the ith matrix in these series for convergence as the summations are built up. This convergence is guaranteed if all the real parts of the eigenvalues of $\mathbf{A}$ are negative. That is, convergence is guaranteed if the system is stable.

The usual test for convergence is that the largest entry in $\mathbf{\Phi}_i$ or $\mathbf{\Gamma}_i$ must be smaller in magnitude than some convergence criteria. Mathematically, the convergence tests are

$$\text{Max}\,[Abs\,(\mathbf{\Phi}_i)] \le \phi_{\text{TEST}}, \tag{8-92}$$

and

$$\text{Max}\,[Abs\,(\mathbf{\Gamma}_i)] \le \Gamma_{\text{TEST}}. \tag{8-93}$$

Thus $\mathbf{\Phi}(\tau_s)$ and $\mathbf{\Gamma}(\tau_s)$ are computed matrix by matrix. Each series computation is halted when its convergence criterion is met.

The analog subsystem model then becomes

$$\boldsymbol{x}((k+1)\tau_s) = \boldsymbol{\Phi}(\tau_s)\,\boldsymbol{x}(k\tau_s) + \boldsymbol{\Gamma}(\tau_s)\,\mathbf{B}\,\boldsymbol{u}_D(k\tau_s), \tag{8-94}$$

$$y_A(k\tau_s) = \mathbf{C}\,\boldsymbol{x}(k\tau_s) + \mathbf{D}\,\boldsymbol{u}_D(k\tau_s). \tag{8-95}$$

Notice that you only have to compute $\boldsymbol{\Phi}(\tau_s)$ and $\boldsymbol{\Gamma}(\tau_s)$ *once* for a given set of **A** and τ_s. The same $\boldsymbol{\Phi}$ and $\boldsymbol{\Gamma}$ may then be used over and over again as the simulation goes from sample time to sample time in its predictions.

One drawback of the numerical method is that it only computes the states and outputs of the analog sub-system at the sample instants. However, to an imbedded digital component, these are the only data that are of interest.

Generalization of the $\boldsymbol{\Phi}$-$\boldsymbol{\Gamma}$ Method. The $\boldsymbol{\Phi}$-$\boldsymbol{\Gamma}$ method may be used when $\boldsymbol{u}_D(t) =$ constant. This occurs when there is a digital component in the system which has zero-order holds for its outputs. However, the method may also be used whenever $\boldsymbol{u}_D(t) \approx$ constant. Consequently, this method is also used to simulate system performance in systems without digital components, provided that their inputs act in such a way that this approximation remains valid.

The primary advantage of the numerical method is that it allows a larger time step than a finite-difference approach. And it may yield an analytic answer if the duration of the constant inputs is long.

HOMEWORK

8-13. Given the SISO system shown in Fig. 8-8, use the $\boldsymbol{\Phi}$-$\boldsymbol{\Gamma}$ method to find an analytic expression for y_A. The input is $u_D = 10u_s(t)$. Assume all initial conditions are zero. The analog subsystem is described by

$$\mathbf{A} = \begin{bmatrix} 0 & 1 \\ -2 & -3 \end{bmatrix},\ \mathbf{B} = \begin{bmatrix} 0 \\ 1 \end{bmatrix},\ \mathbf{C} = [1 \quad 0],\ \mathrm{D} = 0.$$

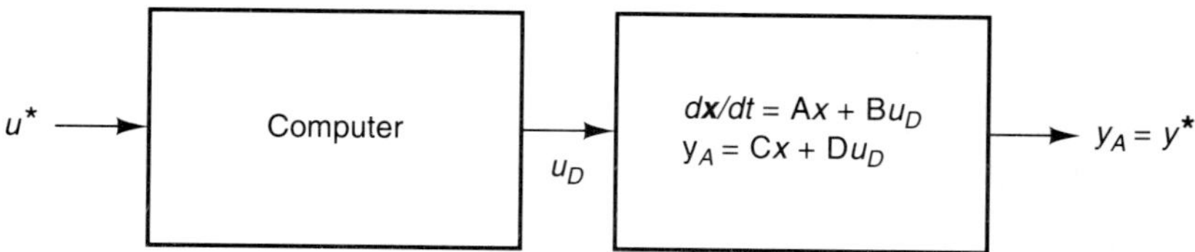

Figure 8-8. Multiport for Prob. 8-13

8-14. a) Use the numerical $\boldsymbol{\Phi}$-$\boldsymbol{\Gamma}$ procedure to compute the $\boldsymbol{\Phi}(\tau_s)$ and $\boldsymbol{\Gamma}(\tau_s)$ matricies for a sample step of $\tau_s = 0.1$ sec., using the system data given in Prob. 8-13. Use the convergence criteria, $\Phi_{\text{TEST}} = \Gamma_{\text{TEST}} = 0.035$. Check your answer against the answer for the earlier problem.

b) What are the simulation equations using the **Φ**, **Γ** matrices for the analog subsystem?

8-15. Use the **Φ-Γ** method to find analytic expressions for y_{A1} and y_{A2} as shown in Fig. 8-9. The inputs are $u_{D1} = 3u_s(t)$, and $u_{D2} = 5u_s(t)$. All initial conditions are zero. The analog subsystem is

$$\mathbf{A} = \begin{bmatrix} 0 & 1 \\ -1 & -2 \end{bmatrix}, \mathbf{B} = \begin{bmatrix} 1 & 0 \\ 0 & 1 \end{bmatrix}, \mathbf{C} = \begin{bmatrix} 4 & 0 \\ 0 & 2 \end{bmatrix}, \mathbf{D} = \begin{bmatrix} 0 & 0 \\ 0 & 0 \end{bmatrix}.$$

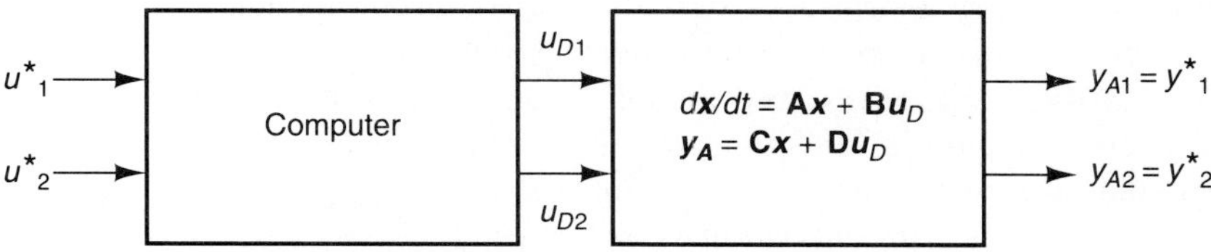

Figure 8-9. Multiport for Prob. 8-15

8-16. a) Use the numerical **Φ-Γ** procedure to compute the $\mathbf{\Phi}(\tau_s)$ and $\mathbf{\Gamma}(\tau_s)$ matrices for a sample step of $\tau_s = 0.1$ sec., using the system data given in Prob. 8-15. Use the convergence criteria, $\Phi_{\text{TEST}} = \Gamma_{\text{TEST}} = 0.03$. Check your answer against the answer for the earlier problem.

b) What are the simulation equations using the **Φ**, **Γ** matrices for the analog sub-system?

8-4 DESIGN PARAMETER SENSITIVITIES

The system output response is determined by the inputs, the initial conditions, and the fundamental physics—and the designer uses these to shape the system response. This is done in two steps, by manipulating the constants which appear in the system model. That is, the first design choice is: What are the system elements, and how are they connected together? This choice leads to a system model of a certain order, with a set of unknown coefficients.

The second design choice is: What are the numerical values of the elements (e.g., the model coefficients) which are necessary to shape the system response? Often, the shaping of the response is accomplished by testing various combinations of the element values for a given set of initial conditions, element connections, and inputs. Consequently, the element values are also called the *design parameters*.

For example, in the simple SISO system

$$\text{RC}\, de/dt + e = \text{R}i, \tag{8-96}$$

the element values of R and C are the design parameters. Recall that the form of Eq. 8-96 comes from the connection and causality of the elements. Once the elements are connected together, the designer must choose numerical values for R and C to shape the system response for the design input.

So, given a numerical design and its simulated performance, the designer is next interested in how sensitive the performance is to changes in the values of the element parameters. If the performance is very sensitive, then the elements will have to be very precisely manufactured. Or the system will have to be redesigned to use less-precise elements. The need for very-precise elements often creates expensive systems. Parameter sensitivities are thus computed to help anticipate this need.

Parameter sensitivities also point the way towards design optimization. This optimization may appear as design for precise steady outputs, and/or design for precise output transient shapes. In this text, you will not study how the optimization is accomplished, that is reserved for follow-on study. Rather, you will study how the parameter sensitivities are computed in order to prepare for design optimization.

The parameter sensitivities of interest are usually of two types: steady-output sensitivities and transient-output sensitivities. In both cases, the inputs, the initial conditions, and the model equations are assumed to be specified. For the steady sensitivities, you change the parameters slightly to study the effect of this change on the steady-system outputs. For the transient sensitivities, you study the effect of small parameter changes on the system eigenvalues.

Steady-output Parameter Sensitivities. To compute this sensitivity, you must be able to find the value of the steady-system outputs. For simple systems, you may use an analytical approach. That is, if you let the n parameters in a system model be designated by P_n, then you must compute the sensitivities of the m outputs, y_m^*, to each of these parameters. So, you must compute the $n \times m$ partial derivatives

$$\partial y_m^*/\partial P_n,$$

which are the sensitivities of interest.

For the simple example given earlier, the steady output is

$$e_s = \mathrm{R}i, \tag{8-97}$$

and the two steady-output sensitivities are

$$\partial e_s/\partial \mathrm{R} = i, \tag{8-98}$$

and

$$\partial e_s/\partial \mathrm{C} = 0. \tag{8-99}$$

Steady-output sensitivities are seldom computed for frequency-domain systems. This is because the eigenvalue sensitivities are used instead for frequency analyses, and these contain all the necessary information. Note that the eigenvalue sensitivities are the same as the pole sensitivities, which also describe the break-frequency sensitivities on the Bode diagram. In other words, the eigenvalues govern both the transient response in the time domain, and the steady frequency response.

For complicated or high-order systems, a computer-based numerical method is often used for computation of sensitivities. This method will be examined following a preparatory study of eigenvalue sensitivities.

Eigenvalue Parameter Sensitivities. In previous sections of this text the eigenvalues, λ, are shown to act as the primary determinants of the system transient response. Consequently, it is quite reasonable to expect designers to be very interested in how sensitive the system's eigenvalues are to variations in the design parameters, as expressed by the $i \times n$ partials

$$\partial\lambda_i/\partial P_n.$$

Note that all complex eigenvalue pairs will have one sensitivity for the real part, and another for the imaginary part.

Again, for simple systems, the eigenvalue sensitivities may be computed with an analytical approach using partial derivatives. For the SISO electrical system presented earlier, there is one eigenvalue with two parameters, and thus two sensitivities,

$$\lambda = -1/\tau = -1/\mathrm{RC}, \tag{8-100}$$

so

$$\partial\lambda/\partial \mathrm{R} = 1/(\mathrm{R}^2\mathrm{C}), \tag{8-101}$$

and

$$\partial\lambda/\partial \mathrm{C} = 1/(\mathrm{RC}^2). \tag{8-102}$$

More complicated systems usually require a computer-based numerical approach.

Numerical Sensitivity Evaluations. Parameter sensitivities are also computed using the digital computer. In this, the partials of interest are computed using the approximations

$$\partial y_m^*/\partial P_n \approx \Delta y_m^*/\Delta P_n, \tag{8-103}$$

and

$$\partial\lambda_i/\partial P_n \approx \Delta\lambda_i/\Delta P_n. \tag{8-104}$$

The basic approach to this is to first compute the steady outputs and the eigenvalues using the design values of the parameters. Each parameter is then increased or decreased a little while all the others are kept at their design value. This parameter variation must be small (say, 10% or less) so that the partial approximations may be valid.

Each time a parameter is varied, the new eigenvalues and the new steady outputs are computed. These values are then input to Eqs. 8-103 and 104 to compute the sensitivities.

A five-part computer program structure such as that below is often used to compute the sensitivities. This program is created to use the simple electrical system discussed previously (Eq. 8-96).

```
C PART I. ENTER THE DESIGN PARAMETERS.
      R = 100.
      C = 0.01
C PART II. ENTER THE MODEL IN TERMS OF THE PARAMETERS.
      A(1) = -1.0(R*C)
      B(1) = 1.0/C
C
C PART III. COMPUTE THE EIGENVALUES.
      RLAMBDA = A(1)
      PRINT RLAMBDA
C PART IV. ENTER THE INITIAL CONDITIONS.
      X(1) = 0.0
C PART V. COMPUTE THE TIME-DOMAIN STEADY OUTPUT. THIS
      EXAMPLE USES RECTANGULAR INTEGRATION OVER 10 τ.
      N = 1000
      TEND = 10.0*R*C
      DELTAT = TEND/N
      DO 100 J = 1,N
      USTAR(J) = 10.0
      YSTAR(J) = X(J)
      X(J+1) = (A(1)*X(J) + B(1)*USTAR(J))*DELTAT + X(J)
100   CONTINUE
      PRINT YSTAR(N)
```

The program begins in Part I where the design parameters are given numerical values.

In Part II, the state-space model is expressed in terms of the parameters of the system.

Part III is the computation of the eigenvalues for the given set of parameters. In general, this computation is a numerical root-finding method which solves the equation

$$\det(\lambda \mathbf{I} - \mathbf{A}) = 0. \tag{8-105}$$

This is the same as finding the roots to the characteristic equation, if the model is given as a differential equation rather than in state-space form.

Frequency-domain systems usually do not go on to the remaining parts of the program.

In Part IV, the initial conditions are entered for the dynamic solution which estimates the steady-system outputs through simulation.

Part V is the simulation of the system outputs. The number of integration steps, N, and the end time, TEND, are chosen to yield an accurate estimate of the new steady end-time conditions.

Thus, for each set of design parameters, the program above computes a set of eigenvalues and a set of steady outputs. By slightly changing the parameters in Part I, you then use Eqs. 8-103 and 104 to compute the sensitivities.

HOMEWORK

8-17. Given the electronic system model

$$\mathrm{RC}\ de/dt + e = \mathrm{R}i,$$

where the design calls for $\mathrm{R} = 1{,}000\ \Omega$ and $\mathrm{C} = 0.0001$ f.

a) Compute an estimate of the eigenvalue sensitivities. Use +10% as the amount of parameter variation. Compare your answers to those using the analytical sensitivities in the text.

b) Repeat (a) above for the steady-output sensitivities.

8-18. Given the structural-rotational system model

$$\mathrm{J}\ d^2\alpha/dt^2 + \mathrm{B}\ d\alpha/dt + \mathrm{K}\alpha = Q,$$

where α is the angle of rotation in radians. The design is $\mathrm{J} = 100$ n-m-sec^2, $\mathrm{B} = 6$ n-m-sec., and $\mathrm{K} = 2$ n-m. Compute the sensitivities of the eigenvalues to the parameters using the analytical approach.

8-19. Given the structural-translational system model

$$(\mathrm{bM_2/k})\ d^3z/dt^3 + (\mathrm{M_1} + \mathrm{M_2})\ d^2z/dt^2 + \mathrm{b}\ dz/dt + \mathrm{k}z = F,$$

where z is the linear translation in ft. The design is $\mathrm{M_1} = 5.455$ slugs, $\mathrm{M_2} = 0.545$ slugs, $\mathrm{b} = 11$ lb_f-sec/ft, and $\mathrm{k} = 6$ lb_f/ft.
Estimate the sensitivities of the eigenvalues to k. Use the numerical approach with a +1% parameter variation (this system is very sensitive to its parameters).

8-5 SUMMARY, BIBLIOGRAPHY, AND REVIEW PROBLEMS

The system response is of critical interest for the evaluation of design performance. This response concerns not only what the system will do with a given set of design parameters, but also what it would do if the design values were changed slightly. In this chapter, a method for the mathematical assembly of the system components was presented, as were methods for the prediction of the system behavior and methods for the computation of the design parameter sensitivities.

The mathematical component assembly method used the multiport diagram and was completely general—it was able to create a system mathematical model from any set of connected components (linear or nonlinear, with or without delay, and with or without digital components). The multiport diagram provided the basis for this assembly through its port definitions.

The bulk of this chapter was then devoted to methods for the prediction of the system response, guided by the various types of components in the system. Sometimes, it was possible to achieve an analytic solution to the system model, but the usual case was that a numerical method must be used.

At the end of the chapter, the sensitivity of the system design to changes in the design parameters (the element values) was investigated. These sensitivities were computed using analytical methods for low-order systems, and numerical methods for high-order systems.

Thus, the modeling is completed. It was begun with a given hardware system design and a desire to translate that design into a mathematical model for design performance and improvement studies, as discussed in Chap. 1. The remainder of the text then roughly followed the tasks of a system modeler.

In Chap. 2, reductionism and the multiport method were used to segment the system into components for simplification. Chapter 3 was an analysis of the component inputs to give direction to the component mathematical modeling effort. Chapters 4 through 7 discussed the component modeling methods. And the last chapter contained the assembly of the component models into a system mathematical model, and the performance of the system design.

All of the methods discussed in this text were presented at an introductory level, suitable to a beginning modeler, and suitable to preliminary design. However, many more useful methods exist, especially in the area of numerical analysis. So, further study in the specialized areas discussed in this text will better equip you for the very difficult modeling problems which you are certain to encounter.

In all cases, you will find that every modeling effort rests on the art of capturing the *essential* performance which is *characteristic* of the design. However, all hardware is nonlinear and of infinite order, so the art is in characterizing the design in the *simplest* model possible. This art can be developed with practice, resulting in good models. Such good models are very useful to designers, while bad models are less than useless. They are error-prone, and they mislead. Bad designs are like this too, since their unnecessary complexity makes them failure-prone.

So, you can avoid many pitfalls in both modeling and design by following this one rule: Keep it simple!

BIBLIOGRAPHY

Astrom, K.J. and B. Wittenmark, *Computer Controlled Systems*, Englewood Cliffs, N.J.: Prentice-Hall Inc., 1984. The book has a good, readable, follow-on discussion of the Φ-Γ method.

DeCarlo, R.A., *Linear Systems*, Englewood Cliffs, N.J.: Prentice-Hall, Inc., 1989. This book has further discussions of state-space modeling, time-variable coefficient modeling, and system stability.

Kreutzer, W., *System Simulation Programming Styles and Languages*, Reading, Mass.: Addison-Wesley Publ. Co, 1986.

Morrison, F., *The Art of Modeling Dynamic Systems*, New York, N.Y.: J. Wiley & Sons, 1991. This book has a lot of discussion with few equations. It is a very readable sequel to an introductory mathematical-modeling study.

Murphy, D.N.P. and N.W. Page, *Mathematical Modeling*, New York, N.Y.: Pergamon Press, 1990.

Ogata, K., *Modern Control Engineering*, Englewood Cliffs, N.J.: Prentice-Hall, Inc., 1990. The block diagram algebra method for SISO system assembly in the frequency-domain is discussed here in some detail.

Westlake, J.R., *A Handbook of Numerical Matrix Inversion and Solution of Linear Equations*, New York, N.Y.: J. Wiley & Sons, 1968.

REVIEW HOMEWORK

8-20. A modeler has found the following models for the two components shown in Fig. 8-4

$$\text{Component A: } 3\, dy_1/dt + y_1 = 2u_1 + 5u^*,$$
$$\text{Component B: } Y^*(s)/U_2(s) = 10/(s+2),$$
$$Y_2(s)/U_2(s) = 3/(s+1).$$

Find the system model in state space.

8-21. A modeler has found the following models for the two components shown in Fig. 8-4

Component A: $3\, d^2y_1/dt^2 + 6\, dy/dt + 7y_1 = 3\, du_1/dt + 8u_1 + 12u^*$,
Component B: $y^* = 0.5u_2$, $5\, dy_2/dt + y_2 = 8u_2$.

Find the system model in state space.

8-22. A system is represented by the equations

$$dx_1/dt = x_2,$$

$$dx_2/dt = -3x_1 - 4x_2 + 5u^*, \text{ and } y^* = x_1.$$

a) Find the general frequency response of the system.

b) Estimate the specific frequency response to the input

$$u^* = 5 \sin 10t + 2 \sin 20t.$$

8-23. Find the analytic time response of the system in the previous problem to the input $u^* = 4u_s(t)$.

8-24. Given the analog subsystem,

$$\mathbf{A} = \begin{bmatrix} 0 & 1 \\ -6 & -5 \end{bmatrix}, \mathbf{B} = \begin{bmatrix} 1 & 0 \\ 0 & 1 \end{bmatrix}, \mathbf{C} = \begin{bmatrix} 2 & 0 \\ 0 & 4 \end{bmatrix}, \mathbf{D} = \begin{bmatrix} 0 & 0 \\ 0 & 0 \end{bmatrix}$$

Use the **Φ-Γ** method to find the analytic output response of this analog subsystem to the finite-digital inputs

$$\mathbf{u}_D = \begin{bmatrix} 4u_s(t) \\ 2u_s(t) \end{bmatrix}.$$

8-25. Compute the $\mathbf{\Phi}$ and $\mathbf{\Gamma}$ matrices for the analog subsystem of the previous problem using the numerical method. Let $\tau_s = 0.1$ sec., and use the convergence tests $\Phi_{\text{TEST}} = 0.03$ and $\Gamma_{\text{TEST}} = 0.03$. Compare your answers to the matrices of the previous problem evaluated at the sample time.

8-26. Given the thermal system model

$$\text{RC}\ d\theta/dt + \theta = \text{R}h,$$

where the design calls for R = 200 °C-sec/Btu and C = 5 Btu/°C.

a) Compute the analytical eigenvalue sensitivities.

b) Use the numerical approximations of the derivatives to compute an estimate of the eigenvalue sensitivities. Use +10% as the amount of parameter variation. Compare your answers to (a).

c) Repeat (a) and (b) above for the steady-output sensitivities.

8-27. Given the fluid system model

$$\text{C}_f\, d^2P/dt^2 + (1/\text{R}_f)\, dP/dt + (1/\text{I}_f)\, P = dZ/dt,$$

with a design that calls for $\text{C}_f = 100\ \text{m}^5\text{-s/n}$, $\text{R}_f = 0.1667\ \text{n/m}^5$, and $\text{I}_f = 0.5\ \text{n-s/m}^5$.

Compute the sensitivities of the eigenvalues to the parameters using the analytical approach.

Appendix

Consistent Unit Sets and Their Conversions

Variable	*Standard International Unit*	*English Unit*	*Conversion (English/SI)*
Angular displacement	radian, r, angular degree, °	same	1 r = (°)(π/180)
Angular speed	radian/second, r/s, revolution/minute, rpm	same	1 rpm = (r/s)(60/π)
Electrical charge	coulomb, c	same	1
Electrical current	ampere, a a = c/s	same	1
Displacement	meter, m	inch, in feet, ft	39.37 3.2808
Energy or Work	newton-meter, n-m = watt-second, w-s = joule, j	inch-pound force, in-lb foot-pound force, ft-lb	8.8512 0.7376
	kilogram-calorie, kg-cal	British thermal unit, Btu	3.9685, 1 kg-cal = 4186 j
	joule, j	British thermal unit, Btu	0.000948
Force	newton, n	pound, lb	0.2248
Frequency	radian/second, r/s cycle/second, cps = Hertz, Hz	same	1 cps = (r/s)2π
Heat	same as energy	same as energy	same as energy
Heat flowrate	kilogram-calorie/sec, kg-cal/s	Br. thermal unit/sec, Btu/s	3.965
Mass	kilogram, kg	slug, sl	0.0684
Mass flowrate	kilogram/second, kg/s	slug/second, sl/s	0.0684

Variable	*Standard International Unit*	*English Unit*	*Conversion (English/SI)*
Power	newton-meter/sec., n-m/s = joule/second, j/s = watt, w	inch-pound/second, in-lb/s foot-pound/second, ft-lb/s horsepower, hp	8.851 0.7376 0.001341, 1 hp = 550ft-lb/s
Pressure	newton/sq meter, n/sq m	pound/sq inch, psi atmosphere, atm	0.0001451 9.88E-6, 1atm = 14.7 psi
Temperature	degree Centegrade, °C	degree Farenheit, °F	°F = 32° + 9/5 °C
Temperature (absolute)	degree Kelvin, °K	degree Rankine, °R	°R = 9/5 °K °R = °F + 459.69°, °K = °C + 273.16°
Torque	newton-meter, n-m	inch-pound, in-lb foot-pound, ft-lb	8.8512 0.7376
Velocity	meter/second, m/s	inch/second, in/s foot/second, ft/s	39.37 3.2808
Voltage	volt, v = watt/sec, w/s	same	1
Volume	cubic meter, cu m	fluid gallon, gal cubic inch, cu in cubic foot, cu ft	264.2 61020, 1 gal = 231 cu in 35.31, 1 gal = 7.481 cu ft
Volume flowrate	cubic meter/second, cu m/s	gallon/second, gps cubic inch/second, cu in/s cubic foot/second, cu ft/s	264.2 61020, 1 gal/min. = 3.85 cu in/s 35.31, 1gal/min = 0.125cu ft/s

Index

JK

L

M

N

O

P

Q

R

S

T

U

V

WXYZ